# Microfluidic Technologies for Miniaturized Analysis Systems

## MEMS Reference Shelf

Series Editor: Stephen D. Senturia
Professor of Electrical Engineering, Emeritus
Massachusetts Institute of Technology
Cambridge, Massachusetts
sds@mit.edu

---

Microfluidic Technologies for Miniaturized Analysis Systems
Steffen Hardt and Friedhelm Schönfeld, eds.
ISBN 978-0-387-28597-9

*Forthcoming Titles*

BioNanoFluidic MEMS
Peter Hesketh, ed.
ISBN 978-0-387-46281-3

Self-assembly from Nano to Milli Scales
Karl F. Böhringer
ISBN 978-0-387-30062-7

Photonic Microsystems
Olav Solgaard
ISBN 978-0-387-29022-5

Micro Electro Mechanical Systems: A Design Approach
Kanakasabapathi Subramanian
ISBN 978-0-387-32476-0

Experimental Characterization Techniques for Micro-Nanoscale Devices
Kimberly L. Turner and Peter G. Hartwell
ISBN 978-0-387-30862-3

Microelectroacoustics: Sensing and Actuation
Mark Sheplak and Peter V. Loeppert
ISBN 978-0-387-32471-5

Inertial Microsensors
Andrei M. Shkel
ISBN 978-0-387-35540-5

Steffen Hardt · Friedhelm Schönfeld
*Editors*

# Microfluidic Technologies for Miniaturized Analysis Systems

Springer

*Edited by:*
Steffen Hardt
Institut für Nano- und Mikroprozesstechnik
Leibniz Universität Hannover
Callinstraße 36
30167 Hannover, Germany

Friedhelm Schönfeld
Institut für Mikrotechnik Mainz, GmbH
Carl-Zeiss-Str. 18-20
55129 Mainz, Germany

*Series Editors:*
Stephen Senturia
Professor of Electrical Engineering, Emeritus
Massachusetts Institute of Technology
Cambridge, MA 02139

Library of Congress Control Number: 2007928398

ISBN 978-0-387-28597-9 e-ISBN 978-0-387-68424-6

Printed on acid-free paper.



9 8 7 6 5 4 3 2 1

springer.com

# Preface

'Microfluidic Technologies for Miniaturized Analysis Systems' – Is a further book in the field of Lab-on-a-Chip technology needed? Yes, we think so! Sure, the area is already covered by a considerable number of review articles, contributed articles and monographs. Yet, microfluidics and Lab-on-a-Chip technology are fast developing branches of research and technology, so in principle there is a continuous demand for reviews covering the latest results. However, after being exposed to our book we hope that the reader is convinced that there is more than only this aspect that distinguishes it from other publications in the field.

A Lab-on-a-Chip is usually a highly heterogeneous object, requiring the interplay of a number of different disciplines such as microfluidics, bioanalytics, or microfabrication. The first Lab-on-a-Chip systems have emerged in the beginning of the 1990s, to a considerable extent spurred by the progress in microfabrication technology previously made. By now the field has reached a certain degree of maturity, reflected by the large number of bioanalytical assays that have been implemented on Lab-on-a-Chip platforms, the large number of competing technological solutions that have been presented, and the increasing number of economic aspects being taken into account when developing corresponding systems. However, the technology's economic success still falls short of the expectations that have been formulated about ten years ago. The reasons for this mismatch are probably rather complex and diverse and are not to be analyzed here. One possible and very straightforward reason might be that the bar for economic sucess had been raised unrealistically high. We hope that, if only to a modest degree, this book may also help to bridge the gap between scientific ideas and their economic implemenation.

In any case, the recent rapid development in Lab-on-a-Chip technology is closely related to progress in microfluidics, being the art of handling fluids and controlling flow in microscale geometries. As often, the economic prospects of a new, emerging technology also drives related areas of fundamental research. In the case of microflow fundamentals, a large number of relevant and important new results have been obtained in the past few years. This development should not let us forget the foundations of microfluidics that have been laid in the late 19th and early 20th century. Already Maxwell, von Smoluchowski and Knudsen have obtained key results for gas flow in small channels. Another area highly relevant to microfluidics are flows with moving interfaces, for which Washburn and Lucas published some pioneering papers in the early 20th century.

The purpose of this volume is to illuminate the microfluidics side of Lab-on-a-Chip technology in detail. The techniques, methods and effects employed to control and handle fluids in microscale geometries described in the literature constitute a kind of “toolbox”. In the early years this toolbox was relatively sparsely equipped and the situation could be characterized by the poor craftsman’s slogan ‘If your only tool is a hammer every problem looks like a nail’, meaning that appropriate miniaturized solutions were not at hand in many cases. By contrast, nowadays a large portfolio of microfluidic technologies does exist. A Lab-on-a-Chip designer is usually spoilt for choice when it comes to deciding which tool to use to solve his problem.

This book describes many of the most important constituents of the microfluidic technology toolbox. In this regard it is different from most other books covering Lab-on-a-Chip technology which give a *complete* overview of the field including microfabrication, biosensor technology, and packaging. As always, there is a tradeoff between broadness and depth, and in many works the key microfluidic technologies are only covered in a rather superficial manner. By contrast, in the following chapters a detailed account of the microfluidic working principles of Lab-on-a-Chip systems is given, thereby sacrificing other areas such as microfabrication or biosensor technology. We believe that an in-detail coverage is justified as an area grows more mature, going along with an increasing demand for comprehensive, in-depth information rather than overview-style writing. The preferred format for such a project is an edited book, guaranteeing that the know-how pool of the contributing authors by far exceeds the knowledge of a few authors who could possibly set up a monograph in the same field.

The chapters of this volume were written by international experts in Lab-on-a-Chip technology or related areas, each of them covering a specific constitutent of the microfluidic technology toolbox. As already mentioned, we regard the in-depth coverage of the different topics a key aspect of the project. Furthermore, it is still important that each chapter gives an overview of the state-of-the-art in the corresponding field. This was taken care of by including a large number of references, thereby informing the reader where further information can be found. Even in the age of internet search engines and databases it is sometimes difficult to find every relevant article on a specific topic, especially in such an interdisciplinary field as Lab-on-a-Chip technology. Therefore we hope that the reader may find the extensive and up-to-date literature lists at the end of most chapters helpful.

We gratefully acknowledge all the chapter authors who scarified a considerably amount of their time for contributing to the present volume. In every single case we were pleased by the good cooperation and by the quality of the work we received. Last but not least, we wish to express our gratitude to the Springer staff, who always made sure that the project was running smoothly and who always replied quickly to our queries. We do hope that the readers will share our enthusiasm on the contributions, and will frequently use and appreciate the compilation.

January 2007

Steffen Hardt
Friedhelm Schönfeld

# Table of Contents

# List of Contributors

**Martin Brinkmann**
Max-Planck Institute for Dynamics and Self-Organization, Göttingen, Germany

**Jennifer L. Christensen**
Texas A&M University, Artie McFerrin Department of Chemical Engineering, College Station, TX 77843, USA

**Eric Cummings**
LabSmith, Inc., 7665 Hawthorne Ave, Livermore, CA 94550, USA

**Petra S. Dittrich**
ISAS – Institute for Analytical Sciences, Bunsen-Kirchhoff-Str. 11, 44263 Dortmund, Germany

**Klaus-Stefan Drese**
Institut für Mikrotechnik Mainz GmbH Fluidics and Simulation Department Carl-Zeiss-Straße 18-20, D-55129 Mainz, Germany

**Bruce K. Gale**
University of Utah, Utah State Center of Excellence for Biomedical Microfluidics, Department of Bioengineering and Department of Mechanical Engineering, 50 S. Central Campus Drive, Rm. 2110, Salt Lake City UT 84112, USA

**Martin A.M. Gijs**
Ecole Polytechnique Fédérale de Lausanne (EPFL), Institute of Microelectronics and Microsystems, CH-1015 Lausanne, Switzerland

**Ómar Gústafsson**
Technical University of Denmark, MIC – Department of Micro and Nanotechnology, Lyngby, Denmark

**Steffen Hardt**
Leibniz Universität Hannover, Institut für Nano- und Mikroprozesstechnik, Callinstraße 36, 30167 Hannover, Germany

**Nicholas R. Harris**
The University of Southampton,
Electronics and Computer
Science,
Southampton, SO17 1BJ, UK

**Martyn Hill**
The University of Southampton,
School of Engineering Sciences,
Southampton, SO17 1BJ, UK

**Krishnacharya Khare**
Max-Planck Institute for
Dynamics and Self-Organization,
Göttingen,
Germany

**Boris Khusid**
New Jersey Institute
of Technology,
University Heights, Newark,
NJ 07102,
USA

**Chang-Jin "CJ" Kim**
University of California at
Los Angeles (UCLA),
Mechanical and Aerospace
Engineering Department,
Los Angeles, CA 90095,
USA

**Jörg P. Kutter**
Technical University
of Denmark,
MIC – Department of Micro
and Nanotechnology,
Lyngby,
Denmark

**Fedor Malik**
Biophysics/BIOTEC,
TU Dresden, Tatzberg 47-51,
01309 Dresden, Germany

**Shoji Maruo**
PRESTO, Japan Science
and Technology Agency
Yokohama National University,
Dept. of Mechanical Engineering,
Yokohama, 240-8501, Japan

**Hyejin Moon**
University of California at
Los Angeles (UCLA),
Mechanical and Aerospace
Engineering Department,
Los Angeles, CA 90095, USA

**Götz Münchow**
Institut für Mikrotechnik Mainz GmbH
Fluidics and Simulation
Department
Carl-Zeiss-Straße 18-20,
D-55129 Mainz, Germany

**Nam-Trung Nguyen**
Nanyang Technological
University
School of Mechanical and
Aerospace Engineering,
Singapore

**Antonio Ramos**
Universidad de Sevilla.
Departamento de Electrónica y
Electromagnetismo.
Avenida Reina Mercedes s/n.
41012-Sevilla, Spain

**Himanshu J. Sant**
University of Utah,
Utah State Center of Excellence for
Biomedical Microfluidics,
Department of Bioengineering
and Department of Mechanical
Engineering,
50 S. Central Campus Drive,
Rm. 2110, Salt Lake City
UT 84112, USA

**Friedhelm Schönfeld**
Institut für Mikrotechnik Mainz GmbH
Fluidics and Simulation
Department
Carl-Zeiss-Straße 18-20,
D-55129 Mainz,
Germany

**Ralf Seemann**
Max-Planck Institute for
Dynamics and Self-Organization,
Göttingen, Germany

**Victor M. Ugaz**
Texas A&M University,
Artie McFerrin Department
of Chemical Engineering,
College Station,
TX 77843, USA

**Petra Schwille**
Biophysics/BIOTEC, TU
Dresden, Tatzberg 47-51,
01309 Dresden,
Germany

# Chapter 1
# Microfluidics: Fundamentals and Engineering Concepts

S. Hardt[1] and F. Schönfeld[2]

[1]Institut für Nano- und Mikroprozesstechnik, Universität Hannover, Callinstraße 36, 30167 Hannover, Germany
[2]Institut für Mikrotechnik Mainz GmbH, Carl-Zeiss-Str. 18-20, 55129 Mainz, Germany

## 1. Introduction

The terms "Lab-on-a-Chip" and "Micro Total Analysis System" stand synonymous for devices that use fluids as a working medium and integrate a number of different functionalities on a small scale. The most important of these functionalities are sample preparation and transport, separation and biosensing/detection. With the exception of the latter they require the handling and processing of small amounts of fluid, mostly liquid, and this is where microfluidics comes into play.

Microfluidics is a comparatively new branch of science and technology in which considerable progress has been made in the past 10–15 years. The reason why we consider it a new discipline is not only the fact that only recently systems have emerged allowing to carry out complex microfluidic protocols, but also due to the different physical regime these systems are based on compared to macroscopic systems for fluid handling and processing. In traditional branches such as chemical process technology fluids are contained, transported and processed in large vessels and ducts. Although the fundamental equations describing the physics and chemistry of such processes are the same as in microfluidics, some effects being important on a macroscopic scale become unimportant on small scales, while other effects that can largely be neglected macroscopically turn out to be dominant in microfluidics. Thus, when it appeared worth while to miniaturize systems for fluid handling and processing, some of the technological developments were heading towards a *terra incognita* where fluids behaved in a way unknown from previous experience with macroscopic systems. By now, much of this *terra incognita* has been explored and technological solutions for many problems in microfluidics have been developed. In fact the development in microfluidics has both achieved such a bandwidth and a depth that we can regard the technological arsenal as a toolbox that offers solutions for most of the specific problems emerging in the context of Lab-on-a-Chip technology. The purpose of this volume is to describe the components of this toolbox, hoping to support a Lab-on-a-Chip designer with an overview of the state-of-the-art of microfluidic technology helpful for attempting to identify a suitable design implementing the desired microfluidic functionalities.

Before proceeding in this direction we wish to bring the specific features of flow phenomena on small scales to the reader's attention. An engineer who designs devices for handling and processing fluids develops a certain intuition about the physics occurring in these devices, and much of this intuition regarding conventional, macroscopic systems has to be abandoned when going over to small-scale systems. The specific features of flow in microdevices are nicely illuminated in two recent review articles [1, 2] that discuss the fundamentals as well as some application perspectives. While these articles are mainly written for readers with a physics background, with this volume a broader audience should be reached, last but not least practitioners from a variety of different fields who have a professional interest in Lab-on-a-Chip technology. However, we believe that everybody involved in this area should have some basic understanding of the mechanisms governing fluidic transport phenomena on small scales. Having had numerous discussions with people getting in contact with microfluidics in one way or the other we know that there is a lot of uncertainty about whether or not the usual laws for fluid flow, heat, and mass transfer are applicable on the micro scale.

Without going into details of the mathematical description, significant insight into the effects governing flow in microchannels can be gained through a scaling analysis. Just by a dimensional analysis it can be shown how different effects, e.g., volumetric and surface forces, scale with the channel diameter or the size of the system under consideration. On this basis, it can be judged which phenomena will govern the behavior of a system and which are of minor importance and can be neglected. A significant part of this chapter is devoted to a scaling analysis of transport phenomena relevant to microfluidics.

In a sense, this chapter lays the foundations for the different contributions to this volume. We highlight the effects and phenomena that are relevant on the microscale and indicate how these may be exploited to establish specific microfluidic functionalities, e.g., fluid transport or sample preconcentration. At the same time, we try to delineate the field of microfluidics from those phenomena occurring at even smaller length scales, which are now beginning to move into the focus of applied research and development. If the scale is further reduced to a few nanometers, the granularity of matter starts to play a role and invalidates the macroscopic (continuum) transport equations that are still applicable on the microscale. Nanofluidics is a developing field in which significant progress has been made on the fundamental level, but which is so far largely lacking the application perspective and degree of maturity existing in microfluidics. For this reason, nanofluidics plays a subordinate role in this volume, and in the current chapter is only touched upon briefly, mainly to explore the limits of validity of the presented scaling analysis on the nanometer scale.

We hope that in the following sections, we can convey some key insights into fluidic transport phenomena on small scales to the practitioner without being too formal. We start with a comparison between the key aspects of flow in macroscopic and microscopic vessels and delineate microfluidics from the phenomena becoming important on the nanoscale. After that, we present a scaling analysis revealing the importance of various transport phenomena as a function of length scale. We conclude with a discussion and overview on a number of engineering and design aspects concerning the realization, design, and modeling of microfluidic systems.

## 2. Essentials of Fluidic Transport Phenomena at Small Scales

### 2.1. Microflow Versus Macroflow

Perhaps the most important quantity determining the flow regime found in a specific system under specific constraints is the Reynolds number, being defined as

$$\mathrm{Re} = \frac{ul\rho}{\mu}, \tag{1}$$

where $u$ is a characteristic velocity, $l$ a characteristic length scale, and $\rho$, $\mu$ the density and dynamic viscosity of the fluid, respectively. For example, $l$ could be the diameter of a channel or duct and $u$ the mean flow velocity. Depending on the Reynolds number, the flow may occur in two distinct regimes: The laminar regime in which the streamline pattern is ordered and regular and the turbulent regime, which is characterized by chaotic velocity fluctuations. At small Reynolds numbers the flow is laminar and all transport processes usually occur at a comparatively slow speed since momentum, heat, and mass have to be spread mainly by diffusion. When a critical Reynolds number $Re_{crit}$ is reached, the flow becomes turbulent and the chaotic velocity fluctuations substantially speed up the transport processes. For flow in a circular pipe $Re_{crit} \approx 2{,}300$, while different geometries are characterized by different critical Reynolds numbers, some being significantly smaller.

An engineer designing macroscopic vessels and devices for fluid processing is well aware of the fact that in the vast majority of cases, the flow will be turbulent. Thus, most of the engineering correlations in daily use for pressure drop, heat, and mass transfer are based on turbulent flow. The turbulent character of the flow brings about a number of advantages and disadvantages. As an advantage, transport processes for heat and mass are by orders of magnitude faster than they are in the diffusion dominated laminar regime. On the other hand, the pressure drop increases beyond the value that would be obtained in laminar flow.

By contrast, on the microscale flow is usually laminar. It is instructive to insert values characteristic for Lab-on-a-Chip systems for the parameters entering (1). As a length scale we can use the hydraulic channel diameter, which is often of the order of 100 μm. Typical flow velocities are of the order of 1 cm $s^{-1}$ or below and typically aqueous solutions are used with a density of about 1,000 kg $m^{-3}$ and a viscosity of about $10^{-3}$ Pa s. Inserting these numbers gives a Reynolds number of 1. This shows that the flow within microchannels usually occurs in the deep laminar regime. Yet, when increasing the channel dimension and/or the flow speed so that Reynolds numbers of the order of 10 are reached, the flow is not solely governed by viscous forces but inertia comes into play. This facilitates the realization of more complex, versatile fluidics, such as (passive) diffuser or Tesla valves or the induction of so-called Dean vortices driven by centrifugal forces. These effects play a role in some microfluidic systems; however, the critical Reynolds number for breakdown of laminar flow is hardly reached.

The laminar character of microflow has important consequences for the design and layout of microfluidic systems. Bearing in mind the increasing capabilities of flow simulation tools and the constant increase of processor speed, the performance of systems for fluid handling and processing can be predicted with increasing reliability before ever developing the first prototype. However, in macroscopic systems the predictive power of flow simulations is limited by the turbulent character of the flow. Turbulent flows can usually only be computed via so-called turbulence models, which are based on certain empirically motivated assumptions about the formation and decay of turbulent eddies on different length scales. By contrast, flow fields in microfluidic systems can often be computed from first principles and no assumptions that are based on shaky foundations have to be made. That way a rational design becomes possible, meaning that many components of Lab-on-a-Chip systems can be designed and optimized using computer simulations, where the predictive power of the simulation results is often high enough to draw reliable conclusions on the performance of components and devices.

An example illustrating this distinction between fluidic systems on the micro and on the macroscale is the modeling of chemical reactions. Usually two or more reagents have to be mixed before undergoing a chemical reaction, which, in turn, depends sensitively on the local concentrations of the reagents achieved in the course of the mixing process. In a turbulent mixing process the complex structure of the eddies leads to complex spatial fluctuations of the concentration fields occurring on scales that can usually not be resolved by the numerical grids used to compute the corresponding mixing and reaction processes. Thus, it becomes very difficult to predict the product distribution resulting from the chemical reaction, and statistical methods are usually employed for this purpose (see, e.g., [3]). By contrast, chemical reactions occurring in microchannels can usually be modeled much more reliably. This is due to the structure of the concentration fields that can often be resolved by the computational grid used in the simulation, at least if no flows with laminar chaos are involved [1]. As a result, the product distribution of reactions occurring in microchannels can usually be predicted with a degree of accuracy unprecedented in macroscopic reactors.

There are, however, a number of factors limiting the applicability of simulation-based rational design approaches in microfluidics. The statements made in the paragraph above are mainly valid for certain classes of flow, especially single-phase flow. Flows with moving three-phase (e.g. gas/liquid/solid) contact lines constitute an example for which the predictive power of simulation approaches is often unsatisfactory. This is due to the fact that on the nanoscale a three-phase contact line can assume a multitude of metastable configurations, a phenomenon manifesting itself as contact-angle hysteresis (see, e.g., [4]). On the other hand, while microflow simulation may be a powerful tool for the design of fluidic components, methods for evaluating the performance of extended systems (Lab-on-a-Chip, μTAS) are largely missing. This stands in contrast to the design tools available in microelectronics, which by now play a key role for the design of microelectronic circuits. Thus, while the laminarity of microflow offers great prospects for the rational design of microfluidic systems, correspondingly powerful tools for system simulation still remain to be developed.

The fundamental difference between micro- and macroflow can also be understood in terms of the influence of the boundary of the flow domain. The boundary, e.g. a solid wall, may be characterized by certain values of the velocity, temperature, and concentration fields. For example, a boundary may move with a certain velocity and may have a fixed temperature. More than 100 years ago a theory was developed, the so-called boundary-layer theory [5], which describes how far the boundary values of these fields penetrate into the fluid. Roughly speaking, there is a comparatively thin layer in which the fluid "feels" the influence of the boundary. Outside of this so-called boundary layer, the fields in the fluid are dominated by values characteristic for the bulk, e.g., velocities and temperatures given by some inflow condition. A boundary layer over a flat plate is depicted on the top of Fig. 1. When the plate is subjected to an external flow field, the region where the velocity field is significantly different from the bulk velocity is quite thin close to the edge of the plate and becomes increasingly thicker when moving further downstream. In a rough analogy, a distinction between micro- and macroflow is the importance of the boundary layer for the flow field. In typical microflows, the boundary layers govern the entire flow field, or in other words, the boundary layers usually extend over the whole flow domain. This is depicted on the bottom of Fig. 1, which shows two parallel plates with boundary layers that finally merge and extend over the whole space between the plates. In macroflows, the boundary layers usually constitute only a comparatively thin shell surrounding the bulk of the fluid.

The dominance of the boundary in microflow can be exploited for flow control and manipulation unparalleled on the macro scale. The boundary can be used to determine flow profiles, temperature, and concentration fields in micro devices in a much better way than in macroscopic vessels. A nice example for determining flow patterns via the structure of the boundary is flow over surfaces containing microscopic grooves. If a fluid is driven by a pressure gradient over a surface containing a multitude of parallel grooves, the flow resistance is smaller if the pressure gradient is parallel to the grooves than if it is orthogonal. Consequently, if the pressure gradient and the groove orientation stand at an inclined angle, the flow close to the surface will tend to follow the grooves and will therefore not be aligned with the pressure gradient. This effect can be used to create a helical flow in a channel, which is, in the simplest version, a parallel-plates arrangement of the surface containing the grooves and a second boundary. The resulting flow pattern is displayed in Fig. 2.

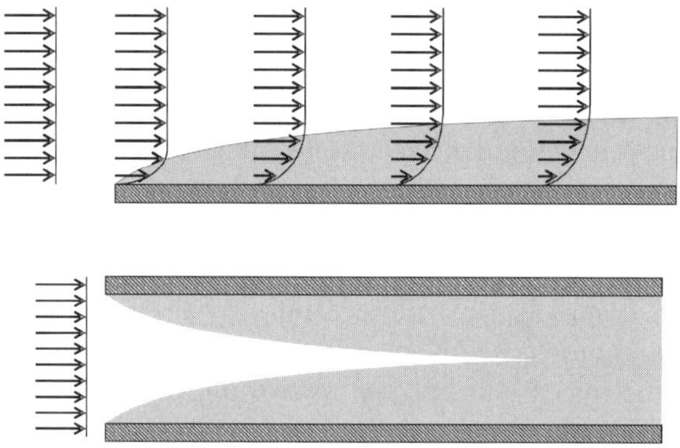

**Fig. 1.** Boundary layer over a flat plate subjected to an external flow field (*top*) and merging boundary layers between two parallel plates (*bottom*)

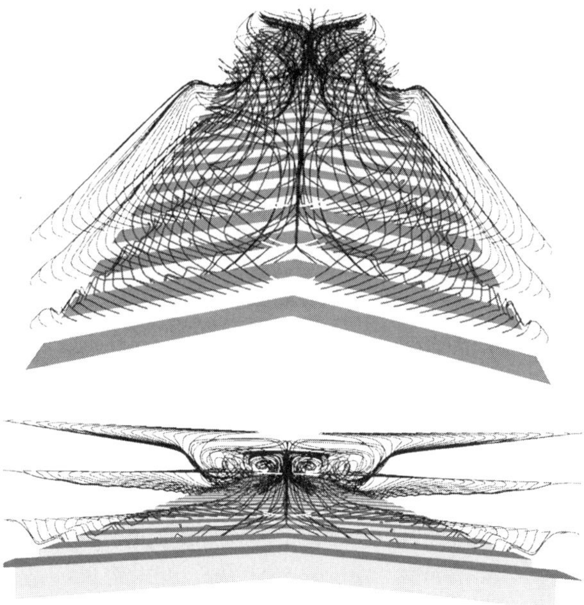

**Fig. 2.** Stripe of a surface structured with herringbone grooves together with streamlines confined between parallel surfaces, viewed from two different angles. Owing to the grooved surface, the streamlines get twisted

The figure shows a stripe of a surface structured with a herringbone pattern of microscopic grooves. When a pressure gradient aligned with the stripe is applied, the streamlines, being reflected from the second, parallel and unstructured surface (top wall, not shown), form a helical pattern. This helical flow pattern extends over the complete domain between the parallel plates. When the distance between the plates is increased (thus moving from the microflow regime towards the macroflow regime), a point will be reached where the groove pattern only influences the flow in the boundary layer above the structured surface and no longer in the full space between the parallel plates. Beyond this point, it is apparently no longer possible to determine the complete streamline pattern by the boundary structure. In a number of examples it has been demonstrated how surfaces with parallel groove patterns may be utilized in microfluidics to design structures performing specific operations, e.g. the rapid mixing of liquids [6].

## 2.2. Nanoflow

In the past few years, significant progress has been made in the field of nanofabrication. This progress has triggered the development of nanofluidic structures in which fluids are transported through domains with a lateral extension of less than a micrometer. Apart from a few exceptions, the application perspective of nanofluidics seems to be more remote than that of microfluidics. Nevertheless, as the fluidic structures become smaller and smaller, it becomes necessary to explore the limits of applicability of the macroscopic laws for transport of momentum, heat, and matter. Such a program is especially important in view of the rational-design approach described in the previous section, which is based on the assumption that the fundamental laws governing the transport processes in small-scale devices are known and may serve as a basis for modeling and simulation studies. A comprehensive overview on modeling and simulation approaches for micro- and nanoflow is given in the book by Karniadakis et al. [7].

The question is as follows: As we further and further reduce the dimension of fluidic devices, when do we arrive at a point where the laws of macroscopic continuum theory are no longer applicable? As described in the previous section, microscopic systems may exhibit a flow behavior significantly different from their macroscopic counterparts. This is mostly due to scaling effects, i.e., a certain phenomenon or effect is present also on the macroscale but does not become apparent because it hides behind other effects, which are dominant on the length scale considered [8]. An example for such scaling effects are free-surface flows, e.g., flows with a gas/liquid interface. Ocean waves are dominated by gravity forces even though surface tension is present also on this macroscopic scale. Free-surface flows in microchannels are often governed by surface-tension forces whereas gravity is negligible. Even if different effects become visible at different length scales, both scenarios are described by the usual continuum equations for transport of momentum, heat, and matter, which are explained in any text book on fluid mechanics. By contrast, when further reducing the length scale of fluidic devices to below 1 nm, from a certain point on the usual continuum description ceases to be valid and a regime is reached where fluids behave in a way that is, at least partly, not yet understood very well and the subject of current research efforts. In the following we will refer to this regime as the molecular flow regime, as opposed to the continuum regime, which is the subject of most classical studies in fluid mechanics.

As it turns out, to explore the molecular flow regime a distinction between gas and liquid flow has to be made. In gas flow, deviations from continuum behavior are observed well before the dimensions of a fluidic device reach the scale of the molecular diameter. By contrast, continuum theory can be used to describe liquid flow even down to a scale of a few molecular diameters. The reason for this difference lies in the intermolecular spacing characteristic for gases and liquids. Owing to the much lower density of gases compared to liquids, gas molecules are well separated in space and the interactions between them are virtually point-like, meaning that the time intervals during which two gas molecules interact are separated by much longer intervals in which the molecules travel freely without undergoing collisions. Molecules in a liquid, however, permanently interact with their neighbors due to their comparatively dense packing.

For gas molecules confined in a micro- or nanochannel this means that when shrinking the channel diameter, at a certain point a regime will be reached where it is equally probable for a molecule to collide with the channel wall as with a neighbor molecule. The two extreme cases for a gas confined between two parallel walls are sketched in Fig. 3. On the top of the figure, transport is dominated by collisions between molecules, whereas the bottom shows a situation where the collisions occur predominantly between the molecules and the walls of the channel. A quantity allowing to differentiate between these two different scenarios is the Knudsen number, defined as

$$\mathrm{Kn} = \frac{\lambda}{d}, \tag{2}$$

where $\lambda$ is the mean free path of the gas molecules and $d$ the channel diameter. The mean free path is the average distance a molecule travels between two collisions.

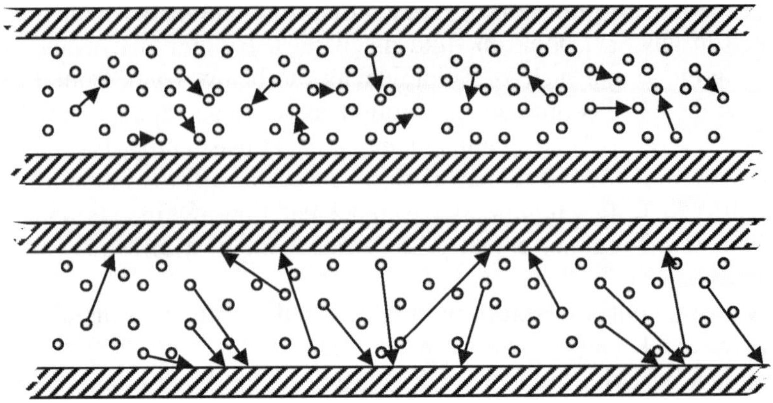

**Fig. 3.** Gas molecules in a parallel-plates channel in the continuum (*top*) and in the free molecular flow regime (*bottom*)

The Knudsen number determines whether flow occurs in the continuum or in the free molecular flow regime. When Kn < 1, it is more probable for a molecule to collide with another molecule than with the channel wall. In the reverse case, collisions with the channel walls occur more frequently. Exactly these collisions with the channel walls limit the validity of the classical continuum transport equations. The properties of a gas are closely connected to the molecular transport mechanisms. If another transport mechanism takes over, in that case wall collisions, one should expect that continuum theory for gases in micro- and nanochannels ceases to be valid. How exactly the continuum description looses its validity is depicted in Fig. 4. The figure shows the channel diameter $d$ as a function of the normalized molecular number density ($n/n_0$) of air for Knudsen numbers of 0.01, 0.1, and 10, where $n_0$ is the density at standard conditions (298 K, 1 atm). The lines of constant Kn delineate regions for which different descriptions have to be applied to model the transport of momentum, heat and matter. For Kn < 0.01, the usual continuum transport equations (Navier-Stokes equation, Fick's law, Fourier's law, etc.) may be used. In the interval 0.01 < Kn < 0.1, the continuum transport equations are still valid; however, the flow boundary conditions at solid surfaces have to be modified. For example, on a macroscopic scale it is often an excellent approximation to assume that all components of the flow velocity have to vanish at a solid surface. This is no longer true if Kn > 0.01, where it becomes important to take into account the slipping of gas at the surface. In other words, it can no longer be assumed that the tangential gas velocity at the surface is identical to zero. For even higher Knudsen numbers, in the so-called transition-flow regime ($0.1 < Kn < 10$), the continuum equations no longer provide an adequate description for the transport processes in the gas. Rather than that, generalized models based on the Boltzmann equation [9] have to be employed, which means that the effort for computing flow fields substantially increases with respect to the continuum description. For $Kn > 10$ the free molecular flow regime is reached. In this regime most of the collisions occur between molecules and channel walls, the collisions between molecules can be neglected. This again leads to a simplification of the transport models and free-molecular flow fields can be computed with moderate effort.

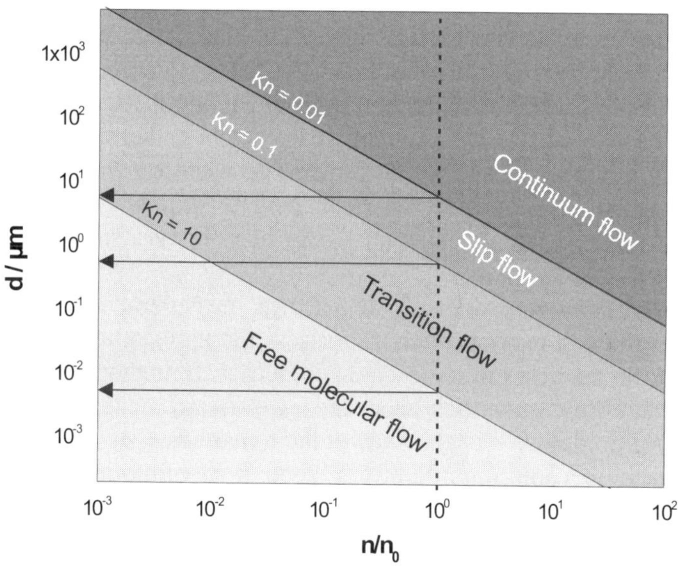


**Fig. 4.** Flow map in a plane spanned by the normalized molecular number density and the channel diameter showing the different regimes of gas flow in micro- and nanochannels

For example, consider the flow of air in small scale channels at standard pressure and temperature. In a channel of about 6 μm diameter it can be expected that flow features and thermal properties such as the friction factor or the Nusselt number start to deviate from the relationships derived from macroscopic systems. This is due to the modified boundary conditions that start to play a role at that scale. When the channel diameter is further reduced to about 600 nanometers, the continuum description ceases to be valid and there is no longer such a comparatively "simple" framework for understanding the transport processes as the Navier-Stokes equation.

In summary, there is a band of Knudsen numbers ($0.1 < Kn < 10$) where the transport processes are extremely complex and are determined both by intermolecular collisions and molecule-wall collisions. In this regime it is difficult to predict velocity, temperature, and concentration fields, and the knowledge on flow behavior is limited. Nevertheless, the specific characteristics of the transition flow regime may be advantageous for a number of applications. It is known that gas molecules slipping over a solid surface lead to a reduced pressure drop compared to flow subjected to a no-slip boundary condition [10]. Furthermore, in the transition flow regime so-called cross-diffusion effects become apparent, which may be exploited for novel driving mechanisms such as fluid pumping by a temperature gradient [11].

Compared to gas flow in micro- and nanochannels, the situation for liquid flow is very different. Liquids such as polymer solutions may exhibit a very complex behavior when confined to a nanoscale geometry (see, e.g., [12]). However, in this chapter we wish to highlight the most elementary phenomena related to flow in small-scale devices, therefore we limit the discussion to liquids consisting of simple molecules such as water or ethanol. Such simple liquids show a behavior that, perhaps surprisingly, conforms very well to predictions from continuum theory even in very small-scale geometries.

Naturally, it is very difficult to perform measurements of velocity or density profiles in nanochannels. To assess the applicability of continuum theory for liquid flows in nanoscale geometries, a method for determining characteristic flow quantities is needed. A standard method for computing nanoscale liquid flows is the molecular dynamics (MD) approach [13]. MD is based on a solution of Newton's equations of motion for a molecule moving in the force field of the other molecules surrounding it. Thus, MD is a very computationally intense approach that requires tracking of every single molecule in a liquid. A comparison of continuum theory with MD results allows to isolate the effects that are due to the granularity of matter and to identify the regime where the continuum description breaks down. Such a comparison reveals that in channels with a width larger than about 10 molecular diameters, continuum theory makes reasonable predictions, at least when allowing the viscosity to deviate from its macroscopic value [7]. Thus, down to a scale which is typically of the order of a few nanometers, simple liquids more or less behave as predicted by the continuum transport equations. It should be noted, however, that again within the realm of continuum theory there are scaling effects that may emerge only in minute channels and devices. An example for such effects are forces in the electric double layer (EDL) of an electrolyte close to a solid surface

[14], which usually become apparent only on the microscale and may be dominant on the nanoscale.

When the dimension of a fluidic channel is further and further reduced and the channel diameter assumes values of a few nanometers, the molecular structure of a liquid shows up in form of density oscillations. Such density oscillations have been observed independently in a large number of computational studies and have also been verified experimentally. Figure 5 shows the density profile in a liquid close to the channel wall as obtained from a MD simulation [15]. The isolated peak on the right side of the diagram represents the density of the wall molecules. The density oscillations occur with a wavelength of the order of the intermolecular distance and are damped within about one nanometer which shows that molecular ordering occurs only in the immediate vicinity of the channel walls.

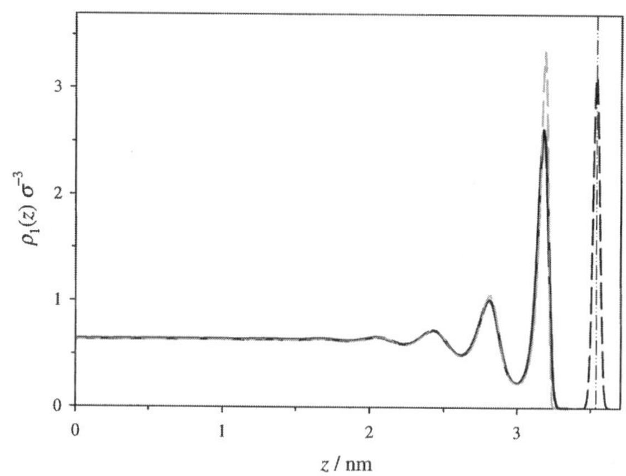


**Fig. 5.** Density oscillations in a liquid close to a solid wall

Similarly as in the case of gas flow, before the regime is reached where the continuum description breaks down, a modification of the flow boundary conditions at the solid surface has to be taken into account. However, while a universal law exists that describes boundary slip of gas flow, a corresponding universal relationship for liquid flow can neither be derived from experiments nor from theory. Several groups have conducted experiments aiming at the determination of the flow boundary conditions for simple liquids at a solid surface [16]. Most of these studies seem to agree in the sense that on a hydrophobic surface the flow velocity does not vanish, i.e., the usual no-slip boundary condition breaks down. The length scale characterizing this slip flow can be significant, often in a range of some 10 nm. By contrast, the experiments are inconclusive with regard to the flow boundary conditions on a hydrophilic surface. While some groups have measured a significant slip velocity, other groups conclude that the no-slip boundary condition is valid at a hydrophilic surface [16].

The reasons and mechanisms for boundary slip of liquids still remain speculative. There is experimental evidence that a hydrophobic surface can be covered with nanoscale bubbles [17], which may be formed by nucleation of gaseous admixtures to a liquid. Such bubbles can, via an internal recirculation flow, allow the liquid to slip along a surface with a considerable velocity. However, also without gas bubbles or an attached gas film there can be a nonzero tangential flow velocity at a solid surface, as has been shown by MD simulations [18]. Therefore, it has to be concluded that the boundary conditions for liquids at solid surfaces remain a matter of ongoing research and no simple, conclusive picture has emerged so far.

## 3. Scaling Analysis

When an engineer develops a solution for a specific problem in microfluidics, he can gain substantial insight by analyzing how changing the length scale affects the various fluidic tools he has at his disposal. Sometimes a physical effect that seems promising for achieving certain functionalities (perhaps due to experience with macroscopic systems) loses its appeal when translated to the microscale and, vice versa, other effects may come into play. Therefore a scaling analysis of forces and effects related to fluidic systems is often beneficial.

Consider a fluid volume with characteristic linear dimension of $L$ on which a number of different forces may act. This fluid volume can be thought of as being part of a fluidic system containing a number of channels, chambers, etc. The forces acting on the molecules in the volume can be classified as imposed and dynamic forces. The former are forces that are imposed from external sources such as electric fields, the latter result from the dynamics of the system and cannot easily be controlled from the outside, e.g. viscous forces. Different forces usually exhibit a different scaling behavior with $L$. As far as the imposed forces are concerned, the scaling behavior originates from the spatial dimension of the regions onto which they apply. Specifically, volume forces scale as $L^3$, whereas surface forces are proportional to $L^2$. In view of their origin depending on the dynamic state of the system, the scaling behavior of dynamic forces is often less obvious.

When performing a scaling analysis, it is extremely important to exactly define which quantities are kept fixed as constraints and which quantities are varied. Scaling a quantity such as a spatial dimension may give different results depending on the constraints that are being imposed. In the following, we wish to consider the scaling sketched in Fig. 6 where a fluid volume of characteristic length scale $L$ moving with velocity $\vec{u}$ is rescaled in all three spatial dimensions while keeping the velocity scale fixed. The scaling operation performed on the fluid volume can be thought of as originating from the scaling of a microfluidic system, i.e., a system including all of its channels and fluidic elements is shrunk by a factor of $L'/L$.

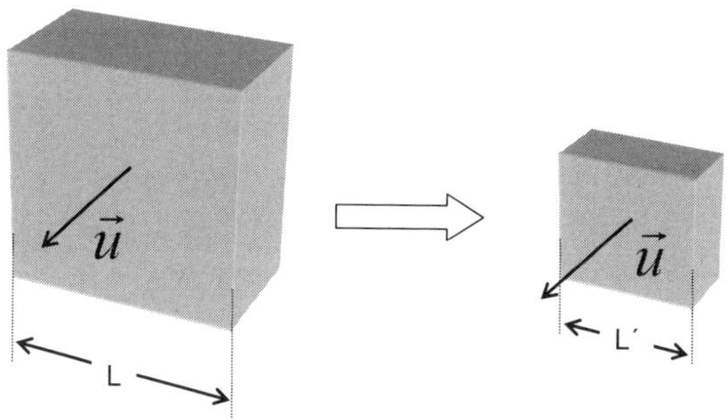


**Fig. 6.** Scaling of a fluid volume of linear dimension $L$ at fixed velocity

Microfluidic systems are often multiphysics systems in the sense that a number of different forces and effects come into play and are superposed to each other. For example, a pressure difference may be used to drive liquid through a channel, while an electric field is employed to separate particles suspended in the liquid. It is worth while to classify the potentially relevant forces in such multiphysics systems in terms of their scaling behavior, which is done in Table 1 for a fluid volume characterized by a linear dimension of $L$. From this table it becomes apparent how the various forces behave when the length scale $L$ is changed, a piece of information that is relevant for those who attempt to find the right dimensions for their microfluidic channels or to select a physical principle suitable for performing a specific microfluidic operation. It should be noted, however, that the entries of Table 1 were derived based on the assumption of a fixed velocity scale and a fixed velocity profile, i.e., it was assumed that in the scaled system the same flow profile exists as in the original one. The latter assumption will be made throughout this chapter if not mentioned otherwise. Therefore the results are not universal and the reader should be advised to carefully check the validity of the invariance of the flow profile, or, alternatively, refer to a scaling analysis with appropriate constraints. An overview of scaling phenomena, similarity solutions, and dimensional analysis in fluid mechanics is given in the book by Sedov [19]. A scaling analysis based on the right constraints can be very helpful in identifying suitable mechanisms for fluid manipulation in a multiphysics context without ever having to solve complex transport equations such as Navier-Stokes.

**Table 1.** Scaling of different forces on a volume element assuming a fixed velocity scale and profile

| Type of force | Relevance | Scaling behavior | Comment |
|---|---|---|---|
| Inertial | Flow with acceleration of fluid volumes, e.g. curved streamlines | $L^2$ | |
| Viscous | Flow with shear forces | $L$ | |
| Pressure | Pressure-driven flow | $L$ | For low Reynolds numbers |
| Gravity | Dense fluids or fluid mixtures with significant differences in density | $L^3$ | |
| Surface tension | Flow of immiscible fluids | $L$ | |
| Electrostatic – single-phase fluid | Electrostatic forces in a single-phase fluid act in those regions where a net charge density exists or the dielectric permittivity changes. Both phenomena usually occur in thin layers, i.e. on two-dimensional manifolds for most practical purposes | $L^2$ | |
| Electrostatic – particle suspension | For many operations in microfluidics it is of interest to manipulate and separate particles suspended in a liquid. An electric field couples to a particle's charge and dipole moment. Here $L$ is assumed to describe the size of the particles | $L^0$<br>$L^3$ | Coupling to charge, thin EDL, small $\zeta$-potential<br>Coupling to dipole moment, fixed polarization density |
| Magnetostatic – single-phase fluid | Magnetostatic forces in a single-phase fluid are usually Lorentz forces exerted on an electric current flowing in a liquid | $L^3$ | Fixed current density |
| Magnetostatic – particle suspension | As in the case of electrostatic forces, $L$ describes the size of the particles suspended in the fluid. There are no magnetic charges, so the magnetic field can only couple to a particle's dipole moment | $L^3$ | Fixed polarization density |

As pointed out above, a scaling analysis can be a simple and appealing method, but there are a number of pitfalls with regard to the constraints imposed on the system. When imposed forces such as electro- or magnetostatic forces are considered, it has to be defined whether or not the sources of these forces should be co-scaled with the fluidic elements of the system under consideration. The analysis presented in Table 1 is based on the assumption that the sources remain fixed when the scale of a fluid volume is changed. A more favorable force scaling when shrinking the dimension of a system can be achieved by scaling the electric or magnetic sources along with the fluidic parts. For example, when the electrode structures producing the electric field acting on particles via dipole forces are scaled along with the particles, the scaling behavior changes from $L^3$ to $L^2$.

Clearly, the scaling analysis presented above has its limitations. In particular the assumptions such an analysis is based on may only be fulfilled for systems with characteristic dimensions lying within a certain range. For systems of a size no longer within this range, the analysis may break down because novel phenomena not yet accounted for come into play. For example, the above analysis is based on the assumption of a fixed velocity profile. Since in microfluidic systems flow phenomena usually occur in the laminar regime with Re $\approx$ 1, this assumption is fulfilled with reasonable accuracy for a significant range of the scale parameter $L$. However, it is clear that by increasing $L$ and keeping the velocity scale fixed the Reynolds number increases and a point will be reached where the velocity profile in the flow domain changes. Depending on the geometry under consideration, evident signatures for a change in flow profile are inertia-driven secondary flows perpendicular to the main flow direction and the formation of recirculation zones separated from the bulk fluid stream. Finally, at a certain Reynolds number there is a transition to turbulent flow. Then, at the latest, the assumptions the scaling analysis presented above is based on break down.

Another scaling limit is reached when the system size becomes so small that the thickness of thin fluid layers, which are the sources of forces acting on the system, can no longer be neglected. For example, in channels made from an electrically nonconducting solid, an electrically charged fluid layer often forms in the close vicinity of the channel walls (see next section). In the fluidic part of this electric double layer an electric field gives rise to a force that can drive a fluid through a channel. In most studies of such electroosmotic flow (EOF) phenomena it is assumed that the thickness of the EDL (typically between 1 and 100 nm) is negligible compared to the channel diameter. However, when further and further shrinking the channel dimensions, at a certain point the thickness of the EDL can no longer be neglected and has to be taken into account when computing the flow profile. Then a scaling analysis based on the usual assumption of an infinitely thin double layer reaches its limits.

## 3.1. Scaling Analysis for Single-Phase Flow

### 3.1.1. Flow Rate

In most microfluidic systems there exists a mechanism for fluid transport. Since in Lab-on-a-Chip systems mostly liquids are processed, the focus of this section will be on liquid pumping. In this context the two most prominent principles for driving a liquid through a channel are based on pressure-driven flow and EOF, as also described in a recent review article [20]. In pressure-driven flow, a pressure difference is generated between the inlet and the outlet of a channel, e.g., by a displacement pump. This pressure difference drives a liquid through a channel, where the characteristic Poiseuille flow profile is formed. By contrast, EOF is generated by an electric field along the channel (see Chap. 2 of this volume). This field acts on charged molecules in a thin fluid layer close to the channel walls, being part of an EDL. An EDL may be formed when molecules at the wall surfaces dissociate, leaving behind charged groups immobilized on the surface and ions in the liquid. The characteristic flow profile of EOF is virtually flat. Only within the EDL the velocity increases from zero at the channel wall to its terminal value characteristic for the bulk of the liquid. In Fig. 7 the velocity profiles of pressure-driven flow and EOF are sketched. Measured by typical dimensions of microchannels, the thickness of the EDL shown in the figure is exaggerated, under realistic conditions its extension can hardly be detected on that scale. Further details on pressure-driven and electroosmotic flow in micro- and nanochannels can be found in the book by Karniadakis et al. [7].

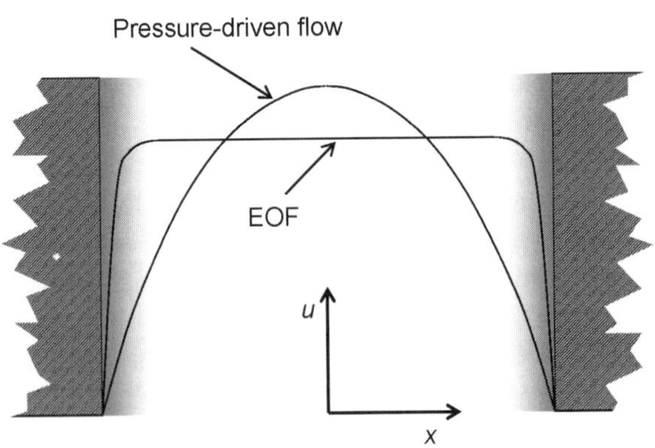


**Fig. 7.** Velocity profiles $u(x)$ for pressure-driven and electroosmotic flow. The extension of the EDL is indicated by shaded areas

When a fluid is flowing through a channel, the resulting stationary-state flow rate is always such that the friction force equals the external driving force. Friction only occurs in those regions where a velocity gradient exists, and Fig. 7 shows that in this respect pressure-driven and electroosmotic flow show a different behavior. While for pressure-driven flow velocity gradients exist virtually everywhere, in EOF the gradients are limited to a thin layer close to the wall. That means that by shrinking the channel diameter in a regime where the EDL thickness is still negligible, scaling does not influence the friction forces in EOF. By contrast, shrinking the channel diameter increases the friction forces in pressure-driven flow.

This different scaling behavior of pressure-driven and electroosmotic flow can be formulated more quantitatively. For that purpose, we assume fully developed flow, i.e., the channels to be considered should be long enough to neglect any effects due to special conditions at the channel entrance. Furthermore, the driving forces for the flow are assumed to be fixed, i.e., the pressure gradient (pressure drop per unit length along the channel) and the electric field are constant. Under these conditions the average flow velocity of pressure-driven flow scales as $L^2$, while the average velocity of EOF stays fixed. Here and in the following $L$ stands for the hydraulic channel diameter, if not stated otherwise. Translated to volume flow this means that in the pressure-driven case the flow rate scales as $L^4$, while for EOF it is proportional to $L^2$. For this reason EOF is much better suited to drive liquid through very narrow channels or constrictions than pressure-driven flow.

### 3.1.2. Heat Generation

Fluid transport through a channel inevitably requires a driving force, which can be due to a pressure difference or an electric field. Multiplying that driving force by the flow velocity gives the power required to drive the fluid. The work input in such a way is usually converted into thermal energy, i.e., the fluid is heated up while being transported through a channel, an effect being referred to as viscous heating. Viscous heating is an undesired phenomenon in most cases and is often not taken into account when designing microfluidic systems. Yet, especially in biofluidic applications a temperature shift can be detrimental, since biomolecules or cells are often sensitive to temperature changes.

When keeping the average velocity of a pressure-driven flow in a channel fixed, the thermal power created per m$^3$ channel volume due to viscous heating scales as $L^{-2}$, i.e., it rapidly increases when the channel becomes smaller. However, it also has to be noted that viscous heating is proportional to the flow velocity, usually being as small as a few mm $s^{-1}$ in Lab-on-a-Chip systems. For this reason it turns out to be unimportant in most practical cases, in spite of its unfavorable $L^{-2}$ scaling. In other applications such as microreactors for chemical production or high-throughput microchannel heat sinks viscous heating can play a more important role, and its significance has been analyzed in more detail in a number of publications [21–24].

As mentioned above, electroosmosis is a principle that is often better suited for fluid pumping in small-scale systems than pressure-driven flow. EOF requires electrodes in contact with an electrolyte where an ion current is created. The power needed to drive this ion current is $I \cdot U$, where $I$ is the electric current and $U$ the voltage difference. Eventually, this power is dissipated into heat, and it turns out that this Joule heating usually creates a much more significant temperature shift in the fluid than viscous heating. When keeping the driving force, in that case $U$, fixed, $I$ scales as $L^2$, which means that the thermal power created per $m^3$ channel volume due to viscous heating scales as $L^0$, i.e. stays constant. Joule heating does not only play a role in systems where an EOF is created, but is also important in electrophoretic separation devices. Electrophoretic transport equally relies on the creation of ionic currents in a liquid, and in the same way as in the case of EOF electrical energy is converted into thermal energy. A number of authors have studied Joule heating in microfluidic systems in more detail [25–31].

The heat balance in a microchannel depends on the heat-generation as well as the heat-removal rate. A large heat source leads to a large temperature shift only if at the same time no efficient mechanism for heat removal exists. Luckily, the heat removal rate scales favorably with the channel diameter $L$, as discussed in the next section. For this reason, the problem of undesired heat generation in a microchannel can usually be solved by designing the microchannel in such a way that a rapid heat transfer is guaranteed.

### 3.1.3. Heat Transfer

A typical heat transfer problem in a microfluidic system is to transfer heat from the walls of a channel to a fluid by forced convection or vice versa. Especially in Lab-on-a-Chip systems the purpose of this is usually to keep the fluid at a well-defined, constant temperature. Such a thermal control is very important especially in biofluidic assays, where biochemical reactions require a well-defined temperature level or cells and biomolecules can only withstand a maximum temperature.

Speeding up heat transfer is an important research objective in many technological disciplines, and a large number of ideas have been developed how heat transfer enhancement can be achieved by special channel geometries or surface structures [32]. In context with Lab-on-a-Chip systems, such ideas have rarely been adopted, most probably due to the fact that in most cases the speed up in heat transfer that can be achieved just by shrinking the channel diameter is sufficient.

The efficiency of heat transfer can be characterized by the time it takes to heat up a fluid volume in contact with a thermal reservoir to a temperature close to the reservoir temperature (the same arguments apply to the reverse process, namely cooling). For heat transfer occurring purely by conduction, this thermal diffusion time scale is given by

$$\tau_{\mathrm{d}} = \frac{L^2}{k} \tag{3}$$

where it should be understood that $\tau_{\mathrm{d}}$ only characterizes the real heat-transfer time scale in an order-of-magnitude sense. In this expression, $L$ is the usual length scale parameter and $k$ the thermal diffusivity, related to the thermal conductivity $\lambda$, the material density $\rho$ and specific heat capacity $c_{\mathrm{p}}$ through $k = \lambda/(\rho \cdot c_{\mathrm{p}})$. The thermal diffusion time scale is also helpful for characterizing the stationary-state temperature field in a system where heat is transported by conduction, in the sense that the bigger $\tau_{\mathrm{d}}$ the bigger the temperature differences. This is simply due to the fact that $\tau_{\mathrm{d}}$ is inversely proportinal to $\lambda$.

Coming back to the heat-transfer problem in microfluidics, forced-convection heat transfer from a channel wall to a fluid, again it has to be stated that in the most general case it is not possible to derive a simple scaling law, since the flow pattern can change with $L$. However, for the case of fully developed laminar channel flow, a situation encountered in many practical situations, the time scale for forced-convection heat transfer $\tau_{\mathrm{c}}$ displays the same dependence on $L$ as the thermal diffusion time scale, i.e. $\tau_{\mathrm{c}} \sim L^2$. Thus, shrinking the channel diameter has a pronounced effect on the heat transfer to the fluid and rapidly diminishes the temperature difference between fluid (the sample processed in the Lab-on-a-Chip system) and solid (the channel walls).

In practice it has to be taken into account that a heat source or sink is usually not directly located at the walls of a microchannel, but outside of the device such that it is separated from the fluid by a layer of wall material. The time scale for heat transfer is then determined by both the time scales for conductive heat transfer in the solid and forced convection to the fluid. Both of them scale as $L^2$, and as a result the total time needed for heat transfer between the heat source/sink and the fluid displays the same scaling.

### 3.1.4. Mass Transfer and Mixing

Figure 8 shows a schematic of the typical mass transfer and the typical mixing problem in microfluidics. In the mass transfer problem molecules are transported to the wall of a solid where they react with other molecules. Mixing occurs between two fluid streams, where at least one molecular species initially in only one of the streams has to be distributed over both streams. Compared to the heat transfer problem described in the previous section, mass transfer and mixing in liquids is much more challenging. The reason for this difficulty is the molecular diffusion constant that is usually much smaller than the thermal diffusivity. Thus, mass diffusion in liquids is much slower than heat conduction, and in the past decade a lot of research effort has been spent to speed up mass transfer and mixing in microdevices. In the course of these activities a large number of methods for mass transfer and mixing enhancement have been developed, usually relying on inducing convective fluxes for distributing the molecular sample (for overviews, see [33–35], and Chap. 3 of this volume). Owing to this variety of approaches, it is impossible to formulate a unique scaling law capturing the size dependence of the mass transfer or mixing speed.

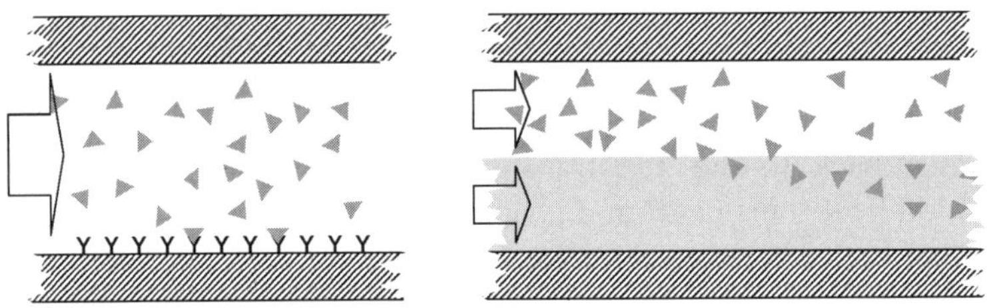

**Fig. 8.** Mass transfer problem with molecules being transported to immobilized reaction partners (*left*) and mixing problem where the transport occurs between two fluid streams (*right*)

In spite of the large number of possible solutions for mass transfer and mixing enhancement, devices that rely on pure diffusion still play an important role in Lab-on-a-Chip technology. This is due to their simplicity and the challenges for microfabrication technology that more advanced devices bring along with them. A typical channel structure for diffusive mass transfer is sketched in the left diagram of Fig. 8. Likewise, a simple mixer that works by diffusion is no more complex than the structure shown on the right side of the figure. In both cases the flow is laminar and the transport of molecules to their destination occurs by diffusion, since there is no component of the velocity field pointing in the direction in which the molecular flux builds up. Accordingly, the same scaling law as for diffusive heat transfer applies. Specifically this means that the time scales necessary for transporting most of the molecules of a sample to their binding partners (mass transfer) or distributing most of the molecules over two different streams (mixing) are proportional to $L^2$, i.e.

$$\tau_{\mathrm{d}} = \frac{L^2}{D} \tag{4}$$

where $D$ is the diffusion constant.

The $L^2$ scaling suggests that a suitable method for speeding up mixing is simply reducing the channel diameter. This is undoubtedly true, but can result in very small diameter channels. In addition, usually the mass flow of the fluids to be mixed is given, which means that shrinking the channel diameter increases the flow speed in the channel. This raises the question for the mixing length scale $\lambda_{\mathrm{d}}$, i.e., the length along a channel after mixing is close to complete. The mixing length and time scales are simply related to each other by the average flow velocity $u$ via $\lambda_{\mathrm{d}} = u \cdot \tau_{\mathrm{d}}$. Consider, for simplicity, the scaling being done only in the direction of the concentration gradient. This means that the walls shown in Fig. 8 move closer together while all other dimensions of the channel are kept fixed. With the constraint of a fixed volume flow $u$ scales as $L^{-1}$, resulting in $\lambda_{\mathrm{d}} \sim L$. Thus, even if the flow speed increases when reducing the channel width, the mixing length decreases linearly with $L$.

### 3.1.5. Hydrodynamic Dispersion

The term "hydrodynamic dispersion" refers to the broadening a concentration tracer suffers while being transported through a channel, as depicted on the left side of Fig. 9. The broadening of the tracer is usually mainly due to the velocity profile in the channel that effects a fast (slow) transport in regions with high (low) velocity. Clearly, hydrodynamic dispersion is detrimental to a number of operations in Lab-on-a-Chip systems, since it dilutes samples of initially high concentration and leads to a significant broadening of the residence–time distribution.

With respect to sample dispersion, pressure-driven flow is much more problematic than EOF, since the velocity profile in the latter is virtually flat. Sample dispersion in Poiseuille flow in circular tubes has first been analyzed by Taylor [36] and Aris [37]. The main result of their analysis is that after a sufficiently long time the cross-section averaged concentration plotted against the coordinate along the channel assumes a Gaussian shape, irrespectively of its initial distribution, as indicated on the right side of

Fig. 9. The broadening of this Gaussian occurs as a diffusive process with an effective diffusivity given as

$$D_e = D + \frac{u^2 L^2}{192 D} \tag{5}$$

where $u$ is the average flow velocity, $L$ the tube diameter, and $D$ the molecular diffusivity. Thus, the higher the velocity, the faster a sample will disperse.

From (5) the width of the sample distribution can be derived without effort, giving $\Delta x = (2D_e t)^{1/2}$. To calculate the scaling behavior with $L$ it is assumed that the first term in (5) can be neglected, an assumption that is valid in many practical cases. The quantity of interest is usually the width of the distribution after it has passed a channel of given length. Interestingly, this quantity displays the same scaling behavior no matter if the flow velocity or the total volume flow is fixed during scaling. In both cases it is found that $\Delta x \sim L$. This shows that reducing the diameter of a microchannel can help to mitigate the undesired effects of hydrodynamic dispersion, an advantage of miniaturization that has so far been less frequently quoted than the speed up of heat and mass transfer.

**Fig. 9.** Poiseuille flow transporting an initially well-localized concentration tracer through a channel (*left*) and evolution of the cross-section averaged concentration (*right*). Irrespectively of the initial concentration profile, the final distribution is a Gaussian

### 3.2. Scaling Analysis for Two-Phase Flow

Two-phase flow has received increasing attention in microfluidics and Lab-on-a-Chip technology in the past few years. Under the genus "two phase" a variety of different systems are subsumed. Examples are cell suspensions, gas–liquid, and liquid–liquid systems, where for the latter it is understood that the liquids are immiscible. In macroscopic devices for fluid processing the phases can often be classified as "disperse" (e.g., cells or bubbles) or "continuous" (the surrounding fluid medium). In microscale gas–liquid or liquid–liquid flows this distinction is often less clear, since the size of droplets or bubbles is usually comparable to the channel diameter. Because of this fact a large variety of microscale two-phase flow patterns and phenomena have been observed. Naturally, the quantities that have been in the focus of the analysis presented in the previous section such as time scales for heat and mass transfer are also of interest for two-phase flows. However, because of their complexity and diversity it is almost impossible to derive any scaling laws that are general enough. Rather than that, in this section a few specific scenarios with high relevance for Lab-on-a-Chip applications are illuminated and analyzed in terms of their scaling behavior.

### 3.2.1. Capillary Filling

Figure 10 depicts a situation that is often encountered in microfluidic systems, even if these are designed for single-phase flow. A liquid–vapor meniscus penetrates into a channel, driven by the Laplace pressure

$$\Delta p = \frac{2\sigma}{R} \tag{6}$$

where $\sigma$ is the surface tension and $R$ the curvature of the meniscus. The reason why capillary filling is also relevant in systems based on single-phase flow is the fact that before use the channel network has to be flooded with liquid, which is often done by capillary forces. It is extremely important that the filling process proceeds in such a way that all channels and cavities are completely flooded. The filling process can be controlled and regulated by using channels with a tailor-made diameter. In turn, this means that the course of filling in a network-structure can be different depending on the channel diameter in the different branches.

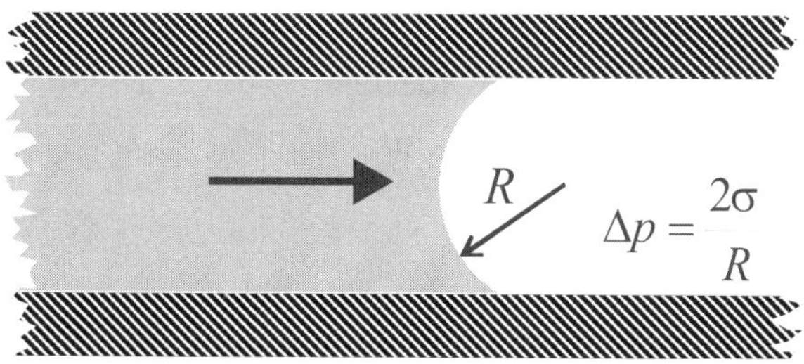


**Fig. 10.** Sketch of capillary meniscus driven into a channel by the Laplace pressure

In the following it is assumed that the liquid–vapor meniscus has a spherical-cap geometry, yielding a well-defined and uniform curvature $R$ in (6). This assumption is clearly fulfilled for a static meniscus in a small-scale channel of circular cross section, but can break down for a highly wetting liquid in a channel of rectangular cross section ([38], Chap. 4 of this volume). In addition, the peculiarities of the initial stages of the filling process when the meniscus enters the channel are disregarded (see, e.g., [39]) and only the situation when the meniscus is sufficiently far away from the channel entrance is considered. Then the position of the meniscus $x(t)$ as a function of time can be obtained from the solution of the Lucas-Washburn equation [40, 41] from which $\dot{x}(t) \propto \sqrt{R/t}$ can be derived. This implies that the capillary filling speed in a channel scales as $L^{1/2}$, where $L$ is the channel diameter. When $L$ is decreased, the Laplace pressure rises, but this increase of the driving force is overcompensated by the increase in viscous friction obtained from Poiseuille's law. Thus, capillary filling in small channels is slower than in large ones, but the decrease in filling speed obeys a square-root law and is only moderate.

### 3.2.2. Droplet Formation

In the past few years an increasing number of microfluidic systems have been described in the literature that rely on droplet transport in small channels. Such droplets can be used as (bio)chemical reactors, with the advantages that the liquid in a droplet being transported through a channel is kept well-stirred and sample molecules stay confined in a small volume. In the context of droplet dispensers and ink-jet systems a lot of research effort has been spent to create monodisperse droplets at a high repetition rate (see [42] and references therein). By contrast, in Lab-on-a-Chip systems the task is to create droplets on chip, which usually means that simpler solutions than those stemming from ink-jet technology are preferred. There are two main flow configurations to create droplets in microchannels (cf. Fig. 11). One configuration relies on two crossing streams of immiscible liquids, where droplets are created from the sidestream. The second configuration is based on coflowing streams, where the sheath fluid flows in the same direction as the droplet-forming fluid.

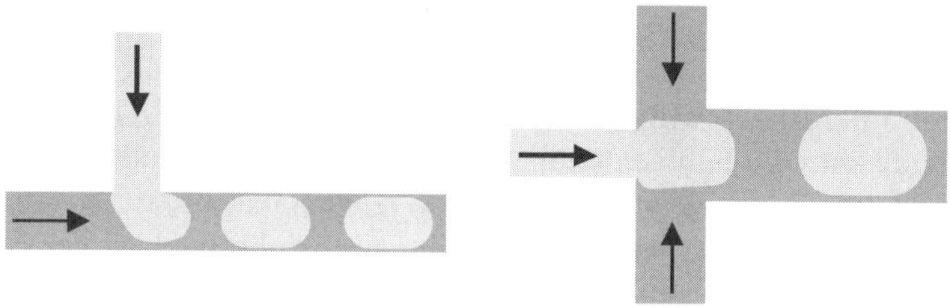

**Fig. 11.** Droplet formation in a channel by crossing (*left*) and coflowing (*right*) streams

A fundamental aspect of microfluidic droplet generators is the size of the droplets that can be created with a specific configuration. In the context of a scaling analysis the question is how the droplet size changes if the channel geometry together with the orifice from which the droplets emerge is scaled. In a general context this question is difficult to answer. An illustration for this difficulty are flow-focusing devices in which very small droplets are produced by narrowing the cross section of a liquid jet which decays into droplets [43]. However, there is a simple "rule of thumb" applicable in a large number of practical situations. This rule of thumb is inspired by early work of Rayleigh on the decay of an inviscid jet into droplets [44]. Rayleigh showed that usually an inviscid jet decays into droplets with a diameter of about 1.9 times the jet diameter, i.e. the droplet diameter is proportional to the jet diameter. This proportionality has been found to hold for a variety of drop generators based of drop formation by capillary forces (as opposed to gravity forces) and usually special efforts are needed to create drops smaller than the orifice diameter [45]. Droplet creation by the methods depicted in Fig. 11 is likewise governed by capillary forces. Thus, it can be concluded that in most practical cases the droplet diameter scales as $L$, where $L$ characterizes the size of the orifice or confluence region where the droplets are created. Such a scaling behavior makes the design of microchannel droplet generators very straightforward: It means that a rescaled system will work in the same way as the original system, with the ratio of droplet and channel diameter being a fixed quantity.

### 3.2.3. Blocking of Channels by Bubbles

Bubbles are often a major problem in microfluidic systems since they can block channels and thereby inhibit liquid flow. Even if a system is constructed in such a way that no air from outside can enter, bubbles can still emerge, e.g. by degassing of the liquid processed in the channels. Bubbles are much more difficult to remove in microfluidic systems than in their macroscopic counterparts. The reason for this problem are surface tension forces that inhibit bubble transport. Two situations that are often encountered in microflows with bubbles are depicted in Fig. 12. The left side of the figure shows a bubble entering a constriction, as considered in [46]. On the right side, a bubble sitting on a solid is depicted, where the contact angles $\theta_L$ and $\theta_R$ on the left and the right side of the bubbles are different. Such a so-called contact-angle hysteresis is a common phenomenon and occurs due to surface roughness or heterogeneities [47]. The situation depicted on the right side of Fig. 12 can occur if a force (e.g., from a pressure difference) acts on the bubble. Instead of inducing a motion of the bubble, the force is balanced by the difference in surface-tension forces, which comes about from the different curvatures of the gas–liquid interface on the left and on the right [48]. A similar thing happens if a bubble enters a constriction, as depicted on the left side of Fig. 12. Owing to the higher interface curvature in the narrow passage, the magnitude of the Laplace pressure $P_R$ is bigger than $P_L$ Eq. (6), thus counteracting the forces driving the bubble into the constriction.

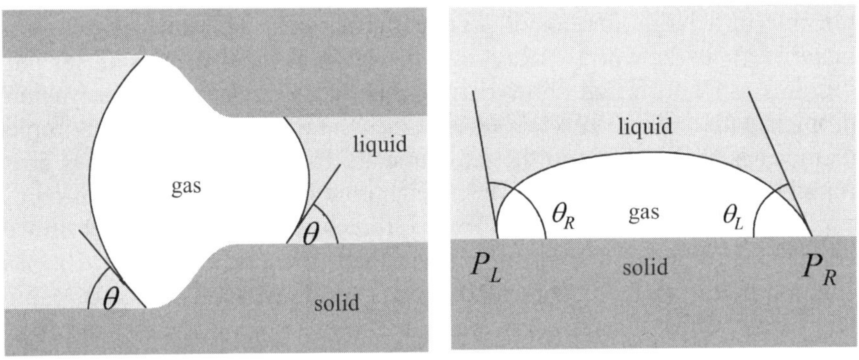


**Fig. 12.** Bubble entering a microchannel constriction (*left*) and bubble attached to a surface displaying contact-angle hysteresis (*right*). In both cases a force has to be applied to deform the gas–liquid interface in the depicted manner

In both cases depicted a critical pressure difference is needed to force the bubble into the constriction or move it along the surface. In the latter case, the critical pressure difference is reached if $\theta_R = \theta_a$ and $\theta_L = \theta_r$, with $\theta_a$, $\theta_r$ being the advancing and receding contact angle, respectively. The form of the solution for the gas–liquid interface shape in this critical state infinitesimally before the bubble moves together with the choice of scaling variables shows that this shape is scale-invariant, i.e., a bubble in the critical state looks the same no matter how big it is [48], given that gravity forces can be neglected. Using the fact that the Laplace pressure is inversely proportional to the radius of curvature of a fluid interface (6) it follows that the pressure difference required to move a bubble on a solid surface is proportional to $L^{-1}$, if $L$ is a length characterizing the size of the bubble. Similarly, for the case of a bubble forced into a constriction, the dependence of the Laplace pressure on the radius of curvature shows that the required pressure difference is proportional to $L^{-1}$, where now $L$ characterizes the channel diameter. The inverse proportionality to $L$ also explains why blocking of channels by bubbles is a typical microscale problem and much less of an issue for larger devices.

### 3.2.4. Particle Trapping by Dipole Forces

The separation and concentration of particles from liquid samples is an important step in many microfluidic assays [49] The term "particle" refers to a variety of different objects that can be suspended in a liquid, such as blood cells, bacteria, or magnetic beads. Such particles can be separated from their surrounding medium using size-exclusion filters, but separation based on dipole forces has many advantages compared to that, such as a higher selectivity and an easy release of the trapped particles. The dipole forces are either due to a magnetic (as described in Chap. 6 of this volume) or electric (as described in Chap. 8 of this volume) field, and the particles' dipole moments can either be permanent or induced. A typical arrangement for trapping particles by dipole forces is sketched in Fig. 13. Interdigitated field sources (either for the electric or the magnetic field) of alternating polarization are arranged periodically on the bottom wall of a flow cell. If the dipole moment per unit volume of the particles is bigger than that of the surrounding medium, they are attracted to the high field strength regions and get trapped at the surface of the field sources.

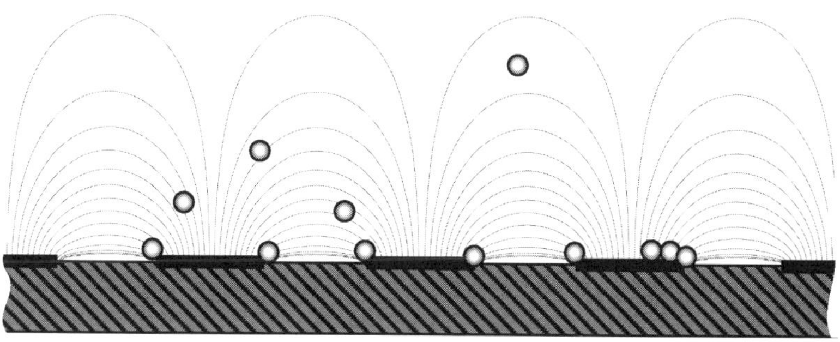

**Fig. 13.** Schematic of a flow cell for particle trapping based on dipole forces. The field lines emerge from sources that are arranged periodically on the bottom wall of the cell

One of the most important figures of merit of a dipole-force particle concentrator is the maximum flow rate with that the suspension can be fed to the system such that still a significant amount of particles gets trapped. If the flow rate is too high, the hydrodynamic drag forces outweigh the dipole forces with the result that the particles move along with the fluid stream instead of being immobilized at the surface. In the following, $L$ should be a measure for the particle size and it is assumed that the size of the field sources is kept fixed. Then, according to Table 1, the force on a particle with a fixed (electric or magnetic) polarization density scales as $L^3$. According to the Stokes formula, the hydrodynamic drag force on a particle is proportional to $L$ and to $v$, the particle's velocity relative to the surrounding medium. The maximum flow velocity $v_{\max}$ at which a particle can still be trapped in an array such as shown in Fig. 13 is obtained by equating the maximum dipole force and the hydrodynamic drag force. From that it follows that $v_{\max}$ is proportional to $L^2$. It is well known that dipole-force particle concentrators have quite severe limitations with respect to the flow velocities at which they operate. The $L^2$ law derived above offers an explanation for this observation.

**Table 2.** Scaling of different quantities related to single- and two-phase microflows

| | | Quantity | Scaling behavior | Assignment of $L$ |
|---|---|---|---|---|
| Single-phase flow | Pressure-driven flow | Flow rate | $L^4$ | Channel diameter |
| | | Viscous heating rate per unit volume | $L^{-2}$ | " |
| | | Sample dispersion width | $L$ | " |
| | EOF | Flow rate | $L^2$ | " |
| | | Joule heating rate per unit volume | $L^0$ | " |
| | Pressure-driven flow & EOF | Time scale for forced convection heat transfer | $L^2$ | " |
| | | Time scale for diffusive mass transfer or mixing | $L^2$ | " |
| | | Mixing length scale | $L$ | " |
| Two-phase flow | | Capillary filling speed | $L^{1/2}$ | " |
| | | Droplet diameter | $L$ | Size of orifice or confluence region |
| | | Pressure difference for bubble mobilization | $L^{-1}$ | Channel diameter or bubble size |
| | | Maximum flow velocity for particle trapping by dipole forces | $L^2$ | Particle size |

### *3.3. Summary of Scaling Laws*

As explained and exemplified in the previous paragraph, studying the scaling behavior of the various phenomena relevant to microfluidics can provide a significant insight to system designers without the need to conduct complex experiments or simulations. While for single-phase flow some quite general scaling laws can be derived, the complexity of two-phase flows only allows illuminating some specific effects and scenarios. In any case, for such a scaling analysis it should be carefully defined which quantities are kept fixed and which are varied. Table 2 summarizes the scaling laws derived for single-phase and two-phase flows.

## 4. System/Engineering Concepts and Design Approaches for Microfluidics

In the subsequent sections system aspects of microfluidic devices as well as various design approaches are discussed. With respect to the former, aspects concerning the functionalities, architecture, and the realization of microsystems are considered. The latter section – design approaches – introduces various modeling approaches appropriate for systems design and the deduction of more comprehensive design rules beyond the aforementioned scaling analysis. Many of the issues highlighted in this section are discussed in more detail in the various chapters of this volume. The purpose of the following sections is mainly to provide the reader with some overview and orientation in the by now vast field of Lab-on-a-Chip technology.

### *4.1. Engineering Concepts for Microfluidic Systems*

Surely, brute force downscaling of macroscopic devices, as cartooned in Fig. 14, is not a promising route for the design of μTAS and Lab-on-a-Chip systems. If we are to design a microfluidic system there are three questions to be answered for the decision on the concept: What functionalities are needed? How can they be transferred into a miniaturized system? And what is the needed or desired integration depth? We start by considering the first two questions and close the section by briefly discussing further important aspects as interfacing, integration, materials, and fabrication. Operations common to almost every μTAS are mixing, separation, detection, fluid control, and actuation. These more or less standard operations are briefly addressed below. For further reading we refer to the following chapters and the given references. Of course there are plenty of other functionalities such as various sample-preparation steps, storage of reagents, lysis and purification, to name a view, the discussion of which is well beyond the scope of the present chapter that has more of an overview character.

**Fig. 14.** After his great success in miniaturization Prof. H.D. Bunsen was a bit too optimistic when it came to pouring in the sample

### 4.1.1. Mixing

Because of the low mass diffusion in liquids micromixing of laminar flows has been an active research area in the past decade. Detailed reviews of the field are given in Chap. 3 and in Hardt et al. [33], Hessel et al. [34], and Nguyen and Wu [35]. An essay on the interesting side-aspect of mixing and irreversibility was recently given by Gollup and Pine [50].

In the context of Lab-on-a-Chip systems, passive micromixers, which do not need any additional energy input other than that needed to pump the fluids, are most common. Besides the scaling discussed above, two dimensionless groups are particularly helpful with respect to the basic design of such systems, the Fourier number (Fo) and the Peclét number (Pe). Fo relates the available time span to achieve mixing, essentially the residence time for flow systems, to the mixing time $\tau_m$, cf. e.g. (4). If the involved times are translated into corresponding lengths by means of the average velocity, Fo may be expressed as the ratio of the channel length to the mixing length, $L_m = \tau_m \cdot u$. Thus, Fo (<) > 1 indicates (in)complete mixing. On the other hand, Pe relates the convective mass transport to the diffusive one, i.e. $Pe = u{\cdot}L/D$.

Passive micro mixers can be categorized by the applied hydrodynamic principles. In the present context the most important ones are multilamination with subsequent geometric or hydrodynamic focusing, splitting and recombination of laminar flows and chaotic advection.

In terms of the efficiency of a passive micro mixer, two further characteristics need to be considered. First, besides an average degree of mixing the homogeneity of the mixing result is often important. For instance, chaotic flows may induce an extremely fast mixing in certain regions of the flow domain while others are hardly affected by the chaotic flow and unmixed islands persist for long times. Second, the pressure drop has to be taken into account. In this regard mixers relying on successive splitting and recombination or chaotic transversal advection are most favorable, since a logarithmic dependence of the mixing length on Pe is achieved, while typical pressure drops are similar to those of corresponding straight pipe flows, i.e. showing a linearly increase with the length of the mixing channel.

Last but not least, problems of reduced peak concentrations caused by the Taylor-Aris dispersion, cf. (5), can be remedied by enhancing lateral species transport by means of efficient mixing approaches (see, e.g., [6]).

### 4.1.2. Separation

Separation is one of the key functionalities incorporated in μTAS, and increasing separation performance was one of the drivers for miniaturization [51, 52]. Moreover, several chapters of the present book are dedicated to detailed discussions of miniaturized separation techniques or touch important aspects thereof. Nowadays almost all separation techniques based on the various variants of electrophoretic or chromatographic effects have been successfully transferred into miniaturized systems. Discrimination of particles and species relies on physicochemical properties such as electromagnetic charge or polarizability, physicochemical affinity, diffusivity, size, mass, mechanical properties, or even on the intrinsic mobility as in the case of motality analysis of sperms by means of a microfluidic H-filter [53]. The complete field of miniaturized separation technologies is much too complex to be reviewed in the present context and the reader is referred to the subsequent chapters and to the references given therein. Yet, from a practitioner's viewpoint it is noteworthy that utilized microstructures do not necessarily have to be fabricated from scratch. Beads [54] and membranes [55] are readily incorporated into microfluidic separation systems. Since membrane technology has been a subject of active research for several decades, versatile products with tailored permeability, selectivity, and geometry are often commercially available.

#### 4.1.3. Sensing and Detection

Sensing in microfluidic systems naturally implies that minute sample volumes and accordingly small amounts of analyte are to be detected. Among the diversity of detection methods, electrochemical methods are most suitable for integration since miniaturized electronic circuits are ubiquitous and can readily be combined with microfluidics. A common sensing method is the direct measurement by means of an electrochemical transducer transferring a target concentration into an electrical signal. For instance the oxygen concentration is often determined relying on an amperometric measurement where a semipermeable membrane is inserted between the electrodes and the sample volume. In context with the detection of heavy metal ions Zhu et al. [56] used mercury droplets as electrodes. And Goral et al. [57] reported the detection of nucleic acid sequences by means of an interdigitated ultramicroelectrode array on glass substrate integrated in a micro channel.

On the other hand, in context with DNA-chip applications, such as the analysis of gene expressions and single nucleotide polymorphisms (SNPs), electrochemical sensing is utilized for the detection of binding events (see, e.g., [58–60]. Furthermore, as shown by Choi et al. [61] different voltammetry modes allow for the detection of immobilization of DNA probes on gold electrodes and for the detection of target DNA.

In the context of optical detection methods, the challenge regarding high sensitivities is often solved using bulky, high-resolution off-chip equipment. In that respect, many microsystems based on external light sources and photo-detectors (see, e.g., [62]) have been realized, while components as lenses, mirrors, and wave guides are often integrated on-chip. For fluorescence detection simplified layouts involve blue light-emitting diodes (LED) and silicon photodiodes instead of a laser source and a photomultiplier tube, respectively (cf., e.g., [63]). Partly along these lines a compact fluorescence sensor has been recently realized by Irawan et al. [64]. Balslev et al. [65] report on a fully integrated sensor system for absorption measurements, comprising a laser, wafeguides, and detectors, fabricated by applying a lithographic step on a polymer thin film subsequent to a spin-on process onto a doped silicon wafer. Zhu et al. [66] report on a highly sensitive integrated microfluidic UV absorbance detector.

The small sample volume is commonly compensated by axial optical pathways with multireflections to increase the optical path length. This may, however, lead to a decrease in the signal-to-noise ratio (SNR) [67] and a larger minimum bandwidth [68]. Miniaturization is challenging with respect to these aspects since downscaling generally implies attenuation of optical signals. Yet, miniaturization reveals its full strength when it comes to single molecule detection [69]. In this case the background noise is drastically reduced by decreasing the sample volume, whereas the signal or source intensity is fixed. Thus, miniaturization enables a considerable increase of the SNR.

### 4.1.4. Pumping

Considering the scaling properties of pressure-driven and electroosmotic flow, it is apparent that electrokinetic pumps (EKPs) suggest themselves for miniaturized applications. If channel diameters are reduced the EOF rates decrease proportional to the cross-sectional area of the channel, whereas pressure-driven flow rates decrease as the cross-sectional area squared, for fixed voltage and pressure drop, respectively (cf. Table 2). Aside from the scaling behavior important advantages of EKPs are virtually plug-flow profiles in straight channels and pulse-free flows. On the other hand there are several disadvantages. Primarily, these are high voltages, electrochemical reactions occurring at the electrodes, and Joule heating. Several studies have addressed these issues and improved pump concepts are developed (see, e.g., [70, 71]). Because of the mentioned disadvantages of EKPs and the fact that only specific fluids can be pumped, which is also true for some alternative methods such as for magnetohydrodynamic and ferrofluidic pumps [72], many microfluidic applications rely on seemingly unfavorable pressure-driven flow induced by mechanical means, such as syringe pumps, peristaltic pumps, pressurized chambers, and membrane or diaphragm pumps. Yet, these pumps are less suitable for miniaturization and much harder to be integrated in microfluidic systems. Concerning further details and the discussion of the multitude of alternative pumping mechanisms, such as acoustic streaming, AC electroosmosis, centrifugal forces, and capillary action, to name a few, the reader is referred to Chap. 2 and to the comprehensive reviews by Laser and Santiago [20], Singhal et al. [73], Woias [74], and Nguyen et al. [75].

### 4.1.5. Valving

Microvalves, as the essential components for flow control, are commonly classified (e.g. [76]) as passive, needing no energy input except for driving the flow, active, with additional energy input, and so-called intelligent valves, where the valve operation is inherently coupled to a stimulus provided by the flow. Among these classes passive valves are typically most easily to be integrated into microfluidic networks. In context with multiphase flows they typically rely upon the minimization of surface energies, such as the balance of surface tension and fluid pressure forces. The generic design of such valves is the microchannel constriction discussed earlier (Fig. 12). A slug or meniscus propelled by a subcritical pressure entering from the left will stop at the constriction. In case of single-phase flows simple passive valves rely on diffuser-type geometries exhibiting a direction-dependent pressure drop for Reynolds numbers beyond the creeping flow regime (see, e.g., [77]). Alternatively, the directional dependence of the fluidic resistance can be achieved using moving parts, such as cantilevers, membranes or even chains of micron-sized dielectric particles [78].

There are a number of requirements miniaturized valves have to fulfill, such as minimum dead volume, low energy consumption, good sealing, short response times, etc. The various different valve concepts have their strength in one or the other field (see, e.g., the recent review by Kwang and Ahn [79]). The choice of possible concepts is, however, severely limited if low-cost solutions are needed. Principles used for macroscopic flow control can be partly adapted to the microscopic needs, but particular attention with respect to sealing and friction has to be paid (e.g., [80]). One realization of an active microfluidic multiway valve is shown in Fig. 15. The core element is an elastomer disk comprising tailored channels, which is pressed onto a microfluidic chip. Depending on the position various functions as closing, connecting, and multiplexing of the incoming channels are possible. Typical channel and disc dimensions for this design are a few hundred microns and a few millimetres, respectively.

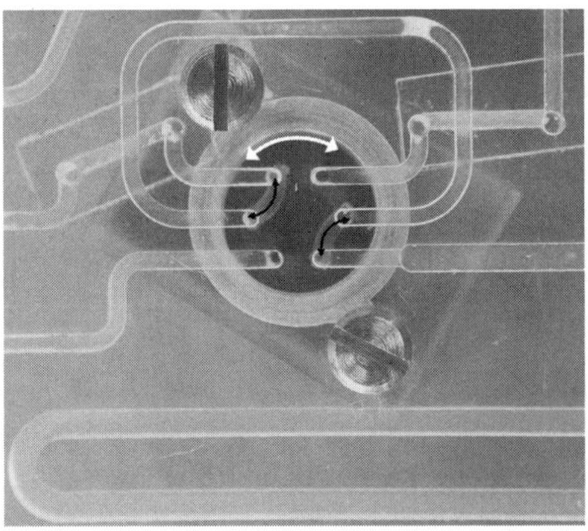

**Fig. 15.** Example of a multiway turning valve attached to a laminated microfluidic polymer chip (courtesy of IMM, Germany). Microfluidic channel inlets can be connected (*black arrows*), closed or multiplexed depending on the position and structuring of the disc (*white arrow*)

### 4.1.6. Manipulation of Bubbles and Slugs

Utilization of immiscible fluid compartments has turned out to be beneficial in a variety of contexts, such as amplification and analysis of nucleic acids (cf., e.g., [81–83]), protein crystallization, and cell encapsulation [84, 85], to name a few. Compared to single-phase flow enhanced fluid-to-wall as well as fluid-to-fluid mass transfer and a minimum signal dispersion are found in Taylor flow due to the circulation within the individual slugs (see, e.g., [86]). In this context curved or wavy channels are especially beneficial since alternating asymmetric internal circulation is induced leading to chaotic advection and rapid mixing within slugs ([87]).

As mentioned above, common methods applied to achieve regular drop formation are crossing of fluid streams or utilization of sheath flows (cf. Fig. 11). Once the droplets are formed, the basic unit operation to be performed are droplet fusion and fission (for a review on experiments and simulation of droplet manipulation see Christini and Tan [88]). Microchannels offer unique approaches for highly regular manipulations on individual drops, as exemplarily shown in Fig. 16. The sequence of micrographs on the left show self-synchronization and joining of two liquid slugs [83]. Both slugs to be mixed are propelled by a well-defined pressure. The slug that reaches the merging section first is stopped by a passive valve. The second slug triggers the valve [89] releasing the first slug and both slugs move synchronously downstream. In case of suspended droplets, Ahn et al. [90] demonstrated that controlled coalescence can be facilitated if droplet pairs are exposed to an electric field.

Prominent examples of two break-up processes mediated by microchannel T-junctions and obstacles are shown on the right hand side of Fig. 16 [91]. Both methods are highly reproducible break-ups and the polydispersity obtained by this sort of controlled emulsification is typically very low.

The positioning of slugs can be achieved or at least supported by means of dedicated constrictions as sketched on the left hand side of Fig. 12 or by locally varying surface energies, such as spotting hydrophilic patches on the channel walls.

Finally, liquid slugs can even be reshuffled using loop channel designs [92] similar to the rearrangement of sample slugs in single-phase EOFs by means of delay loops [93].

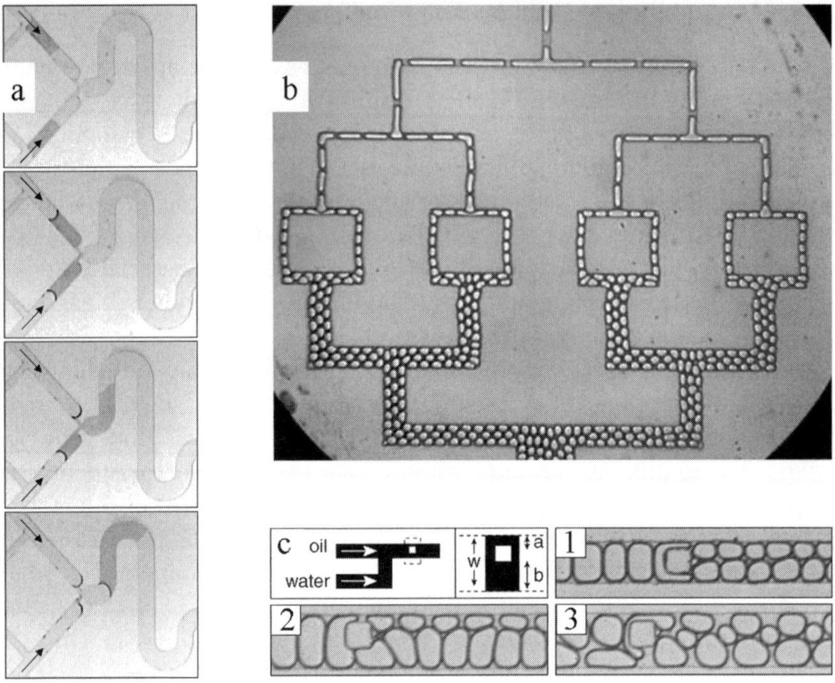


**Fig. 16.** (**a**) Four consecutive micrographs showing synchronization and merging of liquid slugs, (**b**) Microchannel branchings for highly reproducibly droplet fission and (**c**) obstacle-induced droplet break-up within a microchannel. Figures (b) and (c) reprinted with permission from (Link et al. [91]). Copyright (2004) by the American Physical Society

### 4.1.7. Integration and Materials

While a good deal of the pioneering work on microfluidics was mainly focussed on particular aspects and certain small-scale realizations of single operations, most developments are nowadays more application oriented and more and more economically driven. To this end, complete solutions ranging from sample input to signal read out are needed.

Having discussed a number of basic operations and corresponding engineering concepts and bearing in mind that there are numerous further operations on Lab-on-a-Chip systems the problem remains to integrate all these functionalities into one system.

Regarding the integration of functional submodules, two examples of somewhat extreme approaches are shown in Fig. 17. One may decide for an extremely modular system with distributed functionalities as shown on the left side or a highly integrated possibly self-contained and monolithic system (see right side of Fig. 17). While the latter is particular appealing since it transfers the ideas of highly integrated electric circuits to the microfluidics field, a modular system has the advantage that it allows to apply established components where possible and to focus the development effort on a few, most challenging, components. Thus, in particular in the course of development cycles where the solution approaches have to undergo successive revisions, modular concepts are preferable. With respect to the cost pressure present in any commercialization, the general design rule for fluidic systems is the following: Having a maximum functional integration depth while keeping the system complexity as low as possible. In this regard, world-to-chip and chip-to-chip *interfacing* are major issues. While reliable interconnections are generally a mandatory prerequisite, they are of particular importance in context with modular approaches. Interfacing has to provide the transmission of driving forces for microfluidic operations and it has to provide an option for introduction and possibly removal of samples. So far a variety of different solutions have been reported in the literature including "quick-connects" and "breadboards" [95], "plug and play" interconnects [96], and others (see, e.g., [97–100] and the citations given in the μTAS review article by Vilkner et al. [101]). Yet, compared to the reliability of fluid manipulation methods and analysis techniques, interfacing and in particular the availability of standardized solutions lacks

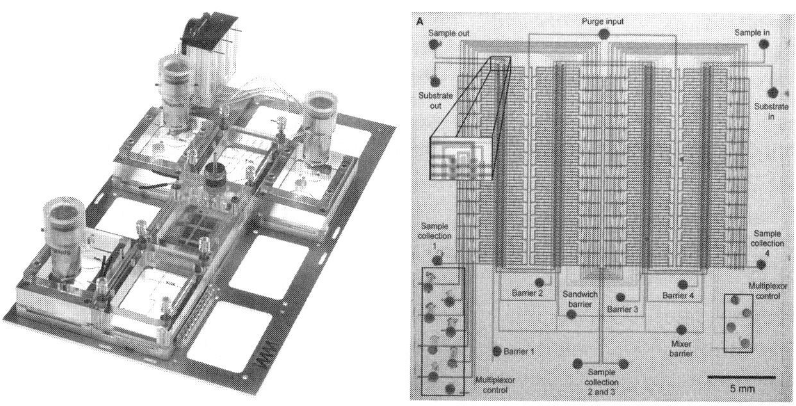


**Fig. 17.** Left: Development platform for microfluidic protocols and modules. Polymer chips are connected among each other and to liquid reservoirs and pumps, which are kept on a separate chip (courtesy of IMM, Germany). Right: Prototype of a highly integrated microfluidic realization of a comparator array. Reprinted with permission from Thorsen et al. [94]. Copyright 2002 AAAS

somewhat behind. One of the principal issues in this regard are dead volumes. Whilst the handling of picoliter quantities of sample is well established within microfluidic systems, dead volumes in an interconnect can easily be several orders of magnitude larger [97]. Standardized solutions for appropriate, simple, and leak tight interconnections are so far not in reach. Therefore, in this regard autonomous, highly integrated, monolithic designs are favorable and architectures of modular microfluidic systems should comprise only a minimum number of (inter-)connections.

Microfluidic systems have an enormous economical potential in the area of life-science as medical diagnostic platforms and are furthermore of high importance with respect to public health care particularly in the developing world [102]. In this regard one-time devices are needed since necessary cleaning protocols accompanied by reliable quality assurance as needed for reusable devices imply high efforts and costs. The problem of equipment sterilization gets even harder for microfluidic systems compared to macroscopic devices. Moreover, the risk of getting false-positive results due to a contamination of a patients sample by its predecessor can not be completely ruled out. The demand for low-cost, disposable systems requires other materials than silicon or glass, which have been used in the early days of microelectromechanical systems (MEMS). In context with exploratory research of microfluidic systems polydimethylsolixane (PDMS) is widely being used. Although PDMS is a rather soft elastomer, microstructures down to the submicron size can be structured by casting [103]. Moreover, highly integrated systems comprising pneumatic pumps, valves, and mixers can be fabricated by soft-lithografic methods in PDMS in much shorter times than are typically needed for classical thin-film technologies (see, e.g., [94, 104, 105]).

Generally, various polymer materials come into question, depending on different factors as biocompatibility, chemical resistivity, optical, mechanical, and thermal properties. With respect to the former, unspecific binding or so-called biofouling of surfaces is a major problem in many biofluidic applications, since sample concentrations, for instance the concentration of proteins, may be thereby reduced below the detection limit. Generally, there are two common ways to address this issue. First, appropriate bulk materials are chosen or bulk material properties are tailored by means of copolymerization of a functional monomer with a commonly used monomer incorporating new properties into the parent polymer [106–108]. Second, the surface of micro devices can be modified while retaining the plastic's bulk properties. Here, covalent binding of coatings is commonly considered to be more efficient compared to physical treatments or adsorbed coatings (see, e.g., Wu et al. [109] and references therein). Prominent examples are poly(ethylene glycol) (PEG) coatings (e.g. onto polymethylmethacrylate, PMMA) [110] or silanisation [111].

With respect to fluorescence measurements of minute samples, the optical properties of the bulk materials have to be considered, since many polymers show (background) fluorescence, particularly when excited with light of wavelengths below about 500 nm (see, e.g., [112–114]). All of the most common polymer materials, PDMS, cycloolefin copolymer (COC), and PMMA show a relatively low or at least tolerable autofluorescence, although generally higher than that of glass. Furthermore, the magnitude of the fluorescence was found to decrease over time and depends on the production process as well on the postproduction age. Specifically, autofluorescence of chips was found to be significantly higher than that of the associated plastic materials [114]. Furthermore, spectral material properties are important, specifically in context with absorbance-based detection systems. In contrast to quartz and certain glasses that are transparent well into the UV range, most polymers are opaque below 300 nm [115] and therefore not suitable for devices based on UV-absorbance [66].

The decision on the material is, of course, also related to the choice of manufacturing techniques. When it comes to the production of larger quantities direct methods such as mechanical polymer micro machining, casting [116, 117], soft lithography [118], or laser micro machining [119] have to be substituted in favor of replication methods such as hot embossing [120, 121] and in particular injection molding [122], which is often applied for mass production. Because of the short cycle times and the high reproducibility this technique is especially appealing for commercialization of microfluidic systems. By tuning of the machine parameters as injection pressure, postpressure, and the temperature ramps, structures with size of the order of microns and below can be replicated with aspect ratios (channel depth over width) up to about 10 [123]. The technique is, however, limited concerning the shape of the replicated structures, since basically planar so-called 2.5-dimensional structures can be realized. However, more complex designs are possible by stacking of microfluidic chips.

Apart from the manufacturing of the polymer chip itself, the cost of back-end processes such as drilling, metallization, and sealing has to be considered. Sealing is most easily achieved for PDMS chips because the material hermetically seals to glass. In case of commonly used thermoplastics, such as polycarbonate (PC), PMMA, or COC, the challenge is to assure a durable tight sealing of the microstructures without affecting their shape, size, and surface properties and without entering of impurities or adhesives into the channels. Common methods are adhesive bonding, thermal bonding, and laser welding (see, e.g., [76, 124]).

### 4.2. Design Methods

Mathematical approaches are generally of great help in the design process of microfluidic systems. Depending on the chosen approach, rough performance estimations, detailed studies, and even design optimizations can be performed without building a single prototype. Moreover, mathematical modeling can help in controlling process parameters and conditions for operating certain assays on microfluidic platforms. If real-time simulations of microfluidic operations are possible they can be performed prior to, or simultaneously with, the actual analysis run alerting the user if problems are to be expected for the particular choice of fluids and operation conditions. For instance, a computer controlled handling system of a microfluidic chip could compute minimum heating or mixing times. If these are not met by the pre-defined operation protocol the user is notified that the corresponding operation, e.g. mixing, heating, or cooling, is expected to be insufficient. Alternatively, within a fully automated environment the protocol could be automatically adjusted to avoid the expected problems.

When it comes to the (computer-aided) design of microfluidic systems a variety of approaches are available. Microfluidic simulations can be performed on different levels concerning the details, which are explicitly included in the simulation approach. The hierarchy of models starts with extremely detailed atomistic models as Molecular Dynamics and the particle-ensemble based Direct Simulation Monte Carlo (DSMC). However, the former method requires typical time resolutions down to the order of femtoseconds, which is many orders of magnitude below the time scales relevant in practice. Because of the large gap between the time scales alternative methods are needed, such as the mesoscopic approach relying on dissipative particle dynamics [125, 126] or lattice-Boltzmann methods [127–129]. The discussion of these methods is beyond the scope of this book and the interested reader is referred to the chapters about multiscale modeling of the book by Karniadakis et al. [7].

On the next – less detailed – level are CFD methods based on continuum models. Here, the generic equations are of advection-diffusion type governing fluid dynamics, mass, and thermal transport when momentum components, concentration, and temperature fields, respectively, are inserted. Finally, on the least detailed level one has coarse grained approaches or reduced order models, such as so-called network or macromodels.

The corresponding computational costs are, of course, strongly model dependent. While high-detail approaches can give fundamental insight into certain features such as the occurrence of density oscillations close to a fluid-solid interface (cf. Fig. 5) they are usually inappropriate for system design. Continuum models, on the other hand, are considerably less laborious with respect to the needed CPU recourses while still providing highly detailed information. Although efficient numerical schemes as the finite-volume method (FVM) or the finite-element method (FEM) are available and are implemented in various commercial solvers (see, e.g., http://www.cfd-online.com/Links/), the simulation of complete microfluidic systems of moderate complexity is still beyond the scope of state-of-the-art soft- and hardware capabilities. Thus, for the computer-aided design of whole systems, simpler, less complex models are needed. A multitude of model-order-reduction techniques have been developed. We proceed by discussing some of the most relevant schemes mainly applied to systems with single phase flow in the following section and subsequently turn to some particular multiphase problems, highlighting methods for reducing the simulation effort.

### 4.2.1. Reduced Order Models for Single-Phase Flow

Typical CFD models often involve multi-million degrees of freedom and in many respects are still too costly using state-of-the-art soft- and hardware. Hence, from the viewpoint of a practitioner methods are needed allowing to reduce the computational effort in generic design problems of microfluidics. These methods should be based on models that incorporate all effects relevant on the microscale, yet keeping the model complexity as low as possible. To this end various approaches come into question. The approaches are classified into so-called black box models and network models. Black box models usually stand for mostly nonphysically motivated formal model-order reduction. For instance, usually one is left with the solution of sets of linear equations with several millions of unknowns, after discretization and linearization of the differential equations governing the system dynamics. To reduce the dimensionality of the problem the solution space may be truncated in such a way that only a restricted space, e.g., the Krylov subspace [7], is considered and the generalized variables usually lack any physical interpretation. In contrast, network models rely on the division of complex system into simple submodules, which can be described by efficient models, for instance by representation of fluidic components by their electric counterparts, as listed in Table 3.

**Table 3.** Analogy between fluidic and electric properties

| Fluidic quantity | Electric equivalence |
|---|---|
| Volume | Charge |
| Flow Rate | Current |
| Pressure | Voltage |
| Inertia | Inductance |
| Compliance | Capacitance |

With respect to the scaling analysis discussed above, it is noteworthy that the mathematical correspondence between a fluidic and an electrical property does not imply that both obey the same scaling. In the general case in fact, they do not. The fluidic resistance for instance scales as $L^{-4}$ (cf. Table 2), whereas the electric resistance scales as $L^{-2}$, with $L$ being the characteristic lateral dimension. The compliance, i.e. a pressure-induced volume change $dV/dP$, e.g. of a membrane covering a square chamber scales as $L^6$, while the electric analogue, the capacitance, scales as $L^2$, where $L$ denotes the characteristic dimension of the membrane and the capacitor surface area, respectively.

The formulation of an equivalent network for a laminar pressure-driven system starts by subdivision of the complete device structure into so-called lumped elements and identification of the fluidic and equivalent electric quantities as specified in Table 3 for each element (see, e.g., [130, 131]). Sets of differential or in the simplest case algebraic equations are derived by application of Kirchhoff's mesh and nodal rules. The mesh rule implies that the sum of pressure drops along a closed loop vanishes indicating that pressure acts as a potential. The nodal rule reflects mass conservation at network nodes. The picture gets a bit more complicated for the superposition of pressure-driven flow and EOF. In the simplest scenario (one-way coupling), where pressure is build up by electrokinetic forces but pressure-driven flow does not affect the EOF, the corresponding models can be solved consecutively starting with the electric part.

If streaming effects have to be considered, i.e. the fluid flow induces an electric current, the model equations for pressure-driven flow and EOF are inherently (two-way) coupled and a matrix representation has to be applied [132]. In the graphical representation within this scenario each port of a network element is equipped with two pins, one carrying the fluidic and one the electric information. However, for the design of microfluidic systems not only the flow itself but also the transport and possibly the separation of species have to be considered. So, the above mentioned model has to be augmented in order to allow for the solution of additional convection-diffusion equations. For steady-state applications, this is most easily achieved by implementation of spectral approaches ([133] and references therein). For instance, in the case of mixing by means of serial multilamination, species concentrations are readily expressed in terms of a cosine Fourier series [134]. In the context of network models, the Fourier coefficients are passed from one element to the other in down-stream direction by introducing additional pins for each species at each port of a network element. For each geometrical element, the dependence of the Fourier coefficients on the residence time is derived, establishing an input–output relationship, which is stored in a library. For the case that species concentrations are time dependent, such as in the case of migrating and broadening bands, the method of moments (cf. [37]) has been combined with the network model outlined above (see [133, 135] and references therein). In this case spectral methods are applied for solving the governing differential equations of the first moments of the cross-section averaged concentrations. As above, the corresponding Fourier coefficients are processed in the downstream direction starting from the first element [133]. By application of these approaches the computational effort can be reduced by orders of magnitude with often tolerable losses in accuracy.

### 4.2.2. Multiphase and Particulate Flows

The modeling of multiphase flows is usually more complex and the derivation of coarse-grained models is often not possible. To illustrate common CFD approaches with various levels of detail we start by considering the modeling of a particle suspension. As above (Sect. 3.2) under "particles" different objects such as cells, bacteria, beads, or a dispersed immiscible phase are subsumed. Yet, one has to keep in mind that the considered particles may be able to move reletive to the surrounding fluid by applying electric or magnetic fields. In more exotic cases microorganisms may be able to move away from unfavorable conditions, such as areas of high mechanical stress or high concentration of toxins [136].

Within approaches motivated by macroscopic scenarios the suspended particles are not resolved numerically. For instance in the case of blood flow simulations it is common to use an effective, e.g. hematocrit dependent, apparent viscosity. Prominent rheological models of blood behavior are the so-called Casson or Quemada liquids (see, e.g., [137, 138] and references therein). Alternatively, liquid and particle concentrations can be modeled as interpenetrating continua.

Within the simplest scenario particles are noninteracting among each other and are just convected with the liquid flow obeying a convection-diffusion equation. In the case of this so-called one-way coupling the liquid velocity field is not altered by the presence of particles and the governing Navier-Stokes equation is decoupled from the particle concentration.

In a slightly more complex scenario, we assume the particles to be propelled by external fields. These may induce magnetic or electric forces on the individual particles (cf. Fig. 13) or may cause sedimentation in the case of gravity. Particles are accelerated until the external force is compensated by the counteracting friction forces (see, e.g., [139]). The relevant time scale for this is given by

$$\tau = \frac{2\rho L^2}{9\eta} \tag{7}$$

where $\eta$ is the dynamic viscosity of the liquid, $\rho$ and $L$ denote the particle's density and radius, respectively. Because of the quadratic size dependence, $\tau$ is of the order of micro seconds for micrometer-sized particles in an aqueous solution. Thus, in contrast to macroscopic bodies microparticles move almost instantaneously with their terminal velocity defined by the balance of the external field-induced and the frictional forces. Here, a case differentiation is appropriate. In the high-concentration limit, the afore mentioned continuum approach is justified if the particle dimensions are well below all other relevant dimensions of the microfluidic system and the mass flow is governed by a Nernst-Planck-type equation [14]. Yet, the force balance implies that momentum is continuously transferred from the particles to the liquid. Thus, apart from the impact of the fluid onto the particle motion as mentioned earlier, the impact of the particles onto the liquid has to be considered, implying a two-way coupling between particle and fluid dynamics [140]. On the other hand, in the low-concentration limit of small particles a continuum approach is no longer justified due to large statistical fluctuations while the momentum transfer to the fluid may be negligible. Furthermore, if Brownian motion can be neglected, as in the case of micron-sized particles in an aqueous solution, deterministic particle trajectories may be computed by direct integration of particle velocities.

Yet, in many cases the simulation of particulate microflows is complex since typical particle dimensions are similar to the characteristic length scale of the system. Thus, the assumption of small particles is no longer justified. Within the so-called force coupling method momentum transfer from finite-size particles to the liquid is explicitly taken into account in form of multipole expansions of source terms [7]. The two-way coupling discussed so far results in an effective fluid-mediated particle–particle interaction, such as a trailing particle is drawn by a leading particle. On the other hand, if for instance the particles carry a charge or dipole a moment, corresponding interactions, as repulsion of charged particles or dipole–dipole interaction have to be taken into account additionally. In case of the latter, dipole moments tend to align causing an attracting force which leads to the so-called pearl chain formation of particles. For details, see, e.g., the books of Pohl [141] and Jones [142]. These phenomena can considerably influence, often enhance, the capturing performance for instance in case of DEP driven cell immobilization (cf. Fig. 18). Moreover, in case of biological suspensions deformability of particles has to be taken into account. This can be achieved using the immersed boundary (IB) method, which is intended for simulation of fluid-structure interaction, especially in biological fluid dynamics [143]. Among others, IB methods have been applied to modeling of swimming of multiple microorganisms in biofilms [144, 145], or to cardiovascular simulations as platelet aggregation during blood clotting (e.g., [146, 147]) and the deformation of red blood cells in shear flow [148].

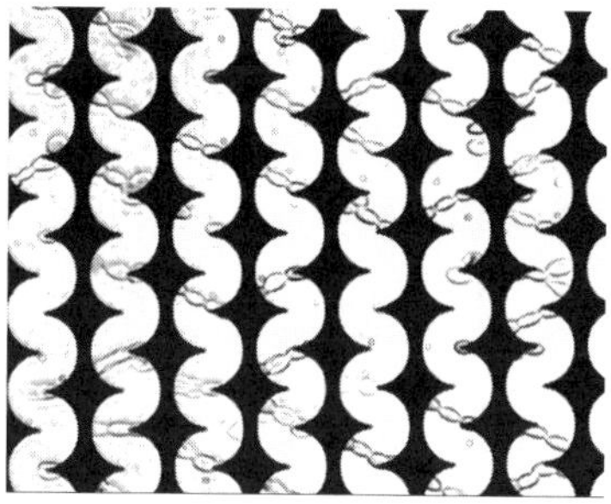

**Fig. 18.** Transmitted light micrograph of dielectrophoretic immobilization of erythrocytes in diluted pork blood. The black areas indicate microelectrodes deposited on a glass wafer. The spacing between the electrodes is partly covered by immobilized erythrocytes showing a pronounced tendency toward pearl-chain formation

Simulation methods of particulate flows, as the IB method, are still an active field of research. Simple and efficient simulations methods in this field are presently not at hand and are not to be expected in the near future.

Yet, less costly CFD approaches are available if only one or a few interfaces instead of a multitude of particles are to be modeled. These multiphase (normally two-fluid) simulations fall either into the class of interface tracking methods or into the class of interface capturing methods. While in the former a sharp fluid–fluid interface is attained by transiently aligning the computational mesh with it, a static computational mesh is used in the latter method allowing for topologically complex deformations of the interface (cf. [88] and references therein). For surface tension dominated flows, an alternative, much simpler energy based modeling can be applied reducing the needed CPU costs by orders of magnitudes compared to the aforementioned approaches [149]. The essential assumption in this case is that typical flow velocities are sufficiently low so that viscous effects are negligible, i.e. $Ca = \mu U/\sigma << 1$, where Ca denotes the capillary number relating viscous to surface tension forces. Within a quasi-steady approximation free-surface dynamics are described as a succession of equilibrium configurations. These configurations are defined by the interface shapes that minimize the potential energy subject to constraints imposed by the boundary conditions and a given amount of fluid [149]. The computation of the constrained equilibrium configurations can be efficiently performed with the help of the Surface Evolver program [150].

Yet, for instance in the case of capillary filling (cf. Fig. 10) even simpler approaches such as application of scaling laws or using steady-state approaches as the Lucas-Washburn law outlined earlier can already give important insight into the meniscus dynamics. The latter is just one particular case of solving Newton's equation of motion within an 1D approach, where the liquid's inertial force is compensated by the Laplace pressure, viscous friction as well as the hydrostatic pressure. The resulting nonlinear second order differential equation is of the type

$$y\ddot{y} + \dot{y}^2 = ay\dot{y} + by + c, \tag{8}$$

where $y$ denotes the axial position of the meniscus and the constants $a, b, c$ depend on the material properties, gravity, and channel dimensions. In the general case (8) has to be solved numerically. It reduces to the aforementioned Lucas-Washburn equation in quasi steady state and to the Quéré equation in case of an inviscid liquid [151]. Along the same line, 1D models, such as equivalent network models, can be derived capturing the filling dynamics in communicating vessels or the dynamics of bubble train flows. Yet, the description is somewhat oversimplified for noncompact menisci in rectangular channels. Especially for small contact angles, less than pi/4, menisci are frayed over time since wetting liquid films tend to flow along geometrical edges [152, 153]. In this case the flow of menisci tips in the interior corners are governed by a Washburn-type dynamics [153, 154], whereas for the complete meniscus dynamics the aforementioned CFD methods have to be applied.

#### 4.2.3. Optimization and System Design

Optimization methods applied in microfluidics are of course dependent on the chosen design and simulation methods. Two modi operandi are commonly distinguished: top-down and bottom-up. The classical top-down strategy begins with the representation of the complete system. The system or "top" level is subsequently divided into basic elements, for which parameterized models have been derived. Ideally, the element representation relies on accurate analytical models based on first principle assumptions [133]. Systematic design variations are straightforward and common optimization routines can be applied as only a limited number of optimization parameters are involved. In contrast, classical bottom-up approaches start with descriptions of the basic building blocks by means of detailed simulations such as CFD. From these macromodels appropriate for system simulation are developed. However, optimizations are not possible on the system level, since parameterized element models are generally not available. Hence, optimizations are to be performed on the detailed level. In this context the optimization approach of Olesen et al. [155] deserves particular attention. The method introduces a porous medium in the overall flow domain. Mathematically, this translates to the introduction of an additional Darcy term in the Navier-Stokes equations. The porosity is iteratively refined to maximize a predefined objective function. In this way designs of microfluidic modules can be optimized from scratch.

As a simple yet instructive example Fig. 19 shows optimization results of a four-terminal device. The optimization criterion is the minimization of the total power dissipation inside the device, subject to the constraint that fluid channels should cover at most 40% of the two-dimensional domain. As to be expected, for low Re the optimal structure turns out to be a pair of U-turns connecting the inlets to the outlets on the same side of the design domain, while at Re = 200 the optimal structure is a pair of parallel channels; however, more complex hybrid structures are found in the intermediate range of Reynold's numbers [155].

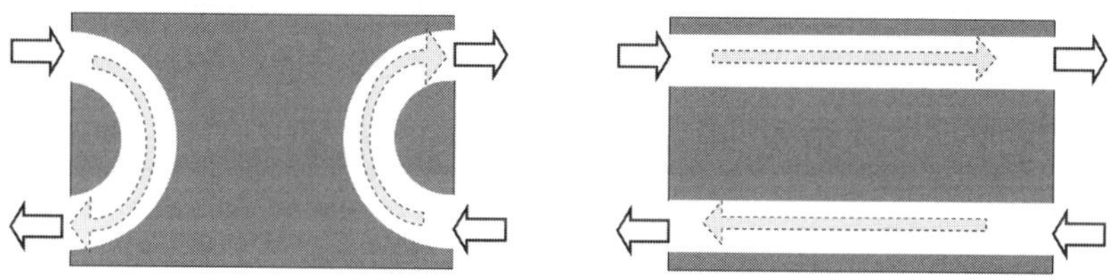

**Fig. 19.** Results of topology optimization with respect to dissipation minimization within a four-terminal fluidic device; adapted from [155].

## References


1. Stone HA, Stroock AD, Ajdari A (2004) Engineering flows in small devices: Microfluidics toward a Lab-on-a-Chip. Annu Rev Fluid Mech 36:381–411
2. Squires TM, Quake SR (2005) Microfluidics: Fluid physics at the nanoliter scale. Rev Mod Phys 77:977–1026
3. Marchisio DL, Barresi AA (2003) CFD simulation of mixing and reaction: The relevance of the micro-mixing model. Chem Eng Sci 58:3579–3587

4. de Gennes PG, Brochard-Wyart F, Quéré D (2004) Capillarity and wetting phenomena. Springer, New York
5. Schlichting H, Gersten K (2003) Boundary-layer theory, 8th Ed. Springer, Berlin
6. Stroock AD, Dertinger SKW, Ajdari A, Mezic I, Stone HA, Whitesides GM (2002) Chaotic mixer for microchannels. Science 295:647–651
7. Karniadakis G, Beskok A, Aluru N (2005) Microflows and nanoflows. Springer, New York
8. Herwig H (2002) Flow and heat transfer in micro systems: Is everything different or just smaller? ZAMM 82:579–586
9. Harris S (2004) An introduction to the theory of the Boltzmann equation. Dover Publications, Mineola
10. Arkilic E, Schmidt MA, Breuer K (1997) Gaseous slip flow in long microchannels. J Microelectromech Syst 6:167–178
11. Sone Y, Waniguchi Y, Aoki K (1999) One-way flow of a rarefied gas induced in a channel with a periodic temperature distribution. Phys Fluid 8:2227–2235
12. Khare R, Graham MD, de Pablo JJ (2006) Cross-stream migration of flexible molecules in a nanochannel. Phys Rev Lett 96:224505
13. Drikakis D, Kalweit M (2005) Computational Modelling of flow and mass transport processes in nanotechnology. In: Rieth M, Schommers W (eds) Handbook of theoretical and computational nanotechnology. American Scientific Publishers, Stevenson Ranch, chapter 34
14. Probstein RF (1994) Physicochemical hydrodynamics. Wiley, New York
15. Sokhan VP, Nicholson D, Quirke N (2001) Fluid flow in nanopores: An examination of hydrodynamic boundary conditions. J Chem Phys 115:3878–3887
16. Lauga E, Brenner MP, Stone HA (2005) Microfluidics: The no-slip boundary condition. In: Foss J, Tropea C, Yarin A, Handbook of experimental fluid dynamics, Springer, New York, chapter 15
17. Tyrrell J, Attard P (2002) Atomic force microscope images of nanobubbles on a hydrophobic surface and corresponding force-separation data. Langmuir 18:160–167
18. Thompson P, Troian S (1997) A general boundary condition for liquid flow at solid surfaces. Nature 389:360–362
19. Sedov LI (1993) Similarity and dimensional methods in mechanics, 10th Ed. CRC Press, Boca Raton
20. Laser DJ, Santiago JG (2004) A review of micropumps. J Micromech Microeng 14:R35–R64
21. Tunc G, Bayazitoglu Y (2001) Heat transfer in microtubes with viscous dissipation. Int J Heat Mass Tran 44:2395–2403
22. Morini GL (2005) Viscous heating in liquid flows in micro-channels. Int J Heat Mass Tran 48:3637–3647
23. Chen CH (2006a) Slip flow heat transfer in a microchannel with viscous dissipation. Int J Heat Mass Tran 42:853–860

24. Morini GL (2006) Scaling effects for liquid flows in microchannels. Heat Transfer Eng 27:64–73
25. Zhao TS, Liao Q (2002) Thermal effects on electroosmotic pumping of liquids in microchannels. J Micromech Microeng 12:962–970
26. Tang GY, Yang C, Chai JC, Gong HQ (2004) Joule heating effect on electroosmotic flow and mass species transport in a microcapillary. Int J Heat Mass Tran 47:215–227
27. Xuan X, Xu B, Sinton D, Li D (2004) Electroosmotic flow with Joule heating effects. Lab Chip 4:230–236
28. Horiuchi K, Dutta P (2004) Joule heating effects in electroosmotically driven microchannel flows. Int J Heat Mass Tran 47:3085–3095
29. Tang GY, Yan DG, Yang C, Gong HQ, Chai CJ, Lam YC (2006) Joule heating and its effects on electroosmotic flow in microfluidic channels. J Phys – Conference Series 34:925–930
30. Tang GY, Yan DG, Yang C, Gong HQ, Chai CJ, Lam YC (2006) Assessment of Joule heating and its effects on electroosmotic flow and electrophoretic transport of solutes in microfluidic channels. Electrophoresis 27:628–639
31. Chein R, Yang YC, Lin Y (2006) Estimation of Joule heating effect on temperature and pressure distribution in electrokinetic-driven microchannel flows. Electrophoresis 27:640–649
32. Hesselgreaves JE (2001) Compact heat exchangers. Elsevier Science, Oxford
33. Hardt S, Drese KS, Hessel V, Schönfeld F (2005) Passive micromixers for applications in the microreactor and μTAS fields. Microfluid Nanofluid 1:108–118
34. Hessel V, Löwe H, Schönfeld F (2005) Micromixers – A review on passive and active mixing principles. Chem Eng Sci 60:2479–2501
35. Nguyen NT, Wu Z (2005) Micromixers – A review. J Micromech Microeng 15:R1–R16
36. Taylor GI (1953) Dispersion of soluble matter in solvent flowing slowly through a tube. Proc Roy Soc A219:186–203
37. Aris R (1956) On the dispersion of a solute in a fluid flowing through a tube. Proc Roy Soc A235:67–77
38. Seemann R, Brinkmann M, Kramer EJ, Lange FF, Lipowsky R (2005) Wetting morphologies at microstructured surfaces. PNAS 102:1848–1852
39. Dreyer M, Delgado A, Rath HJ (1994) Capillary rise between parallel plates under microgravity. J Colloid Interface Sci 163:158–168
40. Lucas R (1918) Ueber das Zeitgesetz des kapillaren Aufstiegs von Fluessigkeiten. Kolloid-Z 23:15–22
41. Washburn EW (1921) The dynamics of capillary flow. Phys Rev 17:273–283
42. Brenn G, Lackermeier U (1997) Drop formation from a vibrating orifice generator driven by modulated electrical signals. Phys Fluids 9:3658–3669
43. Anna SL, Bontoux N, Stone HA (2003) Formation of dispersions using "flow focusing" in microchannels. Appl Phys Lett 82:364–366

44. Lord Rayleigh (1879) On the capillary phenomena of jets. Proc R Soc London 29:71–97
45. Chen AU, Basaran OA (2002) A new method for significantly reducing drop radius without reducing nozzle radius in drop-on-demand drop production. Phys Fluids 14:L1–L4
46. Jensen MJ, Goranović G, Bruus H (2004) The clogging pressure of bubbles in hydrophilic microchannel contractions. J Micromech Microeng 14:876–883
47. de Gennes PG (1985) Wetting: Statics and dynamics. Rev Mod Phys 57:827–863
48. Dussan EB, Chow RTP (1983) On the ability of drops and bubbles to stick to non-horizontal surfaces of solids. J Fluid Mech 137:1–29
49. Andersson H, van den Berg A (2003) Microfluidic devices for cellomics: A review. Sens Act B 92:315–325
50. Gollup J, Pine D (2006) Microscopic irreversibility and chaos. Phys Today 8:8–9
51. Terry SC, Jerman JH, Angel JB (1979) A gas chromatographic air analyzer fabricated on a silicon wafer. IEEE Trans Electron Devices 26:1880–1887
52. Manz A, Graber N, Widmer HM (1990) Miniaturized total chemical analysis systems: A novel concept for chemical sensing. Sens Act B 1(1–6):244–248
53. Cho B, Schuster TG, Zhu X, Chang D, Smith GD, Takayama S (2003) A Passively-driven integrated microfluidic system for separation of motile sperm. Anal Chem 75:1671–1675
54. Verpoorte S (2003) Beads and chips: New recipes for analysis. Lab Chip 3:60N–68N
55. de Jong J, Lammertink RGH, Wessling M (2006) Membranes and microfluidics: A review. Lab Chip 6:1125–1139
56. Zhu XS, Chuan CG, Choi JW, Bishop PL, Ahn CH (2005) On-chip generated mercury microelectrode for heavy metal ion detection. Lab Chip 5:212–217
57. Goral VN, Zaytseva NV, Baeumner AJ (2006) Electrochemical microfluidic biosensor for the detection of nucleic acid sequences. Lab Chip 6:414–421
58. Fodor SPA, Rava RP, Juang XC, Pease AC, Holmes CP, Adams CL (1993) Multiplexed biochemical assays with biological chips. Nature 364:555–556
59. Schena M, Shalon D, Davis RW, Brown PO (1995) Quantitative monitoring of gene expression patterns with a complementary DNA microarray. Science 270:467–470
60. Marrazza G, Chianella I, Mascini M (1999) Disposable DNA electrochemical sensor for hybridization detection. Biosensors Bioelectron 14:43–51
61. Choi YS, Lee KS, Park DH (2005) Single nucleotide polymorphism (SNP) detection using microelectrode biochip array. J Micromech Microeng 15:1938–1946
62. Uchiyama K, Nakajima H, Hobo T (2004) Detection method for microchip separations. Anal Bioanal Chem 379:375–382
63. Schwarz MA, Hauser PC (2001) Recent developments in detection methods for microfabricated analytical devices. Lab Chip 1:1–6

64. Irawan R, Tay CM, Tjin SC, Fu CY (2006) Compact fluorescence detection using in-fiber microchannels – its potential for lab-on-a-chip applications. Lab Chip 6:1095–1098
65. Balslev S, Jorgensen AM, Bilenberg B, Mogensen KB, Snakenborg D, Geschke O, Kutter JP, Kristensen A (2006) Lab-on-a-chip with integrated optical transducers. Lab Chip 1:213–217
66. Zhu L, Lee CS, deVoe DL (2006) Integrated microfluidic UV absorbance detector with attomol-level sensitivity for BSA. Lab Chip 6:115–120
67. Snakenborg D, Mogensen KB, Kutter JP (2003) Optimization of signal-to-noise ratio in absorbance detection by integration of microoptical components. 7th Intern Conf on μTAS pp 841–844
68. Bornhop DJ, Swinney K (2000) Detection in capillary electrophoresis: A review. Electrophoresis 21:1239–1250
69. de Mello AJ (2003) Seeing single molecules. Lab Chip 3:29N–34N
70. Takamura Y, Onoda H, Inokuchi H, Adachi S, Oki A, Horiike Y (2003) Low-voltage electroosmosis pump for stand-alone microfluidics devices. Electrophoresis 24:185–192
71. Brask A, Kutter JP, Bruus H (2005) Long-term stable electroosmotic pump with ion exchange membranes. Lab Chip 7:730–738
72. Pamme N (2006) Magnetism and micro fluidics. Lab Chip 6:24–38
73. Singhal V, Garimella SV, Raman A (2004) Microscale pumping technologies for microchannel cooling systems. Appl Mech Rev 57:191–221.
74. Woias P (2005) Micropumps–Past, progress and future Prospects. Sens Act B 105:28–38
75. Nguyen NT, Huang X, Chuan TK (2002) MEMS-micropumps: A review. J Fluid Eng. 124(2):384–392
76. Geschke O, Klank H, Tellemann P (2004) Microsystem engineering of Lab-on-a-Chip devices. WILEY-VCH, Weinheim
77. Andersson H, Van der Wijngaart W, Nilsson P, Enoksson P, Stemme G (2001) A valve-less diffuser micropump for microfluidic analytical systems. Sens Act B 72(3):259–265
78. Terray A, Oakey J, Marr DWM (2002) Microfluidic control using colloidal devices. Science 296:1841–1844
79. Kwang WO, Ahn CH (2006) A review of microvalves. J Micromech Microeng 16:R13–R39
80. Bitsch L, Kutter JP, Storgaard-Larsen T, Bruus H (2006) A low-energy, turning microvalve with high-pressure seals: Scaling of friction. J Micromech Microeng 16:2121–2127
81. Burns MA, Johnson BN, Brahmasandra SN, Handique K, Webster JR, Krishnan M, Sammarco TS, Man FP, Jones D, Heldsinger D, Namasivayam V, Mastrangelo CH, Burke DT (1998) An integrated microfabricated DNA analysis device. Science 282:484–487
82. Nakano M, Komatsu J, Matsuura SI, Takashima K, Katsura S, Mizuno A (2003) Single-molecule PCR using water-in-oil emulsion. J Biotechnol 102:117–124

83. Münchow G, Dadic D, Doffing F, Hardt S, Drese KS (2005) Automated chip-based device for simple and fast nucleic acid amplification. Expert Rev Mol Diagn 5(4), 613–620
84. Zheng B, Tice JD, Roach LS, Ismagilov RF (2004) A droplet-based, composite PDMS/glass capillary microfluidic system for evaluating protein crystallization conditions by microbatch and vapor-diffusion methods with on-chip X-ray diffraction. Angew Chem Int Ed 43:2508–2511
85. Schaack B, Fouque B, Porte S, Combe S, Hennico A, Filhol-Cochet O, Rehoud J, Balakirev M, Chatelain F (2003) Cell culture in microdrops, a new format for cell on chip technology. 7th Intern Conf on μTAS pp 669–672
86. Bercic G, Pintar A (1997) The role of gas bubbles and liquid slug lengths on mass transport in the Taylor flow through capillaries. Chem Eng Sci 52:3709–3719
87. Bringer M, Gerdts CJ, Song H, Tice JD, Ismagilov RF (2004) Microfluidic systems for chemical kinetics that rely on chaotic mixing in droplets. Phil Trans R Soc Lond A 362:1087–1104
88. Cristini V, Tan YC (2004) Theory and numerical simulation of droplet dynamics in complex flows – A review. Lab Chip 4:257–264
89. Melin J, Roxhed N, Gimenez G, Griss P, van der Wijngaart W, Stemme G (2004) A liquid-triggered liquid microvalve for on-chip flow control. *Sens Act* B 100:463–468
90. Ahn K, Agresti J, Chong H, Marquez M, Weitz DA (2006) Electrocoalescence of drops synchronized by size-dependent flow in microfluidic channels. Appl Phys Lett 88:0264106 1–3
91. Link DR, Anna SL, Weitz DA, Stone HA (2004) Geometrically mediated breakup of drops in microfluidic devices. Phys Rev Lett 95(5):054503-1–054503-4
92. Tan YC, Fisher JS, Lee AI, Cristini V, Lee AP (2004) Design of microfluidic channel geometries for the control of droplet volume, chemical concentration, and sorting. Lab Chip 4:292–298
93. Schönfeld F, Hardt S, Böhm M, Püschl RJ, Walder M, Wenclawiak B (2006) Electroosmotic flow patterning using microfluidic delay loops. Lab Chip 6:1525–1529
94. Thorsen T, Maerkl SJ, Quake SR (2002) Microfluidic large-scale integration. Science 298:580–584
95. González C, Collins SD, Smith RL (1998) Fluidic interconnects for modular assembly of chemical Microsystems. Sens Act B 49:40–45
96. Igata E, Arundell M, Morgan H, Cooper JM (2002) Interconnected reversible lab-on-a-chip technology. Lab Chip 2:65–69
97. Nittis V, Fortt R, Legge CH, de Mello AJ (2001) A high-pressure interconnect for chemical microsystem applications. Lab Chip 1:148–152
98. Fujii T, Sando Y, Higashino K, Fujii Y (2003) A plug and play microfluidic device. Lab Chip 3:193–197

99. Kwang WO, Park C, Namkoong K, Kim J, Ock KS, Kim S, Kim YA, Cho YK, Ko C (2005) World-to-chip microfluidic interface with built-in valves for multichamber chip-based PCR assays. Lab Chip 5:845–850
100. Mair DA, Geiger E, Pisano AP, Fréchet JMJ, Svec F (2006) Injection molded microfluidic chips featuring integrated interconnects. Lab Chip 10:1346–1354
101. Vilkner T, Janasek D, Manz A (2004) Micro total analysis systems. Recent developments. Anal Chem 76:3373–3386
102. Yager P, Edwards T, Fu E, Helton K, Nelson K, Tam MR, Weigl BH (2006) Microfluidic diagnostic technologies for global public health. Nature 442:412–418
103. Whitesides GM, Love JC (2001) The Art of Building Small. Sci Am 285(3): 33–41
104. McDonald JC, Duffy DC, Anderson JR, Chiu DT, Wu H, Schueller OJA, Whitesides GM (2000) Fabrication of microfluidic systems in poly(dimethylsiloxane). Electrophoresis 21:27–40
105. Whitesides GM (2006) The origins and the future of microfluidics. Nature 442:368–373
106. Liu Y, Lee JY, Kang ET, Wang P, Tan KL (2001) Synthesis, characterization and electrochemical transport properties of the poly(ethyleneglycol)-grafted poly(vinylidenefluoride) nanoporous membranes. React Funct Polym 47(3):201–213
107. Chen Y, Ying L, Yu W, Kang ET, Neoh KG (2003) Poly(vinylidene fluoride) with Grafted Poly(ethylene glycol) side chains via the RAFT-mediated process and pore size control of the copolymer membranes. Macromolecules 36:9451–9457
108. Xu FJ, Zhong SP, Yung LYL, Kang ET, Neoh KG (2004) Surface-active and stimuli-responsive polymer-Si(100) hybrids from surface-initiated atom transfer radical polymerization for control of cell adhesion. Biomacromolecules 5:2392–2403
109. Wu D, Zhao B, Dai Z, Qin J, Lin B (2006) Grafting epoxy-modified hydrophilic polymers onto poly(dimethylsiloxane) microfluidic chip to resist nonspecific protein adsorption. Lab Chip 6:942–947
110. Bi H, Meng S, Li Y, Guo K, Chen K, Kong J, Yang P, Zhong W, Liu B (2006) Deposition of PEG onto PMMA microchannel surface to minimize nonspecific adsorption. Lab Chip 6:769–775
111. Hjerte´n S (1985) High-performance electrophoresis – Elimination of electroendosmosis and solute adsorption. J Chromatogr 347:191–198
112. Wang SC, Morris MD (2000) Plastic microchip electrophoresis with analyte velocity modulation. Application to fluorescence background rejection. Anal Chem 72:1448–1452
113. Hawkins KR, Yager P (2003) Nonlinear decrease of background fluorescence in polymer thin-films – A survey of materials and how they can complicate fluorescence detection in μTAS. Lab Chip 3:248–252
114. Piruska A, Nikcevic I, Lee SH, Ahn C, Heineman WR, Limbach PA, Seliskar CJ (2005) The autofluorescence of plastic materials and chips measured under laser irradiation. Lab Chip 12:1348–1354

115. Rohr T, Ogletree DF, Svec F, Frechet JMJ (2003) Surface functionalization of thermoplastic polymers for the fabrication of microfluidic devices by photoinitiated grafting. Adv Funct Mater 13:264–270
116. Duffy DC, McDonald JC, Schueller OJA, Whitesides GM (1998) Rapid prototyping of microfluidic systems in poly(dimethylsiloxane). Anal Chem 70:4974–4984
117. Herzig HP (1997) Micro-optics elements, systems and applications. Taylor & Francis, London
118. Xia Y, Whitesides GM (1998) Soft lithography. Ann Rev Mat Sci 28:153–184
119. Roberts MA, Rossier JS, Bercier P, Girault H (1997) UV Laser machined polymer substrates for the development of microdiagnostic systems. Anal Chem 69(11):2035–2042
120. Kricka LJ, Fortina P, Panaro NJ, Wilding P, Alonso-Amigo G, Becker H (2002) Fabrication of plastic microchips by hot embossing. Lab Chip 2:1–4
121. Rötting O, Röpke W, Becker H, Gärtner C (2002) Polymer microfabrication technologies. Micros Technol 8:32–36
122. Piotter V, Hanemann T, Ruprecht R, Hausselt J (1997) Injection molding and related techniques for fabrication of microstructures. Microsyst Technol 3:129–133
123. Chen SH, Jong WR, Chang YJ, Chang JA, Cin JC (2006b) Rapid mold temperature variation for assisting the micro injection of high aspect ratio micro-feature parts using induction heating technology. J Micromech Microeng 16:1783–1791
124. Becker H, Gärtner C (2000) Polymer microfabrication methods for microfluidic analytical applications. Electrophoresis 21:12–26
125. Hoogerbrugge PJ, Koelman JM (1992) Simulating microscopic hydrodynamic phenomena with dissipative particle dynamics. Europhys Lett 19(3):155–160
126. Espanol P, Warren P (1995) Statistical dynamics of dissipative particle dynamics. Europhys Lett 30(4):191–196
127. Ladd AJ (1994) Numerical simulations of particulate suspensions via a discretized lattice Boltzmann equation. Part 1. Theoretical foundation. J Fluid Mech 271:285–309
128. Ladd AJ (1994) Numerical simulations of particulate suspensions via a discretized lattice Boltzmann equation. Part 2. Numerical results. J Fluid Mech 271:311–389
129. Succi A (2001) The Lattice-Boltzmann equation for fluid dynamics and beyond. Oxford University Press, Oxford
130. Tilmans HAC (1996) Equivalent circuit representation ofelectromechanical transducers: I. Lumped-parameter systems. J Micromech Microeng 6:157–176.
131. Bourouina T, Grandchamp JP (1996) Modeling micropumps with electrical equivalent networks. J Micromech Microeng 6:398–404
132. Ajdari A (2004) Steady flows in networks of microfluidic channels: Building on the analogy with electrical circuits. CR Physique 5:539–546

133. Wang Y, Lin Q, Mukherjee T (2006) Composable behavioral Models and schematic-based simulation electrokinetic lab-on-a-chip systems. IEEE Trans Comp Aided Design ICs Syst 25(2):258–273
134. Hardt S, Pennemann H, Schönfeld F (2006) Theoretical and Experimental Characterization of a low-Reynolds-number split-and-recombine mixer. Microfluid Nanofluid 2:237–248
135. Bedekar AS, Wang Y, Krishnamoorthy S, Siddhaye SS, Sundaram S (2006) System-level simulation of flow-induced dispersion in lab-on-a-chip systems. IEEE Transactions on Computer Aided Design of Inegrated Curcuits and Systems 25(2):294–304
136. Kamm RD (2002) Cellular fluid mechanics. Ann Rev Fluid Mech 34:211–232.
137. Das B, Johnson PC, Pope AS (1998) Effect of nonaxisymmetric hematocrit distribution on non-Newtonian blood flow in small tubes. Biorheology 35(1):69–87
138. Errill EW (1969) Rheology of blood. Physiol Rev 49:863–888
139. Warnke KC (2003) Finite-element modeling of the separation of magnetic microparticles in fluid. IEEE Trans Magn 39:1771–1777
140. Mikkelsen CI, Bruus H (2005) Microfluidic capturing-dynamics of paramagnetic bead suspensions. Lab Chip 5:1293–1297
141. Pohl A (1978) Dielectrophoresis. Cambridge University Press, Cambridge, U.K
142. Jones TB, (1995) Electromechanics of Particles, Cambridge University Press, New York
143. Peskin CS (2002) The immersed boundary method. Acta Numerica 11:479–517
144. Dillon R, Fauci LJ, Fogelson AL, Gaver D (1996) Modeling bio processes using the immersed boundary method. J Comput Phys 129:57–73
145. Dillon R, Fauci LJ (2000) A microscale model of bacterial and biofilm dynamics in porous media. Biotechnol Bioeng 68:536–547
146. Fogelson AL (1984) A mathematical model and numerical method for studying platelet adhesion and aggregation during blood clotting. J Comput Phys 56:111–134
147. Fogelson AL (1992) Continuum models of platelet aggregation: Formulation and mechanical properties. SIAM J Appl Math 52:1089–1110
148. Eggleton CD, Popel AS (1998) Large deformation of red blood cell ghosts in a simple shear flow. Phys Fluid 10:1834–1845
149. Treise I, Fortner N, Shapiro B, Hightower A (2005) Efficient energy based modeling and experimental validation of liquid filling in planar micro-fluidic components and networks. Lab Chip 3:285–297
150. Brakke K (1992) The Surface Evolver. Exp Math 1(2):141–165
151. Zhmud BV, Tiberg F, Hallstensson (2000) Dynamics of capillary rise. J Colloid Interf Sci 228:263–269
152. Concus P, Finn R (1974) On capillary free surfaces in the absence of gravity. Acta Math 132:177–198

# Chapter 2
# Electrohydrodynamic and Magnetohydrodynamic Micropumps

Antonio Ramos

Departamento de Electrónica y Electromagnetismo.
Universidad de Sevilla. Avenida Reina Mercedes s/n. 41012-Sevilla, Spain. ramos@us.es

## 1. Introduction

A great variety of strategies have been developed to pump fluids in microsystems. According to Laser and Santiago [1] micropumps generally fall into one of two classes: (a) displacement pumps, which exert pressure forces on the working fluid by one or more moving boundaries and (b) dynamic micropumps, which exert forces directly on the liquid, without moving parts. From the latter category, we are going to deal with micropumps that exert electric or magnetic forces on liquids, called electrohydrodynamic (EHD) or magnetohydrodynamic (MHD) micropumps, respectively. The main goal of the present work is to describe the physical principles behind the EHD and MHD micropump actuation, and what physical parameters affect the performance of these micropumps.

The transport of small volumes of liquid in microdevices is often achieved by taking advantage of passive mechanisms such as capillarity or gravity. Sometimes microfluidic applications employ macroscale pumps like syringe pumps or pressure/vacuum chambers and valves. However, many microfluidic applications would benefit from an on-chip active pump with size comparable to the small volume of fluid to be pumped, i.e. an integrated micropump. An integrated pump design that has been studied intensively over the years is the piezo-actuated micropump. In this, a membrane is displaced to create a pulsating flow that is rectified using valves. However, the moving parts make the fabrication and operation delicate. In this context, the requirement of an integrated micropump with no moving parts can be fulfilled by using EHD and MHD micropumps.

Micropumps have prominent features with length scales of the order of 0.1 mm or smaller but greater than 100 nm. Therefore, subcontinuum effects, that may be important in nanosystems, are not important here. Subcontinuum effects can also be important for the pumping of gases in microsytems; however, our study is restricted to the pumping of liquids in microsystems. Typically, fluid flows will be laminar and, in many situations, in the low Reynolds number regime.

Possible applications of micropumps [1] cover dispensing therapeutic agents into the body, cooling of microelectronic devices, dispensing liquids in miniature systems for chemical and biological analysis, micropumping of liquids in space exploration where miniaturization is highly desirable, and more.

EHD and MHD dynamic micropumps may be classified into the following:

(a) pumps that exert electric forces in the liquid bulk: the EHD *injection* [2], *conduction* [3], and *induction* [4] *pumps*
(b) those that exert electric forces in the diffuse double layer: the *electroosmotic* [5] and *AC induced-charge electroosmotic* [6] *pumps*
(c) those that exert magnetic forces in the liquid bulk: the *DC* [7] and *AC* [8] *MHD* pumps

In all of them, electric current flows through the working liquid and a variety of pumping techniques is encountered depending on the conductivity of the liquids. We first deal with some basic aspects of electric conduction in liquids and some basic fluid-mechanical aspects that are common for micropumps. Secondly, for each micropump, we analyse the physical principle behind the micropump actuation. We also describe some important properties of pumps like the maximum flow rate $Q_{max}$ and maximum pressure $\Delta p_{max}$ that the pump can generate, the power consumption and energy efficiency, and the typical conductivity range that the pump can actuate. We discuss what kind of problems affect the performance of micropumps such as heat generation, or changes in physical properties of the working fluid and bubble generation both caused by electrochemical reactions. For each type of micropump, we address what are the possible applications. Finally, we compare the performance of the different micropumps and draw some conclusions.

### 1.1. Basic Features of Conduction in Liquids

Charge carriers in liquids are electrons, holes, and ions [9]. Measurements of electron and hole mobility require extremely pure dielectric liquids in order for the electron or hole to survive attachment or charge transfer, respectively [9]. Except for liquid metals and/or ultrahigh purified noble liquids, there exist many situations where free electrons in liquids are quickly trapped by electronegative impurities or molecules of the liquid [10]. In most cases, the charge carriers in liquids will be ions in dissolution, and this is the situation we are going to consider. Under the action of the electric field, these ions can be considered to move at the terminal velocity $\mathbf{v} = \mu\mathbf{E}$ (ion inertia is negligible) [10], where $\mu$ is the mobility of the ion. A good model is to consider the ions as spheres affected by viscous friction. This provides an expression for the mobility as $\mu = q/6\pi a\eta$ (Stokes-Einstein relation), where $q$ is the ion charge, $a$ the hydrodynamic radius of the solvated ion, and $\eta$ the dynamic viscosity of liquid. The ion charge is $q = ze$, where $e$ is the charge of a proton and $z$ is the valence. The mobility of ions in liquids is usually between $10^{-7}$ to $10^{-9}$ $m^2$ $(Vs)^{-1}$. Ion motion is also affected by molecular diffusion and convection. Therefore, the electric current in the liquid bulk can be written as

$$\mathbf{j} = \sum_i q_i(n_i\mu_i\mathbf{E} - D_i\nabla n_i + n_i\mathbf{u}), \tag{1}$$

where $n_i$ is the number density of ionic species, $D_i$ is the diffusion coefficient of an ion, and $\mathbf{u}$ the velocity of the fluid. The first term on the right represents ion migration in the electric field, the second describes transport of charge by diffusion, and the third the transport by convection. The diffusion and mobility coefficients of an ion are linked by Einstein's relation, $D_i/\mu_i = k_B T/q_i$, where $k_B$ is the Boltzmann's constant and $T$ is the absolute temperature.

In many situations common in microsystems, the diffusion current is negligible in front of the electro-migration current. If we compare diffusion and migration, $|D\nabla n|/|\mu n\mathbf{E}| \sim D/\mu El$, where $l$ is a typical distance. From the Einstein's relation, the ratio $D/\mu = k_B T/q \approx 0.025$ V at room temperature, and $El$, the typical increment of voltage in the bulk, is usually much greater than 0.025V. Therefore, $|D\nabla n|/|\mu n\mathbf{E}| \ll 1$. However, the diffusion current can be important close to the electrodes and walls, where double layers are formed.

For electrolytes the convective current is usually much smaller than the conduction current. In effect, the ratio between convective current and electro-migration current is of the order of

$$\frac{\rho u}{\sum_i q_i n_i \mu_i E} \sim \frac{\nabla\cdot(\varepsilon\mathbf{E})u}{\sum_i q_i n_i \mu_i E} \sim \frac{\varepsilon u}{\sum q_i n_i \mu_i l}, \tag{2}$$

where $\rho = \sum_i q_i n_i$ is the charge density. For electrolytes, $\sum_i q_i n_i \mu_i$ is of the order of (or equal to, if $n_i$ are equilibrium concentrations) the conductivity of the liquid, so the ratio is of the order of $u\varepsilon/l\sigma$, which is much smaller than one [11]. For insulating dielectric liquids, the ratio between convection and ion-migration velocities, $u/\mu E$, can be greater or smaller than one.

The electric current expression (1) needs to be modified when dealing with the MHD pumping of electrolytes. In effect, from the Lorentz force on an ion $\mathbf{f} = q(\mathbf{E} + \mathbf{u} \times \mathbf{B})$, the term $\sum_i q_i n_i \mu_i \mathbf{u} \times \mathbf{B}$ should be added to the previous equation. However, it is common that the term $\mathbf{u} \times \mathbf{B}$ is negligible as compared to the electric field $\mathbf{E}$ for electrolytes actuated with magnetic fields in microsystems (see Sect. 4).

Each ionic and neutral species satisfies a conservation equation [12]

$$\frac{\partial n_i}{\partial t} + \nabla \cdot (n_i \mu_i \mathbf{E} - D_i \nabla n_i + n_i \mathbf{u}) = r_i, \tag{3}$$

where $r_i$ is the rate of production due to chemical reactions ($\mu_i = 0$ for neutral species).

In solutions of strong electrolytes, the majority of ions come from the dissociation of neutral salt and this is completely dissociated. In this case, there are no neutral species and no reactions in the bulk ($r_i = 0$ per each ionic species). Electrochemical reactions take place only at the electrodes.

In *unipolar injection*, there is only a single ionic species, and there is no production of charges in the bulk $r_1 = 0$. The ions are produced by reactions at the electrodes.

For a weak binary $z$-$z$ electrolyte, there are three species: the positive ion, the negative ion, and the neutral molecule that dissociates into these ions. The ions are produced (or consumed) by reactions at the electrodes and by homogeneous chemical reactions in the bulk,

$$-r_0 = r_+ = r_- = k_d n_0 - k_r n_+ n_-, \tag{4}$$

where $k_d$ and $k_r$ are the dissociation and recombination rate constants and $n_0$, $n_+$ and $n_-$ are the concentrations of neutral, positive, and negative species, respectively.

In dielectric liquids, the charge carriers originate from dissociation of impurities in the bulk, or of the liquid itself, and/or from ions injected at the surface of the electrodes [10]. The conductivity of a dielectric liquid is highly dependent on its dielectric constant. Ions recombine if the energy of Coulomb attraction is greater than the thermal energy $k_B T$ at the point of closest approach of the ions. The distance at which Coulomb attraction energy equals $k_B T$ is called the Bjerrun distance, $l_B = e^2/4\pi\varepsilon k_B T$ [10]. If the solvated ions have a diameter greater than the Bjerrun distance, the ions do not recombine. For instance, $l_B = 0.7$ nm for water at room temperature, and solvated ions have a greater radius. Therefore, in water impurities tend to be dissociated. On the other hand, if the Bjerrun distance is greater than the diameter of the ions, an equilibrium is established between dissociation and recombination. Because the probability to occur dissociation is proportional to the exponential factor $\exp(-l_B/a)$, the conductivity is highly dependent on the dielectric constant $\varepsilon_r$ of the liquid. In polar liquids ($\varepsilon_r \gg 2$), all impurities tend to be dissociated. In contrast, in nonpolar liquids ($\varepsilon_r \sim 2$) there is an equilibrium between dissociation and recombination.

To establish a DC current in a liquid, electrons should be transferred from the metal electrodes to the molecules in the liquid, and viceversa. This can happen by electron emission from surface irregularities, from electrochemical redox reactions at the surface or by other means. From the mathematical point of view, the different mechanisms are described by similar parameters [13]. A simple picture is to consider that neutral molecules can be charged at the electrodes: on the cathode $X + e^- = X^-$, on the anode $X - e^- = X^+$; and that ions can be discharged at the electrodes: on the cathode $X^+ + e^- = X$, on the anode $X^- - e^- = X$. This together with the homogeneous reactions of ion production in the bulk form a simple dissociation-injection picture of conduction in liquids [13].

In the case of AC currents, the charging of the double layer at the electrodes should be considered [14] (see also Sect. 3). A simple model is to consider the electric double layer of the electrodes as a capacitor and a resistor in parallel and both connected to a resistive liquid bulk. The current through the capacitor represents the charging of the double layer, while the current through the resistor represents the Faradaic current. In the AC case, if the voltage drop across the double layer is not high, the electrochemical reactions at the electrodes can be very much reduced. In this case, the continuity of electric current can be guaranteed with no electrochemical reactions (no faradaic current) by the charging of the double layer.

153. Goldschmidtboeing F, Rabold M, Woias P (2006) Strategies for void-free liquid filling of micro cavities. J Micromech Microeng 16:1321–1330
154. Weislogel MM, Lichter S (1998) Capillary flow in an interior corner J Fluid Mech 373:349–378
155. Olesen LH, Okkels F, Bruus H (2006) A high-level programming-language implementation of topology optimization applied to steady-state Navier–Stokes flow. Int J Numer Meth Eng 65:975–1001

The equations that govern the electric field are those of Electro-quasi-statics. In effect, the electric field that drives the current can be considered to be irrotational if the magnetic energy density is much smaller than the electric energy density [10]. When the electric currents in the liquid are the sources of $\mathbf{B}$, the magnetic field intensity can be estimated from Ampère-Maxwell law

$$\nabla \times \mathbf{B} = \mu_0 \mathbf{j} + \mu_0 \varepsilon \frac{\partial \mathbf{E}}{\partial t}, \tag{5}$$

as $B \sim \max(\mu_0 jl, \mu_0\, \varepsilon\, \omega El)$, where $\omega$ is a typical frequency. Putting $j = \sigma E$, the ratio of energies is

$$\frac{B^2/\mu_0}{\varepsilon E^2} \sim \max(\mu_0 \sigma^2 l^2/\varepsilon, \mu_0 \varepsilon \omega^2 l^2). \tag{6}$$

For liquid conductivity $\sigma < 10\,\mathrm{S\,m^{-1}}$ and a typical system length $l \sim 100\,\mu\mathrm{m}$, $\mu_0 \sigma^2 l^2/\varepsilon < 0.14/\varepsilon_r \ll 1$, where $\varepsilon_r$ is the relative permittivity of the liquid. For frequencies smaller than 10 MHz and system length $l < 1\,\mathrm{mm}$, $\mu_0 \varepsilon \omega^2 l^2 < 3\times 10^{6-}$. Under these assumptions our system is in the quasi-electrostatic limit of Maxwell equations [10,15]:

$$\nabla \cdot (\varepsilon \mathbf{E}) = \rho = \sum_i q_i n_i, \tag{7}$$

$$\nabla \times \mathbf{E} = 0, \tag{8}$$

$$\nabla \cdot \mathbf{j} + \frac{\partial \rho}{\partial t} = 0. \tag{9}$$

The electric field that drives the current is also very approximately irrotational in the case of electrolytes actuated by MHD micropumps (see Sect. 4).

### *1.2. Mechanical Aspects of Micropumps*

Liquid motion is governed by the Navier-Stokes equations for an incompressible fluid

$$\nabla \cdot \mathbf{u} = 0, \tag{10}$$

$$\rho_m \frac{\partial \mathbf{u}}{\partial t} + \rho_m (\mathbf{u} \cdot \nabla)\mathbf{u} = -\nabla p + \eta \nabla^2 \mathbf{u} + \mathbf{f}, \tag{11}$$

where $\mathbf{f}$ represents any body force density applied to the liquid. For microsystems the Reynolds number is usually very small and the convective term in the Navier-Stokes equations can be neglected. In effect, the Reynolds number is a measure of the ratio $|\rho_m \mathbf{u}\cdot\nabla\mathbf{u}|/|\eta\nabla^2\mathbf{u}|$ and is given by $Re=\rho_m ul/\eta$, where $l$ is a typical length of the system. Taking typical values in microsystems, $u\sim10^{-3}$ m s$^{-1}$, $l\sim10^{-4}$ m, it is obtained $Re\sim0.1$ in the case of water. Therefore, viscous forces are usually dominant in front of inertial forces in microsystems [16]. Under steady-state conditions, the momentum equation is then

$$0=-\nabla p+\eta\nabla^2\mathbf{u}+\mathbf{f}\,. \tag{12}$$

To the equations of motion, we must add the boundary condition of no-slip for the velocity.

The equations relating fluid velocities to forces and pressure gradients are linear, and this linearity allows for the usual representation of microfluidic networks in terms of equivalent electrical circuits. The applied force will commonly be concentrated in a certain region of the microfluidic system, this is the micropump region, and it is equivalent to the generator of an electrical circuit. For a channel of the network outside the micropump, the volume flow rate $Q$ is linearly related to the pressure difference $\Delta p$ between the inlet and outlet of the channel. For a straight channel of circular cross section we have

$$Q=\Delta p\frac{\pi D^4}{128\eta L}, \tag{13}$$

where $D$ is the diameter and $L$ length of the channel. The channel behaves as a resistor in an electrical circuit, where $Q$ and $\Delta p$ are equivalent to the current intensity and potential drop, respectively.

Two important parameters of a pump are the maximum pressure $\Delta p_{\max}$ and maximum flow rate $Q_{\max}$ that the pump can generate. The maximum flow rate is obtained when the pump actuates without an external load, that is, when the pressure difference between the inlet and outlet of the pump is zero. The maximum pressure is the pressure difference required to stop the flow that the pump generates. In many situations, the flow rate $Q$ that a pump is able to produce against a back pressure $\Delta p$, under constant operating conditions, can be approximated by a linear relation of the form $Q=Q_{\max}(1-\Delta p/\Delta p_{\max})$.

Power consumption and energy efficiency are important operational parameters of pumps [1]. When comparing different pumps, we would prefer the one that consumes less power for obtaining a given flow rate or a given pressure. According to Laser and Santiago [1], the most useful definition of energy efficiency for a pump is the hydraulic power divided by the power consumption, $eff = \Delta p\, Q / P_{\text{in}}$. Here $P_{\text{in}}$ is the power that goes into the pump. For instance, the total power consumption of a mechanical pump that delivers the hydraulic power $Q\Delta p$ is $Q\Delta p$ plus the friction losses of its moving parts. Therefore, the efficiency of a frictionless mechanical pump is unity. When, under constant operating conditions, there is a linear relationship between $\Delta p$ and $Q$, the maximum energy efficiency can be calculated as

$$eff = \frac{1}{4} \frac{Q_{\max} \Delta p_{\max}}{P_{\text{in}}}. \tag{14}$$

We will use this expression in order to estimate the efficiencies of micropumps, even in cases where the linear relationship between $\Delta p$ and $Q$ does not hold.

## 2. Electric Forces in the Bulk: Injection, Conduction, and Induction EHD Pumps

When there is an electric field in a liquid, electric forces are exerted on it. The electric body force density is [17]

$$\mathbf{f}_E = \rho \mathbf{E} - \frac{1}{2} E^2 \nabla \varepsilon + \frac{1}{2} \nabla \left[ E^2 \rho_{\text{m}} \left( \frac{\partial \varepsilon}{\partial \rho_{\text{m}}} \right)_T \right], \tag{15}$$

where the first term is the Coulomb force, the second term is the dielectric force, and the third is the electrostriction. The latter can be incorporated into the pressure for an incompressible fluid. A simple way of seeing that the electric field exerts forces on the fluid is to consider that the forces on the free charges and dipoles are transmitted directly to the liquid [12]. The Coulomb term represents the force on the free charges. For a free charge $q$, the force is $q\mathbf{E}$ and from the definition of $\rho$ as charge density, we obtain the Coulomb force density $\rho\mathbf{E}$. The dielectric term represents the force on the dipoles. The dielectric term in (15) can be rewritten as $-\frac{1}{2} E^2 \nabla \varepsilon = -\nabla \left( \frac{1}{2} (\varepsilon - \varepsilon_0) E^2 \right) + (\varepsilon - \varepsilon_0)(\mathbf{E} \cdot \nabla)\mathbf{E}$, taking into account $(\mathbf{E} \cdot \nabla)\mathbf{E} = \nabla E^2/2$ for irrotational fields. For an incompressible fluid, the first term on the right-hand side can be incorporated into the pressure. The second term, $(\varepsilon - \varepsilon_0)(\mathbf{E}\cdot\nabla)\mathbf{E} = (\mathbf{P}\cdot\nabla)\mathbf{E}$, represents the force density on $n$ dipoles per unit volume, taking into account the force on an individual dipole $\mathbf{p}$ as $(\mathbf{p}\cdot\nabla)\mathbf{E}$ and the definition of the polarization vector as $\mathbf{P} = n\mathbf{p}$.

The dielectric force in a homogeneous fluid (spatially constant properties) does not produce continuous flow; its influence is important in the hydrostatic equilibrium [18]. The dielectric siphon is a good example of how liquid flow can be achieved with dielectric forces [18]. However, the dielectric siphon is not a pump that can move fluid in a loop or between two points at equal pressure. The term responsible for generating pumping in Electrohydrodynamics is primarily the Coulomb force $\rho\mathbf{E}$. There are three main mechanisms of generating free charge in the bulk of a dielectric liquid: (a) injection of ions in a very insulating liquid from electrodes; (b) generation of nonequilibrium layers of free charge close to the electrodes because the rate of dissociation of neutral molecules exceeds the rate of recombination of ions [3]; and (c) induction of charges in a conducting liquid due to gradients of electrical conductivity. These three mechanisms give rise to three kinds of pumps.

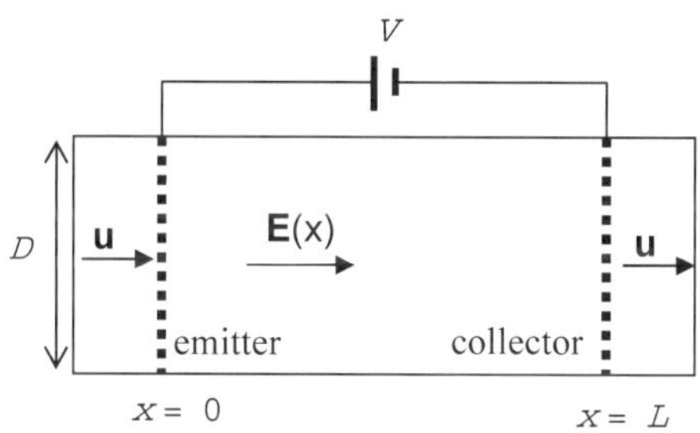


**Fig. 1.** Injection pump configuration

### *2.1. Injection Pump*

The injection pump, also known as ion-drag pump, uses the interaction of an electric field with electric charges injected into a dielectric fluid. The electric field is imposed between an electrode called emitter and another called collector. The ions, travelling from the emitter to the collector, drag by friction the working fluid.

This pumping mechanism has been known for quite some time. Stuetzer [19, 20] and Pickard [21, 22] were among the first who studied theoretically and experimentally this EHD pump. In microsystems, Richter and coworkers [2, 23] fabricated an ion-drag EHD micropump, consisting of pairs of metallic planar grids through which the pumped fluid moves. Ahn and Kim [24] experimented with an ion-drag pump that consisted of pairs of coplanar microelectrodes. Darabi et al. [25] tested different designs of coplanar microelectrodes for ion-drag pumping: emitter electrodes of planar shapes, of saw-tooth shapes or of saw-tooth shapes with bumps.

### 2.1.1. Pump Principle

Let us consider two plane-parallel perforated electrodes placed perpendicular to the axis of a pipe of circular cross section along which the liquid is pumped (see Fig. 1). For instance, this is a first approximation of two parallel wire grids that function as emitter and collector [2]. The emitter supplies ions that are driven by the electric field toward the collector placed at a distance $L$ from the emitter. We restrict here to the simple case of unipolar injection (only one kind of ion is present). The body force applied to the liquid is the Coulomb force $\mathbf{f} = \rho\mathbf{E}$. As a simplification, we are going to consider the problem one-dimensional. The force per unit area generated by the pump is

$$\Delta p = \int_0^L \rho E \,\mathrm{d}x \,. \tag{16}$$

From the Gauss's law in this 1D-problem, i.e. $\partial \varepsilon E / \partial x = \rho$,

$$\Delta p = \int_0^L \frac{\partial \varepsilon E}{\partial x} E \,\mathrm{d}x = \frac{\varepsilon E_L^2}{2} - \frac{\varepsilon E_0^2}{2}, \tag{17}$$

where $E_\mathrm{L} = E(L)$ is the field at the collector and $E_0 = E(0)$ is the field at the emitter. The 1D current density is, neglecting diffusion,

$$j = \rho(\mu E + u) = (\mu E + u)\frac{\partial \varepsilon E}{\partial x}. \tag{18}$$

In this simplified 1D-problem, the fluid velocity $u$ is the average velocity in a cross section of the pipe. In the steady state, the current density and fluid velocity are constant along $x$. Equation (18) can be integrated and the electric field strength is

$$E = \left[\frac{2jx}{\varepsilon\mu} + \left(E_0 + \frac{u}{\mu}\right)^2\right]^{1/2} - \frac{u}{\mu}. \tag{19}$$

The charge density is

$$\rho = \frac{j/\mu}{\left[\left(E_0 + u/\mu\right)^2 + 2jx/\varepsilon\mu\right]^{1/2}}. \tag{20}$$

Here $u > 0$, the liquid moves in the same direction as the ions, which is the case when the flow is generated by the pump. From the emitter to the collector, the charge density decreases while the field strength $E$ increases. The injection of charge decreases the electric field at the emitter, since each additional unipolar charge already present in the pipe repels the incoming charge. When the field is too low, charge can no longer be removed from the emitter, and the pump reaches its space charge limit [26].

Let us analyse first the case when the fluid velocity is much smaller than the ion-migration velocity, $u \ll \mu V/L$. For space-charge-limited emission (SCLE), the electric field at the emitter is much smaller than the average field, $E_0 \ll V/L$. The boundary condition is, therefore, $E_0 \approx 0$. This happens under strong injection, $\rho \to \infty$ at the emitter. In this case, the generated pressure is maximized because $E_0 \approx 0$, see Eq. (17). The electric field is

$$E(x) = \left[\frac{2jx}{\varepsilon\mu}\right]^{1/2}. \tag{21}$$

From $V = \int_0^L E\,\mathrm{d}x$, the current density can be obtained,

$$j = \frac{9}{8}\frac{\mu\varepsilon V^2}{L^3} \tag{22}$$

and the generated pressure is

$$\Delta p_{max} = \frac{9}{8}\frac{\varepsilon V^2}{L^2}. \tag{23}$$

Under these conditions, the maximum average fluid velocity and flow rate that the pump can deliver are

$$u_{\max} = \frac{9}{8}\frac{\varepsilon V^2}{L^3}\frac{D^2}{32\eta} \qquad Q_{\max} = \frac{9}{8}\frac{\varepsilon V^2}{L^3}\frac{\pi D^4}{128\eta}, \tag{24}$$

where $D$ is the diameter of the pipe. Here we have used expression (13) since laminar flow is expected to be valid in microsystems.

For the dynamic case ($u \neq 0$) and under SCLE condition, $E_0 = 0$, the equations provide voltage and generated pressure as functions of $u$ and $j$ [19]:

$$V = \frac{\varepsilon\mu}{3j}\left\{\left[\frac{2jL}{\varepsilon\mu} + \left(\frac{u}{\mu}\right)^2\right]^{3/2} - \left(\frac{u}{\mu}\right)^3\right\} - \frac{uL}{\mu}, \tag{25}$$

$$\Delta p = \frac{\varepsilon}{2}\left\{\left[\frac{2jL}{\varepsilon\mu} + \left(\frac{u}{\mu}\right)^2\right]^{1/2} - \frac{u}{\mu}\right\}^2. \tag{26}$$

Voltage drop and current density can be expressed as functions of the generated pressure $\Delta p$ and flow velocity $u$

$$j = \frac{\varepsilon\mu}{2L} E_L \left[E_L + \frac{2u}{\mu}\right], \tag{27}$$

$$V = \frac{2L}{3} E_L \frac{E_L + \frac{3}{2}u/\mu}{E_L + 2u/\mu}, \tag{28}$$

where $E_L = \sqrt{2\Delta p/\varepsilon}$. The generated pressure by the pump is used in overcoming the internal hydrodynamic resistance of the pump plus the external load: $\Delta p = R_{in}Q + \Delta p_{out}$, where $R_{in} = 128\eta L/\pi D^4$. It is now possible to obtain the flow rate $Q$ versus the external back pressure $\Delta p_{out}$ at different applied potentials $V$. Here we can see that if ion migration velocity $\mu E_L$ is much smaller than the fluid velocity $u$, the voltage drop is $V = E_L L/2$ and the pressure generated is $\Delta p = 2\varepsilon V^2/L^2$. Therefore, the maximum pressure at negligible flow rate and the maximum flow rate at zero pressure are, respectively,

$$\Delta p_{max} = 2\varepsilon V^2/L^2 \quad Q_{max} = 2\varepsilon V^2/(R_{in}L^2), \tag{29}$$

which are 16/9 times greater than for the case $u \ll \mu E$.

Up to here the theory has taken into consideration that there is only one kind of charge, i.e. unipolar injection. In practice, there are always other ions different from those injected that come from dissociation of impurities or of the liquid itself. All these ions should be considered when dealing with a real experiment. According to Pickard [21] dissociation will produce equal amounts of positive and negative ions and their contribution to the space charge should be negligible. The pressure heads can be relatively well accounted for by emission theory, while the total current is not so well explained. This is because the electric force comes from the space charge originated by emission processes, while the measured current is the sum of the emission current and the conduction current.

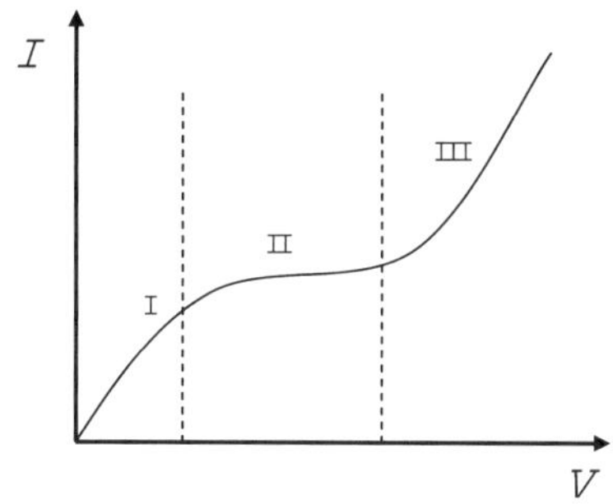


**Fig. 2.** Typical current-versus-voltage characteristic for a dielectric liquid

The current-versus-voltage characteristic for a dielectric liquid will typically have a form as shown in Fig. 2 [9,13]. Generally, different behaviors can be distinguished. At low voltages (region I), the current is caused by the dissociation of impurities in the liquid, and ohmic behavior is observed. As the voltage increases (electric fields around $10^5$ V m$^{-1}$), ions generated by the dissociation of impurities can not be created as fast as they are neutralized at the electrodes and the current density reaches saturation (region II). This region II is always present in gases but is ill-defined or may not be present in dielectric liquids. At higher voltages, a steep increase in the current is observed (region III). The beginning of region III can be due to enhancement of the dissociation process of impurities by the electric field strength. Nevertheless, at further increase of the electric field strength, ion injection from the electrodes becomes the dominating process. The current then tends to the $I \propto V^2$ law of the space-charge-limited emission. Finally, at electric fields of sufficient high strength (of the order of 100 MV m$^{-1}$), breakdown may take place. The breakdown electric field is very dependent on the liquid, impurities, and geometry.

To reduce the voltage needed to observe the emission, the emitter is usually designed to produce a local, highly intense electric field. The emitters usually have sharp edges or points while the collectors should be as smooth as possible. For instance, in the ion-drag pump experiments using arrays of coplanar microelectrodes by Darabi et al. [25], the micropump with saw-tooth-shaped emitters with bumps obtained the best pumping performance.

### 2.1.2. Characteristics

#### *2.1.2.1. Efficiency*

According to Crowley et al. [26], the electrical power consumption of the ion-drag pump, which is the product of voltage and current, can be approximated by

$$P_{\text{in}} = jSV \sim \rho(u + \mu E)SV + \sigma ESV, \tag{30}$$

where $S$ is cross-sectional area, and $E$ and $\rho$ are here the mean electric field and mean charge density, respectively. The term $\sigma E$ is the ordinary bipolar conduction, in which charge carriers are not produced by injection but by dissociation of impurities in the bulk. We may think of this as residual currents that do not contribute to the pumping. Crowley et al. defined the useful output power of the pump as all the mechanical work, rather than the mechanical work the pump delivers to a external load. Under this consideration, the output power is $P_{\text{out}} \sim \rho EuSL = \rho uSV$, giving an efficiency of

$$eff^* = \frac{P_{\text{in}}(\mu = \sigma = 0)}{P_{\text{in}}} = \frac{1}{1+\alpha}, \tag{31}$$

where $\alpha$ is the coefficient,

$$\alpha = \frac{\mu E}{u} + \frac{\sigma E}{\rho u}. \tag{32}$$

From our definition of efficiency, (14), we obtain $eff \sim eff^*/4$, considering $Q_{\max} = u_{\max} S$ and $\Delta p_{\max} \sim \rho EL$. The absolute maximum efficiency is $eff = 0.25$, when $\alpha = 0$, which can also be obtained from the values of maximum pressure and flow rate given in (29). Since $\alpha = 0$ implies that the ion mobility must vanish, the pump is more efficient if the fluid velocity is much greater than the ion-migration velocity $\mu E$.

The second term due to the losses originated by residual conductivity can be written as

$$\frac{\sigma E}{\rho u} \sim \frac{L/u}{\varepsilon/\sigma}, \tag{33}$$

where the charge density estimate $\rho \sim \varepsilon E/L$ comes from Gauss's law. It can be seen that the expression is the ratio of two characteristic times: the numerator is the transit time and the denominator is the charge relaxation time of the fluid. For high efficiency, this ratio should be small, i.e., the fluid must go from emitter to collector before the charge has time to relax.

The maximum efficiency obtained in the experiments of Darabi and Wang [27] was around 0.0015 for an applied voltage of 180 V using refrigerant HFE-7100 as working fluid.

*2.1.2.2. Liquids*

Typically, the liquids that are candidate to be pumped by ion-drag micropumps are highly insulating liquids. An upper bound for the conductivity of the liquid to be pumped comes from comparison between the time needed for the ion to go from the emitter to the collector and the charge relaxation time $\varepsilon/\sigma$. If the charge relaxation time is too short, the charge is screened before enough force is transmitted to the liquid. Therefore, we do not expect that ion-drag pumps can actuate liquids with conductivities $\sigma \gg v\,\varepsilon/L$, where $v$ is the ion velocity $v = u + \mu E$. Crowley et al. [26] presented a table of liquids that were good candidates to be used as working fluids for ion-drag pumps. The highest conductivity in that table of liquids was around $10^{-7}$ S m$^{-1}$. In microsystems, Richter and Sandmaier [2] were able to pump several organic solvents such as ethanol, methanol, acetone, and propanol. Richter et al. [23] and Ahn and Kim [24] presented results with ethanol as working fluid, which, according to Crowley et al. [26], has a conductivity around $10^{-7}$ S m$^{-1}$ if it is not deionized. The liquids used in the micropump of Darabi and Wang [27] had conductivities lower than $10^{-9}$ S m$^{-1}$. The highest fluid velocities are predicted for liquids with high dielectric constant and low viscosity. However, high dielectric constant also means that they have conductivities too large to allow efficient operation. Liquids of high dielectric constant should be deionized, in order to be used for ion-drag pumping.

#### *2.1.2.3. Problems*

The high electric field strengths used can change the liquid electrical properties, and can cause corrosion of the electrodes. These two effects can affect run-to-run repeatability and reduce pump life. However, some liquids like liquid nitrogen can be ideal working fluids for ion drag pumping. According to Darabi and Wang [27] slight electrode degradation was observed using the HFE-7100 fluid; however, no noticeable electrode degradation was observed using liquid nitrogen. It seems that the good performance of liquid nitrogen is due to its very stable molecular structure. According to Zhakin [13], the choice of a liquid, electrode material, and ionizing groups can be important in order to have reproducible results.

#### *2.1.2.4. Applications*

A potential application of an ion-drag pumping system is its use in pumping fluids in cryogenic cooling microsystems. Cryogenic cooling has become a widely adopted technique to improve the performance of electronics and sensors. Super-conducting devices, which are used to increase signal-to-noise ratio in communications, must also be maintained at cryogenic temperatures. Micropumps capable of pumping liquid nitrogen would enable the development of compact and lightweight chip-integrated cooling systems [27, 28]. In [28], the tested micropumps were able to pump sufficient liquid nitrogen to cool super-conducting sensors and detectors in a cryogenic chamber.

Other promising applications for ion-drag micropumps include fuel injection loops, and gas and liquid pumping where small quantities of dielectric fluids need to be pumped.

### 2.2. Conduction Pump

Under the weak electric field regime (field much less than $10^7$ V m$^{-1}$), the conduction in a dielectric liquid is carried out by ions generated by dissociated molecules. When an electric field exceeds a certain threshold (on the order of $10^5$ V m$^{-1}$), there appear layers in the vicinity of the electrodes where the rate of dissociation exceeds that of the recombination. These nonequilibrium layers are charged, with the opposite sign from that of the adjacent electrode [3, 29]. The thickness of these heterocharge layers is much greater than the Debye length and increases with the electric field. The Coulomb force acting on these charge layers produce the pumping mechanism. Because the thickness of the charge layer is much greater than the Debye length, we have included this kind of pump among those whose mechanism is due to electric forces in the bulk.

Micropumps based on conduction EHD pumping have not yet been reported, although Atten and Seyed-Yagoobi [3] and Jeong and Seyed-Yagoobi [30] reported conduction EHD pumps in the centimeter scale.

### 2.2.1. Pump Principle

The EHD conduction pumping mechanism can be illustrated considering two parallel perforated electrodes immersed in a dielectric liquid [3, 31]. The analysis here presented considers that no injection from electrodes takes place (the ions reaching the electrodes neutralize). Electric conduction in a pure dielectric liquid can be explained with a simple model, which considers a reversible process of dissociation-recombination of a neutral species (denoted by AB) into univalent positive $A^+$ and negative $B^-$ ions. These ions are the charge carriers between the parallel electrodes.

Let us assume that the electrode placed at $x = 0$ is subjected to a potential $V$ and the electrode placed at $x = L$ is grounded (geometry as in Fig. 1). The governing 1D-equations for the steady-state behavior of the ionic and neutral species are

$$\frac{\partial \mu_+ n_+ E}{\partial x} = k_\mathrm{d} n_0 - k_\mathrm{r} n_+ n_- , \tag{34}$$

$$\frac{\partial \mu_- n_- E}{\partial x} = -k_\mathrm{d} n_0 + k_\mathrm{r} n_+ n_- , \tag{35}$$

and

$$\frac{\partial E}{\partial x} = e\frac{n_+ - n_-}{\varepsilon} , \tag{36}$$

where $k_\mathrm{d}$ and $k_\mathrm{r}$ denote the dissociation and recombination rates, respectively. Several assumptions are made: (a) the dissociation rate is very small so that the neutral species concentration $n_0$ is constant and (b) diffusion and convection are neglected. At equilibrium $k_\mathrm{d} n_0^\mathrm{eq} = k_\mathrm{r} n_+^\mathrm{eq} n_-^\mathrm{eq}$ and $n_+^\mathrm{eq} = n_-^\mathrm{eq} = n^\mathrm{eq}$.

The boundary conditions for the previous equations are the following: (a) $n_+ = 0$ at $x = 0$ and $n_- = 0$ at $x = L$, because of the Coulomb repulsion between electrode and co-ions and (b) $\int_0^L E(x)\,dx = V$, where $V$ is the applied potential difference between electrodes. These equations were solved by Thomson and Thomson [32]. They provided an approximate solution valid for the quasi-ohmic regime, i.e., when the thickness of the heterocharge layer, $\lambda$, is much smaller than the distance between electrodes $L$. In this case, the liquid is electroneutral in the bulk except for the heterocharge layers. The approximation consists on neglecting the recombination in the heterocharge layers. For the layer adjacent to the positive electrode ($x = 0$), the approximate solution for $x \le \lambda_+$ is

$$\mu_+ n_+ E = k_\mathrm{d} n_0 x\,. \tag{37}$$

The electric current is constant along $x$

$$\sigma E_\mathrm{b} = e(\mu_+ n_+ + \mu_- n_-)E = ek_\mathrm{d} n_0 x + e\mu_- n_- E\,, \tag{38}$$

where $E_\mathrm{b}$ is the electric field in the bulk and $\sigma = e(\mu_+ + \mu_-)n^\mathrm{eq}$ is the bulk conductivity. For $x > \lambda_+$, the liquid is assumed to be electroneutral, $n_+ = n_- = n^\mathrm{eq}$. The thickness $\lambda_+$ can be obtained from $\sigma E_\mathrm{b} = ek_\mathrm{d} n_0 \lambda_+ + e\mu_- n^\mathrm{eq} E_\mathrm{b}$ giving

$$\lambda_+ = \frac{\mu_+ n^\mathrm{eq} E_\mathrm{b}}{k_\mathrm{d} n_0} = \frac{\mu_+ E_\mathrm{b}}{k_\mathrm{r} n^\mathrm{eq}}\,. \tag{39}$$

Equivalently, for the charge layer adjacent to the negative electrode,

$$\lambda_- \quad \frac{\mu_- E_\mathrm{b}}{k_\mathrm{r} n^\mathrm{eq}}\,. \tag{40}$$

The thickness of the charge layer can be seen as the distance that an ion, under the action of the electric field, travels without recombination, taking into account that $t_c = 1/k_\mathrm{r} n^\mathrm{eq} \approx \varepsilon/\sigma$ [33] is the charge relaxation time.

Solving Gauss's law in the charge layer $0 \le x \le \lambda_+$ leads to [34]

$$E = E_b \sqrt{1 + (1 - x/\lambda_+)^2 \mu_+/\mu_-}\,. \tag{41}$$

The value of the electric field in the bulk is close to $V/L$ for $\lambda_+/L \ll 1$ and $\lambda_-/L \ll 1$, typically, when $\varepsilon V/en^{eq}L^2 \ll 1$.

In the two charge layers, the Coulomb force is directed towards the electrodes. In the absence of motion, the force balance equation gives

$$\frac{\partial p}{\partial x} = e(n_+ - n_-)E = \frac{\partial \varepsilon E}{\partial x} E = \frac{\partial \varepsilon E^2/2}{\partial x}. \tag{42}$$

Therefore, the pressure generated in the layer close to $x = 0$ is

$$p_0 - p_b = \frac{\varepsilon}{2}(E_0^2 - E_b^2) = \frac{\varepsilon}{2} E_b^2 \frac{\mu_+}{\mu_-}, \tag{43}$$

where $p_0$ is the pressure at $x = 0$ and $p_b$ is the pressure in the bulk. An equivalent expression can be obtained for the pressure drop across the layer adjacent to the negative electrode:

$$p_b - p_L = \frac{\varepsilon}{2}(E_b^2 - E_L^2) = -\frac{\varepsilon}{2} E_b^2 \frac{\mu_-}{\mu_+}, \tag{44}$$

where $p_L$ is the pressure at $x = L$. The total pressure generated across the pump is, therefore,

$$p_0 - p_L = \frac{\varepsilon}{2} E_b^2 \left( \frac{\mu_+}{\mu_-} - \frac{\mu_-}{\mu_+} \right). \tag{45}$$

Here, contrary to [3, 31], the net pressure is different from zero because of the difference in mobilities between positive and negative ions.

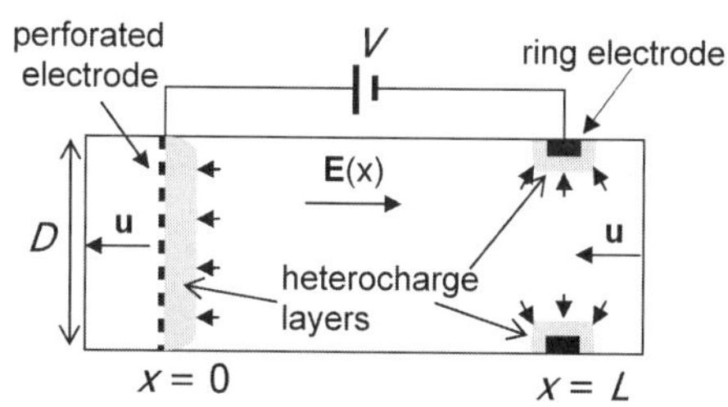


**Fig. 3.** Scheme of the conduction pump of ref. [31]

To have a pressure difference even in the case of equal mobilities, the electrode pair is made nonsymmetric [3, 30, 31]. For instance, a perforated disc electrode against a ring electrode was used in [31] (see Fig. 3). The Coulomb force in the charge layer adjacent to the perforated disc electrode is mainly axial while the electric force in the charge layer adjacent to the ring electrode is mainly radial. The radial force is balanced by the pipe wall while the axial force drives the dielectric liquid along the pipe. According to [31], the pressure generated at zero flow rate for this configuration is $\Delta p_{\max} = 0.85 \varepsilon V^2 / L^2$, which is of the order of the expressions previously obtained for only one electrode, (43). For finite flow rate, if the convection current can be neglected in front of the conduction current, the previous expression for the maximum pressure is the total force per unit area that the pump exerts to the fluid. In this case, the maximum average fluid velocity and flow rate that this pressure can generate are

$$u_{\max} = 0.85 \frac{\varepsilon V^2}{L^3} \frac{D^2}{32\eta} \quad Q_{\max} = 0.85 \frac{\varepsilon V^2}{L^3} \frac{\pi D^4}{128\eta}. \tag{46}$$

Experimentally, the conduction pump drives the liquid in the opposite direction as the ion-drag pump for the same geometry. For instance, in the point-plane configuration of [3], the liquid moves towards the point in the low voltage regime (below 17.5 kV) and moves from the point to the plane for higher voltages. This indicates that in this geometry the effect of direct injection of charges overcomes the corresponding effect of the heterocharge layers at approximately 17.5 kV.

## 2.2.2. Characteristics

### *2.2.2.1. Efficiency*

For $\lambda \ll L$, we can estimate the electric input power $IV \approx \sigma V^2 S / L$. The energy efficiency is then

$$eff = \frac{\Delta p_{\max} Q_{\max}}{4IV} \approx 0.2 \frac{u_{\max} \varepsilon}{\sigma L} \approx 0.2 \frac{\varepsilon^2 V^2 D^2}{32 L^4 \eta \sigma}. \tag{47}$$

The combination of parameters $u\varepsilon / \sigma L$ is called the electric Reynolds number [35]. It represents the ratio between convection current and conduction current. In the conduction regime, the convection of current is usually negligible, so that this number is typically very small. The efficiency decreases as $\sigma$ increases. For the experiments presented in [31] at $V = 10$ kV, the maximum efficiency was $eff \sim 10^{-3}$.

### *2.2.2.2. Liquids*

Liquids of low conductivity can be used as working fluids. An upper bound for $\sigma$ comes from the thickness of the heterocharge layer as compared to the Debye length. If the conductivity is high, both layers (diffuse and heterocharge layers) can have similar thicknesses. In this case, the diffusion processes are not negligible and the approximations are not applicable. This condition $\lambda \gg \lambda_D$ implies that the conductivity should be $\sigma \ll \mu\varepsilon e E^2/k_B T$. A lower bound for $\sigma$ comes from the comparison between the heterocharge layer thickness $\lambda$ and the distance between electrodes $L$ [3]. If $\lambda \gg \mathrm{L}$, the recombination is negligible also in the bulk. This gives rise to the saturation regime for the current where the ions are removed from the bulk at the rate of production. In this situation, the electric current increases very slightly with voltage approaching the asymptotic value $j_{sat} = ek_d n_0 L$. The electric field does not vary significantly between electrodes [34] and the pressure generated is small (because the pressure comes from the increments in $\varepsilon E^2$). Therefore, in order to have significant pressure, the conductivity should not be much smaller than $\mu E\varepsilon/L$. In [3], the pressure obtained for the working liquid n-hexane was smaller than predicted ($\sigma = 7\times10^{-11}\,\mathrm{S\,m^{-1}}$). For this liquid, the authors estimated $\lambda \sim 4L$ ($C_0 \sim 0.25$ in their notation), and the conclusion was that the pressure generated will take much smaller values than the prediction $\Delta p_{max} \sim \varepsilon V^2/L^2$ when $\lambda \gg L$. Tentatively, we can say that a lower bound for $\sigma$ is $0.01\mu E\varepsilon/L$. Therefore, the conduction pump will mainly operate for liquids with conductivities between

$$0.01\frac{\mu\varepsilon V}{L^2} < \sigma < \frac{\mu\varepsilon V}{L^2}\frac{eV}{k_B T}. \tag{48}$$

In [3, 30, 31], the primary working fluid was HCFC refrigerant R-123 with conductivity $\sigma = 2.8\times10^{-8}\,\mathrm{S\,m^{-1}}$, which is greater than $\mu\varepsilon V/L^2 \sim 10^{-9}\ \mathrm{Sm^{-1}}$ for their experimental conditions.

#### *2.2.2.3. Problems*

The voltages can not exceed the threshold voltage for charge injection, which may limit the pressure and flow generated. To limit the charge injection, the use of electrodes with small radii of curvature should be avoided. As an advantage, the working fluid and electrodes are not subjected to the degradation that ion injection usually generates.

*2.2.2.4. Applications*

As for the ion-drag pump, the main application may be the pumping of refrigerant for chip-integrated cooling systems.

## 2.3. Induction Pump

In this pump, charge is induced in an inhomogeneous liquid by the electric field, and the electric force upon this induced charge is transmitted to the liquid. Usually the inhomogeneity in the liquid is achieved by imposing a gradient of temperature. Liquid conductivity depends on temperature mainly because ionic mobilities are inversely proportional to viscosity (that depends on temperature) and because the dissociation rate of neutral molecules also depends on temperature. The common induction pump consists of an array of coplanar electrodes subjected to a travelling wave potential together with a vertical gradient of temperature (see Fig. 4) [36]. The temperature gradient may be obtained by the Joule heating generated by the imposed electric field instead of by an externally imposed temperature gradient. Typically, this has been the case in the application of this pump in microfluidics [4, 37, 38]. Recently, an induction pump was realized that employs an external gradient of temperature [39]. Unidirectional motion can also be obtained with an array of electrodes subjected to a single phase signal with an imposed longitudinal temperature gradient [40, 41].

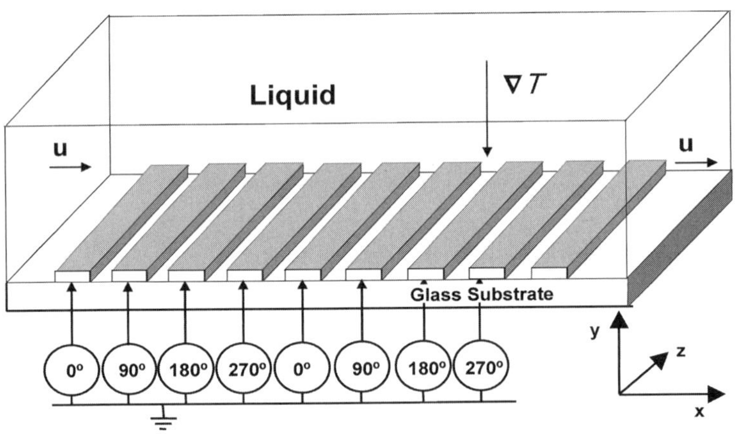


**Fig. 4.** Scheme of travelling wave induction pump

### 2.3.1. Pump Principle

Let us consider a simple geometry like the one shown in Fig. 1: two parallel perforated electrodes subjected to a potential difference inside a circular capillary. The electrode at $x = 0$ is at temperature $T_0$ and the electrode at $x = L$ is at temperature $T_\mathrm{L}$. Solving the diffusion equation for temperature, a linear temperature profile is achieved between electrodes in the steady state. Because the conductivity and permittivity of the liquid are functions of $T$, gradients of conductivity and permittivity will appear. The increment of temperature between electrodes is such that, in good approximation, there are also linear variations for $\sigma$ and $\varepsilon$ between the electrodes:

$$\sigma(x) = \sigma_1 + ax \qquad \varepsilon(x) = \varepsilon_1 + bx. \tag{49}$$

where $a$ and $b$ are given by

$$a = \frac{\mathrm{d}\sigma}{\mathrm{d}T}\frac{\mathrm{d}T}{\mathrm{d}x} \qquad b = \frac{\mathrm{d}\varepsilon}{\mathrm{d}T}\frac{\mathrm{d}T}{\mathrm{d}x}. \tag{50}$$

We assume that any convection of current is negligible in front of the conduction of current. In this case the continuity of current leads to

$$\frac{\partial \sigma E}{\partial x} = 0 \quad j = \sigma E = constant. \tag{51}$$

And from $\int_0^L E\,\mathrm{d}x = V$ we obtain

$$j = V\left(\int_0^L \frac{\mathrm{d}x}{\sigma}\right)^{-1} = \frac{aV}{\ln(1 + aL/\sigma_1)}. \tag{52}$$

An induced charge density appears in the bulk

$$\rho = \frac{\partial \varepsilon E}{\partial x} = \sigma E \frac{\partial}{\partial x}\left(\frac{\varepsilon}{\sigma}\right) = E\frac{\sigma b - \varepsilon a}{\sigma}. \tag{53}$$

In this case that the permittivity varies spatially, the electric force expression has two contributions, the Coulomb term and the dielectric term

$$f = \rho E - \frac{1}{2}E^2\frac{\partial \varepsilon}{\partial x} = E\frac{\partial \varepsilon E}{\partial x} - \frac{1}{2}E^2\frac{\partial \varepsilon}{\partial x} = \frac{1}{2}\frac{\partial \varepsilon E^2}{\partial x}. \tag{54}$$

The pressure drop that is generated at zero flow rate is then

$$\Delta p_{\max} = \frac{1}{2}\varepsilon(L)E^2(L) - \frac{1}{2}\varepsilon(0)E^2(0) = \frac{j^2}{2}\left(\frac{\varepsilon(L)}{\sigma(L)^2} - \frac{\varepsilon(0)}{\sigma(0)^2}\right). \quad (55)$$

For the case $bL/\varepsilon_1 \ll 1$ and $aL/\sigma_1 \ll 1$ the pressure drop is

$$\Delta p_{\max} = \frac{1}{2}\varepsilon_1 \frac{V^2}{L^2}\left(\frac{bL}{\varepsilon_1} - \frac{2aL}{\sigma_1}\right) = \frac{1}{2}\varepsilon_1 \frac{V^2}{L^2}(\beta - 2\alpha)(T_L - T_0), \quad (56)$$

where $\alpha = (1/\sigma)(\mathrm{d}\sigma/\mathrm{d}T)$ and $\beta = (1/\varepsilon)(\mathrm{d}\varepsilon/\mathrm{d}T)$. For water-saline solutions, $\alpha \sim 0.02\,K^{-1}$ and $\beta \sim -0.004\,K^{-1}$. If $T_0 > T_L$ the pump drives the fluid from $x = 0$ to $x = L$, as should be for a positive induced charge (see (53)). The liquid motion modifies the linear profile of temperature because of heat convection. To neglect heat convection in front of heat diffusion, the Péclet number should be small, i.e. $uL/\chi < 1$, where $\chi$ is the thermal diffusivity. It can be seen, by solving the 1D convection-diffusion equation for temperature, that the linear profile is not very much distorted even if $uL/\chi = 1$. For a fluid velocity of 1 mm s$^{-1}$, length $L = 10^{-4}$ m and water thermal diffusivity $\chi = 1.4\times10^{-7}$ m$^2$ s$^{-1}$, the Péclet number is close to 0.7, and the previously calculated expressions for $T$ and $\Delta p$ are approximately correct. Assuming that the convection of temperature and the convection of charge are negligible, the maximum average fluid velocity and flow rate that this pressure can generate are

$$u_{\max} = \frac{1}{2}\frac{\varepsilon_1 V^2}{L^3}\frac{D^2}{32\eta}(2\alpha - \beta)(T_0 - T_L) \quad Q_{\max} = u_{\max}\frac{\pi D^2}{4}. \quad (57)$$

With these simple expressions, we can see that the induction pump generates much smaller flow rate and pressure than the conduction or injection pumps for the same applied voltage and geometry (between 5 and 10 times less). In effect, the temperature increment will be on the order of 10 K or less, and the pressure generated is on the order of $\Delta p \sim \varepsilon(V^2/L^2)\alpha\Delta T \le 0.2\varepsilon(V^2/L^2)$ in the case of water.

To avoid faradaic reactions, AC voltages of high enough frequency should be applied. In this case, the time-average force is different from zero because it is quadratic with voltage. The expressions are the same substituting $V$ by the rms value $V_{rms}$. The theoretical results presented here are valid in the AC case if the angular frequency of the applied signal $\omega$ is much smaller than $\sigma/\varepsilon$. For frequencies $\omega \gg \sigma/\varepsilon$, the liquid behaves as a pure dielectric liquid and no free charge is induced. However, the dielectric force is still operating at these frequencies, giving a generated pressure of the order of $\Delta p \sim \varepsilon(V^2/L^2)\beta\Delta T$, which is $\alpha/\beta$ times smaller than the force for $\omega \ll \sigma/\varepsilon$ (around 5 times smaller if the liquid is water).

As already mentioned, the usual experimental case (see Fig. 4) consists of an array of coplanar electrodes subjected to a travelling wave signal $V_0 \cos(\omega t - kx)$ and to a vertical gradient of temperature that is generated by Joule heating [4]. In this case, the vertical component of the electric field induces free charge in the liquid bulk, while the longitudinal component acts on the induced charge driving the liquid along the array. The dependence of the electric force with frequency is very different from the case we have just analysed. Here, the force is maximized at $\omega = \sigma/\varepsilon$, going to zero at high and low frequencies. For a long channel of rectangular cross section (where $h$ is the height, $w$ the width and $w \gg h$), the expressions for the average velocity and flow rate at zero pressure can be obtained from the results of [36]

$$u_{\max} = \frac{\varepsilon V_0^2}{16h\eta}(\alpha - \beta)\Delta T \frac{\omega\tau}{1+(\omega\tau)^2} \qquad Q_{\max} = u_{\max} hw, \tag{58}$$

where $\tau = \varepsilon/\sigma$ is the charge relaxation time of the liquid and $\Delta T$ is the temperature difference between top and bottom of the channel. Here it was supposed that $kh \gg 1$ and that the change in permittivity and conductivity are small, $\Delta\varepsilon/\varepsilon \ll 1$ and $\Delta\sigma/\sigma \ll 1$. The pressure that can be generated per wave-length $L = 2\pi/k$ is

$$\Delta p_{\max} = \frac{12\eta u L}{h^2} = \frac{3}{4}\frac{\varepsilon V_0^2(\alpha - \beta)\Delta T L}{h^3}\frac{\omega\tau}{1+(\omega\tau)^2}. \tag{59}$$

If $L \sim h \sim w$, these expressions for $\Delta p_{\max}$ and $Q_{\max}$ at frequency $\omega = 1/\tau$ are of the same order as the previously obtained expressions (56, 57).

When the origin of the temperature gradient is due to Joule heating, the boundary conditions for the temperature are essential in order to know the generated velocity or pressure. Nevertheless, an estimation of the increment of temperature can be obtained from the temperature equation neglecting heat convection [42]

$$\sigma E^2 + \kappa\nabla^2 T = 0, \tag{60}$$

where $\kappa$ is thermal conductivity. The estimate for the increment of temperature is $\Delta T \sim \sigma V_{rms}^2/\kappa$, which leads to pressure and flow proportional to $V^4$ when Joule heating governs the temperature profile.

## 2.3.2. Characteristics

### *2.3.2.1. Efficiency*

The efficiency of the EHD induction micropump can be estimated as

$$eff = \frac{1}{4}\frac{\Delta p_{max} Q_{max}}{IV + \dot{q}}, \tag{61}$$

where $\dot{q}$ is the power that is consumed in order to generate the temperature difference. When the temperature profile is generated by Joule heating, $\dot{q}$ is already included in the term $IV$. For an imposed temperature gradient in the system of parallel electrodes studied before, the minimum value of $\dot{q}$ should be $\dot{q} \sim Sk\Delta T / L$, where $S$ is the area of the pipe and $\Delta T$ the difference of temperature between electrodes. For small increments of temperature ($\Delta \varepsilon / \varepsilon \ll 1$ and $\Delta\sigma / \sigma \ll 1$), the maximum energy efficiency is

$$eff \approx \frac{1}{4}\frac{(u_{max}\varepsilon/\sigma L)\,\alpha\Delta T}{1+\dot{q}/IV} \approx \frac{\varepsilon^2 V^2 D^2\,(\alpha\Delta T)^2}{128\sigma\eta L^4\,(1+\dot{q}/IV)}. \tag{62}$$

We can see that the efficiency is very small, proportional to the product of two small numbers: the electric Reynolds number and $\alpha\,\Delta T/(1+\dot{q}/IV)$.

### *2.3.2.2. Liquids*

The liquids that can be pumped should have some conductivity. In the previous expressions, the convection of charge was neglected in front of the ohmic current. This is a fair assumption when the electric Reynolds number is small $u\varepsilon/\sigma L \ll 1$ [35]. When the convection current is not small, the pattern of induced charge can be very different and the pump may not work. In addition, any charge that could be injected from the electrodes should relax in a distance much shorter than the typical length, i.e. $\mu E\varepsilon/\sigma L \ll 1$. Therefore, the condition is that the conductivity should be greater than $\sigma > v\varepsilon / L$, where $v = u + \mu E$. For a micropump where the velocity is $u \sim 10^{-3}$ ms$^{-1}$, typical length is $L \sim 10^{-5}$ m, the condition is $\sigma > 10^{-7}$ S m$^{-1}$ in the case of water, and $\sigma > 10^{-9}$ S m$^{-1}$ for non-polar liquids. An upper bound for $\sigma$ comes from the generated temperature due to Joule heating. Typically the voltages needed to produce enough flow are in the range from 10 to 50 V peak-to-peak [38]. We expect that the increment of temperature will be excessive (reaching boiling temperature) for conductivities $\sigma \geq 1$ S m$^{-1}$ [11]. Therefore, a wide range of conductivities can be used for this micropump: $10^{-9} < \sigma < 1$ S m$^{-1}$.

*2.3.2.3. Problems*

If the electric field is very high, especially for liquids with conductivity around $\sigma \sim 1\ \mathrm{S\,m^{-1}}$ or more, the temperature rise can be excessive. Also, this may be a limiting factor if we are interested in pumping liquids with bio-particles (cells, DNA,...) where the temperature must be controlled. The expected temperature rise for $0.1\ \mathrm{S\,m^{-1}}$ ($0.01$ S/m) at $V_0 = 10$ V is smaller than 10 K (1 K).

If the electric field is too high, there exists the possibility that EHD flow instabilities appear [36], destroying the pumping mechanism. The formation of gas bubbles is also a limiting factor at high electric fields, and especially at low frequencies. EHD instabilities and bubbles were observed at voltages around $40$ V in the device studied in [38].

The problems related to the formation of bubbles by Faradaic currents can be eliminated by coating the electrodes with a thin dielectric layer. At high frequencies, the electric fields can penetrate the dielectric coating easily. As an advantage, avoiding Faradaic currents using dielectric coating makes the performance of the pump very reproducible: accurate pumping over eight months was observed in [38].

*2.3.2.4. Applications*

Since saline solutions with $\sigma < 1\ \mathrm{S\,m^{-1}}$ can be pumped, the possible applications range from bio-medical to chemical analysis. To use the micropump for liquids with low conductivity, the temperature gradient should be imposed externally. The energy efficiency of this pump is low, nevertheless there are many applications where this is not a limiting issue. For example, we may have a macroscopic power generator to feed our microscopic device. Although the generated pressure is typically low, many applications do not require high pressure like the generation of secondary flows for mixing [43] or the generation of controlled circular streaming in microfluidic systems for cooling of special elements, or for supply of living cells.

## 3. Electric Forces in the Diffuse Layer: Electroosmotic and AC/IC Electroosmotic Pumps

At the interface between solid walls and electrolytes, double layers are formed [44] due to differences in electrochemical potentials of both phases. The solid surface becomes charged and counterions coming from the bulk liquid shield this surface charge. At equilibrium, the electrostatic attraction between the charged surface and the counterions is balanced by thermal agitation. The liquid is electrically neutral except for a charged layer adjacent to the solid surface. The characteristic thickness of this layer is the Debye length, $\lambda_{\rm D}$, given by

$$\lambda_{\rm D} = \sqrt{\frac{\varepsilon k_B T}{e^2 \sum_i z_i^2 n_i^\infty}} \,. \tag{63}$$

Here $z_i$ and $n_i^\infty$ are the valence number and number density of ionic species $i$ in solution. For the case of a symmetric electrolyte, the Debye length can be written as $\lambda_D = \sqrt{\varepsilon D/\sigma}$, where $D$ is here the ion diffusion coefficient. The application of an electric field along the surface produces a Coulomb force on the diffuse charge and the liquid is set into motion. Because the thickness of the Debye layer is very small (1–100 nm in water) the Coulomb force is considered as a superficial force.

We can distinguish two ways of generation of charge at the solid surface: (a) a chemical mechanism such as ionization of bound surface groups or as ion adsorption and (b) an electrostatic mechanism in which the solid surface (typically a metal) gets charged because it is subjected to an electric potential difference with respect to the bulk electrolyte. In both cases, a double layer is formed in order to screen the charge of the solid surface. We can identify two kinds of pumps based on these charging mechanisms.

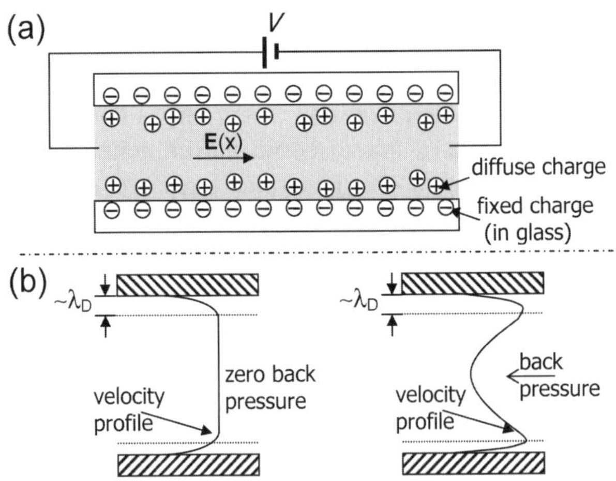


**Fig. 5.** Electroosmosis in a capillary. (**a**) The electric field acts on the counterions in the diffuse double layer. (**b**) Velocity profiles generated in the capillary with (*right*) and without (*left*) a back pressure

### 3.1. Electroosmotic Pump

A DC Electroosmotic (EO) pump makes use of the electric double layer that spontaneously develops between a liquid in contact with a solid. Generally, the chemical state of the solid surface is altered because of the presence of the electrolyte. For example, in the case of silica-based ceramics, like glass, with $SiOH$ groups at the surface, a fraction of the $Si-OH$ bonds change into $Si-O-$ releasing the proton when immersed in water. It results in a net negatively charged surface. The resulting redistribution of ions in the liquid keeps the liquid bulk electrically neutral, while a positive diffuse charge layer appears close to the surface. Application of an electric field along the surface causes the mobile ions of the double layer to move towards the electrode of opposite sign dragging the fluid with them. In this case, bulk flow moves in the direction of the applied electric field. The negatively charged $SiO-$ groups are bound to the solid surface by covalent bonds and hence will be stationary. Electroosmosis thus refers to the bulk movement of a liquid past a stationary solid surface due to externally applied electric field. The phenomenon is illustrated in Fig. 5.

The EO phenomena have been known for almost 200 years (F.F. Reuss discovered electroosmosis in 1809). Flows generated by EO pumping are used in a range of applications, including soil remediation or contaminant removal from groundwater, and have been used in chemical and biological analysis since a long time. Pretorius et al. [5] proposed electroosmosis for high-speed chromatography in 1974. A number of important techniques used in miniaturized systems for chemistry and life sciences make use of EO flow: flow injection analysis [45], on-chip electrophoretic separation [46], and on-chip liquid chromatography [47]. Yao and Santiago [48] presented a detailed description of the history and development of EO pumps that are able to generate high pressure.

#### 3.1.1. Pump Principle

The most basic EO pump is simply a capillary or micro-pipe with electrodes submerged within end-channel reservoirs. Let us consider a cylindrical capillary of radius $a$ [49], with the wall of the capillary at potential $\zeta$, the zeta potential. Zeta potential is a key parameter in the theory of electrokinetics. Classical theory describes the electrical double layer as divided into the Stern layer and Gouy-Chapman diffuse layer [44]. The Stern or compact layer is formed of counterions absorbed onto the wall, while the ions of the Gouy-Chapman layer are diffuse. The plane separating these two layers is called the shear plane, and the potential at this plane is the zeta potential.

Let us assume that inside the capillary there is a symmetric 1-1 electrolyte of bulk ionic concentration $n^{\infty}$. The Poisson-Boltzmann equation in this cylindrical geometry is

$$\frac{1}{r}\frac{\partial}{\partial r}\left(r\frac{\partial \Phi}{\partial r}\right)=-\frac{\rho(r)}{\varepsilon}=\frac{2en^{\infty}}{\varepsilon}\sinh\left(\frac{e\Phi}{k_B T}\right). \tag{64}$$

In the Debye-Hückel approximation [44], $e\Phi/k_B T \ll 1$, and the Poisson-Boltzmann equation becomes

$$\frac{1}{r}\frac{\partial}{\partial r}\left(r\frac{\partial \Phi}{\partial r}\right)=\frac{\Phi}{\lambda_D^2}, \tag{65}$$

where the Debye length is $\lambda_D = \varepsilon k_B T/2e^2 n^{\infty}$. The general solution for this equation, which is finite at $r=0$, is $\Phi(r)=AI_0(r/\lambda_D)$, where $I_0$ is the zero-order modified Bessel function of the first kind. Taking into account the boundary condition at $r=a$, the expressions for $\Phi$ and $\rho$ are

$$\Phi(r)=\zeta\frac{I_0(r/\lambda_D)}{I_0(a/\lambda_D)} \qquad \rho=-\frac{\varepsilon}{\lambda_D^2}\Phi. \tag{66}$$

Under an applied electric field in the axial direction, $E_z$, the free charge is subjected to a Coulomb force. The equation of liquid motion for an infinite cylindrical tube under the influence of an axial applied electric field and no applied pressure gradient is

$$\frac{\eta}{r}\frac{\partial}{\partial r}\left(r\frac{\partial u}{\partial r}\right)+E_z\rho=0. \tag{67}$$

With the boundary no-slip condition $u(a)=0$,

$$u(r)=-\frac{\varepsilon\zeta E_z}{\eta}\left(1-\frac{I_0(r/\lambda_D)}{I_0(a/\lambda_D)}\right) \tag{68}$$

is obtained. In the case that the radius is much greater than the Debye length, $a \gg \lambda_D$, the velocity varies from zero at the wall to the asymptotic value $u_s = -\varepsilon\zeta E_z/\eta$ in a very thin layer of the order of $\lambda_D$. Furthermore, the bulk of the liquid in the capillary moves with velocity $u_s$, known as the electroosmotic slip velocity. For this case of $a \gg \lambda_D$, the slip velocity expression is valid in the general case where $\zeta$ is not restricted to be smaller than $k_B T/e$.

The average velocity and flow rate at zero pressure are

$$u_{max} = u_s F(a^*) \quad Q_{max} = \pi a^2 u_{max}, \tag{69}$$

where $a^* = a/\lambda_D$ and

$$F(a^*) = \frac{2}{a^2}\int_0^a \left(1 - \frac{I_0(r/\lambda_D)}{I_0(a/\lambda_D)}\right) r\,\mathrm{d}r = 1 - \frac{2I_1(a^*)}{a^* I_0(a^*)}. \tag{70}$$

The pressure difference between the capillary ends required to balance this flow rate is the maximum pressure that can be generated, and this is

$$\Delta p_{max} = -\frac{8\varepsilon\zeta\Delta\Phi}{a^2} F(a^*). \tag{71}$$

Here $\Delta\Phi = E_z L$ is the potential difference applied between the ends of the capillary separated a distance $L$. In the thin double layer limit where $a \gg \lambda_D$, $F(a^*) \to 1$. Substitution yields $Q_{max} = -\pi a^2 \varepsilon\zeta E_z/\eta$ and $\Delta p_{max} = -8\varepsilon\zeta\Delta\Phi/a^2$. For capillary radii much smaller than the Debye length, $F(a^*) \to (a/\lambda_D)^2/8$.

To increase the pressure drop, the size of the micropipe can be reduced. This, however, reduces the flow rate. A way of increasing the pressure while not reducing much the flow rate is to use a bundle of parallel micropipes with small radius. In practice, EO pumps incorporate porous structures so that each pore acts as a tortuous capillary [48]. These porous structures can be visualized as a large number of tortuous micropipes in parallel. From this model, the theoretical maximum pressure is given by expression (71) and the maximum flow rate by [48]

$$Q_{max} = -\frac{\psi}{\tau}\frac{\varepsilon\zeta A}{\eta} F(a^*), \tag{72}$$

where $A$ is the cross-sectional area of the structure, $\psi$ is its porosity (void volume divided by total volume of the porous medium), and $\tau$ is its tortuosity defined as $(L_e/L)^2$ ($L_e$ is the characteristic length of the tortuous path of the pores and $L$ the length of the porous structure). From these expressions the number $\psi A/\tau\pi a^2$ can be seen as the effective number of EO capillaries with radius $a$ connected in parallel which give the same flow rate as the porous structure.

### 3.1.2. Characteristics

#### *3.1.2.1. Efficiency*

Let us analyze the efficiency of a single capillary. The power consumption is given by the applied voltage multiplied by the electric current through the pump $P_{\text{in}} = IV$. The current transported by the liquid consists of two contributions [44]: that due to the bulk conductivity and that due to the surface conductivity including the convective current due to the electro-osmotic flow itself. Thus, one can write for a capillary

$$I = \pi a^2 \sigma E_z + 2\pi a \sigma_s E_z, \tag{73}$$

where $\sigma_s$ is a specific surface conductivity. For a capillary we have

$$\frac{Q_{\max}}{I} = \frac{\varepsilon\zeta}{\eta(\sigma + 2\sigma_s/a)} F(a^*). \tag{74}$$

Therefore, our estimation for the maximum efficiency is

$$eff = \frac{1}{4}\frac{\Delta p_{\max} Q_{\max}}{IV_{\text{app}}} = \frac{\Delta\Phi}{V_{\text{app}}}\frac{2(\varepsilon\zeta F(a^*))^2}{a^2\eta(\sigma + 2\sigma_s/a)} \tag{75}$$

where $V_{\text{app}}$ is the potential applied to the main circuit and, in general, is different from the potential difference at the ends of the capillary $\Delta\Phi$. In particular, the difference $V_{\text{app}} - \Delta\Phi$ should account for the threshold voltage needed to start the Faradaic reactions at the electrodes and the junction voltages at inlet and outlet of the capillary. The efficiency of a porous structure will be of the same order as that of a capillary since in the model, both $Q$ and $I$ are proportional to the number of capillaries in parallel and $\Delta p$ and $V$ are the same for all the capillaries.

In the case $\lambda_{\text{D}} \ll a$, the current will be given by the bulk contribution and, assuming an ideal situation where $V_{\text{app}} = \Delta\Phi$, the efficiency is

$$eff = \frac{2\varepsilon^2\zeta^2}{a^2\eta\sigma}. \tag{76}$$

We can see that, in this limit, the efficiency decreases with pore size $a$ and with the product $\sigma\eta$, which is proportional to ionic concentration. It does not decrease with viscosity because the product $\sigma\eta$ is, in fact, almost independent of viscosity. In the general case, the efficiency is a maximum when $a$ is of the order of $\lambda_D$ [50, 48].

If the working liquid is relatively conductive, electroosmotic pumps can be very inefficient. Taking $\zeta = 0.1$ V, $a = 1\,\mu$m and an aqueous solution of conductivity $\sigma$, we obtain $eff = 10^{-5}/\sigma$, with $\sigma$ in SI units. The reason for this is that the two main origins for energy losses, Joule heating and viscous losses, both increase with conductivity. Joule heating is directly proportional to $\sigma$, given by $\int \sigma E^2\,\mathrm{d}\tau$. Viscous losses increase with $\sigma$ because the shear stresses that balance the Coulomb force are confined to the thin diffuse layer, which is a decreasing function of $\sigma$. The viscous dissipation is greater in this case than when the viscous stresses are distributed across the entire capillary. The viscous dissipation is, for the thin diffuse layer, of the order of

$$P_v \sim \eta \frac{u_s^2 2\pi a L}{\lambda_D}. \tag{77}$$

This is of the order of $a/\lambda_D$ times greater than for the case of viscous stresses evenly distributed in the capillary (as in the Poiseuille flow). However, the losses due to Joule heating are usually much greater, and Joule heating is the primary mechanism of power consumption at typical conditions of EO pumps [50]. The ratio between power consumption due to viscous stresses and due to Joule heating is of the order of

$$\frac{P_v}{P_J} \sim \frac{\eta u_s^2 2\pi a L/\lambda_D}{\sigma E^2 \pi a^2 L} = \frac{2\varepsilon^2\zeta^2}{\eta\sigma\lambda_D a} = eff\,\frac{a}{\lambda_D} \qquad (\lambda_D \ll a). \tag{78}$$

In this limit $\lambda_D \ll a$, we can see that $eff \sim (P_v/P_J)(\lambda_D/a)$, the product of two numbers smaller than one.

The EO micropump experiments presented in [51] show a maximum efficiency of around 0.004 for deionized water of conductivity $\sigma \sim 10^{-4}$ S m$^{-1}$, while the efficiency was around 0.022 for acetonitrile of

$\sigma \sim 10^{-6}\,\mathrm{S\,m^{-1}}$. The higher efficiency of acetonitrile is mainly due to its lower conductivity.

*3.1.2.2. Liquids*

Electrokinetic phenomena tend to disappear at high ionic concentrations. When the electrolyte conductivity is high, the Debye length becomes very small and comparable or smaller than the thickness of the Stern layer. Since the total double-layer voltage is shared between the compact and the diffuse layers, the reduction of the diffuse part leads to a reduction of the potential across it, i.e., a reduction of the zeta potential $\zeta$. Applying a simple model of two capacitors in series (the compact-layer capacitor in series with the diffuse-layer capacitor) leads to the conclusion that $\zeta$ decreases with the squared-root of electrolyte concentration $n^{\infty}$. For values of $\sigma \sim 1\,\mathrm{S\,m^{-1}}$, both capacitances are of the same order, (the Stern layer capacitance is of the order of $C_s \sim 0.3\,\mathrm{F\,m^2}$ [44]). The prediction that $\zeta \propto 1/\sqrt{n^{\infty}}$ fails when the pH of the solution controls the surface charge [48]. However, $\zeta$ is still a decreasing function of ion concentration.

A lower bound for the conductivity of the liquid comes from a comparison between the electric field inside the double layer and the applied electric field. If the liquid conductivity is very small, the electric field inside the double layer (of the order of $\zeta/\lambda_{\mathrm{D}} \sim k_{\mathrm{B}}T/e\lambda_{\mathrm{D}}$) becomes of the same order as the applied field required to have significant EO velocity. In this case, the distributions of ions in the capillary would be very distorted by the applied electric field. Therefore, it is not expected that electroosmosis can work properly at very low conductivity. For an applied field $E_z \sim 10^4\,\mathrm{V\,m^{-1}}$, $\zeta \sim 0.025\,\mathrm{V}$, the condition implies $\sigma > 10^{-7}\,\mathrm{S\,m^{-1}}$ for water (or $\sigma > 10^{-8}\,\mathrm{S\,m^{-1}}$ for nonpolar liquid with $\varepsilon_{\mathrm{r}} = 2$). Also, when the charge on the surface is governed by the process of deprotonation, as in the case of silica surfaces, the pH of the liquid is another parameter to be taken into consideration. The zeta potential is a strong function of pH, typically saturating at high and low pH [1]. According to Yao et al. [52], low buffer concentrations (below 0.1 mM) are impractical because pump operation generates prohibitively large fluctuations in pH.

Liquids that have been pumped using electroosmosis have conductivities that range from $10^{-6}$ to $1\,\mathrm{S\,m^{-1}}$ [50].

*3.1.2.3. Problems*

EO pumps operated continuously in DC mode suffer from inherent problems. Electrolysis generates gas bubbles, which must be prevented from coming into the pump channel. Redox reactions at the electrodes will eventually change the pH, which is a problem because the EO flow depends on it. For instance, under conditions of $Q = 0$ in order to obtain the maximum pressure, the liquid is recirculating in the pumping channels. The pH change introduced at the electrodes will stop the EO flow after a while when the buffer is depleted [53]. Traditionally, some of these problems are avoided by separating the electrodes from the EO pumping channels with an ion exchange system [45]. For example, in [54] the electrodes are placed in reservoirs separated by the channels by a porous glass disk. The disk allows for exchange of ions but not of bulk flow. The reservoirs are filled with buffers to reduce the changes in pH due to operation. Yao et al. [52] recombined the products of the reaction, where electrolytic gases were converted to water and reused by the system. Brask et al. [53] presented an EO pump that worked in AC mode in order to prevent the bubble formation and reduce the pH change problem. The pump was energized by square signal of ±30 V and produced a pulsating flow. To rectify this oscillating flow, a system of valves was employed.

The high voltage usually required by EO pumps (traditionally in the kilovolt range) can be a drawback, especially for portable chips. A low voltage EO pump that produces high pressure (around 25 kPa at 10 V) has been realized by using a cascade pump design [55, 56]. The idea is to place several pumps in series in order to increase the pressure and each pump operates at low voltage (from 0 to 10 V) (see Fig. 6). Each pump consists of a narrow channel section, containing 10 parallel channels, followed by a wide single channel. The narrow channel section acts as a high pressure pump with forward electric field. In the wide channel section, the electric field is reversed, but here the channel is so wide that the generated back pressure is small compared to the previous pressure. Also, in order to reduce the voltage, the electrodes were in the flow path (gel salt bridge electrodes were used). Another way of reducing the voltage needed is to employ pumping based on transverse electrokinetic effects [57, 58]. It is based on the anisotropic response of a channel with obliquely oriented grooves to an applied field. The electric field is applied perpendicularly to the principal axis of the channel, resulting in flow in the axial direction. This allows for the generation of fluid velocities around 100–150 $\mu m\ s^{-1}$ for low applied voltages, around 10 V. Since the electrodes are inside the flow path, low ionic strength solutions are preferable to reduce generation of bubbles and to reduce local pH changes at the electrodes.

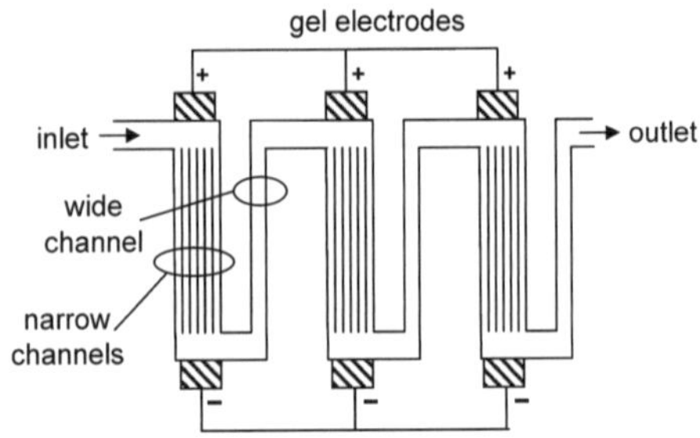


**Fig. 6.** Scheme of cascade pump by Takamura et al. [55]

The increment in temperature due to Joule heating can also be a problem. An order of magnitude estimate for the temperature increment is obtained by equating the generated heat in a capillary $\sigma E^2 \pi a^2 L$ to the heat flow through the lateral wall $\kappa(\partial T/\partial r)2\pi a L$. The estimation for the increment of temperature gives

$$\Delta T \sim a\frac{\partial T}{\partial r} \sim \frac{\sigma V^2 a^2}{2\kappa L^2}. \qquad (79)$$

The smaller the radius of the cross-sectional area of the pump, the smaller the increment in temperature. As an example, the increment of temperature was around 20°C for $\sigma \sim 0.5$ S m$^{-1}$ at an applied electric field of $6\times10^4$ V m$^{-1}$ in the experiments of Tang et al. [59]. Therefore, the temperature rise can be excessive for conductivities equal or greater than 1 S m$^{-1}$.

*3.1.2.4. Applications*

Saline solutions with $\sigma \leq 1$ S m$^{-1}$ can be pumped; therefore, possible applications range from bio-medical to chemical analysis, such as, the previously mentioned, on-chip liquid chromatography and on-chip electrophoresis separation. The energy efficiency can be very low for conductivities around $0.1\,\mathrm{S\,m^{-1}}$, nevertheless there are many applications where this is not a limiting issue. Potential applications include the replacement of high-pressure pumps in micro-total-analysis-systems: for drug delivery, sample analysis, separation, and mixing processes. Another application is the use of closed-loop electroosmotic microchannels as cooling systems for microelectronics [60]. Arrays of micropumps, temperature sensors, and fluid flow networks could be used to redistribute heat from hot spots that arise during microchip operation.

### 3.2. AC/IC Electroosmotic Pump

An alternating-current/induced-charge electroosmotic pump makes use of the Coulomb force on the induced-charge in the double layer [61, 62, 63]. Typically, the normal component of the electric field charges the double layer at the electrode/electrolyte interface, while the tangential electric field component produces a force on the induced charge in the diffuse layer that results in fluid motion [64, 65, 66]. Unidirectional fluid flow is typically obtained in two ways: arrays of asymmetric pairs of electrodes subjected to a single AC signal [6] or arrays of symmetric electrodes subjected to travelling-wave signals [67].

#### 3.2.1. Pump Principle

Let us analyse the case of a travelling-wave potential that generates EO flow (see Fig. 7). We are going to apply the linear approximation of the double layer (Debye-Hückel approximation). Also, we assume that the applied potential is low enough so that Faradaic currents from the electrodes to the liquid are absent, i.e., we assume perfectly polarizable electrodes. At the level of the electrodes, $z = 0$, a travelling potential is applied that we model as $V(x,t) = V_0 \cos(kx - \omega t)$, where $\omega$ is the frequency of the applied signal and $k$ is the wave-number that characterizes the spatial periodicity of the array. At the interface between electrodes and electrolyte, an electric double layer is formed. The frequency $\omega$ is low enough so that the double layer is in quasi-equilibrium. We assume that the Debye length is much smaller than any other length in the system and, in particular, $k\lambda_D \ll 1$. Above the array, the bulk electrolyte is electro-neutral characterized by its conductivity $\sigma$. The electric current follows Ohm's law $\mathbf{j} = \sigma \mathbf{E}$ and the electric potential holds Laplace's equation $\nabla^2 \Phi = 0$. The generated electric potential in the bulk is of the form (in phasor notation) $\Phi = Re[A e^{-i(kx-\omega t)-kz}]$, where $i = \sqrt{-1}$. Here we have assumed that $kh \gg 1$, where $h$ is the height of the upper wall of the device. The potential in the bulk at the limit of the Debye layer ($z = 0^+$) is related to the applied potential by the following boundary condition

$$\sigma \frac{\partial \Phi}{\partial z} = C_{\mathrm{DL}} \frac{\partial (\Phi - V)}{\partial t} = i\omega C_{\mathrm{DL}} (\Phi - V), \tag{80}$$

where $C_{\mathrm{DL}}$ is the double layer capacitance given by the series combination of the compact layer capacitance $C_{\mathrm{s}}$ and the Debye layer capacitance $C^{-1}{}_{\mathrm{DL}} = C_{\mathrm{s}}^{-1} + C_{\mathrm{d}}^{-1}$. The previous boundary condition expresses the conservation of charge: the current arriving at the double layer charges the double-layer capacitor. From this equation, the constant $A$ is obtained as $A = i\Omega V_0/(1+i\Omega)$, where $\Omega = \omega C_{DL}/\sigma k$. The Debye-Hückel theory gives the value of the diffuse double layer capacitance as $C_{\mathrm{d}} = \varepsilon/\lambda_{\mathrm{D}}$.

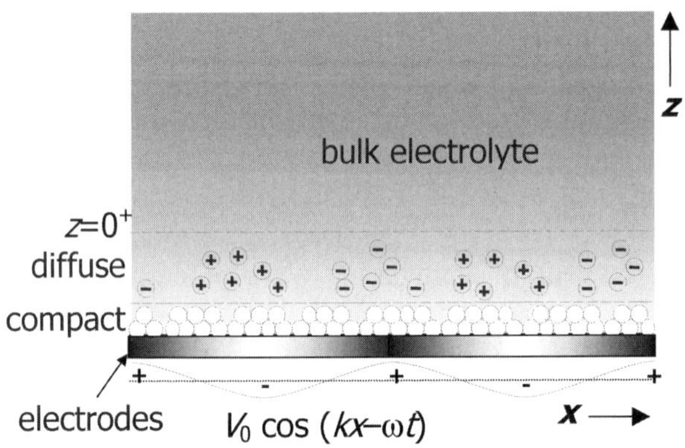


**Fig. 7.** Travelling wave electroosmotic pump. The dimension of the double layer is exaggerated.

The tangential electric field $E_{\mathrm{x}}$ produces a force on the induced charge in the diffuse layer $q_{\mathrm{s}} = -C_{\mathrm{d}}\zeta$. The potential drop across the diffuse layer $\zeta$ can be related to the total double-layer potential drop by $\zeta = C_{\mathrm{s}}(V - \Phi)/(C_{\mathrm{s}} + C_{\mathrm{d}})$. Although both electric field and induced charge are oscillating functions in time, the Coulomb force has a nonzero time-average that leads to a nonzero slip velocity. For electrical double layers in quasiequilibrium on perfectly polarizable electrodes, the classical calculation [68] gives for the electroosmotic slip velocity $u_{\mathrm{s}} = -\varepsilon E_{\mathrm{x}}\zeta/\eta$. The time-averaged slip velocity is given by [67, 69]

$$\langle u_s \rangle = \frac{1}{2} Re\left[ \frac{-\varepsilon\zeta E_x^*}{\eta} \right] = \frac{C_s}{C_s + C_d} \frac{\varepsilon k V_0^2}{2\eta} \frac{\Omega}{1+\Omega^2}, \tag{81}$$

where $E_x^*$ is the complex conjugate of $E_x$. The EO slip velocity is a function of frequency with a maximum at $\Omega = 1$, and this maximum velocity is

$$\langle u_s \rangle_{max} = \frac{C_s}{C_s + C_d} \frac{\varepsilon k V_0^2}{4\eta}. \tag{82}$$

Let us assume that there is also an array of electrodes subjected to the same travelling-wave in the upper wall. The fluid velocity from $z = 0$ to $z = h$ is constant under conditions of zero back pressure. The maximum average velocity is then equal to $u_{max} = \langle u_s \rangle_{max}$.

Now we can obtain the maximum flow rate and pressure generated by this pump. For a channel of rectangular cross section (height $h$ and width $w$, $w \gg h$), and axial dimension $L$, the maximum flow rate and pressure are

$$Q_{max} = \frac{C_s}{C_s + C_d} \frac{\varepsilon k V_0^2}{4\eta} wh \;\; \Delta p_{max} = \frac{C_s}{C_s + C_d} \frac{3\varepsilon k V_0^2 L}{h^2}. \tag{83}$$

According to these expressions, both pressure and flow rate increase if the wave-length of the periodic array is reduced ($k$ is increased). The maximum pressure can be increased by reducing the height of the device, although at expenses of reducing the flow rate.

In experiments, the quadratic dependence on voltage is only observed at very low voltages. The velocity tends to a lower increment with voltage, which may be explained because of, at least, two mechanisms acting at the same time: (a) The potential $\zeta$ can not grow without limit because the charge in the diffuse double layer has an upper limit given by the size of the ions. Therefore, $\zeta$ saturates and the velocity becomes proportional to $V$. (b) If the potential is beyond the threshold voltage for the appearance of Faradaic currents, the double layer starts to leak charge. Experimentally and numerically, saturation of the velocity with voltage is observed [69, 70]. At still higher voltages, the flow changes direction, and generation of bubbles start to appear easily [71, 69, 72]. It seems that experimentally the AC EO mechanism provides a maximum slip velocity around 500 $\mu$m s$^{-1}$.

### 3.2.2. Characteristics

*3.2.2.1. Efficiency*

In the theoretical model, no influence of the fluid velocity in the electric field is taken into account. In this limit, the mechanical power is negligible as compared to the power consumed by Joule heating. Therefore, our estimation for the total power consumption is $P_{\mathrm{J}} \approx \int \sigma E_{\mathrm{rms}}^2 \, \mathrm{d}\tau$. The estimation for the efficiency is

$$eff = \frac{1}{4}\frac{Q_{max}\Delta p_{max}}{P_J} = \frac{3\varepsilon^2 V_0^2 k}{4(1+C_d/C_s)^2 \sigma\eta h} \tag{84}$$

obtained at the optimum frequency that gives $\Omega = 1$. The expression for the efficiency is of the order of $\varepsilon u_{\mathrm{max}}/\sigma h$, an electric Reynolds number, multiplied by the factor $C_{\mathrm{s}}/(C_{\mathrm{s}}+C_{\mathrm{d}})$. The efficiency increases for decreasing conductivity $\sigma$, decreasing height $h$, and increasing wave number $k$.

*3.2.2.2. Liquids*

The conductivity of the liquid to be pumped can not be very high. As the conductivity increases, the potential across the diffuse layer becomes much smaller than the applied potential across the total double layer. If we increase the total voltage in order to increase the potential $\zeta$, Faradaic reactions can easily appear. Saline solutions in water that have been pumped using the AC/IC electroosmotic mechanism have conductivities below or equal to $0.1\,\mathrm{S\,m^{-1}}$.

A lower bound for the conductivity of the liquid must also exist because the mechanism is based on the induced charge in the diffuse double layer. The mechanism can only work if there exists a clear distinction between a charged layer and an electro-neutral bulk liquid. Under the action of an electric field, the ion concentration should be greater than a certain value in order to have electro-neutrality in the bulk. Electro-neutrality does not apply if the parameter $\Lambda = \varepsilon E/eln^{\infty}$ is of the order one or greater [12]. When $\Lambda \sim 1$, the electric field can create a charge layer of thickness $\sim l$ and charge density $\sim en^{\infty}$. For a typical electric field of $10^5\ \mathrm{V\,m^{-1}}$ and a typical length of $10^{-5}\,\mathrm{m}$, the electric field destroys the electro-neutrality if the ion concentration is of the order of $4\times10^{19}$ m$^{-3}$ ($\sigma\sim3\times10^{-7}$ S m$^{-1}$) for a polar liquid like water and it is of the order of $10^{18}$ m$^{-3}$ ($\sigma\sim7\times10^{-9}$ S m$^{-1}$) for a nonpolar liquid of $\varepsilon\sim2\varepsilon_0$.

Liquids that have been pumped using the AC/IC electroosmotic pump had conductivities in the range $10^{-4}$ to 0.1 S m$^{-1}$.

#### *3.2.2.3. Problems*

The applied voltage can not be very high in order to avoid Faradaic currents since the electrodes are in contact with the liquid. Therefore, low ionic strength solutions are preferable. The Faradaic reactions can degrade the electrodes after some time of operation. A way of reducing this problem is to use a dielectric coating on the electrodes, although this reduces the slip velocity for a given applied voltage. The use of titanium electrodes can be advantageous because titanium spontaneously form an oxide passivation layer that preserves them from degradation. As an advantage, because the pump is supposed to operate at voltages below the ionization potentials, the working liquid is not expected to suffer changes in its electrical properties.

#### *3.2.2.4. Applications*

Water saline solutions with $\sigma\leq0.1$ S m$^{-1}$ can be pumped, therefore, possible applications range from biomedical to chemical analysis. Typically, the generated pressure is small. However, these kinds of pumps have many potential applications for manipulating particles and fluids in closed microdevices, which do not require high pressure. For example, it may have applications in circular chromatography [73].

## 4. Magnetic Forces: DC and AC MHD Pumps

In a magnetohydrodynamic (MHD) pump, the force acting on the liquid is the Lorentz force. It appears when an electric current is subjected to a magnetic field. The magnetic force density is given by $\mathbf{f}=\mathbf{j}\times\mathbf{B}$. A simple way of looking at this expression is to consider that the magnetic force on a moving free charge, $\mathbf{F}=q\mathbf{v}\times\mathbf{B}$, is transmitted directly to the fluid, giving the previous body force density. We can distinguish two basic strategies of operation: a DC electric current is actuated by a static magnetic field (DC MHD micropump [7, 74, 75]) and an AC electric current is actuated by an alternating magnetic field (AC MHD micropump [8, 76]).

### 4.1. DC MHD Micropump

The Lorentz force is generated by applying a static magnetic field, usually created by a permanent magnet, upon a DC electric current. The magnetic field generated by the current is usually negligible as compared to the applied magnetic field. In effect, from Ampere's law $\nabla \times \mathbf{B} = \mu \mathbf{j}$, where $\mu$ is the magnetic permeability, the magnetic field created by the current is of the order of $B_j \sim \mu_0 jl$ ($l$, typical distance and for nonmagnetic materials $\mu \approx \mu_0$, the permeability of vacuum). The ratio between the generated field $B_j$ and the applied field $B_0$ is then $B_j/B_0 \sim \mu_0 jl/B_0$. An estimation for the current comes from the Lorentz force needed to generate a certain fluid velocity. From the Stokes equations, valid for low Reynolds number that are typical of microsystems, the comparison of the viscous term and the force density leads to a fluid velocity $u \sim jBl^2/\eta$. Therefore, the ratio is given by

$$\frac{B_j}{B_0} \sim \frac{\mu_0 \eta u}{B_0^2 l}, \tag{85}$$

which is usually very small. For example, taking $u \sim 10^{-3}\ \mathrm{m\,s^{-1}}$, $B_0 \sim 0.1T$, $l \sim 10^{-4}\ \mathrm{m}$, $\eta \sim 10^{-3}\ \mathrm{Pa\,s}$, the ratio is of the order of $10^{-6}$. Therefore, the applied magnetic field is almost unperturbed by the electric current.

The electric current density for a moving conductor of conductivity $\sigma$ is given by $\mathbf{j} = \sigma(\mathbf{E} + \mathbf{u} \times \mathbf{B})$. Except for liquid metals, the electric current driven by the electric field is much greater than that driven by the Lorentz term. In effect, the ratio is given by

$$\frac{|\mathbf{u} \times \mathbf{B}|}{E} \sim \frac{\sigma u B_0}{j} \sim \frac{\sigma B_0^2 l^2}{\eta}, \tag{86}$$

using the previous estimation for the liquid velocity $u \sim jBl^2/\eta$. The ratio is equal to the Hartmann number squared, $H = Bl\sqrt{\sigma/\eta}$, and it is a measure of the ratio of the magnetic body force and the viscous force [74]. The value of $H^2$ is very small for liquids with conductivities

$\sigma \ll 10^6\,\mathrm{S\,m^{-1}}$, such as electrolytic solutions. For instance, taking $B_0 \sim 1\,\mathrm{T}, l \sim 10^{-4}$ m, $\eta \sim 10^{-3}$ Pa s and $\sigma = 1\,\mathrm{S\,m^{-1}}$ we obtain $H^2 \sim 10^{-5}$. Therefore, the electric current is given by the applied electric field for nonmetallic liquids in microsystems. For liquid metals, where the Hartmann number can be much greater than unity, the actual electric current distribution should be taken into account. The continuity of electric current $\nabla . \mathrm{j} = 0$ must be considered and new induced electric fields appear. We restrict ourselves to the case of negligible Hartmann number.

#### 4.1.1. Pumping Principle

Let us consider a straight conduit of length $L$ with a rectangular cross section of height $h$ and width $w$ filled with an electrolyte, see Fig. 8. Perfectly conducting electrodes are placed on the two opposite sides of the conduit. The electrodes are subjected to a potential difference $V$, so there is an electric field $\mathbf{E} = E_0\,\mathbf{e}_\mathrm{x}$, with $E_0 = V/w$. There is also an applied homogeneous magnetic field in the vertical direction $B = B_0\,\mathbf{e}_\mathrm{y}$. The liquid inside the conduit is subjected to the Lorentz force density $\mathbf{j} \times \mathbf{B}$ and fluid flow is generated. The conduit is considered very long so that fringe effects are neglected and the fluid flow is considered to have velocity only in the $\mathrm{z}$-direction. The equation for the 1D velocity under a constant pressure gradient is

$$0 = \eta \nabla^2 u - \frac{\partial p}{\partial z} + \sigma E_0 B_0, \tag{87}$$

where we have considered the term $-\sigma u B_0^2$ to be negligible (Hartmann number much less than one). The flow has a profile similar to that of pressure-driven flow, and obeys Poiseuille law in microchannels [8, 75].

The pressure difference needed to stop the flow is

$$\Delta p_{\max} = \sigma E_0 B_0 L, \tag{88}$$

and the maximum average velocity and flow rate at zero pressure are approximately given by

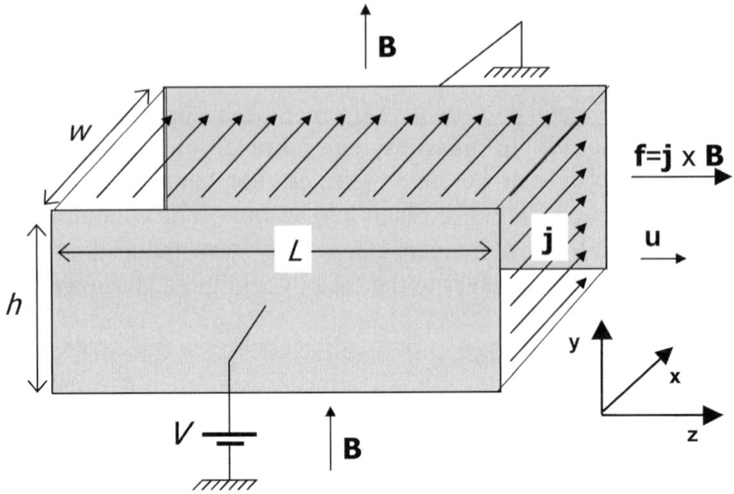


**Fig. 8.** Scheme of a magnetohydrodynamic pump

$$u_{\max} = \sigma E_0 B_0 \frac{D_h^2}{32\eta} \quad Q_{\max} = \frac{\sigma E_0 B_0 \pi D_h^4}{128\eta}, \tag{89}$$

where $D_h = 2wh/(w+h)$ is the hydraulic diameter of the conduit.

For liquid metals, the electric current density will surely be determined by the resistance of the total circuit and the applied voltage to it. It is usually the case that the applied voltage is much greater than $uBw$ (the electromotive force generated by fluid motion). In this case, an estimation of the maximum pressure and maximum flow rate can be obtained from the previous expressions by substituting $\sigma E_0$ for $j$, the mean current density in the channel.

### 4.1.2. Characteristics

#### *4.1.2.1. Efficiency*

Let us obtain an expression for the efficiency in the case when the Hartmann number is very small, such as the pumping of electrolytes. The power consumption is $IV_{app}$, where $I = \sigma E_0 hL$ is the current intensity through the conduit and $V_{app}$ the potential applied to the circuit. The estimation for the maximum efficiency is

$$eff = \frac{1}{4}\frac{\Delta p_{\max} Q_{\max}}{IV_{\mathrm{app}}} = \frac{V}{V_{\mathrm{app}}}\frac{\sigma B_0^2 \pi D_{\mathrm{h}}^4}{512\eta h w} \sim \frac{V}{V_{\mathrm{app}}}\frac{\sigma B_0^2 D_{\mathrm{h}}^2}{128\eta}, \tag{90}$$

where $V$ is the potential difference between the channel walls. It can be seen that the efficiency is proportional to the Hartmann number squared. The efficiency of the pump is proportional to the conductivity and inversely proportional to the viscosity. It is inversely proportional to the square of the hydraulic diameter, which makes miniaturization more demanding. For $D_{\mathrm{h}} = 100\,\mu\mathrm{m}$, $B = 1\,\mathrm{T}$, $\sigma = 1\,\mathrm{S\,m^{-1}}$, $\eta \sim 10^{-3}\,\mathrm{Pa\,s}$, $V/V_{\mathrm{app}} \sim 1$, the maximum efficiency is $eff \sim 10^{-7}$. The estimated efficiency for the micropump experiments presented by Homsy et al. [75] is of the order of $10^{-9}$ for an applied voltage of 20 V and magnetic field of 0.4 T.

*4.1.2.2. Liquids*

The liquids that can be pumped by the MHD micropumps are typically conducting. For magnetic field intensities available using miniature permanent magnets (up to 1 T), the electric current needs to be of a certain value, which makes the required electric field to be inversely proportional to $\sigma$. A lower bound for the conductivity can be obtained assuming a limiting electric field around $10^7$ V m$^{-1}$. With a required mean velocity of $10^{-3}\,\mathrm{m\,s^{-1}}$, hydraulic diameter $D_{\mathrm{h}} \sim 10^{-3}$ m, applied magnetic field $B_0 \sim 1$ T and viscosity $\eta \sim 10^{-3}\,\mathrm{Pa\,s}$, we obtain a lower bound for the liquid conductivity of $\sigma \sim 3\times10^{-6}$ S m$^{-1}$ ($\sigma \sim 3\times10^{-4}$ S m$^{-1}$ for $D_h \sim 10^{-4}$ m). Another limitation working with liquids of low conductivity is the appearance of electroosmotic flow generated by the applied electric field that may affect adversely the MHD flow [75]. According to Homsy et al. electrolytes with ionic strengths from 0.01–0.1 M presented this problem, this is, electrolytes with conductivities around $0.1-1$ S m$^{-1}$.

Mercury slugs, seawater, saline solutions and deionized water have been pumped using MHD micropumps [7, 74].

*4.1.2.3. Problems*

When aqueous solutions are actuated with DC voltage, bubble formation due to water electrolysis appears. Most of the MHD micropumps have integrated the electrodes directly into the pumping channel and, therefore, the formation of bubbles is problematic. To avoid that these bubbles get into the channel, the electrodes were placed in reservoirs outside the channel in [75]. Frit-like structures connected the reservoirs to the channel in a similar way as is done in some electroosmotic pumps. The degradation of the electrodes due to Faradaic reactions is another issue [75]. The proper selection of electrode material and redox species can reduce this problem. Another way of reducing the problems due to Faradaic reactions is to use AC electric currents as in the AC MHD micropumps.

Joule heating can also be problematic. The generation of heat may increase the temperature in the channel up to values that are not appropriate. An estimate of the temperature rise can be done in a similar manner as was performed to obtain expression (79). The expression using the current density is

$$\Delta T \sim l\frac{\partial T}{\partial l} \sim \frac{j^2 l^2}{2\sigma\kappa}, \tag{91}$$

where $l$ is a typical distance from the center of the channel to the lateral walls. It can be seen that, since a certain current density is required to generate a certain velocity, the increment of temperature is inversely proportional to $\sigma$. For a current density of $4000\,\mathrm{A\,m^{-2}}$, $l \sim 5\times10^{-5}$ m, the temperature rise can be around 30 K for $\sigma \sim 10^{-3}\,\mathrm{S\,m^{-1}}$.

#### *4.1.2.4. Applications*

The performance of the pump increases with conductivity. Homsy et al. [75] showed that their MHD micropump was suitable for use with saline solutions of relatively high ionic strength, $\geq 0.1$ M (conductivity around $\sigma \geq 1\,\mathrm{S\,m^{-1}}$). This is advantageous for the handling of physiological liquids that require higher buffer concentrations. MHD micropumps can also be used to produce secondary flows that may be beneficial for stirring, mixing [77] or to direct fluid into any desired path in a fluidic network [78]. Circular chromatographic applications may benefit from MHD micropumps [76].

Other applications for MHD micropumps include those where small quantities of liquid metals need to be pumped.

## 4.2. AC MHD Micropump

The AC MHD micropump uses both oscillating current and magnetic field to produce a continuous flow. The use of AC voltages reduces or avoids the electrolytic reactions needed to sustain the electric current and the associated formation of gas bubbles and electrode degradation [8]. As in the DC MHD micropump, for electrolytic solutions the Hartmann number is small and the electric current is driven by the applied electric field, $\mathbf{j} = \sigma\mathbf{E}$. However, from Faraday's law there is an electric field generated by the oscillating magnetic field, $\nabla \times \mathbf{E}' = -\partial\mathbf{B}/\partial t$, that produces eddy currents in the channel. This electric field is negligible as compared to the electric field needed to drive significant current in electrolytes if

$$\frac{E'}{E_0} = \frac{\omega Bl}{E_0} = \frac{\sigma\omega Bl}{j_0} \ll 1, \tag{92}$$

where we have taken $E_0 = j_0/\sigma$. For $\omega \leq 10^6\ \mathrm{s}^{-1}$, $B \leq 1$ T, $l = 10^{-4}$ m and $\sigma \leq 1\,\mathrm{S\,m}^{-1}$, we obtain $E'/E_0 \leq 0.05$ for a current density of $j_0 = 2000\,\mathrm{A\,m}^2$, typical in experiments [76, 75]. Therefore, the electric field is given by the applied AC voltage. For liquid metals the situation is very different and the eddy currents dominate. We restrict again our analysis to the case of nonmetallic liquids.

#### 4.2.1. Pump Principle

An electromagnet is placed under the channel and generates the oscillating magnetic field. The situation is equivalent to that of Fig. 8 with the applied voltage and magnetic field varying periodically in time with a single frequency $\omega$.

To have continuous motion, the time-averaged Lorentz force $\langle \mathbf{f} \rangle = \int_0^T \mathbf{j} \times \mathbf{B}\, \mathrm{d}t/T$ must be different from zero (here, $T = 2\pi/\omega$). The generated flow is actually pulsed; however, at high frequencies, only the time-averaged motion is observed.

The oscillating electric and magnetic fields are given by $\mathbf{E} = E_0 \cos\omega t\,\mathbf{e}_x$ and $\mathbf{B} = B_0 \cos(\omega t + \phi)\mathbf{e}_y$, respectively. The equation for the time-averaged velocity is still given by (87), but the driving time-averaged force density is now $(1/2)\sigma E_0 B_0 \cos\phi$. The control of the phase allows for controlling both the speed and flow direction [8]. The expressions for the maximum pressure, maximum velocity and maximum flow rate are equivalent to the DC case, changing $E_0 B_0$ by $E_0 B_0 \cos\phi/2$.

In experiments, the AC magnetic field amplitude is not constant but it is a function of the frequency of the electromagnet circuit [8, 76], typically decreasing for frequencies greater than several kHz. It is desirable that the frequency is high so that electrolysis and bubble formation is avoided. Electric currents can be higher at higher frequencies without bubble formation. Therefore, there is an optimum frequency of operation in order to maximize $|\mathbf{j} \times \mathbf{B}|$.

### 4.2.2. Characteristics

*4.2.2.1. Efficiency*

The efficiency is less than in the case of DC MHD pump for similar electric and magnetic field amplitudes because the power consumption includes, in addition to the power consumed by the electric current in the channel, the power needed to feed the electromagnet. The estimate for the maximum efficiency is, therefore,

$$eff \sim \frac{V}{V_{\mathrm{app}}} \frac{\sigma B_{\mathrm{rms}}^2 D_{\mathrm{h}}^2}{128\eta(1+\alpha)}, \tag{93}$$

where $\alpha$ is the ratio between the power consumed by the electromagnet and the power consumed by the electric current in the channel.

*4.2.2.2. Liquids*

As for the DC MHD pump, liquids that can be pumped are typically conducting. The lower bounds for the conductivity previously obtained for the DC MHD pump are also valid here. The actuation of liquid metals can be problematic for the AC MHD pump because the generated eddy currents in the liquid metal are not negligible and might have negative effects. They can heat the sample and the fluid flow might be quite different from what is expected. The liquids that were pumped by the AC MHD micropump of Lemoff and Lee [8] have conductivities from around $0.1$ to around $10\ \mathrm{S\ m^{-1}}$.

*4.2.2.3. Problems*

The generation of heat in the channel due to Joule heating is also present here as for the DC MHD pump. However, new sources of heat appear such as induction heating in the magnetic core material of the electromagnet and heating by eddy currents in the electrodes of the channel [76]. These effects limit the intensity of the AC magnetic field that can be obtained at high frequencies and, as a result, the fluid flow. The problem of the eddy currents in the electrodes can be reduced by laminating the electrodes, in a similar ways as is done in transformers [76].

*4.2.2.4. Applications*

The applications are similar to those of the DC MHD pump. The pumping of small samples of liquid metals by using AC MHD micropumps might be problematic.

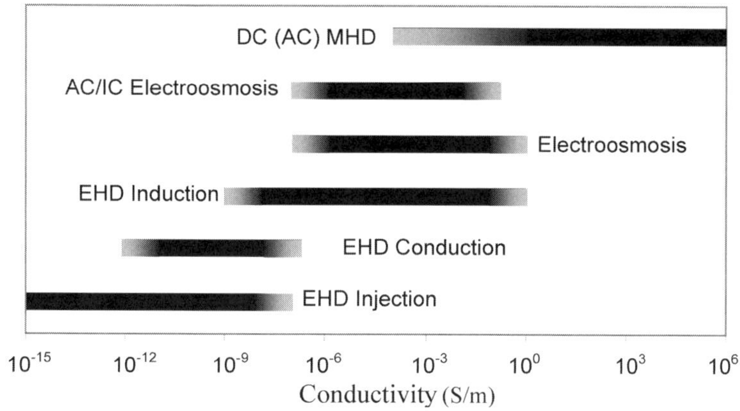


**Fig. 9.** Estimated conductivity ranges of the working fluid for operation of the different micropumps

## 5. Comparisons and Conclusions

In a first approach, each micropump can actuate liquids within a certain range of conductivities. The choice of a micropump should take into account this fact. For instance, very insulating liquids can be actuated by the EHD injection micropump, while liquid metals can be actuated by the DC MHD micropump. Figure 9 shows estimated conductivity ranges of the working fluid that each micropump can actuate. These ranges have been obtained using either values from theory or from experiments for microsystems. The limits of these ranges are necessarily fuzzy, since they depend on several parameters and geometry.

With respect to the performance of the different pumps, Table 1 summarizes theoretical expressions for the estimated maximum pressure and maximum flow rate for each micropump, and Table 2 shows experimental values of some selected works. For the theoretical expressions, we consider a given channel where the electric or magnetic forces are acting on the working fluid. The channel has the following dimensions: axial length equal to $L$, and cross-section of width $w$ and height $h$. We have used in some expressions the hydraulic diameter $D_h = 2wh/(w+h)$. The injection pump presents two limits: (a) electro-migration is negligible in front of convection, $\mu E \ll u$, and (b) convection is negligible in front of electro-migration $u \ll \mu E$. The second case ($u \ll \mu E$) leads to values for the maximum pressure and flow rate that are 9/16 smaller than the first case ($\mu E \ll u$). For the induction pump, we have neglected $\beta = (1/\varepsilon)\partial\varepsilon/\partial T$ in front of $\alpha = (1/\sigma)\partial\sigma/\partial T$ because it is usually much smaller. For the electroosmotic pump, we have written the expressions for the case $\lambda_D \ll D_h$. In the case of AC/IC EO pump we have written the expression corresponding to $w \gg h$ as an approximation. For the AC magnetohydrodynamic micropump, the root-mean-square values of the AC magnetic field $B_{rms}$ and voltage $V_{rms}$ should be employed.

**Table 1.** Estimated maximum pressure and flow rate generated for the different mechanisms in a channel of length $L$, width $w$, and height $h$.

| Mechanism | $\Delta p_{max}$ | $Q_{max}$ |
|---|---|---|
| EHD injection | $\Lambda 2\varepsilon V^2/L^2$ | $\Lambda \pi D_h^4 \varepsilon V^2/64\eta L^3$ |
| EHD induction | $0.85\varepsilon V^2/L^2$ | $0.85\pi D_h^4 \varepsilon V^2/128\eta L^3$ |
| EHD induction | $\varepsilon V^2 \mid \alpha\Delta T \mid /L^2$ | $\pi D_h^4 \varepsilon V^2 \mid \alpha \Delta T \mid /128\eta L^3$ |
| AC/IC electroosmosis | $32\varepsilon \mid \zeta \mid V/D_h^2$ | $wh\varepsilon \mid \zeta \mid V/\eta L$ |
| AC/IC electroosmosis | $6\varepsilon k V_{rms}^2 L/h^2(1+C_d/C_s)$ | $\varepsilon k V_{rms}^2 wh/2\eta(1+C_d/C_s)$ |
| (DC or AC) MHD | $\sigma VBL/w$ | $\pi D_h^4 \sigma VB/128\eta w$ |

For the injection case, $\Lambda = 1$ if $\mu E \ll u$, and $\Lambda = 9/16$ if $u \ll \mu E$.

With respect to the experimental values of Table 2, some $\Delta p_{max}$ values have been obtained by multiplying the flow rates by estimations of the hydraulic resistance of the system. The systems would be able to provide maximum pressures of the order of the estimated ones. We have also included the length $L$ and cross-sectional area $S$ of the estimated liquid volumes actuated by the micropumps because the hydraulic power delivered by a micropump is usually an increasing function of volume. In fact, connecting micropumps in series increases the maximum pressure and connecting micropumps in parallel increases the maximum flow rate.

**Table 2.** Selected experimental values of micropumps.

| Reference | Type | Working liquid | $\Delta p_{max}$(Pa) | $Q_{max}$(mm$^3$s$^{-1}$) | $V$(V) | $S$(mm$^2$) | $L$(mm) |
|---|---|---|---|---|---|---|---|
| Darabi and Wang [27] | Injection | HFE-7100 | 500 | 75 | 180 | 0.5 | 20 |
| Richter Sandmaier [2] | Injection | Ethanol | 2,500 | 200 | 700 | 9 | 0.35 |
| Ahn Kim[24] | Injection | Ethyl alcohol | 175 | 0.25 | 100 | 0.3 | 15 |
| Feng Seyed-Yagoobi [31] | Conduction | R-123 | 270 | 4,000 | $10^4$ | 80 | 90 |
| Fuhr et al. [38] [a] | Induction | Water | 6 | $7.5 \times 10^{-3}$ | 14 | 0.025 | 4 |
| Felten et al. [39] [a] | Induction | Water | 2.7 | $3 \times 10^{-4}$ | 3.5 | 0.003 | 0.24 |
| Yao et al. [52] | Electroos. | Borate buffer | $1.3 \times 10^5$ | 550 | 100 | 1300 | 1 |
| Chen Santiago [50] | Electroos. | Water | $3.3 \times 10^4$ | 0.25 | 1,000 | 0.034 | 1 |
| Takamura et al. [55] | Electroos. | Phosphate buffer | 800 | $7 \times 10^{-3}$ | 10 | 0.003 | 1.2 |
| Wang et al. [51] | Electroos. | Water | $3 \times 10^5$ | 0.05 | 6,000 | 0.008 | 60 |
| Studer et al. [71] [a] | AC/IC EO | Water | 90 | $1.4 \times 10^{-3}$ | 3.5 | 0.002 | 6 |
| García et al. [72] [a] | AC/IC EO | Water | 0.14 | 0.04 | 2.1 | 0.2 | 2 |
| Homsy et al. [75] [a] | DC MHD | 1 M KCl solution | 28 | $8.3 \times 10^{-3}$ | 20 | 0.01 | 16 |
| Eijkel et al. [76] [a] | AC MHD | 1 M KCl solution | 26 | $1.2 \times 10^{-4}$ | 2.8 | 0.006 | 60 |

[a] The value $\Delta p_{max}$ is estimated from flow rate and hydraulic resistance.

To have a rough idea of the pressure difference needed to obtain a given flow rate in a microfluidic system, let us consider that our external load is a microchannel of cross section $100\mu\text{m} \times 100\mu\text{m}$ and length $L$. The channel is folded and compacted in a small area (of the order of 1 cm$^2$) so that the length $L$ can be something between 1 mm and 1 m. We would like to obtain a liquid velocity around 1 mm s$^{-1}$, that can be sufficient for a microsystem ($Q = 10^{-11}$ m$^3$ s$^{-1}$ $= 0.01$mm$^3$ s$^{-1}$ in this example). From (13), the pressure required if the liquid is water is of the order of $\Delta p \sim 4000L$ Pa ($L$ in meters). Therefore, we have a range of pressures from 4 to 4000 Pa that are required to be generated by the micropumps.

We can see in Table 2 that the EO pumps can provide these pressures and flow rates. We remember here that a single capillary of small radius ($\sim 1\mu$ m) can generate high pressure, although the generated flow rate will be small. The EO pumps that generate high pressure and flow rate are typically made of porous structures [1]. These pumps can be visualized as a bundle of tortuous capillaries in parallel.

The injection pump of Richter and Sandmaier [2] could also work in all the pressure range of our example provided that some micropumps are placed in series in order to increase the pressure. According to the theoretical expressions, the maximum pressure that the EHD injection micropump can generate is of the order of $\varepsilon E^2$, for only one stage. Electric breakdown will limit the maximum attainable pressure, and this is a factor to be taken into account. The breakdown field can vary from 10 to 100 MV m$^{-1}$ depending on liquid, impurities, and geometry. For the experiments of Darabi and Wang [27], a breakdown electric field of 11 MV m$^{-1}$ was observed for HFE-7100 liquid. Taking a breakdown electric field of around 10 MV m$^{-1}$, a maximum attainable pressure of $\Delta p_{\mathrm{max}} = \varepsilon_{\mathrm{r}} 900\,\mathrm{Pa}$ is obtained ($\varepsilon_{\mathrm{r}}$, relative permittivity). Of course, the maximum attainable pressure generated by one stage can be much higher if the breakdown field is higher: it may be very high for very pure dielectric liquids ($>100$ MV m$^{-1}$).

For the conduction pump, the upper bound for the generated pressure is governed by the maximum electric field before ion injection reverses the flow direction [3]. The pressure generated (around 300 Pa at 10 kV) by the conduction pump of Feng and Seyed-Yagoobi [31] may be enough for some microfluidic applications. Although the dimensions of this pump are in the milimeter to centimeter scale, it seems possible to build a real micropump based on this mechanism.

In the case of the induction micropump, the theoretical value for the generated pressure is much smaller than $\varepsilon E^2$ because of the factor $\alpha \Delta T$, that can be very small. The maximum electric field is also limited by EHD instabilities [36, 38]. From the theoretical expressions for the travelling-wave device, these induction micropumps may provide higher pressures if the height of the device is of the order of the gap and electrode width. This procedure, of course, reduces the flow rate.

The AC/IC EO micropump generates very little pressure (see Table 2). However, looking at the theoretical expressions, reducing the height $h$ of the actuated microchannel would increase the generated pressure (proportional to $1/h^2$). This procedure reduces the flow rate and, in order to maintain it, the strategy should be to use as many micropumps connected in parallel as possible. However, this may not be easy or possible to realize experimentally.

The MHD micropumps also provide little pressure. The electric current, $\sigma E$, can be increased but it is limited by Joule heating effects. It is also very difficult to increase the magnetic field beyond 1 T.

Looking at Table 2, we can see that, at present, only the electroosmotic and EHD injection micropumps perform well in terms of generating enough pressure and flow rate for a microfluidic system like the one described in the example. However, there are many possible applications where strong pumping is not required and where the microfluidic system can be actuated by electric or magnetic forces distributed in all, or a significant part, of the device: cooling of special elements [28], circular chromatographic applications [76,73], pumping and local control of flow in microfluidic networks [57,78], or pumping of secondary flows that can enhance stirring and mixing [43,77].

The voltage required to operate a micropump is also an important factor to be considered, especially for portable microsystems where batteries would be used. The voltage used is included in table 2. The use of pumps connected in series like the cascade pump of Takamura et al. [55] reduces the required voltage. The use of electrodes inside the microchannel together with techniques like transverse electrokinetic effects [57], or AC/IC electroosmosis [6], also requires a low voltage.

To conclude, the choice of a particular integrated micropump should consider, among other factors, the properties of the working fluid (in particular the conductivity), the voltage required to operate the micropump, the generation of gases by the electrochemical reactions and the generation of heat by the Joule effect.

## Acknowledgments

The author is grateful to Prof. Antonio Castellanos and Dr. Alberto T. Pérez for their critical readings of the manuscript and their valuable comments. This work was supported by the Spanish government agency DGCyT under contract FIS2006-0364 and the regional government Junta de Andalucía under project FQM-421.

## References

1. Laser D.J. and Santiago J.G., A review of micropumps, *J. Micromech. Microeng.*, **14**, R35–R64 (2004)
2. Richter A. and Sandmaier H., An electrohydrodynamic micropump. In: *Proceedings of Micro Electro Mechanical Systems, 1990.* 'An Investigation of

Micro Structures, Sensors, Actuators, Machines and Robots', IEEE, pp. 99–104 (1990)
3. Atten P. and Seyed-Yagoobi J., Electrohydrodynamically induced dielectric liquid flow through pure conduction in point/plane geometry, *IEEE Trans. Dielectr. Electr. Insul.*, **10**, 27–36 (2003)
4. Fuhr G., Hagedorn R., Müller T., Benecke W. and Wagner B., Microfabricated electrohydrodynamic (EHD) pumps for liquids of higher conductivity, *J. Microelectromech. Syst.*, **1** , 141–146 (1992)
5. Pretorius V., Hopkins B.J. and Schieke J.D., Electro-osmosis: A new concept for high-speed liquid chromatography, *J. Chrom.*, **99**, 23–30 (1974)
6. Brown A.B.D., Smith C.G. and Rennie A.R., Pumping of water with AC electric fields applied to asymmetric pairs of microelectrodes, *Phys. Rev. E*, **63**, 016305 (2000)
7. Jang J. and Lee S.S., Theoretical and experimental study of MHD (magnetohydrodynamic) micropump, *Sensors Actuators*, **80**, 84–89 (2000).
8. Lemoff A.V. and Lee A.P., An AC magnetohydrodynamic micropump, *Sensors Actuators B*, **63**, 178–185 (2000)
9. Schmidt W.F., Conduction mechanisms in liquids. In: Bartnikas R. (ed) *Electrical Insulating Liquids. Engineering Dielectrics, Vol III.* American Society for Testing and Materials, Philladelphia, PA, (1994)
10. Castellanos A., Basic concepts and equations in Electrohydrodynamics. Part I. In: Castellanos A. (ed) *Electrohydrodynamics*. Springer-Verlag, New York, (1998)
11. Castellanos A., Ramos A., González A., Green N.G. and Morgan H., Electrohydrodynamics and dielectrophoresis in microsystems: Scaling laws. *J. Phys. D: Appl. Phys.*, **36**, 2584–2597 (2003)
12. Saville D.A., Electrohydrodynamics: The Taylor-Melcher leaky dielectric model. *Annu. Rev. Fluid Mech*, **29**, 27–64 (1997)
13. Zhakin A.I., Conduction phenomena in dielectric liquids. Part II. In: Castellanos A. (ed) *Electrohydrodynamics*. Springer-Verlag, New York (1998)
14. Bard A.J. and Faulkner L.R., *Electrochemical Methods: Fundamentals and Applications, 2nd Edition*. John Wiley and Sons, New York (2001)
15. Haus H.A. and Melcher J.R., *Electromagnetic Fields and Energy*, Prentice Hall, Englewood Cliffs, NJ (1989)
16. Stone H.A., Stroock A.D. and Ajdari A., Engineering flows in small devices: Microfluidics toward a Lab-on-a-Chip, *Annu. Rev. Fluid Mech.*, **36**, 381–411 (2004)
17. Stratton J.A., *Electromagnetic Theory*, McGraw Hill, New York (1941)
18. Jones T.B., Liquid dielectrophoresis on the microscale, *J. Electrostatics*, **51–52**, 290–299 (2001)
19. Stuetzer O.M., Instability of certain electrohydrodynamic systems, *Phys. Fluids.*, **2**, 642–648 (1959)
20. Stuetzer O.M., Ion drag pumps, *J. Appl. Phys.*, **31**, 136–146 (1960)
21. Pickard W.F., Ion drag pumping. I. Theory, *J. Appl. Phys.*, **34**, 246–250 (1963)

22. Pickard W.F., Ion drag pumping. II. Experiment, *J. Appl. Phys.*, **34**, 251–258 (1963)
23. Richter A., Plettner A., Hofmann K.A. and Sandmaier H., Electrohydrodynamic pumping and flow measurement. In: *Proceedings of Micro Electro Mechanical Systems*, MEMS '91. 'An Investigation of Micro Structures, Sensors, Actuators, Machines and Robots', IEEE, pp. 271–276 (1991)
24. Ahn S.H. and Kim Y.K., Fabrication and experiment of planar micro ion drag pump, *International Conference on Solid State Sensors and Actuators, 1997.* Transducers'97, **1**, 373–376 (1997)
25. Darabi J., Rada M., Ohadi M. and Lawler J., Design, fabrication, and testing of an electrohydrodynamic ion-drag micropump, *J. Microelectromech. Syst.*, **11**, 684–690 (2002)
26. Crowley J.M., Wright G.S. and Chato J.C., Selecting a working fluid to increase the efficiency and flow rate of an EHD pump, *IEEE Trans. Ind. Appl.*, **26**, 42–49 (1990)
27. Darabi J. and Wang H., Development of an electrohydrodynamic injection micropump and its potential application in pumping fluids in cryogenic cooling systems, *J. Microelectromech. Syst.*, **14**, 747–755 (2005)
28. Foroughi P., Benetis V., Ohadi M., Zhao Y. and Lawler J., Design, testing and optimization of a micropump for cryogenic spot cooling applications. In: *Proceedings of Semiconductor Thermal Measurement and Management Symposium* Twenty First Annual IEEE Conference, pp. 335–340 (2005)
29. Seyed-Yagoobi J., Electrohydrodynamic pumping of dielectric liquids, *J. Electrostatics*, **63**, 861–869 (2005)
30. Jeong S. and Seyed-Yagoobi J., Experimental study of electrohydrodynamic pumping through conduction phenomenon, *J. Electrostatics*, **56**, 123–133 (2002)
31. Feng Y. and Seyed-Yagoobi J., Understanding of electrohydrodynamic conduction pumping phenomenon, *Phys. Fluid.*, **16**, 2432–2441 (2004)
32. Thomson J.J. and Thomson G.P., *Conduction of Electricity Through Gases, 3rd edition*, Cambridge University Press, Cambridge (1928)
33. Langevin P., Recombinaison et mobilites des ions dans les gaz. *Annales de Chimie et de Physique*, **28**, 433 (1903)
34. Pontiga F., Sobre la estabilidad de una capa de líquido sometida a un campo eléctrico y un gradiente térmico, PhD Thesis, University of Seville (1992)
35. Melcher J.R. and Taylor G.I., Electrohydrodynamics: A review of the role of interfacial shear stresses *Ann. Rev. Fluid Mech.*, **1**, 111–146, (1969)
36. Melcher J.R. and Firebaugh M.S., Travelling-wave bulk electroconvection induced across a temperature gradient, *Phys. Fluids.*, **10**, 1178–1185, (1967)
37. Müller T., Arnold W.M., Schnelle T., Hagedorn R., Fuhr G. and Zimmermann U., A traveling-wave micropump for aqueous solutions: Comparison of 1 g and μ g results. *Electrophoresis*, **14**, 764–772 (1993)
38. Fuhr G., Schnelle T. and Wagner B., Travelling wave-driven microfabricated electrohydrodynamic pumps for liquids, *J. Microelectromech. Syst.*, **4**, 217–226 (1994)

39. Felten M., Geggier P., Jäger M. and Duschl C., Controlling electro-hydrodynamic pumping in microchannels through defined temperature fields, *Phys. Fluids.*, **18**, 051707 (2006)
40. Green N.G., Ramos A., González A., Castellanos A. and Morgan H., Electric field induced fluid flow on microelectrodes: The effect of illumination. *J. Phys. D: Appl. Phys.*, **33**, L13–L17 (2000)
41. González A., Ramos A., Castellanos A., Green N.G. and Morgan H., Electrothermal flows generated by alternating and rotating electric fields in microsystems, *J. Fluid Mech.*, **564**, 415–433 (2006)
42. Ramos A., Morgan H., Green N.G. and Castellanos A., AC electrokinetics: A review of forces in microelectrode structures, *J. Phys. D: Appl. Phys.*, **31**, 2338–2353 (1998)
43. Sigurdson M., Wang D. and Meinhart C.D., Electrothermal stirring for heterogeneous immunoassays, *Lab Chip*, **5** , 1366–1373 (2005).
44. Hunter R.J., *Zeta Potential in Colloid Science.* Academic Press, San Diego (1981).
45. Dasgupta P.K. and Liu S., Electroosmosis: A reliable fluid propulsion system for flow injection analysis, *Anal. Chem.*, **66**, 1792–1798 (1994)
46. Manz A., Effenhauser C.S., Burggraf N., Harrison D.J., Seiler K. and Fluri K., Electroosmotic pumping and electrophoretic separations for miniaturized chemical analysis systems, *J. Micromech. Microeng.*, **4**, 257–265 (1994)
47. Jacobson S.C., Hergenroder R., Koutny L.B. and Ramsey J.M., Open-channel electrochromatography on a microchip, *Anal. Chem.*, **66**, 2369–2373 (1994)
48 Yao S. and Santiago J.G., Porous glass electroosmotic pumps: Theory, *J. Colloid Interface Sci.*, **268**, 133–142 (2003)
49. Rice C.L. and Whitehead R., Electrokinetic flow in a narrow cylindrical capillary, *J. Phys. Chem.*, **69**, 4017–4024 (1965)
50. Chen C.H. and Santiago J.G., A planar electroosmotic micropump, *J. Microelectromech. Syst.*, **11**, 672–683 (2002)
51. Wang P., Chen Z. and Chang H.H., A new electro-osmotic pump based on silica monoliths, *Sensors Actuators B*, **113**, 500–509 (2006)
52. Yao S., Hertzog D.E., Zeng S., Mikkelsen J.C. and Santiago J.G., Porous glass electroosmotic pumps: Design and experiments, *J. Colloid Interface Sci.*, **268**, 143–153 (2003)
53. Brask A., Snakenborg D., Kutter J.P. and Bruus H., AC electroosmotic pump with bubble-free palladium electrodes and rectifying polymer membrane valves, *Lab Chip*, **6**, 280–288 (2006)
54. Lazar I.M. and Karger B.L., Multiple open-channel electroosmotic pumping system for microfluidic sample handling, *Anal. Chem.*, **74**, 6259–6268 (2002)
55. Takamura Y., Onoda H., Inokuchi H., Adachi S., Oki A. and Horiike A., Low-voltage electroosmosis pump for stand-alone microfluidics devices, *Electrophoresis*, **24**, 185–192 (2003)
56. Brask A., Goranovic G. and Bruus H., Theoretical analysis of the low-voltage cascade electro-osmotic pump, *Sensors Actuators B*, **92**, 127–132 (2003)

57. Gitlin I., Stroock A.D., George M., Whitesides G.M. and Ajdari A., Pumping based on transverse electrokinetic effects, *Appl. Phys. Lett.*, **83**, 1486–1488 (2003)
58. Ajdari A., Transverse electrokinetic and microfluidic effects in micro-patterned channels: Lubrication analysis for slab geometries, *Phys. Rev. E*, **65**, 016301 (2002)
59. Tang G., Yan D., Yang C., Gong H., Chai J.C. and Lam Y.C., Assessment of Joule heating and its effects on electroosmotic flow and electrophoretic transport of solutes in microfluidic channels, *Electrophoresis*, **27**, 628–639 (2006)
60. Jiang L., Mikkelsen J., Koo J.-M., Huber D., Yao S., Zhang L., Zhou P., Maveety J.G., Prasher R., Santiago J.G., Kenny T.W. and Goodson K.E., Closed-loop electroosmotic microchannel cooling system for VLSI circuits, *IEEE Trans. Compon. Packag. Technol.*, **25**, 347–355 (2002)
61. Ramos A., Morgan H., Green N.G. and Castellanos A., AC electric-field-induced fluid flow in microelectrodes, *J. Colloid Interface Sci.*, **217**, 420–422 (1999)
62. Ajdari A., Pumping liquids using asymmetric electrode arrays, *Phys. Rev. E,* **61**, R45–R48 (2000)
63. Bazant M.Z. and Squires T.M., Induced-charge electrokinetic phenomena: Theory and microfluidic applications, *Phys. Rev. Lett.*, **92**, 066101 (2004)
64. Green N.G., Ramos A., González A., Morgan H. and Castellanos A., Fluid flow induced by non-uniform AC electric fields in electrolytes on microelectrodes I: Experimental measurements. *Phys. Rev. E*, **61**, 4011–4018, (2000)
65. González A., Ramos A., Green N.G., Castellanos A. and Morgan H., Fluid flow induced by non-uniform AC electric fields in electrolytes on micro-electrodes II: A linear double-layer analysis. *Phys. Rev. E*, **61**, 4019–4028, (2000)
66. Green N.G., Ramos A., González A., Morgan H. and Castellanos A., Fluid flow induced by non-uniform AC electric fields in electrolytes on microelectrodes III: Observation of streamlines and numerical simulations. *Phys. Rev. E*, **66**, 026305, (2002)
67. Cahill B.P., Heyderman L.J., Gobrecht J. and Stemmer A., Electro-osmotic streaming on application of traveling-wave electric fields, *Phys. Rev. E*, **70**, 036305 (2004)
68. Levich V.G., *Physicochemical Hydrodynamics*. Prentice-Hall, Englewood Cliffs, NJ (1962)
69. Ramos A., Morgan H., Green N.G., González A. and Castellanos A., Pumping of liquids with traveling-wave electroosmosis, *J. Appl. Phys.,* **97**, 084906 (2005)
70. Olesen L.H., Bruus H. and Ajdari A., AC electrokinetic micropumps: The effect of geometrical confinement, Faradaic current injection, and nonlinear surface capacitance, *Phys. Rev. E*, **73**, 056313 (2006)

71. Studer V., Pepin A., Chen Y. and Ajdari A. An integrated AC electrokinetic pump in a microfluidic loop for fast and tunable flow control. *Analyst*, **129**, 944–949 (2004)
72. García-Sánchez P., Ramos A., Green N.G. and Morgan H., Experiments on AC electrokinetic pumping of liquids using arrays of microelectrodes, *IEEE Trans. Dielectr. Electr. Insul.*, **13**, 670–677 (2006)
73. Debesset S., Hayden C.J., Dalton C., Eijkel J.C.T. and Manz A., An AC electroosmotic micropump for circular chromatographic applications, *Lab Chip* **4**, 396–400 (2004)
74. Zhong J., Yi M. and Bau H.H., Magneto hydrodynamic (MHD) pump fabricated with ceramic tapes, *Sensors Actuators*, **96**, 59–66 (2002)
75. Homsy A., Koster S., Eijkel J.C., van den Berg A., Lucklum F., Verpoorte E. and de Rooij N.F., A high current density DC magnetohydrodynamic (MHD) micropump, *Lab Chip*, **5**, 466–471 (2005)
76. Eijkel J.C.T., Dalton C., Hayden C.J., Burt J.P.H. and Manz A., A circular AC magnetohydrodynamic micropump for chromatographic applications, *Sensors Actuators B*, **92**, 215–221 (2003)
77. Bau H.H., Zhong J. and Yi M., A Minute magneto hydro dynamic (MHD) mixer, *Sensors Actuators B*, **79**, 205–213 (2001)
78. Bau H.H., Zhu J., Qian S. and Xiang Y., A Magneto-hydrodynamically controlled fluidic network, *Sensors Actuators B*, **88**, 205–216 (2003)

# Chapter 3
# Mixing in Microscale

Nam-Trung Nguyen

Dr. Ing., Dr. Ing. habil.,
Associate Professor, School of Mechanical and Aerospace Engineering
Nanyang Technological University, Singapore

## 1. Introduction

Mixing in microscale is one of the key technologies for miniaturized analysis systems. Miniaturization brings advantages to a wide range of traditional fields such as chemical industry, pharmaceutical industry, analytical chemistry, biochemical analysis, and high-throughput synthesis. Due to the dominant surface effects, miniaturization also brings in challenges that do not occur in macroscale. For instance, turbulence used in macroscale to improve mixing is not possible in microscale. The laminar, deterministic flow caused by the dominant viscous effects makes turbulence impractical in microscale. Molecular diffusion may be improved by the shorter mixing path. However, the higher flow velocity leads to dominant advective effects and consequently the longer residence time and the longer mixing channel. These challenges will be discussed later with the help of nondimensional numbers. Looking at the evolution of micromixer designs in the last decade, it is apparent that a number of design techniques has been employed to improve mixing. This chapter will review these several concepts – from diffusion-based micromixers in the early stage to the recently reported micromixers based on chaotic advection. In general, this chapter classifies micromixers into two types: passive and active.

Passive micromixers do not need moving parts and actuators, the mixing concept is only based on molecular diffusion as well as chaotic advection. Thus, the two major subgroups of the passive concept are micromixers based on molecular diffusion and micromixers based on chaotic advection. For mixing of large molecules with small diffusion coefficients, chaotic advection is the more favorable concept. Mixing concepts based on molecular diffusion can be further categorized based on the arrangement of the mixed phases as parallel lamination, serial lamination, serial segmentation, and injection.

Active micromixers require external disturbance to improve mixing. Based on the source of the disturbance, active mixing is categorized here as pressure-driven, temperature-induced, electrohydrodynamic, dielectrophoretic, electrokinetic, magnetohydrodynamic, and acoustic concepts.

## 2. Mass Transport in Microscale

### *2.1. Transport Effects*

#### 2.1.1. Diffusive Transport

Diffusive transport is caused by the random motion of molecules, also called the Brownian motion. The mixing rate is determined by the flux of diffusion $j_{\text{diff}}$:

$$j_{\text{diff}} = -D\frac{dc}{dx}, \tag{1}$$

where $D$ is the diffusion coefficient in $\text{m}^2\,\text{s}^{-1}$ and $c$ is the species concentration in kg $\text{m}^{-3}$, called the mass concentration. The relationship in (1) is called the Fick's law. The diffusion coefficient for diluted, large spherical molecules was derived by Einstein as [6]:

$$D = \frac{kT}{6\pi\mu r}, \tag{2}$$

where $k$ is the Boltzmann constant, $T$ is the absolute temperature, $\mu$ is the dynamic viscosity, and $r$ is the radius of the molecule. At a constant temperature, the diffusion coefficient $D$ of a molecule is inversely proportional to the dynamic viscosity $\mu$ of the solvent. Equation (2) also applies well to dilute suspension of spherical colloid particles. The diffusion of dilute polymer molecules in a low-molecular-weight solvent is inversely proportional to the square root of the molecular weight $M$:

$$D \propto \frac{1}{\sqrt{M}}. \tag{3}$$

Figure 1 depicts the range of diffusion coefficients of different materials. Common diffusion coefficients found in microfluidics may range from $10^{-5}\,\text{cm}^2\,\text{s}^{-1}$ for small molecules to $10^{-7}\,\text{cm}^2\,\text{s}^{-1}$ for large molecules.

### 2.1.2. Advective Transport

A mass flux can be caused by advection:

$$j_{\text{conv}} = uc\,, \tag{4}$$

where $u$ is the velocity. The three-dimensional diffusive/advective transport can therefore described as a balance equation of mass flux change:

$$\frac{\partial c}{\partial t} + u\frac{\partial c}{\partial x} + v\frac{\partial c}{\partial y} + w\frac{\partial c}{\partial z} = D\left(\frac{\partial^2 c}{\partial x^2} + \frac{\partial^2 c}{\partial y^2} + \frac{\partial^2 c}{\partial z^2}\right) + q\,, \tag{5}$$

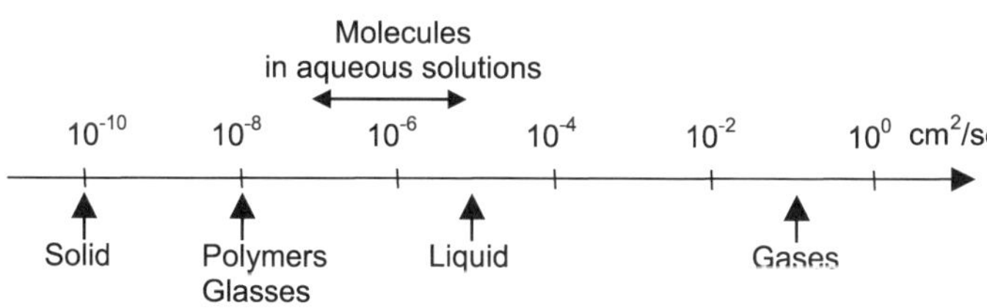


**Fig. 1.** Diffusion coefficient range

where $u$, $v$, $w$ are the velocity components in $x$, $y$, and $z$ direction. The first term on the left of (5) is the transient change of the concentration. The other three terms on the left represents the gradient of the advective mass flux. The first term on the right represents the gradient of the diffusive mass flux. The source term $q$ describes the generation of the solute.

### 2.1.3. Taylor–Aris Dispersion

A pressure-driven flow results in a distributed velocity profile across the microchannel. Convective transport at the fluid layers with different velocities causes an apparently higher axial dispersion then pure molecular diffusion. For a cylindrical capillary, the long-time asymptotic solution for the axial dispersion coefficient is [1,72]:

$$D^* = D\left[1 + \frac{1}{48} Pe(d_c)^2\right], \tag{6}$$

where $Pe(d_c) = \left(Ud_c/D\right)$ is the Peclet number based on the capillary diameter $d_c$ as the characteristic length. As discussed in Sect. 2.2, Peclet number in typical microfluidic applications may reach 100,000 causing a several order of magnitude higher axial dispersion coefficient $D^*$ then molecular diffusion coefficient $D$. For the special case of flow between two parallel plates with a gap of $H$, the dispersion coefficient is

$$D^* = D\left[1 + \frac{1}{210} {Pe_H}^2\right], \tag{7}$$

where $Pe_H$ is the Peclet number based on the gap $H$. For a rectangular channel of width $W$ and height $H$ ($W > H$), the dispersion coefficient can be described as ([15]):

$$D^* = D\left[1 + \frac{1}{210} Pe(W)^2 f\left(\frac{H}{W}\right)\right], \tag{8}$$

where $Pe_W$ is the Peclet number based on the channel width and $f(H/W)$ is a function of the aspect ratio $H/W$ [15].

### 2.1.4. Chaotic Advection

Chaotic advection refers to stretching, folding, and breakup processes caused by transverse flows. These transverse flows can be caused either actively by external disturbances or passively by spatially periodic structures. When the flow passes through each of the structure, also called the advection cycle, the cross-sectional concentration distribution is transformed (or mapped) into the next distribution. Repeating these advection cycles stretches, folds, and breaks up the fluids and leads to complete mixing. Reader may refer to Ottino's book The kinematics of mixing: stretching, chaos, and transport for more details about this topic [59]. Recently, Wiggins and Ottino reviewed this concept for mixing in microscale in [81].

### 2.2. Dimensionless Numbers and Scaling Laws

Diffusion coefficient $D$, the kinematic viscosity $\nu$, and the thermal diffusivity $\alpha = k/\rho c$ – where $k$, $\rho$, and $c$ are thermal conductivity, density, and specific heat, respectively – are transport properties and all have the same unit of $\mathrm{m}^2\ \mathrm{s}^{-1}$. The ratios between these properties represent a group of nondimensional numbers which are characteristic for the ratio between the competing transport processes. These nondimensional numbers help to compare molecular diffusion with other transport process in microfluidics.

Schmidt number is the ratio between momentum transport and diffusive mass transport:

$$Sc = \frac{\text{momentum transport}}{\text{diffusive mass transport}} = \frac{\nu}{D} = \frac{\mu}{\rho D}. \tag{9}$$

Lewis number is the ratio between heat transport and diffusive mass transport:

$$Le = \frac{\text{heat transport}}{\text{diffusive mass transport}} = \frac{\alpha}{D} = \frac{k}{\rho c D}. \tag{10}$$

The ratio between advective transport and momentum transport is called the Reynolds number:

$$Re = \frac{\text{advective mass transport}}{\text{momentum transport}} = \frac{\rho U D_h}{\mu} = \frac{U D_h}{\nu}, \tag{11}$$

where $U$ is the mean velocity in the flow direction and $D_\mathrm{h}$ is the hydraulic diameter. In many cases, the hydraulic diameter of the channel is taken as the characteristic length $L_\mathrm{ch}$. For typical values of $L_\mathrm{ch} = 100\ \mu\mathrm{m}$, $U = 1\ \mathrm{mm\ s}^{-1}$, $\nu = 10^{-5}\ \mathrm{cm}^2\ \mathrm{s}^{-1}$, the typical Reynolds number is $Re = 0.01$. This small number means laminar flow exists in almost all microfluidic applications.

Peclet number is the ratio between advective mass transport and diffusive mass transport:

$$Pe = \frac{\text{advective mass transport}}{\text{diffusive mass transport}} = \frac{UL_{\text{ch}}}{D}, \tag{12}$$

where $L_{\text{ch}}$ is the characteristic mixing length.

The ratio between the Peclet number and the Reynolds number is actually the ratio between momentum transport and diffusive mass transport or the Schmidt number above. For diffusion coefficients ranging from $10^{-5}\,\text{m}^2\,\text{s}^{-1}$ to $10^{-7}\,\text{cm}^2\,\text{s}^{-1}$, the Peclet numbers for the typical values in the above example are $100 < Pe < 10{,}000$. That means, advective mass transport dominates over diffusive transport in almost all microfluidic applications.

The average diffusion time $t$ over the characteristic mixing length $L_{\text{mixing}}$, also called the striation thickness, is represented by the Fourier number [11]:

$$Fo = \frac{Dt_{\text{diff}}}{L^2_{\text{mixing}}}. \tag{13}$$

The Fourier number is usually in the range between 0.1 and 1. For a simple T-mixer with two streams in a microchannel of a length $L_{\text{mixer}}$ and a width of $W$, the residence time should be the same as the average diffusion time:

$$\begin{aligned} t_{\text{res}} &= t_{\text{diff}} \\ \frac{L_{\text{mixer}}}{U} &= \frac{Fo\,L^2_{\text{mixing}}}{D} = \frac{Fo\,W^2}{D}. \end{aligned} \tag{14}$$

Thus, the ratio between the channel length and channel width is:

$$\frac{L_{\text{mixer}}}{W} = Fo\,\frac{UW}{D} = Fo\,Pe_w. \tag{15}$$

For the above typical values of Fourier number $0.1 < Fo < 1$ and Peclet number based on the channel width $10 < Pe_w < 10{,}000$, the range of this ratio is $10 < L_{\text{mixer}}/W < 10{,}000$. For some applications, the required mixing channel is unacceptably long.

If the inlets are split and rejoined as $n$ pairs of solute/solvent streams, the mixing length is reduced to $L_{\text{mixing}} = W/n$. The ratio of the required channel length and channel width then becomes:

$$\frac{L_{\mathrm{mixer}}}{W} = Fo\frac{UW}{D} = \frac{1}{n^2} Fo\,Pe_W\,. \tag{16}$$

This concept is called parallel lamination, where the channel length can be reduced by a factor of $n^2$.

If the inlets are stretched and folded in $n$ cycles, the mixing length is reduced to $L_{\mathrm{mixing}} = W/b^n$. The base $b$ depends on the type of the mixer. In case of sequential lamination as discussed in Sect. 3, the base is for instance $b = 2$. The base could have a different value in case of chaotic advection. The ratio between the required channel length and the channel width is:

$$\frac{L_{\mathrm{mixer}}}{W} = \frac{1}{b^{2n}} Fo\,Pe_W\,. \tag{17}$$

The above equation reveals that a very compact micromixer can be designed using sequential lamination or chaotic advection.

In general, fast mixing can be achieved with smaller mixing path and larger contact surface. If the channel geometry is very small, the fluid molecules collide most often with the channel wall and not with other molecules. In this case, the diffusion process is called Knudsen diffusion [11]. The ratio between the distance of molecules and the channel size is characterized by the dimensionless Knudsen number (*Kn*):

$$Kn = \frac{\lambda}{D_{\mathrm{h}}}, \tag{18}$$

where $\lambda$ is the mean free path and $D_{\mathrm{h}}$ is hydraulic diameter of the channel structure. The mean free path for gases is given by:

$$\lambda = \frac{KT}{\sqrt{2}\pi d_{\mathrm{m}}^2 p}, \tag{19}$$

where $k$ is the Boltzmann constant ($k = 1.38066 \times 10^{-23}$ J K$^{-1}$), $T$ is the absolute temperature, $p$ is the pressure, and $d_{\mathrm{m}}$ is the molecular diameter of the diffusing species. The Knudsen number for liquid is small, because the mean free path of liquid is on the order of a few angstroms. Thus, Knudsen diffusion may occur only in pores with nanometer sizes. In gases, the mean free path is on the order of a hundred nanometers to several microns. For example, at room condition, the mean free path of hydrogen is 0.2 $\mu$m. Knudsen diffusion may occur in microchannels with diameters on the order of a few microns.

Among the above dimensionless numbers, Reynolds number (*Re*) represents the flow behavior in the microchannel, while Peclet number (*Pe*) represents the ratio between advection and diffusion. Thus, these two numbers are suitable for characterizing the operation point of a micromixer. From the definitions (11) and (12), the relation between *Pe* and *Re* is:

$$\frac{Pe}{Re} = \frac{uL_{\mathrm{ch}}/D}{uD_{\mathrm{h}}/\nu} = \frac{L_{\mathrm{ch}}}{D_{\mathrm{h}}}\frac{\nu}{D} = \frac{L_{\mathrm{ch}}}{D_{\mathrm{h}}}Sc, \tag{20}$$

where $u$, $D$, $\nu$, and *Sc* are the bulk velocity, the diffusion coefficient, and the kinematic viscosity, and the Schmidt number (9), respectively. The hydraulic diameter $D_{\mathrm{h}}$ and the mixing path $L$ are usually on the same order; therefore, we can assume $L_{\mathrm{ch}}/D_{\mathrm{h}} \approx 1$. The kinematic viscosity of liquids and the diffusion coefficient is on the order of $\nu = 10^{-6}\ \mathrm{m}^2\ \mathrm{s}^{-1}$, while the diffusion coefficient ranges from $D = 10^{-9}$ to $10^{-11}\ \mathrm{m}^2\ \mathrm{s}^{-1}$. The Schmidt number is about $10^3 < Sc < 10^5$. On a *Pe-Re* diagram, the area between the two lines $Pe \approx 1{,}000Re$ and $Pe \approx 100{,}000Re$ represents the operation range of micromixers, Fig. 2. Operation points of micro-mixers are expected to be in this area.

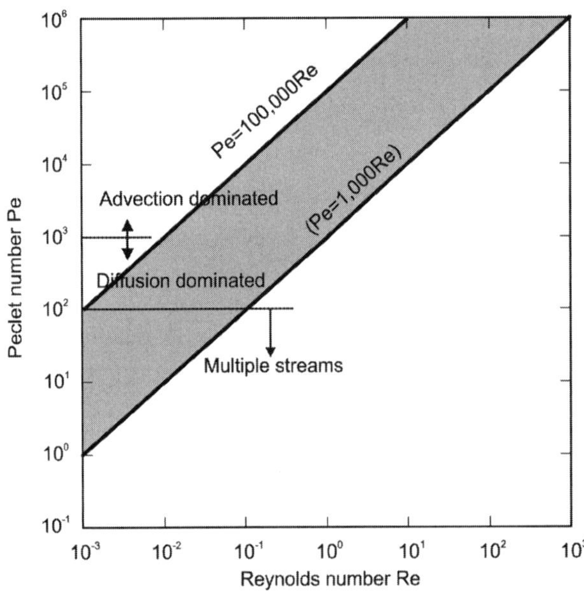


**Fig. 2.** The *Pe–Re* diagram of micromixers, the typical operation of micromixers are located in the gray area

## 3. Micromixers Based on Molecular Diffusion

In general, the final stage of all mixing concepts is molecular diffusion. This section discusses mixing concepts that rely only on diffusive transport. At a given flux (1), a large interfacial area can improve diffusive transport. From (1), a large gradient and a large diffusion coefficient also contribute to a high diffusive flux. Compared to the macroscale, the small mixing length in microscale actually leads to a higher concentration gradient. The relationship in (2) shows that larger diffusion coefficients can be achieved with a higher temperature and a lower viscosity. However, the following analysis assumes that diffusion coefficient is a material constant. Mixing is only optimized by the geometrical designs and the different transport effects. For micromixers based on molecular diffusion, the basic concepts for decreasing the mixing length are parallel lamination, sequential lamination, sequential segmentation, segmentation based on injection, and focusing.

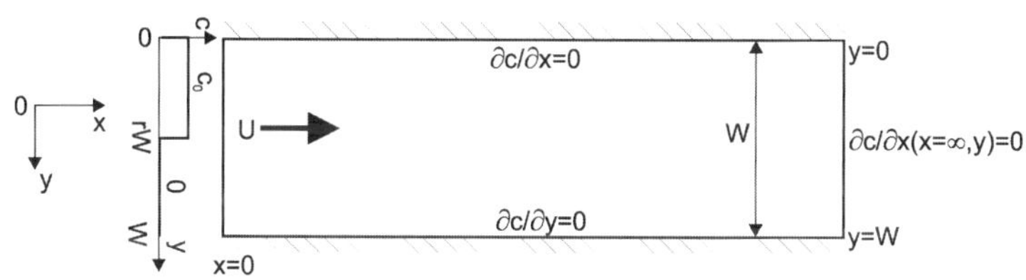


**Fig. 3.** The steady-state two-dimensional model of concentration distribution in a parallel lamination with two substreams

### *3.1. Parallel Lamination*

#### 3.1.1. Mixers Based on Pure Molecular Diffusion

Parallel lamination splits solute and solvent each into $n$ substreams and rejoining them later in a single stream. According to (16), compared to a parallel mixer with two substreams, the mixing time or the required channel length can be reduced by a factor of $n^2$. To start with, let us consider the model of a mixer with two substreams at the inlet as depicted in Fig. 3. The two-dimensional model assumes a flat microchannel. The mixing streams have the same viscosity and velocity $u$. The mixer is a long channel with a width $W$, two inlets, and one outlet. The solute stream has a concentration of $c = c_0$ and a mass flow rate of $\dot{m}_1$. The solvent stream has a concentration of $c = 0$ and a mass flow rate of $\dot{m}_2$. With the above assumptions, the transport equation (5) can be reduced to the steady-state two-dimensional form:

$$u\frac{\partial c}{\partial x} = D\left(\frac{\partial^2 c}{\partial x^2} + \frac{\partial^2 c}{\partial y^2}\right). \tag{21}$$

Assuming the same viscosity and fluid density, the dimensionless interface location $r$ is equal to the mass fraction $\alpha$ of the solvent in the final mixture $\alpha = \dot{m}_1/(\dot{m}_1 + \dot{m}_2)$ $(0 \le \alpha \le 1)$. The mixing ratio of the solute and the solvent is therefore $\alpha : (1-\alpha)$.

Introducing the dimensionless variables for the coordinates system $x^* = x/W$, $y^* = y/W$, the dimensionless concentration $c^* = c/c_0$, and the Peclet number $Pe = uW/D$, the transport equation has the dimensionless form:

$$Pe\frac{\partial c^*}{\partial x^*} = \frac{\partial^2 c^*}{\partial x^{*2}} + \frac{\partial^2 c^*}{\partial y^{*2}}. \tag{22}$$

The corresponding boundary conditions for the inlets are:

$$c^* |_{(x^*=0, 0 \le y^* < r)} = 1 \text{ and } c^* |_{(x^*=0, r \le y^* \le 1)} = 0. \tag{23}$$

Full mixing is assumed at ($x^* = \infty$):

$$\frac{\partial c^*}{\partial x^*} |_{(x^*=\infty, 0 \le y^* \le 1)} = 0. \tag{24}$$

The wall conditions are:

$$\frac{\partial c^*}{\partial y^*} |_{y^*=0,1} = 0 \tag{25}$$

The wall condition (25) is also the symmetry condition in case of mixing with multiple streams. Thus the same model can also be used for the later case. Using separation of variables and the corresponding boundary conditions (23–25), the dimensionless concentration distribution in the mixing channel can be expressed as:

$$c^*(x^*,y^*) = \alpha + \frac{2}{\pi}\sum_{n=1}^{\infty}\frac{\sin\alpha\pi n}{n}\cos(n\pi y^*) \times \exp\left(-\frac{2n^2\pi^2}{Pe+\sqrt{Pe^2+4n^2\pi^2}}x^*\right), \quad . \quad n=1,2,3 \tag{26}$$

As mentioned above, the above solution can also be extended to the case of multiple mixing streams. The wall boundary condition is identical to the symmetry condition in the middle of each stream. Thus, (26) can be extended periodically along the channel width direction. Considering the distance between two neighboring concentration extrema $W_{\text{min, max}}$, the Peclet number is evaluated as $Pe = UW_{\text{min, max}}/D$. Figure 4 shows the typical theoretical and experimental results in a parallel lamination micromixer with two or three streams, respectively.

As discussed above, fast mixing is achieved in parallel lamination micromixers by decreasing the mixing path and increasing the contact surface between the solvent and the solute. The simplest parallel lamination micromixer consists of a long microchannel and two inlets. Because of their characteristic shapes, these designs are simply called the T-mixer or the Y-mixer. If the aspect ratio of the mixing channel is small $W \gg H$, the inlet streams of a T-mixer can be twisted and laminated as two thin liquid sheets to reduce the mixing path from the channel width $W$ to the channel height $H$ [27]. The simple design and the available theoretical models make T-mixer and ideal platform for investigation of Taylor's dispersion [30, 37] and other nonlinear effects [86].

Smaller mixing paths can be achieved by splitting the solvent and the solute in multiple streams and rejoining them or by interdigitating them three dimensionally [24]. For instance, the design of Bessoth et al. can achieve full mixing with 32 streams only after a few milliseconds [4, 36].

Another concept for reducing the mixing path in parallel lamination micromixers is hydrodynamic focusing. Knight et al. [39] reported a simple mixer with three inlets. The width of the solvent stream in the

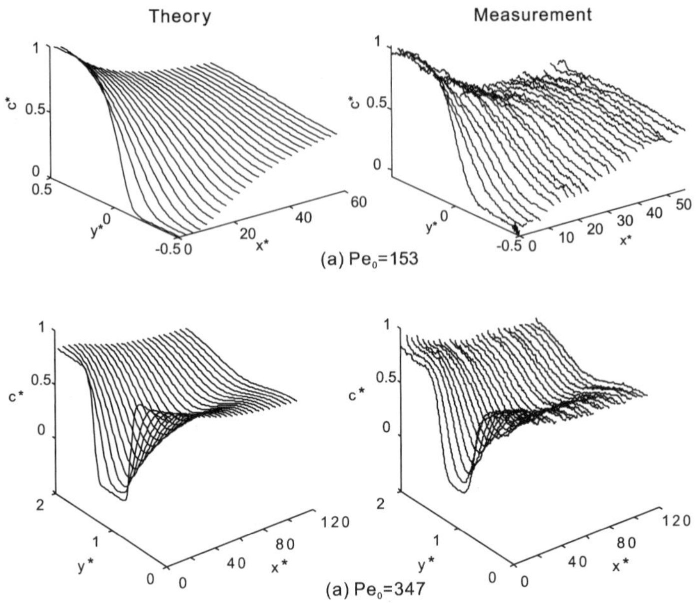


**Fig. 4.** Concentration distribution in a parallel lamination micromixer: (**a**) two streams and (**b**) three streams

middle was focused by adjusting the pressure ratio between the sample flow and the sheath flow. Due to the very small focused width of the solvent stream, mixing time can be reduced to a few microseconds [32]. The same configuration of hydrodynamic focusing and mixing was used for cell infection [77]. Combining parallel lamination with geometric focusing can be a powerful concept to improve micromixer's performance [26].

### 3.1.2. Mixers Based on Inertial Instabilities

According to the nondimensional analysis (15), the channel length is proportional to the Peclet number. As in most cases, the required channel lengths are not practical for implementation in miniaturized platforms. A shorter mixing channel can achieve full mixing at extremely high Reynolds numbers (more than 100) [84, 90]. Secondary flow and chaotic advection improve mixing by further reducing the mixing path. Because of the required high Reynolds number and the high pressure, this mixing concept can only be implemented in mechanically rigid materials such as silicon, glass [84], and stainless steel. For instance, the mixer reported in [84] needs a flow velocity as high as 7.6 m $s^{-1}$ at a pressure up to 7 bar to achieve Reynolds numbers up to 500. The huge velocity gradient in microscale at high Reynolds number leads to the formation of extremely fast vortices. Improved mixing can be achieved with such vortices. Lim et al. [43] created vortices in a diamond-shaped cavity next to a straight microchannel with a flow velocity of $45\ \mathrm{m\ s^{-1}}$. The corresponding Reynolds number in this case $Re = 245$ is very large for a typical microscale application. A similarly high shear stress can be created by multiple inlet streams entering a circular mixing chamber in tangential directions [7]. Such vortex-based mixers requires velocities as high as $10\ \mathrm{m\ s^{-1}}$ and pressure up to 15 bar.

Vortices caused by inertial effects can be generated at moderate Reynolds numbers with turns and geometrical obstacles. For instance, a simple 90° bend in the mixing channel can generate vortices at Reynolds numbers above 10 [90]. Mixing is achieved with a single bend at Reynolds numbers higher than 30. Obstacles such as structures on channel wall [83] or throttling the channel entrance [20] can also induce inertial instabilities into the standard T-mixer design.

### *3.2. Sequential Lamination*

Sequential lamination segregates the joined stream into two channels and rejoins them for the next transformation stage. Sequential lamination is also called split-and-recombine (SAR) concept [63]. Using $n$ such transformation stages in serial, one is able to laminate $2^n$ layers, which causes $4^n$ times faster mixing (17). Figure 5 illustrates this mixing concept, with two transformation. This concept is also called the baker's transformation, or Bernoulli transformation [81].

The implementation of the concept depicted in Fig. 5 is referred to as sequential lamination with vertical lamination and horizontal splitting. A similar transformation can be achieved with horizontal lamination and vertical splitting. This mixing concept requires relatively complicated three-dimensional fluidic structures that can only be made in a bulk system. According to (16) and (17), the sequential concept results in much faster mixing compared to the parallel concept with the same device area. Due to their complex geometry, most of the reported serial lamination mixers were fabricated in silicon, using bulk micromachining technologies such as wet etching in KOH [8, 64] or deep reactive ion etching (DRIE) technique [21]. Polymeric micromachining is another alternative for making serial lamination micromixers. Lamination of multiple polymer layers also allows making complex three-dimensional channel structures. Schoenfeld et al. [63] realizes this mixing concept on PMMA.

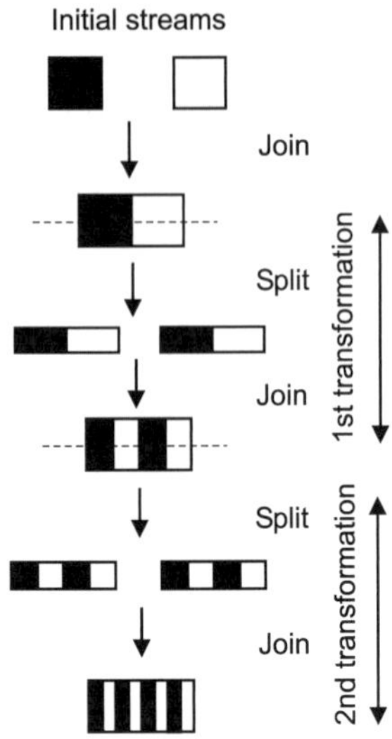


**Fig. 5.** Concept of sequential lamination

He et al. extended the concept of serial lamination to electrokinetic flows as reported in [25]. Serial lamination was realized through electroosmosis flows between the multiple intersecting microchannels. Melin et al. reported a similar design for a pressure-driven flow [49]. However, this design only works for a plug of the two mixed liquids.

### *3.3. Sequential Segmentation*

Sequential segmentation is referred to as the breakup of the solvent and the solute streams axially along the mixing channels. Each segment occupies the entire channel width. Dispersion in the axial direction results in faster mixing. According to (6–8) the axial dispersion coefficient may be of several order of magnitudes higher than pure molecular diffusion. Sequential segmentation is implemented by alternate switching of the inlet flows, using controlled valves for instance, Fig. 6a. The mixing ratio, can be adjusted by the switching ratio, Fig. 6b.

The concentration distribution along the axial direction $x$ can be described with the following one-dimensional model. Given are a flow velocity $u$ and a switching period $T$. The characteristic mixing length is the segment length $L = uT$. The transport equation (5) can then be reduced to the transient one-dimensional form:

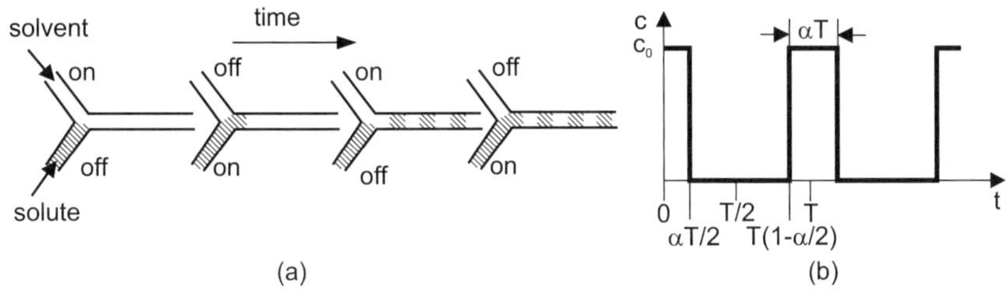


**Fig. 6.** Sequential segmentation: (**a**) alternate switching of the inlet streams and (**b**) switching ratio $\alpha$ determines the mixing ratio

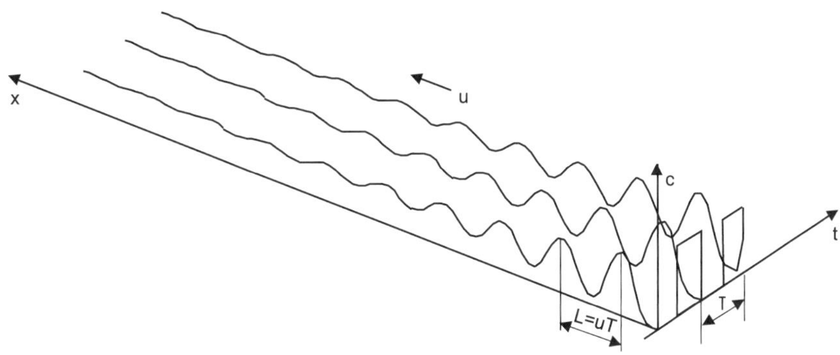


**Fig. 7.** Transient one-dimensional model of sequential segmentation

$$\frac{\partial c}{\partial t} + u\frac{\partial c}{\partial x} = D^* \frac{\partial^2 c}{\partial x^2}, \tag{27}$$

where $D^*$ is the dispersion coefficient determined by (6), (7), or (8). The periodic boundary condition at the inlet ($x = 0$) as depicted in Fig. 6b is

$$c(t,0) = \begin{cases} c_0 & 0 \le t \le \alpha T/2 \\ 0 & \alpha T/2 < t \le T - \alpha T/2 \\ c_0 & T - \alpha T/2 < t \le T \end{cases}, \tag{28}$$

where $c_0$, $T$, and $\alpha$ are the initial concentration of the solute, the period of the segmentation, and the mixing ratio, respectively.

Nondimensionalizing the concentration by $c_0$, the spacial variable by $L$, and the time by $T$ results in the dimensionless form of (27):

$$\frac{\partial c^*}{\partial t^*} = \frac{1}{Pe}\frac{\partial^2 c^*}{\partial x^{*2}} - \frac{\partial c^*}{\partial x^*}, \tag{29}$$

where $^*$ denotes the dimensionless variables. The Peclet number is defined based on the characteristic mixing length $L$ and the dispersion coefficient $D^*$as $Pe = uL/D^*$. The corresponding dimensionless boundary condition is:

$$c^*(t^*,0) = \begin{cases} 1 & 0 \le t^* \le \alpha/2 \\ 0 & \alpha/2 \le t^* \le 1-\alpha/2 \\ 1 & 1-\alpha/2 < t^* \le 1 \end{cases}. \tag{30}$$

Solving (29) with (30) and $c^*(\infty) = \alpha$ results in the transient concentration distribution:

$$c^*(x^*,t^*) = \Re\left\langle \alpha + \sum_1^\infty \frac{2\sin(\alpha\pi n)}{\pi n} \times \left\{ \exp\left[\frac{1}{2}\left(Pe - \sqrt{Pe^2 + 8\pi n\, Pe\, i}\right)x^*\right] \times \exp\left(2\pi t^* i\right)\right\}\right\rangle, \tag{31}$$

where $i$ is the imaginary unit and $\Re$ indicates the real component of a complex number.

The effectiveness of sequential segmentation is clearly shown in Fig. 8. At the same flow rate and Reynolds number, the Peclet number of parallel lamination is about two order of magnitude higher than the Peclet number of a sequential segmentation. The two streams clearly do not mix in a parallel lamination micromixer, Fig. 8a. Mixing appears to be much better with sequential segmentation, Fig. 8b. The analytical model (31) agrees relatively well with experimental data, Fig. 8a, b.

Wang et al. mentioned for the first time the utilization of Taylor–Aris dispersion for mixing in microscale. The Y-mixer was fabricated in polycarbonate by excimer laser [78]. Tan et al. [70] used piezoelectric actuators as active valves for switching the inlet streams [68]. The mixer was fabricated in PDMS. Nguyen and Huang used piezoelectric discs as active valves [55]. The mixer was fabricated by $CO_2$ laser and hot lamination of the multiple polymer sheets. Hydrodynamic focusing was used to eliminate the side effect of Taylor–Aris dispersion in a rectangular to microchannel [54].

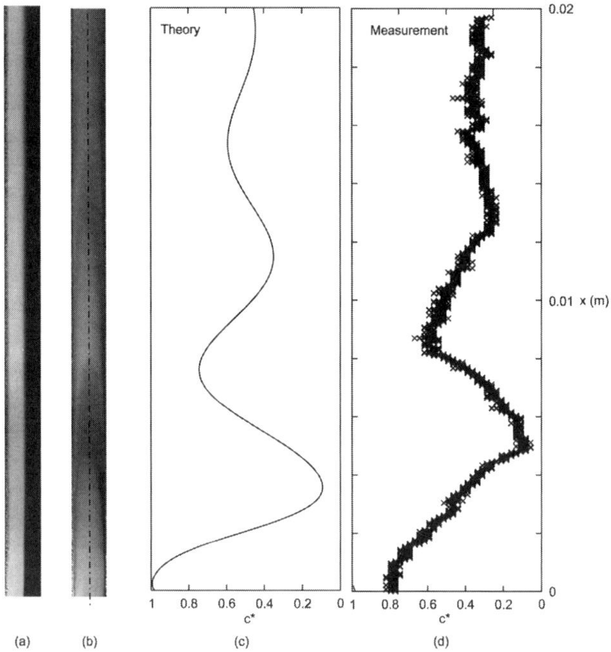


**Fig. 8.** Concentration distribution in a micromixer: (**a**) parallel lamination mixer with two inlets ($Re = 0.662$, $Pe_W = 6{,}173$); (**b**) sequential segmentation mixer at a switching frequency of 1 Hz ($Re = 0.662, Pe = 37.7$); (**c**) theoretical results with the parameters of (**b**); (**d**) measured concentration distribution at the centerline of the mixing channel as indicated in (**b**)

### 3.4. Segmentation Based on Injection

The concept of injection introduces the solvent into the solute flow through a nozzle array. This concept decreases the mixing path and increases the interfacial area between the solvent and the solute. Figure 9 shows a simple two-dimensional model of the concentration distribution around a single circular nozzle.The nozzle has a radius of $R$. The mass flow rate of the solute is $\dot{m}_1$ kg s$^{-1}$, while the mass flow rate of the solvent is $\dot{m}_2$ and an inlet concentration of $c_0$. The solvent flow has a uniform velocity of $u$ and a concentration of $c = 0$.

Considering a steady-state, two-dimensional problem and neglecting the source term, the transport equation (5) reduces to the form:

$$\frac{u}{D}\frac{\partial c}{\partial x} = \frac{\partial^2 c}{\partial x^2} + \frac{\partial^2 c}{\partial y^2}. \tag{32}$$

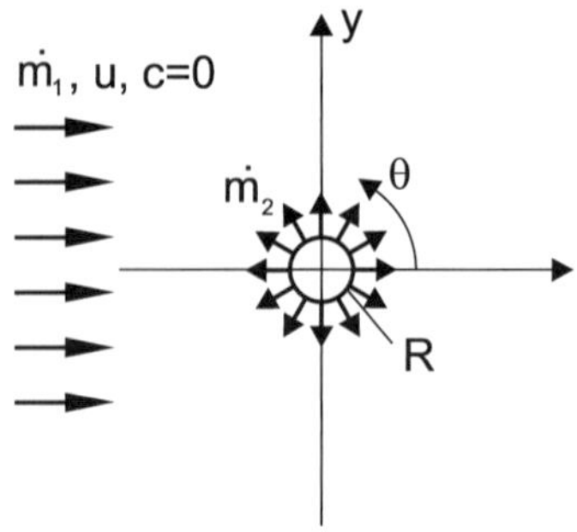


**Fig. 9.** Two-dimensional model of the concentration distribution around a circular nozzle

The general solution of (32) is the product of a velocity-dependent term and a symmetric term $\psi$:

$$c = \exp\left(\frac{ux}{2D}\right)\psi(x, y). \tag{33}$$

Substituting (33) in (32) results in:

$$\left(\frac{u}{2D}\right)\psi = \frac{\partial^2 \psi}{\partial x^2} + \frac{\partial^2 \psi}{\partial y^2}. \tag{34}$$

Defining the variable $r = \sqrt{x^2 + y^2}$ in the polar coordinate system shown in Fig. 9, the boundary condition of the solute flow is after the Fick's law (1):

$$\frac{\mathrm{d}c}{\mathrm{d}r}\Big|_{r=R} = -\frac{j}{D} = -\frac{\dot{m}_2}{2\pi RHD}, \tag{35}$$

where $H$ is the height of the mixing chamber above the injection nozzle. Assuming a small mixing ratio ($\dot{m}_2 \ll \dot{m}_1$), the following boundary condition is acceptable for (34):

$$\frac{\mathrm{d}\psi}{\mathrm{d}r}\Big|_{r=\pm\infty} = 0. \tag{36}$$

Equation (34) can be rewritten for the polar coordinate system as:

$$\frac{\mathrm{d}^2\psi}{\mathrm{d}r^2}+\frac{1}{r}\frac{\mathrm{d}\psi}{\mathrm{d}r}-\left(\frac{u}{2D}\right)^2\psi=0. \quad (37)$$

The solution of (37) is the modified Bessel function of the second kind and zero order:

$$\psi = K_0[ur/(2D)]. \quad (38)$$

The solution of (32) with the above-mentioned boundary conditions is:

$$c(r,\theta)=\frac{\dot{m}_2 D}{\pi RH}u^{-1}\frac{K_0[ur/(2D)]}{K_1[uR/(2D)]-K_0[uR/(2D)]\cos\theta}\frac{\exp[ur\cos\theta/(2D)]}{\exp[uR\cos\theta/(2D)]} \quad (39)$$

where $\theta$ is the angular variable of the polar coordinate system.

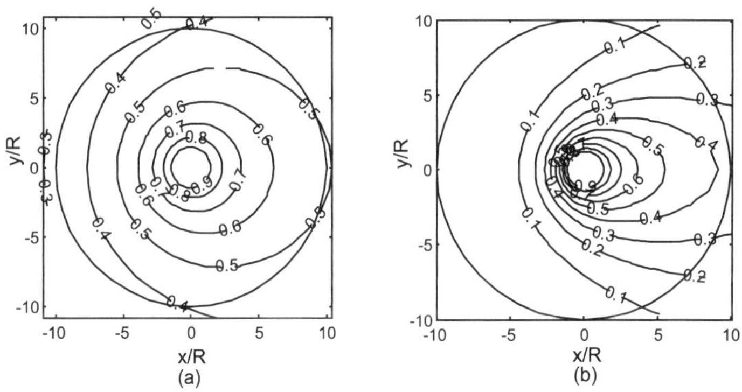

**Fig. 10.** Dimensionless concentration distribution around the injection nozzle: (**a**) $Pe = 0.1$ and (**b**) $Pe = 1$

Introducing the Peclet number $Pe = 2uR/D$, the dimensionless radial variable $r^* = r/R$, and the dimensionless concentration:

$$c^* = \frac{c}{2\dot{m}_2/(\pi H)} \quad (40)$$

the dimensionless form of (39) is:

$$c^*(r^*,\theta)=\frac{K_0(Pe\, r^*/4)/Pe}{K_1(Pe/4)-K_0(Pe/4)\cos\theta}\{\exp[Pe(r^*-1)/4]\}^{\cos\theta}, \quad (41)$$

where $K_1$ is the modified Bessel function of the second kind and first order. Figure 10 shows the typical dimensionless concentration distribution around a single injection nozzle at different Peclet numbers.

Miyake et al. [51, 52] reported a micromixer based on injection. The mixer have 400 nozzles arranged in a square array, which is located in a mixing chamber. The device was fabricated in silicon using DRIE. Larsen et al. [40] reported a similar concept with a different nozzle shape and array arrangement. Seidel et al. [65] and Voldman et al. [76] used passive microvalves for releasing one of the two fluids.

### 3.5. Focusing of Mixing Streams

Another solution for reducing the mixing path is focusing of the mixing streams. Focusing of mixing streams can be realized geometrically as shown in Fig. 11a [26] or hydrodynamically as depicted in Fig. 11(b).

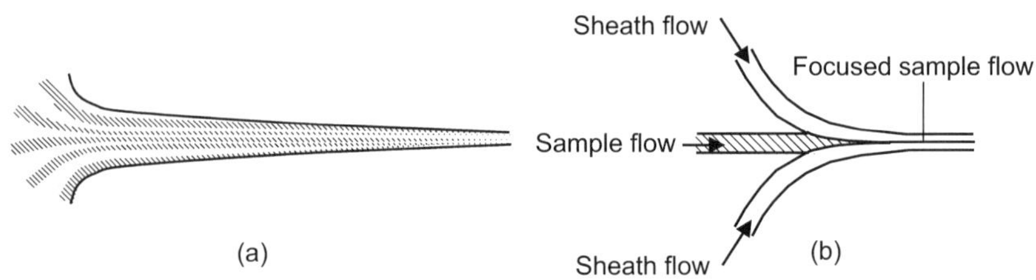


**Fig. 11.** Focusing concepts: (**a**) geometric focusing and (**b**) hydrodynamic focusing

The following model analyzes the effect of hydrodynamic focusing for reducing the width of the mixing streams. In the model, the sample stream is sandwiched between two identical sheath streams (Fig. 12). The sample stream and the sheath stream are assumed to be immiscible. Fig. 12 shows the geometry of the channel cross section with the above two phases. The channel has a width $2W$ and a height $H$. The position of the interface is $rW$. Since the model is symmetrical regarding $y$-axis and $z$-axis, only one-fourth of the cross section needs to be considered [85].

The velocity distribution $u_1$ and $u_2$ in the channel can be described by the Navier–Stokes equations:

$$\begin{cases} \dfrac{\partial^2 u_1}{\partial y^2} + \dfrac{\partial^2 u_1}{\partial z^2} = \dfrac{1}{\mu_1}\dfrac{\partial p}{\partial x} \\ \dfrac{\partial^2 u_2}{\partial y^2} + \dfrac{\partial^2 u_2}{\partial z^2} = \dfrac{1}{\mu_2}\dfrac{\partial p}{\partial x} \end{cases}, \tag{42}$$

where indices 1 and 2 describe the sample flow and the sheath flow, respectively. In (42) $\mu_1$ and $\mu_2$ are the viscosities of the sample fluid and of the sheath fluid. Nondimensionalizing the velocity by a reference velocity $u_0$, and the coordinates by $W$ leads to the dimensionless model:

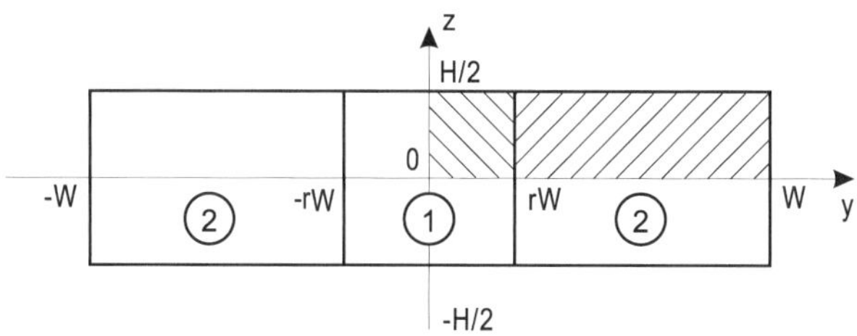


**Fig. 12.** Analytical model for two-phase focusing

coordinates by $w$ leads to the dimensionless model:

$$\begin{cases} \dfrac{\partial^2 u_1^*}{\partial y^{*2}} + \dfrac{\partial^2 u_1^*}{\partial z^{*2}} = P' \\ \dfrac{\partial^2 u_2^*}{\partial y^{*2}} + \dfrac{\partial^2 u_2^*}{\partial z^{*2}} = P' \end{cases}. \tag{43}$$

With $P' = \dfrac{W}{\mu_1 u_0}\dfrac{\partial p}{\partial x^*}$, $\beta = \mu_2/\mu_1$, and $\theta = (2n-1)\pi/h$, the solutions of (43) have the forms ($0 < y^* < 1$, $0 < z^* < h/2$)

$$\begin{cases} u_1^*(y^*,z^*) = P'\left[\dfrac{z^{*2}-h^2/4}{2}+\sum_{n=1}^{\infty}\cos\theta z^*(A_1\cosh\theta y^* + B_1\sinh\theta y^*)\right] \\ u_2^*(y^*,z^*) = \dfrac{P'}{\beta}\left[\dfrac{z^{*2}-h^2/4}{2}+\sum_{n=1}^{\infty}\cos\theta z^*(A_2\cosh\theta y^* + B_2\sinh\theta y^*)\right] \end{cases}. \tag{44}$$

The no-slip conditions at the wall are

$$u_2^*(1,z^*) = 0. \tag{45}$$

The symmetry condition at the $z^*$-axis is:

$$\frac{\partial u_1^*}{\partial y^*}\Big|_{y^*=0} = 0. \tag{46}$$

At the interface between the sample flow and the sheath flow, the velocity and the shear rate are continuous

$$\begin{cases} u_2^*(r,z^*) = u_1^*(r,z^*) \\ \dfrac{\partial u_1^*}{\partial y^*}\Big|_{y^*=r} = \beta\dfrac{\partial u_2^*}{\partial y^*}\Big|_{y^*=r} \end{cases}. \tag{47}$$

For a flat channel ($h \ll 1$), the position of the interface can be estimated as:

$$r = \frac{1}{1+2\beta\kappa}, \tag{48}$$

where $\kappa = \dot{m}_2/\dot{m}_1$ is the flow rate ratio between the sheath streams and the mixing streams. A Fourier analysis with the above boundary conditions results in the coefficients of (44):

$$
\begin{aligned}
A_1 = {} & A_0[\beta \sinh\theta r \cosh^2\theta r \cosh\theta - \sinh^3\theta r \cosh\theta - \\
& (\beta-1)\sinh^2\theta r \sinh\theta\cosh\theta - (\beta-1)\sinh^2\theta r(\cosh\theta - \cosh\theta r) \\
& +(\beta-1)\sinh\theta r\cosh\theta r\cosh\theta(\cosh\theta-\cosh\theta r)]/ \\
& [\beta\cosh^2\theta r\sinh\theta r\cosh^2\theta - \sinh^3\theta r\cosh^2\theta - \\
& (\beta-1)\sinh^2\theta r\cosh\theta r\sinh\theta\cosh\theta], \\
A_2 = {} & A_0[\beta\cosh^2\theta r\cosh\theta - \sinh^2\theta r\cosh\theta \\
& -(\beta-1)\sinh\theta r\cosh\theta r\sinh\theta - \\
& (\beta-1)\sinh\theta r(\cosh\theta-\cosh\theta r)]/[\beta\cosh^2\theta r\sinh^2\theta - \\
& \sinh^2\theta r\cosh^2\theta - (\beta-1)\sinh\theta r\cosh\theta r\sinh\theta\cosh\theta], \\
B_1 = {} & 0, \\
B_2 = {} & A_0[(\beta-1)\sinh\theta r(\cosh\theta-\cosh\theta r)]/[\beta\cosh^2\theta r\cosh\theta - \\
& \quad \sinh^2\theta r\cosh\theta - (\beta-1)\sinh\theta r\cosh\theta r\sinh\theta],
\end{aligned}
\tag{49}
$$

where $A_0 = (-1)^{n+1}\dfrac{4h^2}{(2n-1)^3\pi^3}$. Figure 13 shows the typical dimensionless velocity distribution inside the flow channel for the same flow rate in all streams ($\kappa = 1$). The theoretical model is depicted with the solid lines in Fig. 13. The velocity of the sample flow is lower, if the sample flow is more viscous than the sheath flow ($\beta < 1$), Fig. 13a. If the sheath flows are more viscous ($\beta > 1$), the sample flow is faster, Fig. 13b.

## 4. Micromixers Based on Chaotic Advection

### 4.1. Chaotic Advection in a Continuous Flow

As discussed above, advection is another important form of mass transfer, especially in flows with a low Reynolds number. In most passive micromixers based on molecular diffusion, advection is parallel to the main flow direction. Transversal transport of species relies entirely on molecular diffusion. Thus, advection with a three-dimensional obit could improve mixing significantly [81]. The basic design concept for the generation of advection is the modification of the channel shape for stretching, folding, and breaking of the laminar flow. According to [81] these processes are chaotic. The necessary condition for chaos is that the streamlines should cross each other at different times. This effect occurs in a time-periodic flow or a spatially periodic flow. While the first flow type can be implemented by motions of boundaries, periodic changes in geometry can induce the second flow type. Depending on the geometrical configuration, chaotic advection was observed at different Reynolds numbers. Therefore, micromixer designs based on chaotic advection are categorized by their operation range of Reynolds numbers. A range of $Re > 100$ is considered in this section as high. The ranges of $10 < Re < 100$ is intermediate. The range of $Re < 10$ is regarded as low. In the following, only micromixers based on passive chaotic advection are discussed.

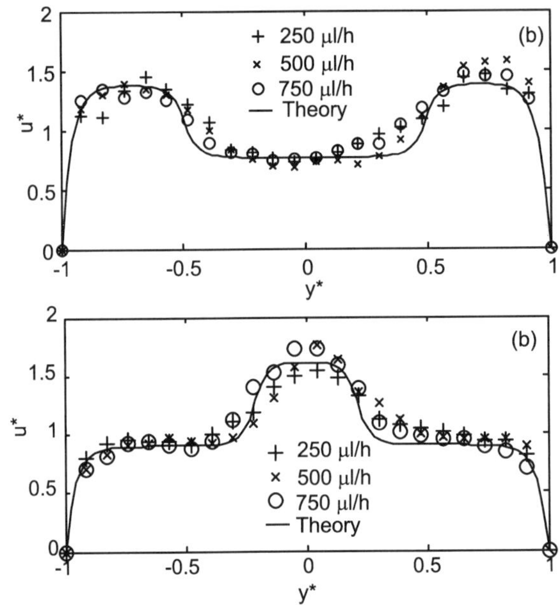


**Fig. 13.** The dimensionless velocity profile ($h = 0.1$, $\kappa = 1$): **(a)** $\beta = 1/1.8$ and **(b)** $\beta = 1.8$

### 4.1.1. Chaotic Advection at High Reynolds Numbers

The simplest way to induce spatially periodic flow is placing periodic structures in the mixing channel. These obstacles can be implemented on the wall, Fig. 14a. Wang et al. [79] numerically investigated the role of cylindrical pillars in a mixing channel at high Reynolds numbers. The results showed that these obstacles can improve mixing at high Reynolds numbers. The obstacles and inertial forces alter the flow directions and creates transversal mass transport. Lin et al. [44] used cylinders placed in a narrow channel to enhance mixing, Fig. 14b. The micromixer was fabricated in silicon. Seven cylinders of 10-$\mu$m diameter were arranged in a $50 \times 100 \times 100$ $\mu$m mixing chamber. Mixing time of 50 $\mu$s was achieved at Reynolds numbers ranging from 200 to 2,000.

Secondary flow in curved channel induces the so called Dean-vortex. Chaotic advection can be achieved in a planar channel by repeating the curved design [33]. Sharp turns or meandering mixing channels can also induce spatially periodic flows at high Reynolds numbers, Fig. 14c. Mengeaud et al. [50] used spatial period of the meandering microchannel as the optimization parameter. The micromixers were made of polyethylene terephthalate (PET). Microchannels with a width of 100 $\mu$m, a depth of 48 $\mu$m, and a length of 2 mm were fabricated using excimer laser. Chaotic mixing was observed in flows above a critical Reynolds number of 80.

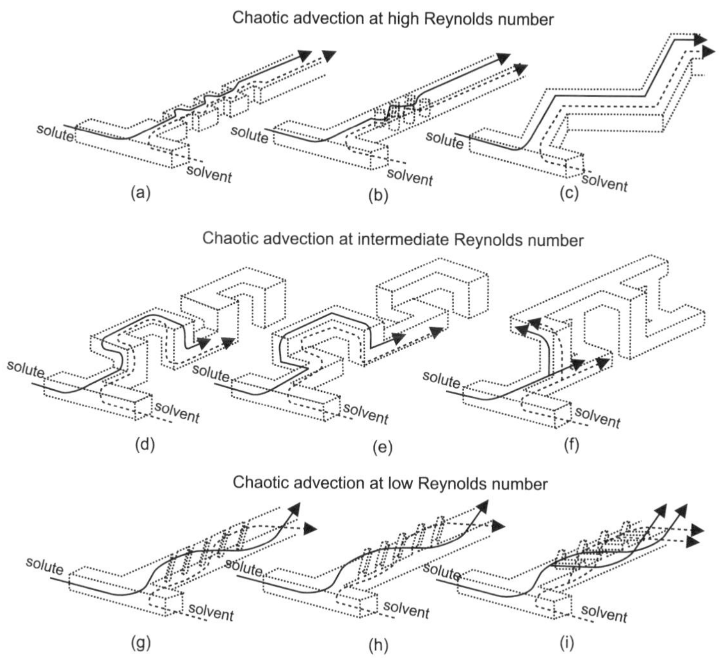


**Fig. 14.** Micromixers based on chaotic advection: (**a**) obstacles on walls [83]; (**b**) obstacles in the channel [44, 80]; (**c**) meandering channel [50]; (**d**) C-shape [45]; (**e**) L-shape [74]; (**f**) connected out-of-plane L-shapes [10]; (**g**) slanted ribs; (**h**) slanted grooves [67, 68]; and (**i**) staggered-herringbone grooves [67, 68]

### 4.1.2. Chaotic Advection at Intermediate Reynolds Numbers

A three-dimensional, periodic channel design allows chaotic advection at intermediate Reynolds numbers. Hong et al. [28] demonstrated an in-plane micromixer with modified Tesla structures. Chaotic advection was induced by the Coanda effect. The planar fluidic structures was embossed in cyclic olefin copolymer (COC) and sealed by thermal direct bonding. Mixing was achieved at Reynolds numbers above 5. Liu et al. [45] reported a three-dimensional serpentine mixing channel fabricated in silicon and glass. The channel was constructed as a series of C-shaped segments positioned in perpendicular planes, Fig. 14d. Chaotic advection was induced by the twisted intersecting streamlines. Mixing was observed at relatively high Reynolds numbers above 25. Vijayendran et al. [74] reported a three-dimensional serpentine mixing channel fabricated in PDMS . The channel was designed as a series of L-shaped segments in perpendicular planes, Fig. 14e. The mixer was tested at the Reynolds numbers of 1, 5, and 20. Better mixing was achieved at higher Reynolds numbers. Chen and Meiners [10] reported another complex channel design realized in PDMS. The periodic structure is formed by two connected out-of-plane L-shapes and measures about $400 \times 300\ \mu m^2$, Figure 14f. Effective mixing was achieved with a purely laminar flow at Reynolds numbers ranging from 0.1 to 2. Park et al. [61] reported a more complex three-dimensional channel structure. The channel rotates and separates the solvent and the solute by partitioned walls. The device was fabricated in PDMS and then bonded to glass. At intermediate Reynolds number, chaotic advection in a mixing chamber can be induced actively by a planar pulsed source–sink system [34]. Based on this concept, Evan et al. fabricated a micromixer in silicon and glass. The mixing chamber measures $1{,}500 \times 600\ \mu m^2$ with a height of $100\ \mu m$ [17].

### 4.1.3. Chaotic Advection at Low Reynolds Numbers

Similar to Kenics static mixer in macroscale [59], rips (Fig. 14g) or grooves (Fig. 14h,i) on the channel wall can twist the flow and generate chaotic advection. Jonson et al. [35] used excimer laser to ablate slanted grooves on the bottom wall of the mixing channel. This structure twists an electrokinetically driven flow into helical stream lines, that allows chaotic mixing at a relatively low velocity of $300\ \mu m\, s^{-1}$. The substrate material of this mixer is polycarbonate (PC), which is sealed by a cover made of polyethylene terephthalate glycol (PETG). The mixing channel was 72-$\mu m$ wide at the top, 28-$\mu m$ wide at the bottom, and 31-$\mu m$ in depth. The width of an ablated groove was 14 μm, the center-to-center spacing between the grooves was 35 μm. The length of the region occupied by the wells from the T-junction was 178 μm. The same configuration with slanted groves was investigated later by Strook et al. [68] for pressure driven flows. Two different groove patterns were considered. The so-called staggered herringbone mixer shown in Fig. 14i can work well at Reynolds numbers ranging from 1 to 100. The concept of twisted flows can be applied to electrokinetic flow by modifying the surface charge [67]. Electrokinetic mixing [5] with only patterned surface modification [23] can also enhance mixing. Compared to the conventional T-configuration, Biddiss et al. reported an improvement of mixing efficiencies from 22 to 68% at Peclet numbers ranging from 190 to 1,500. Kim et al. [38] improved the design of [68] with embedded barriers parallel to the flow direction. This barrier forces the flow to change the original elliptic mixing pattern [68] to a hyperbolic pattern [38]. The mixing channel of this design is 240 μm in width, 60 μm in depth, and 21 mm in length. The barriers have a cross section of 40 μm × 30 μm.

Bertsch et al. [3] reported a miniaturized version of the conventional Kenics static mixer with helical flow-twisting elements. Two designs were tested for this concept. The first design was formed by four mixing elements, which was made of 24 rectangular bars placed at 45°. The four mixing elements were arranged at 45° in the channel. The second design consists of right-handed and left-handed helical elements containing six small-helix structures. Because of this complex geometry, the micromixer was fabricated by stereomicrophotography. The complex structure was built up layer by layer.

### 4.2. Chaotic Advection in Multiphase Flow

In macroscale, chaotic advection is often induced by moving walls [59]. The implementation of moving walls in microscale would require actuators and complex fabrications steps. Moving microdroplets offer an unique alternative to moving walls. Chaotic advection can be achieved inside a microdroplet. Fast mixing can therefore be achieved by forming and driving droplets of solvent and solute through a microchannel network. The basic cross configuration is similar to that of hydrodynamic focusing, Fig. 15a. For the formation of droplets, both solvent and solute need to be immiscible to the carrier fluid. Solvent and solute enter from the two side channels. The formation behavior of droplets depend on the capillary number:

$$Ca = \frac{u\mu}{\sigma}, \tag{50}$$

where $u$ is the average velocity, $\mu$ is the dynamic viscosity, and $\sigma$ the interfacial tension between the immiscible phases. Another important parameter of the formation process is the sample fraction:

$$r = \frac{\dot{Q}_{\text{solvent}} + \dot{Q}_{\text{solute}}}{\dot{Q}_{\text{solvent}} + \dot{Q}_{\text{solute}} + \dot{Q}_{\text{carrier}}}, \tag{51}$$

and $\dot{Q}_{\text{solvent}}$, $\dot{Q}_{\text{solute}}$, and $\dot{Q}_{\text{carrier}}$ are the flow rates of solvent, solute, and carrier fluid.

The capillary number represents the ratio between viscous force and interfacial force. Interfacial force dominates over inertial force at low capillary numbers. At a low capillary number, the solvent and solute can merge into a sample droplet and mix rapidly due to chaotic advection inside the droplet. Increasing the capillary number at the same fraction $r$, viscous force slowly controls the formation process. The droplets form separately and are not able to merge and mix. Increasing further the capillary number, the alternate droplets become smaller and unstable. At a high capillary number, the three streams flow side-by-side as in the case of miscible fluids [92]. Alternatively, droplets can be formed with the T-configuration shown in Fig. 15b. A separation phase avoids the solvent and solute to merge immediately at the entrance. The droplet train formed with this concept can be stored over a long time, because the carrier fluid, such as an oil, can protect the aqueous sample from evaporation. The long-term stability of the sample allows protein crystallization to be carried out in the microscale [93].

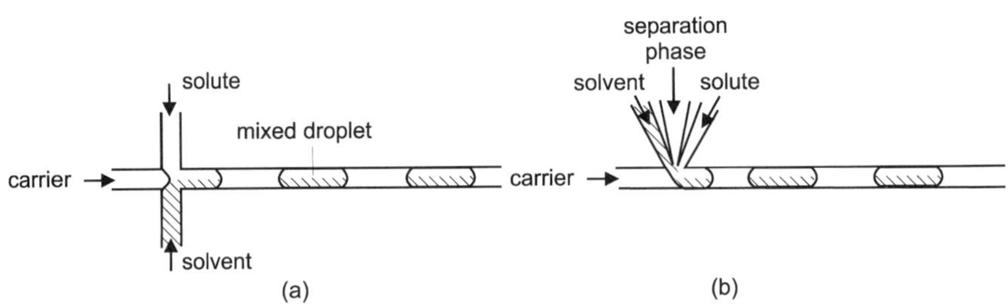


**Fig. 15.** Formation of multiphase droplets: **(a)** cross configuration and **(b)** T-configuration with a separation phase

Chaotic advection in the microdroplets cannot be achieved in a straight microchannel. As depicted in Fig. 16a, the flow pattern inside the droplet is steady and symmetrical. The two vortices keeps the solvent and solute separately in each half of the droplet. Mixing relies on diffusion similar to the case of parallel lamination. If the microchannel has turns as depicted in Fig. 16b, periodic boundary condition can be created for the droplet. Each side of the droplet takes turn to have a faster vortex. This time periodic change of droplet boundary leads to chaotic advection and consequently faster mixing. Compared to diffusive mixing with parallel lamination, the improvement of mixing time can be described as [9]:

$$\frac{t_{\text{chaotic}}}{t_{\text{diffusive}}} = \frac{Pe}{L^* \log(Pe)}, \tag{52}$$

where $Pe_W = uW/D$ is the Peclet number based on the channel width $W$, and $L^* = L/W$ is the dimensionless droplet length normalized by the channel width.

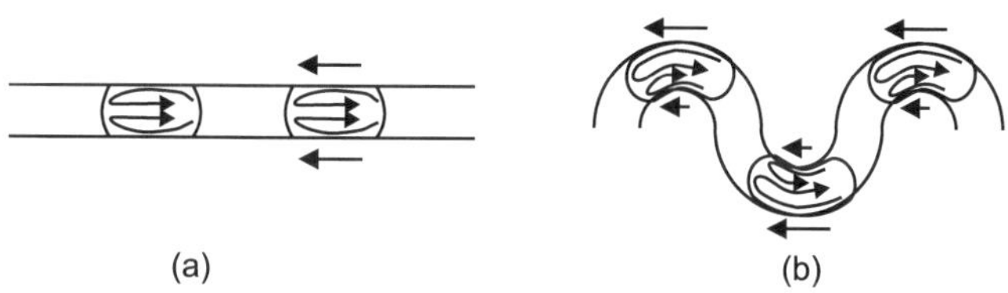


**Fig. 16.** Flow pattern in a microdroplet: (**a**) Schematic flow patterns inside a droplet in a straight channel and (**b**) in a channel with turns

Besides the passive formation process described above, droplets can be generated and transported actively by hydrodynamic force [29] or surface effects such as thermocapillary [22] and electrowetting [60]. Droplets can be formed in microchannels with other immiscible phases, such as liquid and air [66]. Hosokawa et al. [29] reported the earliest droplet-based micromixer, which was fabricated in PDMS. The concept utilized a hydrophobic microcapillary vent, which join the solute with the solvent droplet. Paik et al. [60] reported different droplet-based mixing schemes with the electrowetting concept. The active actuation allows droplets to be merged and split repeatedly. Subsequently, the merged droplet can be transported with different motion patterns to induce chaotic mixing.

# 5. Active Micromixers

## 5.1. Pressure-Driven Disturbance

Pressure-driven disturbance is relatively simple to realize. Therefore, this disturbance type was implemented in the earliest active micromixers. Deshmukh et al. [13] used micropumps to induce pressure disturbance in a T-mixer 14, Fig. 17a. The microfluidic system was fabricated in silicon using DRIE. An integrated planar micropump drives and stops the flow in

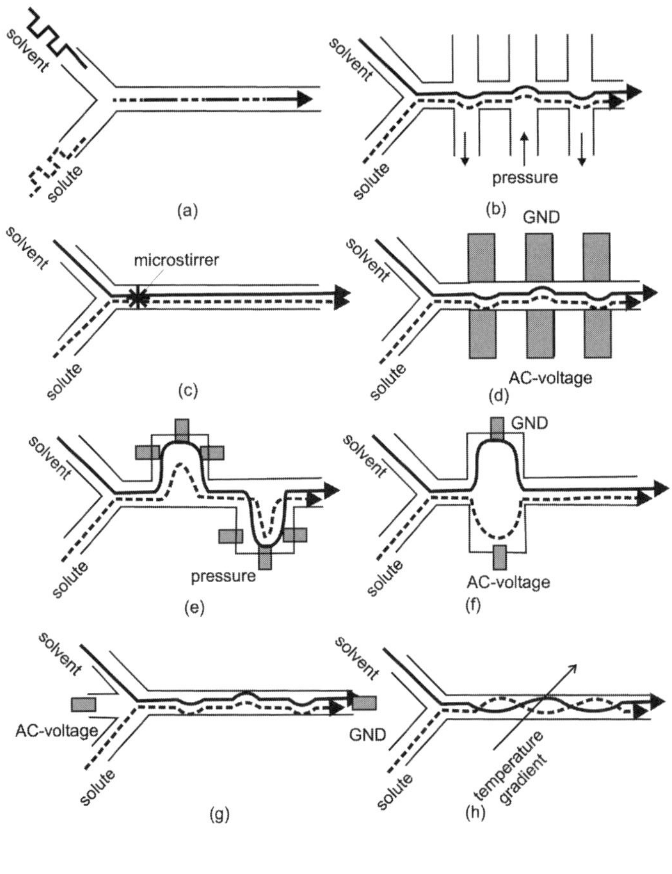


**Fig. 17.** Active micromixers: (**a**) serial segmentation; (**b**) pressure disturbance along mixing channel; (**c**) integrated microstirrer in the mixing channel; (**d**) electro-hydrodynamic disturbance; (**e**) dielectrophoretic disturbance; (**f**) electrokinetic disturbance in a mixing chamber; (**g**) electrokinetic disturbance in a mixing channel; and (**h**) disturbance caused by thermocapillary advection induced by a tranverse temperature gradient

the mixing channel causing sequential segmentation with improved axial dispersion. This concept can also be realized by an external micropump [18].

Equivalent to pressure disturbance is the generation of pulsing velocity [19, 56], Fig. 17b. Niu and Lee [56] generated pressure disturbance with a computer controlled source–sink system, which is partly similar to that of Evans et al. [17]. The pulse frequency and the number of mixing units determine the performance of the mixing process. The mixing pattern in a chamber under a transient pressure-driven disturbance was investigated numerically by Okkels et al. [58].

Moving parts can actively cause pressure and flow disturbance in the mixing channel. Suzuki and Ho [69] demonstrated the miniaturized version of macroscale magnetic stirrer. An electrical conductor generates a magnetic field, which in turn moves magnetic beads of 1–10 μm in diameter. The chaotic movement induced by the magnetic beads improves mixing significantly. Lu et al. [48] integrated a magnetic microstirrer inside the mixing channel, Fig. 17c. The micromachined stirrer is placed at the interface between two liquids in a T-mixer. Stirring speeds between 100 and 600 rpm were achieved with an external magnetic field.

## 5.2. Electrohydrodynamic Disturbance

El Moctar et al. [16] reported an active micromixer with eletrohydrodynamic disturbance, which is similar to the design with pressure-driven disturbance of [56]. Electrodes are placed along the mixing channel, which is 30 mm long, 250 μm wide and 250 μm deep. A number of titanium electrodes are placed in the direction perpendicular to the mixing channel, Fig. 17d. Transversal flows were induced by changing the voltage and frequency on the electrodes. Good mixing was achieved at a Reynolds number as low as 0.02.

## 5.3. Dielectrophoretic Disturbance

Dielectrophoresis (DEP) is the polarization effect of a particle relative to its surrounding medium in a nonuniform electrical field. This dielectrophoretic force causes the particle to move to and from an electrode. Chaotic advection was induced by the motion of these particles. Both periodic electrical actuation and channel geometry are the key parameters for this concept. Micromixers based on dielectrophoretic disturbance were reported by Deval et al. [14] and Lee et al. [41], Fig. 17e.

### 5.4. Electrokinetic Disturbance

Electroosmotic transport is used in microfluidics as an alternative to pressure-driven transport. Thus, flow instabilities can also be induced by electroosmotic forces. Jacobson et al. [31] used electrokinetic flow in a conventional T-mixer. Lettieri et al. [42] utilized electroosmotic forces to create instabilities in a pressure-driven micromixer. The pressure-driven flow becomes unstable. Chaotic advection occurs in the mixing chamber, Fig. 17f. Oscillating electroosmotic flow generated by an ac-voltage was observed by Oddy et al. [57]. Mixing was improved significantly, Fig. 17g. The concept of sequential segmentation can also be implemented in an electrokinetic flow. Tang et al. [71] utilized switching of electrokinetic flow to generate short segments of solvent and solute in the mixing channel. Electrokinetic segmentation was capable of creating stable fluid segments in the mixing channel at a frequency between 0.01 and 1 Hz.

### 5.5. Magnetohydrodynamic Disturbance

The next candidate for transversal transport of species in microchannels is magnetohydrodynamic effect [2]. An applied dc-voltage on an electrode pair generates in the presence of an external magnetic field Lorentz forces. These forces induce transversal movement of solutes in the mixing channel. The Lorentz force can stretch and fold the liquids in the mixing channel. The only drawback of this concept is that active mixing only works with an electrolyte solution. The mixer of Bau et al. was fabricated from co-fired ceramic tapes. The electrodes are printed with a gold paste [2].

### 5.6. Acoustic Disturbance

Acoustic waves can be used in micromixers to stir fluids and induce mixing. Acoustically induced flow, or acoustic streaming, can be used to create transversal transport in a mixing channel. Moroney et al. [53] reported for the first time the proof of concept of acoustic mixing in microscale. The acoustic field was induced from the channel wall by flexible plate wave. Zhu and Kim [94] reported active mixing with focused acoustic waves in a mixing chamber. The mixing chamber was fabricated in silicon and measures 1 mm × 1 mm × 10 μm. A piezoelectric membrane was located at the bottom of the mixing chamber. Rife et al. [62] used acoustic streaming to improve mixing. Vivek et al. [75] reported a micromixer with focused acoustic streaming using different electrode patterns. Acoustic stirring at high frequency can also be implemented by an external micropump [82].

Despite its effectiveness, acoustic mixing may have complications for biological samples. Temperature rise caused by acoustic energy is one of the concerns. Furthermore, ultrasonic waves around 50 kHz are harmful to biological samples because of the possible cavitations. Yasuda et al. [89] reported an acoustic micromixer with loosely focused acoustic waves to generate stirring movements. The wave is generated by a thin zinc oxide film. The actuator was driven by sinusoidal waves with frequencies corresponding to the thickness-mode resonance (e.g., 240 and 480 MHz) of the piezoelectric film. This mixer operated without any significant temperature increase and could be used for temperature-sensitive fluid. Yang et al. reported further acoustic devices for mixing water and ethanol, as well as water and uranine [87, 91].

Liu et al. utilized acoustic streaming induced around an air bubble for mixing. Air pockets with 500-μm diameter and 500-μm depth trap the air bubbles. Acoustic streaming was generated by an integrated PZT actuator [46, 47]. Yaralioglu et al. [88] utilized acoustic streaming to disturb the flow in a conventional Y-mixer. While the channel is made of PDMS, the piezoelectric actuator is integrated into the cover quartz wafer.

### 5.7. Thermal Disturbance

According to (2), the diffusion coefficient highly depends on the temperature. Thus, thermal energy can be used to enhance mixing [12, 73]. Darhuber et al. [12] proposed and numerically analyzed the thermocapillary advection induced by transverse temperature gradient, Fig. 17h. The temperature gradient induces helical streamlines, which is representative for chaotic advection. Tsai et al. [73] utilized a thermal bubble bubble to generate disturbance in a mixing channel.

## References

1. Aris, R. (1956). On the dispersion of a solute in a fluid flowing through a tube. *Proceeding of the Royal Society of London, Series A (Mathematical and Physical Sciences)*, 235:67–77
2. Bau, H. H., Zhong, J., and Yi, M. (2001). A minute magneto hydro dynamic (MHD) mixer. *Sensors and Actuators B*, 79:207–215

3. Bertsch, A., Heimgartner, S., Cousseau, P., and Renaud, P. (2001). Static micromixers based on large-scale industrial mixer geometry. *Lab on a Chip*, 1:56–60
4. Bessoth, F. G., de Mello, A. J., and Manz, A. (1999). Microstructure for efficient continuous flow mixing. *Analytical Communications*, 36:213–215
5. Biddiss, E., Erickson, D., and Li, D. (2004). Heterogeneous surface charge enhanced micromixing for electrokinetic flows. *Analytical Chemistry*, 76:3208–3213
6. Bird, R. B., Stewart, W. E., and Lightfoot, E. N. (2001). *Transport Phenomena*. Wiley, New York
7. Boehm, S., Greiner, K., Schlautmann, S., De Vries, S., and Van Den Berg, A. (2001). *A Rapid Vortex Micromixer for Studying High-Speed Chemical Reactions*. pp. 25–27
8. Branebjerg, J., Gravesen, P., Krog, J. P., and Nielsen, C. R. (1996). *Fast Mixing by Lamination*. pp. 441–446
9. Bringer, M., Gerdts, C. J., Song, H., Tice, J. D., and Ismagilov, R. F. (2004). Microfluidic systems for chemical kinetics that rely on chaotic mixing in droplets. *Philosophical Transactions of the Royal Society of London, Series A*, 362:1087–1104
10. Chen, H. and Meiners, J. C. (2004). Topologic mixing on a microfluidic chip. *Applied Physics Letters*, 84:2193–2195
11. Cussler, E. L. (1996). *Diffusion Mass Transfer in Fluid Systems*, 2nd edition. Cambridge University Press, New York
12. Darhuber, A. A. (2003). Microfluidic actuation by modulation of surface stresses. *Applied Physics Letters*, 82:657–659
13. Deshmukh, A. A., Liepmann, D., and Pisano, A. P. (2000). Continuous micromixer with pulsatile micropumps. In *Technical Digest of the IEEE Solid State Sensor and Actuator Workshop, Hilton Head Island, SC, June 4–8*, pp. 73–76
14. Deval, J., Tabeling, P., and Ho, C. M. (2002). A dielectrophoretic chaotic mixer. In *Proceedings of MEMS'02, 15th IEEE International Workshop Micro Electromechanical System, Las Vegas, NV, Jan. 20–24*, pp. 36–39
15. Dutta, D. and Leighton, D. T. (2001). Dispersion reduction in pressure driven flow through microetched channels. *Analytical Chemistry*, 73:504–513
16. El Moctar, A. O., Aubry, N., and Batton, J. (2003). Electro-hydrodynamic micro-fluidic mixer. *Lab on a Chip*, 3:273–280
17. Evans, J., Liepmann, D., and Pisano, A. P. (1997). *Planar Laminar Mixer*. pp. 96–101
18. Fujii, T., Sando, Y., Higashino, K., and Fujii, Y. (2003). A plug and play microfluidic device. *Lab on a Chip*, 3:193–197
19. Glasgow, I. and Aubry, N. (2003). Enhancement of microfluidic mixing using time pulsing. *Lab on a Chip*, 3:114–120
20. Gobby, D., Angeli, P., and Gavriilidis, A. (2001). Mixing characteristics of T-type microfluidic mixers. *Journal of Micromechanics and Microengineering*, 11:126–132

21. Gray, B. L., Jaeggi, D., Mourlas, N. J., Van Drieenhuizen, B. P., Williams, K. R., Maluf, N. I., and Kovacs, G. T. A. (1999). Novel interconnection technologies for integrated microfluidic systems. *Sensors and Actuators A*, 77:57–65
22. Handique, K. and Burns, M. A. (2001). Mathematical modeling of drop mixing in a slit-type microchannel. *Journal of Micromechanics and Microengineering*, 11:548–554
23. Hau, W. L. W., Trau, D. W., Sucher, N. J., Wong, M., and Zohar, Y. (2003). Surface-chemistry technology for microfluidics. *Journal Micromechanics and Microengineering*, 13:272–278
24. Haverkamp, V., Ehrfeld, W., Gebauer, K., Hessel, V., Loewe, H., Richter, T., and Wille, C. (1999). The potential of micromixers for contacting of disperse liquid phases. *Fresenius Journal of Analytical Chemistry*, 364:617–624
25. He, B., Burke, B. J., Zhang, X., Zhang, R., and Regnier, F. E. (2001). A picoliter-volume mixer for microfluidic analytical systems. *Analytical Chemistry*, 73:1942–1947
26. Hessel, V., Hardt, S., Loewe, H., Schoenfeld, F. (2003). Laminar mixing in different interdigital micromixers: I. Experimental characterization. *AICHE Journal*, 49:566–577
27. Hinsmann, P., Frank, J., Svasek, P., Harasek, M., and Lendl, B. (2001). Design, simulation and application of a new micromixing device for time resolved infrared spectroscopy of chemical reactions in solutions. *Lab on a Chip*, 1:16–21
28. Hong, C. C., Choi, J. W., and Ahn, C. H. (2004). A novel in-plane microfluidic mixer with modified tesla structures. *Lab on a Chip*, 4:109–113
29. Hosokawa, K., Fujii, T., and Endo, I. (1999). Droplet-based nano/picoliter mixer using hydrophobic microcapillary vent. In *Proceedings of the IEEE International Workshop Micro Electromechanical System, Piscataway, NJ, USA*, pp. 388–393
30. Ismagilov, R. F., Stroock, A. D., Kenis, P. J. A., Whitesides, G., and Stone, H. A. (2000). Experimental and theoretical scaling laws for transverse diffusive broadening in two-phase laminar flows in microchannels. *Applied Physics Letters*, 76:2376–2378
31. Jacobson, S. C., McKnight, T. E., and Ramsey, J. M. (1999). Microfluidic devices for electrokinematically driven parallel and serial mixing. *Analytical Chemistry*, 71:4455–4459
32. Jensen, K. (1998). Chemical kinetics: Smaller, faster chemistry. *Nature*, 393:735–736
33. Jiang, F., Drese, K. S., Hardt, S., Küpper, M., and Schönfeld, F. (2004). Helical flows and chaotic mixing in curved. Micro channels. *AICHE Journal*, 50: 2297–2305
34. Johnes, S. and Aref, H. (1988). Chaotic advection in pulsed source–sink systems. *Physics of Fluids*, 31:469–485
35. Johnson, T. J., Ross, D., and Locascio, L. E. (2002). Rapid microfluidic mixing. *Analytical Chemistry*, 74:45–51

36. Kakuta, M., Jayawickrama, D. A., Wolters, A. M., Manz, A., and Sweedler, J. V. (2003). Micromixer-based time-resolved NMR: applications to ubiquitin protein conformation. *Analytical Chemistry*, 75:956–960
37. Kamholz, A. E., Weigl, B. H., Finlayson, B. A., and Yager, P. (1999). Quantitative analysis of molecular interaction in microfluidic channel: The T-sensor. *Analytical Chemistry*, 71:5340–5347
38. Kim, D. S., Lee, S. W., Kwon, T. H., and Lee, S. S. (2004). A barrier embedded chaotic micromixer. *Journal Micromechanics and Microengineering*, 14:798–805
39. Knight, J. B., Vishwanath, A., Brody, J. P., and Austin, R. H. (1998). Hydrodynamic focusing on a silicon chip: Mixing nanoliters in microseconds. *Physical Review Letters*, 80:3863–3866
40. Larsen, U. D., Rong, W., and Telleman, P. (1999). *Design of Rapid Micromixers Using CFD*. pp. 200–203
41. Lee, Y. K., Deval, J., Tabeling, P., and Ho, C. M. (2001). Chaotic mixing in electrokinetically and pressure driven micro flows. In *Proceedings of MEMS'01, 14th IEEE International Workshop Micro Electromechanical System, Interlaken, Switzerland, Jan. 21–25*, pp. 483–486
42. Letteri, G. L., Dodge, A., Boer, G., Lichtenberg, J., Verpoorte, E., and de Rooij, N. F. (2000). *Consequences of Opposing Electrokinetically and Pressure-Induced Flows in Microchannels of Varying Geometries*. pp. 351–354
43. Lim, D. S. W., Shelby, J. P., Kuo, J. S., and Chiu, D. T. (2003). Dynamic formation of ring-shaped patterns of colloidal particles in microfluidic systems. *Applied Physics Letters*, 83:1145–1147
44. Lin, Y., Gerfen, G. J., Rousseau, D. L., and Yeh, S. R. (2003). Ultrafast microfluidic mixer and freeze-quenching device. *Analytical Chemistry*, 75:5381–5386
45. Liu, R. H., Stremler, M. A., Sharp, K. V., Olsen, M. G., Santiago, J. G., Adrian, R. J., Aref, H., and Beebe, D. J. (2000). Passive mixing in a three-dimensional serpentine microchannel. *Journal of Microelectromechanical Systems*, 9:190–197
46. Liu, R. H., Yang, J., Pindera, M. Z., Athavale, M., and Grodzinski, P. (2002). Bubble-induced acoustic micromixing. *Lab on a Chip*, 2:151–157
47. Liu, R. H., Lenigk, R., Druyor-Sanchez, R. L., Yang, J., and Grodzinski, P. (2003). Hybridization enhancement using cavication microstreaming. *Analytical Chemistry*, 75:1911–1917
48. Lu, L. H., Ryu, K. S., and Liu, C. (2002). A magnetic microstirrer and array for microfluidic Mixing. *Journal of Microelectromechanical Systems*, 11:462–469
49. Melin, J., Gimenez, G., Roxhed, N., Van Der Wijngaart, W., and Stemme, G. (2004). A fast passive and planar liquid sample micromixer. *Lab on a Chip*, 3:214–219
50. Mengeaud, V., Josserand, J., and Girault, H. H. (2002). Mixing processes in a zigzag microchannel: Finite element simulation and optical study. *Analytical Chemistry*, 74:4279–4286

51. Miyake, R., Lammerink, T. S. J., Elwenspoek, M., and Fluitman, J. H. J. (1993). *Micro Mixer with Fast Diffusion.* pp. 248–253
52. Miyake, R., Tsuzuki, K., Takagi, T., and Imai, K. (1997). *A Highly Sensitive and Small Flow-Type Chemical Analysis System with Integrated Absorbtionmetric Micro-Flowcell.* pp. 102–107
53. Moroney, R. M., White, R. M., and Howe, R. T. (1991). Ultrasonically induced microtransport. In *Proceedings of MEMS'91, 3th IEEE International Workshop Micro Electromechanical System, Nara, Japan, Jan. 30–Feb. 4.*, pp. 277–282
54. Nguyen, N. T. and Huang, X. Y. (2005). Mixing in microchannels based on hydrodynamic focusing and time-interleaved segmentation: Modelling and experiment. *Lab Chip*, 5:1320–1326
55. Nguyen, N. T. and Huang, X. Y. (2006). Modeling, fabrication and characterization of a polymeric micromixer based on sequential segmentation. *Biomedical Microdevices Journal*, 8:133–139
56. Niu, X. Z. and Lee, Y. K. (2003). Efficient spatial–temporal chaotic mixing in microchannels. *Journal of Micromechanics and Microengineering*, 13:454–462
57. Oddy, M. H., Santiago, J. G., and Mikkelsen, J. C. (2001). Electrokinetic instability micromixing. *Analytical Chemistry*, 73:5822–5832
58. Okkels, F. and Tabeling, P. (2004). Spatiotemporal resonances in mixing of open viscous fluids. *Physical Review Letters*, 92:228–301
59. Ottino, J. M. (1989). *The Kinematics of Mixing: Strectching, Chaos, and Transport.* Cambridge University Press, Cambridge
60. Paik, P., Pamula, V. K., and Fair, R. B. (2003). Rapid droplet mixers for digital microfluidic systems. *Lab on a Chip*, 3:253–259
61. Park, S. J., Kim, J. K., Park, J., Chung, S., Chung, C., and Chang, J. K. (2004). Rapid three-dimensional passive rotation micromixer using the brackup process. *Journal of Micromechanics and Microengineering*, 14:6–14
62. Rife, J. C., Bell, M. I., Horwitz, J. S., Kabler, M. N., Auyeung, R. C. Y., and Kim, W. J. (2000). Miniature valveless ultrasonic pumps and mixers. *Sensors and Actuators A*, 86:135–140
63. Schoenfeld, F., Hessel, V., and Hofmann, C. (2004). An optimized. Split-and-rexombine micro-mixer with uniform 'chaotic' mixing. Lab chip, 4:65–69
64. Schwesinger, N., Frank, T., and Wurmus, H. (1996). A modular microfluid system with an integrated micromixer. *Journal of Micromechanics and Microengineering*, 6:99–102
65. Seidel, R. U. (1999). *Capilary Force Mixing Device as Sampling Module for Chemical Analysis.* pp. 438–441
66. Song, H., Bringer, M. R., Tice, J. D., Gerdts, C. J., and Ismagilov, R. F. (2003). Experimental test of scaling of mixing by chaotic advection in droplets moving through microfluidic channels. *Applied Physics Letters*, 83:4664–4666
67. Stroock, A. D. and Whitesides, G. M. (2003). Controlling flows in microchannels with parterned surface charge and topography. *Accounts of Chemical Research*, 36:597–604

68. Stroock, A. D., Dertinger, S. K. W., Ajdari, A., Mezic, I., Stone, H. A., and Whitesides, G. M. (2002). Chaotic mixer for microchannels. *Science*, 295:647–451
69. Suzuki, H. and Ho, C. M. (2002). A magnetic force driven chaotic micro-mixer. In *Proceedings of MEMS'02, 15th IEEE International Workshop Micro Electromechanical System, Las Vegas, NV, Jan. 20–24*, pp. 40–43
70. Tan, C. K. L., Tracey, M. C., Davis, J. B., and Johnston, I. D. (2005). Continously variable mixing-ratio micromixer with elastomer valves. *Journal of Micromechanics and Microengineering*, 15:1885–1893
71. Tang, Z., Hong, S., Djukic, D., Modi, V., West, A.C., Yardley, J., and Osgood, R. M. (2002). Electrokinetic flow control for composition modulation in a microchannel. *Journal of Micromechanics and Microengineering*, 12:870–877
72. Taylor, G. (1953). Dispersion of soluble matter in solvent flowing slowly through a tube. *Proceeding of the Royal Society of London, Series A (Mathematical and Physical Sciences)*, 219:186–203
73. Tsai, J. H. and Lin, L. (2002). Active microfluidic mixer and gas bubble filter driven by thermal bubble pump. *Sensors and Actuators A*, 97–98:665–671
74. Vijiayendran, R. A., Motsegood, K. M., Beebe, D. J., and Leckband, D. E. (2003). Evaluation of a three-dimensional micromixer in a surface-based biosensor. *Langmuir*, 19:1824–1828
75. Vivek, V., Zeng, Y., and Kim, E. S. (2000). Novel acoustic-wave micromixer. In *Proceedings of MEMS'00, 13th IEEE International Workshop Micro Electromechanical System, Miyazaci, Japan, Jan. 23–27*, pp. 668–673
76. Voldman, J., Gray, M. L., and Schmidt, M. A. (2000). An integrated liquid mixer/valve. *Journal of Microelectromechanical Systems*, 9:295–302
77. Walker, G. M., Ozers, M. S., and Beebe, D. J. (2004). Cell infection within a microfluidic device using virus gradients. *Sensors and Actuators B*, 98:347–355
78. Wang, H., Iovenitti, P., Harvey, E., and Masood, S. (2002a). *Mixing of Two Fluid Streams in a Microchannel Using the Taylor–Aris Dispersion Effect*. pp. 158–163
79. Wang, H., Iovenitti, P., Harvey, E., and Masood, S. (2002b). Optimizing layout of obstacles for enhanced mixing in microchannels. Smart materials and structures, 11: 662–667
80. Wang, H., Iovenitti, P., Harvey, E., and Masood, S. (2003). Numerical investigation of mixing in microchannels with patterned grooves. *Journal of Micromechanics and Microengineering*, 13:801–808
81. Wiggins, S. and Ottino, J. M. (2004). Foundation of chaotic mixing. *Philoshopical Transactions of the Royal Scciety of London, Series A*, 362:937–970
82. Woias, P., Hauser, K., and Yacoub-George, E. (2000). *An active silicon micromixer for mTAS applications. In Micro Total Analysis Systems 2000, A. van den Berg et al. (eds.), Boston: Kluwer Academic Publishers*, pp. 277–282

83. Wong, S. H., Bryant, P., Ward, M., and Wharton, C. (2003). Investigation of mixing in a cross-shaped micromixer with static mixing elements for reaction kinetics studies. *Sensors and Actuators B*, 95:414–424
84. Wong, S. H., Ward, M. C. L., and Wharton, C. W. (2004). Micro T-mixer as a rapid mixing micromixer. *Sensors and Actuators B*, 100:365–385
85. Wu, Z. and Nguyen, N. T. (2005). Rapid mixing using two-phase hydraulic focusing in microchannels. *Biomedical Microdevices*, 7:13–20
86. Wu, Z., Nguyen, N. T., and Huang, X. Y. (2004). Non-linear diffusive mixing in microchannels: theory and experiments. *Journal of Micromechanics and Microengineering*, 14:604–611
87. Yang, Z., Matsumoto, S., Goto, H., Matsumoto, M., and Maeda, R. (2001). Ultrasonic micromixer for microfluidic systems. *Sensors and Actuators A*, 93:266–272
88. Yaralioglu, G. G., Wygant, I. O., Marentis, T. C., and Khuri-Yakub, B. T. (2004). Ultrasonic mixing in microfluidic channels using integrated transducers. *Analytical Chemistry*, 76:3694–3698
89. Yasuda, K. (2000). Non-destructive, non-contact handling method for biomaterials in micro-chamber by ultrasound. *Sensors and Actuators B*, 64:128–135
90. Yi, M. and Bau, H. H. (2003). The kinematics of bend-induced mixing in micro-conduits. *International Journal of Heat and Fluid Flow*, 24:645–656
91. Zhen, Y., Goto, H., Matsumoto, M., and Maeda, R. (2000). Active micromixer for microfluidic systems using lead-zirconate-titanate (PZT)-generated ultrasonic vibration. *Electrophoresis*, 21:116–119
92. Zheng, B., Tice, J. D., and Ismagilov, R. F. (2004a). Formation of droplets of alternating composition in microfluidic channels and applications to indexing of concentrations in droplet-based assays. *Analytical Chemistry*, 76:4977–4982
93. Zheng, B., Tice, J. D., Roach, L. S., and Ismagilov, R. F. (2004b). A droplet-based, composite PDMS/glass capillary microfluidic system for evaluating protein crystallization conditions by microbatch and vapor-diffusion methods with on-chip X-ray diffraction. *Angewandte Chemie-International Edition*, 43:2508–2511
94. Zhu, X. and Kim, E. S. (1997). Acoustic-wave liquid mixer. In *Microelectromechanical Systems (MEMS) American Society of Mechanical Engineers, Dynamic Systems and Control Division (Publication) DSC., ASME, Fairfield, NJ, Vol. 62*, pp. 35–38

# Chapter 4
# Control of Liquids by Surface Energies

Martin Brinkmann, Krishnacharya Khare, and Ralf Seemann

Max-Planck Institute for Dynamics and Self-Organization,
Göttingen (Germany)
martin.brinkmann@ds.mpg.de
krishnacharya.khare@ds.mpg.de
ralf.seemann@ds.mpg.de

## 1. Introduction

Our every day experience of handling liquids is strongly linked to the presence of gravity. In a zero gravity environment, as realized during parabola flights or in space crafts, it is almost impossible to pour water from a bottle into a glass. A close look at a glass of water already reveals the influence of interfacial energies: the curvature of the water meniscus bends up close to its rim due to the high wettability of clean glass surfaces by water. When compartmentalized into structures smaller than a characteristic length being typically on the order of 1 mm gravity becomes irrelevant and interfacial energies dominate both the statics and dynamics of a liquid down to a length scale of about 10 nm [1].

The strong impact of interfacial energies on the morphology of submillimeter-sized wetting droplets can be utilized to control the position and the shape of tiny amounts of liquid on specifically designed substrates. Figure 1 shows a comparison of droplet shapes on a plane substrate decorated with a pattern of wettable stripes (left) and a structure of rectangular grooves on a chemically homogeneous substrate (right). Wettability patterns on the substrate or surface topographies may act as "anchoring points" for the contact line of the wetting liquid.

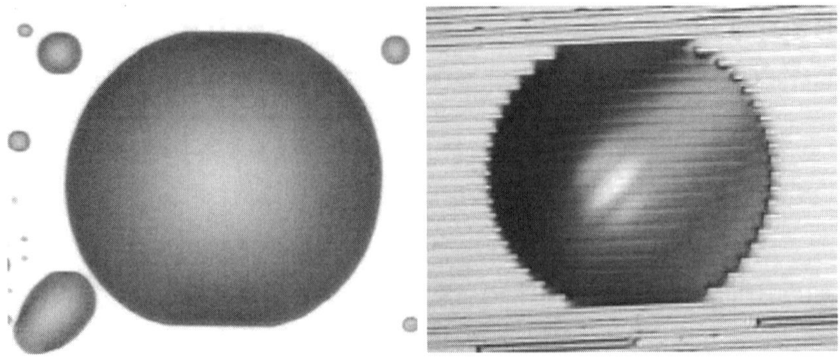

**Fig. 1.** Hexaethylene glycol droplet on a plane substrate bearing horizonal wettable stripes (left) with a periodicity of 400 nm in comparison to a polymer droplet on a topographically structured substrates with rectangular grooves (right). Pinning of the three phase contact line is clearly visible from the non-spherical shape of the droplets perimeter (left image taken from Herminghaus et al. [2] with kind permission of Springer Science and Business Media)

Highly wettable stripes on plane substrates can provide the means to guide a liquid on a surface [3–6]. Changes in the adhesion energy of the liquid can be utilized to fill linear surface topographies such as rectangular grooves [7, 8]. Therefore, chemical patterns and topographic structures such as grooves represent potential building blocks of open microfluidic systems [9], alternative to closed systems fabricated in a solid matrix.

For fundamental physical understanding, but as well for potential use in printing techniques [4, 10, 11], soldering [12–14], biotechnological applications [15] or open microfluidics [3, 8, 9, 16, 17], wetting of chemically or topographically structured surfaces by nonvolatile liquids has become a topic of intense research in recent years. The proper functioning of wettability-based systems requires a full understanding of the liquid behavior for a variety of different substrate designs. Here, in particular, the pioneering work in the group of Lipowsky on the theoretical framework of morphological wetting on chemically structured substrates has to be mentioned [1, 3, 5, 18–21]. Experimental and theoretical studies of basic liquid morphologies have been performed for various wettability patterns and surface topographies. This includes circular domains [18–20], stripes [3, 5], rings of different wettability [21, 22] as well as topographic steps [7] and channels of triangular [23, 93] or rectangular cross section [8, 23].

Wettability patterns and substrate topographies imprinted to the surface are typically static. Here, the morphology can be controlled by liquid volume. Potential applications in open microfluidics require an active control of liquid morphologies of given volume. This can be achieved with techniques to switch the wettability by external stimuli. Here, the electrowetting effect provides a versatile tool to reversibly switch contact angles of conductive wetting liquids [9, 24].

In the following Sect. 2, we will give an outline of the fundamental wetting phenomena involved in the description of liquid morphologies and their transitions on chemically and topographically structured substrates. Section 3 is devoted to plane substrates decorated with basic patterns of differently wettable domains. A short overview of available experimental preparation techniques is included in this section. Results for wetting morphologies on topographically structured substrates are summarized in the subsequent Sect. 4, which, again, includes a short introduction to experimental techniques. We present an application of a particular morphological transition that can be utilized to control the imbibition of liquid into topographic structures.

## 2. Capillary Model

Small insects like the water strider *Gerris remigis* have the astonishing ability to move on a water surface [25]. The weight of the water strider is balanced by interfacial forces of the water–air interface. An increase of the area of the liquid–vapor interface requires work against interfacial tension. The water strider could walk on any liquid provided the strider's weight is small enough. If the deformations of the water interface around the strider's feet are small, the accompanying increase of interfacial free energy scales with the square of the indentation depth in contrast to the gravitational energy that decreases linearly, leading to a stable mechanical equilibrium.

Interfacial free energies can be obtained, in principle, as the excess of total free energy of the system over the free energy of the bulk phases [26]. This surface excess of the free energy can be easily explained from the microscopic interactions between liquid molecules at or close to the interface. Creating new interfacial area between two fluid phases requires to break bonds between neighboring molecules and is therefore related to an energy increase. A negative sign of the interfacial energy would contradict the mechanical stability of the two fluid phases in coexistence.

Details of the molecular interactions, however, are not relevant on macroscopic length scales where the interfacial energies and the spatial configuration of the liquid–vapor interface contains all necessary information about the system. These details are essential only in microscopic models of the liquid interface, for instance, in a description of the density profile between the liquid and an adjacent bulk phase. For all practical purposes, the intrinsic width of the interface and the interfacial roughness caused by thermal fluctuations are many orders of magnitude smaller than the size of the droplets [27]. Hence, in the framework of the following discussion, we regard all interfaces as mathematical surfaces, i.e., ideal interfaces of zero width. As shown experimentally, this assumption stays valid even for mesoscopic liquid droplets with lateral dimensions below 100 nm [6, 28, 29].

### 2.1. Equilibrium Conditions

For nonvolatile and incompressible liquids, the mechanically stable droplet shapes are determined by a local minimum of interfacial and gravitational energy. Contributions of gravitational energy are negligibly small if the droplet size is well below the capillary length

$$\ell_c = \sqrt{\frac{\gamma_{lv}}{g\,|\Delta\rho|}}, \tag{1}$$

which is determined from a balance of gravitational and interfacial energies. In expression (1), $\gamma_{lv}$ is the interfacial tension of the liquid–vapor interface, $\Delta\rho = \rho_l - \rho_v$ the density difference between the liquid l and vapor v, and $g$ the normal acceleration. Typically, $\ell_c$ is on the order of 1 mm. Because of the large interfacial tension between water and air, the related capillary length has a comparably large value $\ell_c \approx 2.7$ mm. For special systems, the capillary length $\ell_c$ can assume values on the order of centimeters, or may become even larger if the ambient phase is an immiscible liquid with a density close to the density of the wetting liquid. In this way, the effect of gravity or, to be precise, of buoyancy on the wetting liquid can be drastically reduced in capillary experiments.

The liquid–vapor interface of a wetting morphology with dimensions well below $\ell_c$ is a surface of constant mean curvature in mechanical equilibrium. The mean curvature $H$ and the interfacial tension $\gamma_{lv}$ of the liquid–vapor interface are related to the difference $\Delta P = P_l - P_v$ between the pressure $P$ in the liquid l and the ambient vapor phase v by the Laplace equation

$$\Delta P = 2H\gamma_{lv}. \tag{2}$$

The local mean curvature $H = (C_1 + C_2)/2$ is the arithmetic mean of the principal curvatures $C_1$ and $C_2$ in the respective point of the surface [30], as illustrated in the sketch Fig. 2a. The liquid–vapor interface of a spherical droplet found on plane substrates has a normal curvature $C = 1/R$ irrespective of the direction and the point on the surface. Hence, a spherical surface is a surface of constant mean curvature with $H = 1/R$, where $R$ is the radius of the sphere.

The equation of Laplace (2) is still valid for liquid–vapor interfaces with a vertical extension on the order or larger than $\ell_c$. Here, the pressure difference in the respective point of the liquid–vapor interface includes a contribution of the hydrostatic pressure in the fluid phases. Corrections to the Laplace pressure have to be introduced for liquid structures of size smaller than a mesoscopic length $\ell_m$. This disjoining pressure is mostly due to long ranged van der Waals or electrostatic forces, which become relevant for liquid structures on the substrate with a vertical extension smaller than roughly $\ell_m$ 10 nm [6]. For these two cases, i.e., in the range of droplet dimensions above $\ell_c$ and droplet dimensions below $\ell_m$, the liquid–vapor interface cannot be approximated by a surface of constant mean curvature.

In mechanical equilibrium, the liquid–vapor interface of a wetting droplet meets the surface of the substrate at a constant angle $\theta$, which is solely

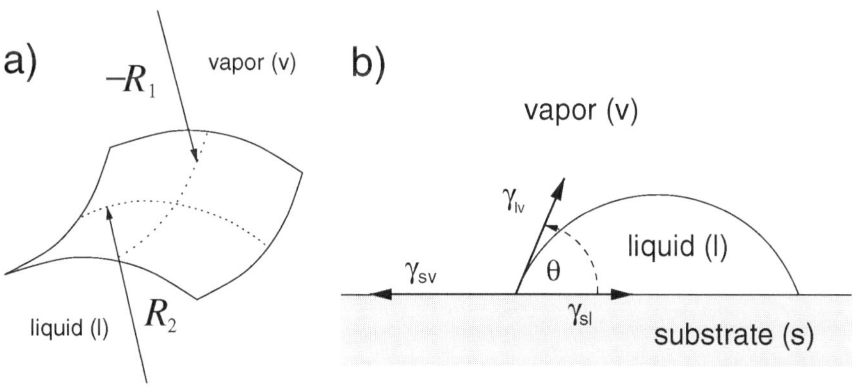


**Fig. 2.** (**a**) The mean curvature $H$ of the liquid–vapor interface is the sum divided by two of the principal curvatures $C_1 = 1/R_1$ and $C_2 = 1/R_2$. Note, that the sign of the normal mean curvatures depends on the orientation of the surface. By definition, the curvature of a sphere is positive. (**b**) Sketch of the balance of interfacial forces at the three phase contact line

determined by interfacial tensions $\gamma_{\mathrm{lv}}, \gamma_{\mathrm{sl}}, \gamma_{\mathrm{sv}}$ of the respective interfaces, cf. Fig. 2b. The equation of Young-Dupré

$$\cos\theta = \frac{\gamma_{\mathrm{sv}} - \gamma_{\mathrm{ls}}}{\gamma_{\mathrm{lv}}}, \tag{3}$$

expresses the mechanical equilibrium of the three phase contact line between the substrate, the vapor, and the wetting liquid. The equilibrium contact angle $\theta$, or the "material" contact angle, is a direct measure of the substrate wettability. According to the Greek terms for "water loving" and "water hating" contact angles $\theta < 90°$ have been termed hydrophilic while contact angles with $\theta > 90°$ are referred to as hydrophobic, respectively. In the sequel, these terms are not restricted to water and will be used for any wetting liquid.

### 2.2. Contact Line Pinning

The equation of Young-Dupré, (3), may become ill-defined on plane surfaces with a discontinuous dependence of the surface tensions $\gamma_{\mathrm{ls}}$ and $\gamma_{\mathrm{vs}}$ on the position. The contact angle $\theta$ in points of the contact line coinciding with such a discontinuity can attain any value between the smaller value $\theta_{-}$ on the more hydrophilic side and the larger value $\theta_{+}$ on the more hydrophobic side of the discontinuity, cf. Fig. 3a. As a consequence, the contact line is immobilized and its position is fixed to the line of discontinuity as long as the contact angle falls into the range between

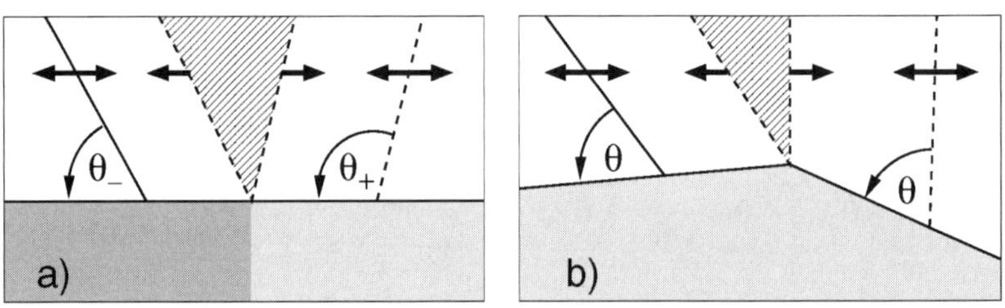


**Fig. 3.** Sketch of a wetting front moving over a wettability step or a kink in the substrate topography. (**a**) The three phase contact line becomes pinned to a boundary between two differently wettable domains with contact angles $\theta_{-}$ and $\theta_{+}$ as long as the position of the liquid–vapor interface remains in the hatched sector. (**b**) Pinning of the contact line to a convex edge on a homogeneous substrate

$\theta_-$ and $\theta_+$. Hence, the contact angle is not defined by the local wettability of the substrate but depends on the global shape of the liquid–vapor interface in equilibrium.

The sketch in Fig. 3b illustrates the pinning of a moving contact line at a convex edge of a topography on a substrate with homogeneous wettability. Any configuration of the liquid–vapor interface that forms a contact angle larger than the equilibrium contact angle $\theta$ according to (3) with respect to the left side of the kink and a contact angle smaller than the contact angle $\theta$ with respect to the right side of the kink can be in equilibrium. Because of its enlarged free range of contact angles, an acute convex edge of the substrate pins the contact line more effectively than less acute edges. Concave corners, contrary to convex edges, cannot immobilize the contact line. Configurations of the liquid–vapor interface with a contact line that is fully or partially aligned with a concave corner of the topography are unstable.

As we will see in the following, constraining the contact line to certain positions on chemically or topographically structured substrates allows the liquid to form a large variety of equilibrium shapes.

### 2.3. Computation of Droplet Shapes

Surfaces decorated with wettability patterns or substrate topographies lead to complex equilibrium morphologies of the wetting liquid due to the pinning mechanisms mentioned above. According to the Laplace equation (2), the liquid–vapor interface of small droplets substrates will always be a surface of constant mean curvature even on heterogeneous substrates. The size of the droplets shall be small compared to the capillary length $\ell_c$ but still large compared to the mesoscopic length $\ell_m$. The vast majority of these equilibrium shapes cannot be parameterized explicitly, even for substrates with regular wettability patterns or surface topographies. Hence, the main features of the liquid–vapor interface may be partially captured through a substitution of the real configuration by known surfaces of constant mean curvature if the dimensions of these liquid structures fall into this range of length scales. Possible shape "primitives" are segments of cylinders [7, 8], parts of a sphere [18], or even suitable pieces of a Delaunay surface [20, 21]. The family of Delaunay surfaces comprises all surfaces of revolution with constant mean curvature being periodic along the axis of symmetry. The surface of a sphere and the catenoid are obtained as limiting shapes while the cylinder is a special Delaunay surface [21].

Generically, the condition of Young-Dupré will not be matched in all points of the contact line. The shape of a primitive is fixed by a low dimensional set of parameters like, for instance, the radius and the contact angle of a spherical cap. Within such an approximation, the interfacial energy of the model droplet is a function of these parameters and can be analytically minimized while respecting the constraint of a constant liquid volume. This method has been successfully applied to several problems, calculating the equilibrium shapes of liquid filaments on wettable stripes [5], topographic steps [7], and rectangular grooves [8]. In some rare cases one may obtain exact analytic solution in this way, i.e., surfaces of constant mean curvature whose boundaries are consistent with the equation of Young-Dupré (3) or the condition of a pinned contact line [20].

Alternative to approximations by shape primitives, the liquid–vapor interface may be represented by a mesh of small triangles spanning between neighboring nodes. A large number of theoretical investigations in the field of wetting on structured surfaces have been performed using this numerical model. The interfacial energy of such a discrete, polyhedral surface is a function of $3N$ coordinates where $N$ is the number of nodes which make up the mesh. Global constraints like a constant droplet volume or local constraints that have to be enforced, for instance, on those nodes gliding on the substrate walls are respected during minimization. Flipping of edges changes the topology of the mesh which is necessary to adapt to changes in the configuration during morphological transitions. New edges can be introduced to the mesh while short edges are constantly removed. In this way, the distribution of edge lengths on the polyhedral surfaces is kept in a certain desired range. Numerical minimizations using such a dynamically triangulated model of the liquid–vapor interface can be performed with the public domains software "surface evolver," developed by K. Brakke [31]. This program offers a whole script language to define the substrate geometry and required energies, methods for mesh manipulation, and a number of minimization schemes.

## 3. Plane Substrates with Wettability Patterns

On plane and chemically homogeneous substrates, small droplets attain the shape of spherical caps, as discussed in Sect. 2. Qualitatively different droplet shapes emerge on surfaces decorated with patterns of differently wettable domains. Driven by the minimization of interfacial energy, the liquid tries to maximize the contact area with the highly wettable surface domains and to avoid the poorly wettable parts of the substrate. In a certain range of liquid volumes and wettabilities, the wetting liquid follows the geometry of the wettable structures and is restricted by the boundaries of the highly wettable domains. Here, contact line pinning is the central feature that governs the morphology of liquid on chemically structured substrates.

Before we discuss liquid morphologies on several fundamental wettability patterns, we briefly summarize in the following Sect. 3.1 common experimental techniques by which wettability patterns and liquid morphologies can be created and analyzed. Contact line pinning and equilibrium morphologies of liquid on chemically structured surfaces are illustrated in Sect. 3.2 for the particular geometry of an array of highly wettable circular domains embedded in a poorly wettable matrix. We will discuss the equilibrium liquid morphologies depending on control parameter such as the contact angles $\theta_-$ and $\theta_+$ on the highly and poorly wettable domains, respectively, and the liquid volume $V$. Furthermore, we extend the complexity of the chemical patterns to lines and rings, and consider morphological wetting transition on these substrates in Sect. 3.3.

### *3.1. Experimental*

The smoothest surfaces with chemical contrast can be found in semiconductor industry. If two crystalline materials are combined or troughs of a crystalline material are “filled” by epitaxial growth, these surfaces can be polished to an almost atomically smooth surface. Unfortunately this technique is very expensive and the range of available materials, i.e., wettabilities and pattern shapes is very limited.

Hence, for scientific experiments, the wettability of surfaces is typically tailored by grafting self assembly monolayers onto surfaces. This preparation technique allows to vary the contact angle within a fairly large range up to about 115° for water. Self assembly monolayers of certain molecules form a closed monomolecular coating, which might be covalently bonded to the respective surface [32]. By the nature of a monolayer, its thickness is given by the extension of the molecules, which is typically on the order of one nanometer. The head groups of the molecules, which bind to glass or silicon oxide surfaces are often mono- and trichlorosilanes. In case of gold surfaces, thiols are used. Generally, the trichlorosilanes layer are more stable than monochlorosilanes as expected by the larger number of possible binding sites, but have the tendency to form multilayers. For the typically hydrophobic wetting properties, the tail group of the molecules usually consists of alkyl chains, which might be perfluorinated.

Basically, there are two possibilities to fabricate chemically patterned substrates. The first strategy is to start with a homogeneous SAM-layer and remove or oxidize the molecules in order to achieve the desired wettability pattern, e.g., by exposure to UV/ozone or oxygen plasma [33–35], through dewetting patterns [36], or by electrochemistry using patterned electrodes [37]. Very fine wetting structures can be achieved by locally oxidizing the substrate or removing the SAM-layer by electrochemistry applying a voltage between the substrate and the tip a scanning force microscope (SFM) [38–42].

The second strategy is to directly graft a SAM-layer only to specific areas of a substrate. One may start with a prepatterned substrate where the SAM-layer binds selectively to parts of the surface with a certain chemical composition. As prepatterns, often thin gold layers are vapor deposited through masks like transmission electron microscopy (TEM) grids [3, 43], or created by colloidal monolayer [44–46], breath figures [47], and dewetting or microphase separation patterns [48]. These masks can furthermore be used to create hydrophilic patterns by vapor deposition of hygroscopic salts like $CaCl_2$ or $MgF_2$ [3, 43].

Alternatively, SAM-layers can be patterned directly, e.g., by domain formation in Langmuir-Blodget monolayers transferred to solid substrates [49] or unstable monolayers produced by such a monolayer transfer [50]. Unfortunately, the spectrum of possible wetting patterns is very limited. The only patterns, for instance, which can be created by the latter technique, are parallel stripes. A much more flexible method, but technically demanding, is the dip pen lithography where lines are written onto a substrate by an SFM tip similar to a fountain pen [51–53]. A straight forward and easy to use method is the so-called microcontact printing [54–57]. Here, molecules that are able to form a SAM-layer are printed to a substrate by the raised parts of a poly(dimethyl siloxane) rubber (PDMS) stamp. To avoid multi-layer formation in microcontact printing preferably monochlorosilanes are used. This soft lithography technique works well for structure sizes down to about 100 nm.

The liquid morphologies are typically created by condensing liquid from the vapor phase onto the patterned surface [3, 43, 50, 59], dip coating the substrates [4, 10, 11], or dewetting liquid from a substrate [60, 61]. Often, organic fluids or water are used as wetting liquids. The liquids might be cooled or osmotically stabilized allowing for an extended observation time [3, 59]. Liquids that can be solidified without remarkable volume change allow for a long term stability of the formerly liquid morphologies and an exact three-dimensional analysis of its shape. Polymers like polystyrene can be solidified by temperature [8], UV-curable prepolymers can be solidified by exposure to light [56, 62], and metal oxide morphologies can be created from an organic sol–gel solution [19, 63].

Optical microscopy techniques are the main investigation tool to characterize the emerging liquid morphologies. These optical techniques are limited in lateral resolution and their ability to analyze the structures in three-dimensions. A quantitative three-dimensional analysis of the wetting morphologies is possible up to a contact angle of about 25°, only. The surface of a solidified liquid droplet can be imaged by atomic force microscopy (AFM) or scanning electron microscopy (SEM), which overcome the limitations of the optical techniques regarding the lateral resolution and the maximal contact angle. Even liquid structures can be imaged by AFM in tapping mode™ [59] or environmental SEM [64].

### 3.2. Circular Surface Domains

It is instructive to start our consideration with a surface pattern of circular hydrophilic domains on a hydrophobic substrate matrix. This ensures that the equilibrium droplet shapes on the patterned substrate are still segments of a sphere which, in return, allows an analytical description of the system. Morphological transitions between different equilibrium configurations of liquid distributed to an array of hydrophilic discs were first addressed by Lenz and Lipowsky [18]. Bechinger et al. performed experiments for the inverse geometry of wetting layers on a hydrophobic matrix that is perforated by an array of circular hydrophobic domains [19]. The morphology of the liquid film and its instabilities were analyzed theoretically by Lenz et al. [20].

A sufficiently small amount of liquid placed onto a circular hydrophilic domain forms a spherical cap with a contact angle given by the material contact angle $\theta_- < 90°$. In this regime (I), the contact line of the small droplet has a circular shape and is completely located on the highly wettable domain, cf. Fig. 4. While adding more liquid, the droplet grows until it covers the entire domain. At this point, the contact line coincides with the circular domain boundary. This regime of pinned droplet configurations is referred to as regime (II). During a further increase of the droplet volume, the position of the contact line remains fixed to the boundary while the contact angle $\theta$ grows until it reaches the limiting value given by the contact angle $\theta_+$ on the less wettable substrate matrix. Beyond this point, the contact line of the spherical droplet detaches from the boundary and slides onto the hydrophobic, respectively, less hydrophilic matrix, cf. regime (III) in Fig. 4.

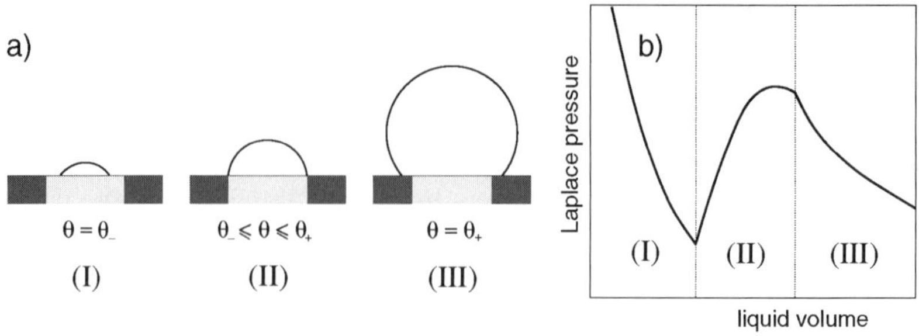


**Fig. 4.** Spherical droplet on a highly wettable domain of circular shape for different liquid volumes (**a**) and the corresponding Laplace pressure (**b**) according to Lenz and Lipowsky [18]. In regime (I) and (III), the Laplace pressure is a strictly monotonously decreasing function of the liquid volume. The Laplace pressure increases with growing volume in regime (II) if the contact angle $\theta$ is below $90°$

The Laplace pressure $\Delta P$ of a liquid droplet on a homogeneous and plane surface is a monotonously decreasing function of the liquid volume $V$. This applies to droplets in regime (I) and (III), as sketched in Fig. 4. The Laplace pressure of a pinned droplet in regime (II) with $\theta < 90°$, however, is a monotonously growing function of the droplet volume. The Laplace pressure of a droplet in regime (II) attains a local maximum either at the boundary between regime (II) and (III), i.e., at a contact angle $\theta = \theta_+$ for $\theta_+ \leq 90°$, or at a contact angle $\theta = 90°$ for $\theta_+ \geq 90°$.

### 3.2.1. Array of Hydrophilic Discs

An array of $N$ droplets on $N$ circular hydrophilic domains has been considered for a full wettability contrast $\theta_- = 0°$ and $\theta_+ = 180°$ by Lenz and Lipowsky [18]. For this particular choice of material contact angles, the droplets are always found in the pinned regime (II). It is assumed that the droplets can exchange liquid volume, e.g., via the vapor phase. A homogeneous configuration of equally sized droplets with a contact angle $\theta < 90°$ is found for small volumes, cf. Fig. 5. At a total volume $V^* = 2\pi N a^3/3$, the contact angle of the droplets reaches the critical value $90°$, as shown in Fig. 5. Beyond $V^*$, the homogeneous droplet configuration is unstable with respect to an exchange of liquid volume. Since the Laplace pressure is a monotonously decreasing function of $V$ for $\theta > 90°$, any small perturbation in the distribution of volume is amplified and leads to a heterogeneous droplet configuration. In equilibrium and, hence, in the globally stable configuration for $V > V^*$, the Laplace pressure has to be the same in all droplets. Because there are only two possible droplet shapes for a given Laplace pressure, a small droplet with $\theta < 90°$ and a large droplet with $\theta > 90°$, the final droplet configuration has to be bimodal. But since two or more large droplets with $\theta > 90°$ cannot be mechanically stable with respect to an exchange of volume, the only possible locally stable equilibrium configuration is one large droplet in coexistence with $N-1$ small droplets.

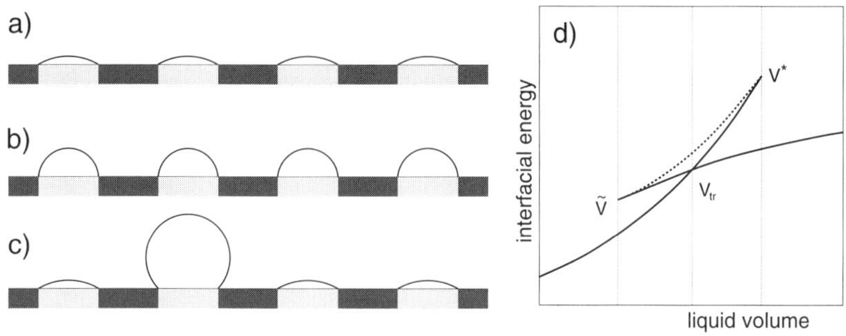


**Fig. 5.** Morphological transition between a homogeneous configuration of four small droplets (**a**) and (**b**), and a heterogeneous configuration (**c**) according to Lenz and Lipowsky [18]. The configuration in (**b**) with $\theta = 90°$ corresponds to the largest total volume $V^*$ where the homogeneous configuration represents a local minimum of the interfacial energy. (**d**) Schematic hysteresis loop for the morphological transition between (**a**) and (**c**). The *solid line* on the lower left side of the diagram corresponds to the homogeneous configuration, the full line on the upper right side to the heterogeneous configuration. A branch of mechanically unstable solutions (*dashed line*) connects the branches of stable (or metastable) solutions. Note that the real hysteresis loop for $N = 4$ is extremely narrow and asymmetric

The transition between the homogeneous and heterogeneous droplet configurations is continuous for $N = 2$ [65]. In case of $N > 2$, however, the transition is hysteretic with respect to changes in the total volume $V$, i.e., the droplet configuration is not uniquely defined by $V$. An inverse transition from the heterogeneous to the homogeneous configuration occurs at a volume $\tilde{V} < V^*$ that depends on the total number of droplets $N$. The interfacial energy of the droplet configuration at its respective point of instability jumps to a smaller value during the shape transitions, cf. Fig. 5. Thus, there exists a liquid volume $V^{\text{tr}}$ with $\tilde{V} < V^{\text{tr}} < V^*$, where the interfacial free energies of both droplet configurations are identical. Many features of this simple system are generic for transitions between different wetting morphologies, as we will demonstrate in the following.

### 3.2.2. Array of Hydrophobic Discs

The "inverse" geometry of a periodic array of circular hydrophobic surface domains on a hydrophilic substrate matrix has been investigated experimentally by Bechinger et al. [19] and later theoretically described by Lenz et al. [20]. Figure 6 shows a solidified wetting layer on a hexagonal lattice of circular hydrophobic domains after dewetting in comparison to numerical minimization of the interfacial energy. In their theoretical analysis, Lenz et al. assumed a perfectly wettable substrate and a separation distance between the circular domains, which is on the order of their diameter. The circular hydrophobic domains are arranged in a periodic lattice structure. It is useful to exploit the various symmetries of the system in numerical minimizations of the interfacial energy. Hence, one may consider only a unit cell of the surface pattern with appropriate periodic boundary conditions or even a smaller piece of the system by virtue of the discrete symmetries of the unit cell [20]. Possible instabilities of the liquid–vapor interface, which are accompanied by a redistribution of liquid on length scales larger than the size of a unit cell, are effectively suppressed.

If the thickness of the wetting layer is large compared to the distance between the holes, a plane liquid–vapor interface represents the configuration of the smallest interfacial energy. For small layer thicknesses, it is

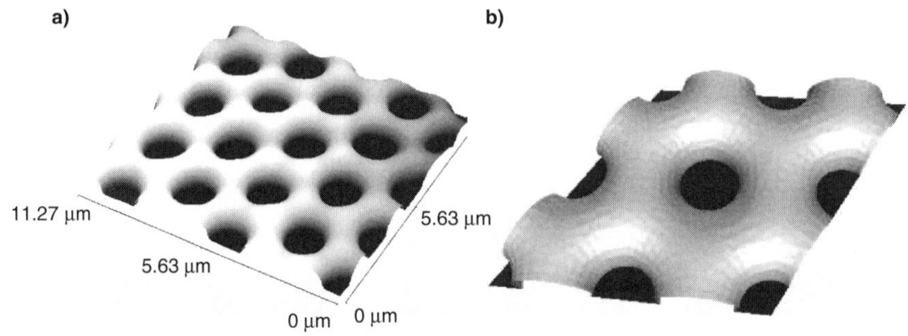


**Fig. 6.** (**a**) AFM micrograph showing a perforated wetting layer of gel-like hydrated $WO_3$ on a hexagonal array of hydrophobic circular discs (reprinted from Bechinger et al. [19], with permission from Elsevier). (**b**) Configuration of the liquid–vapor interface obtained by numerical minimization of the interfacial energy in the same geometry

energetically more favorable for the liquid to dewet from the circular hydrophobic domains and to form a perforated wetting layer. Here, the contact line is pinned to the boundary between the hydrophobic domains and the surrounding hydrophilic matrix. The contour of the liquid–vapor interface close to the contact line can be described by nodoids, a member of the family of Delaunay surfaces [20, 21], see also Sect. 2.3. Numerical minimizations and the analytic model show that the Laplace pressure of a "thin" perforated wetting film is a monotonously growing function of the liquid volume $V$ per unit cell. The Laplace pressure reaches a maximum value at a volume $V^{\max}$ and decreases as the liquid volume is further increased. Finally, at a critical volume $V^* > V^{\max}$, the "thick" perforated film becomes unstable and forms a homogeneous wetting layer.

## 3.3. Striped Surface Domains

A wettability pattern of periodic and parallel stripes represents, in addition to circular domains, the second fundamental geometry which has been the subject of several experimental and theoretical studies [3–5]. Linear wettability domains may provide the means to guide liquid in open microfluidic systems or to control the droplet growth during condensation of liquid on the surface. Complete spreading of the wetting liquid on surface patterns is important in printing techniques, which are based on the transfer of ink from the highly wettable domains onto another surface. Printing techniques with a resolution on the micrometer scale would open, e.g., the possibility to print electrical circuits directly onto plane substrates [4]. In any of those applications, the wetting behavior on linear domains like highly wettable stripes will be essential.

We begin our considerations with liquid morphologies on a single wettable stripe and will extend the discussion to related wetting geometry of annular domains as well as to liquid morphologies wetting two or more hydrophilic stripes.

### 3.3.1. Perfectly Wettable Stripe

Hydrophilic stripes with a material contact angle $\theta_-$ close to 0° have been realized by Gau et al. in a hydrophobic silicone rubber (PDMS) matrix where pure water has a contact angle $\theta_+$ of about 110° [3], see the optical micrograph in Fig. 7. As hydrophilic material, $MgF_2$ has been deposited from the vapor phase through electron microscopy grids forming a layer of less than 20 nm thickness. When cooling the substrate below the dew point, water condenses onto the hydrophilic $MgF_2$ pattern of parallel stripes. Each stripe with a length $l$ of several millimeters has a width $w = 30\ \mu m$ and is separated by a hydrophobic stripe of the same width. Since water perfectly wets $MgF_2$ surfaces, the condensing water spreads along the whole stripe. If the length of the stripe is finite, the liquid forms a cylindrical filament, which grows in size as more liquid condenses. At a certain degree of filling, the liquid–vapor interface of such a homogeneous liquid filament suddenly becomes unstable and decays into a bulge-like droplet, which extends in the form of flat filaments wetting the remaining part of the stripe.

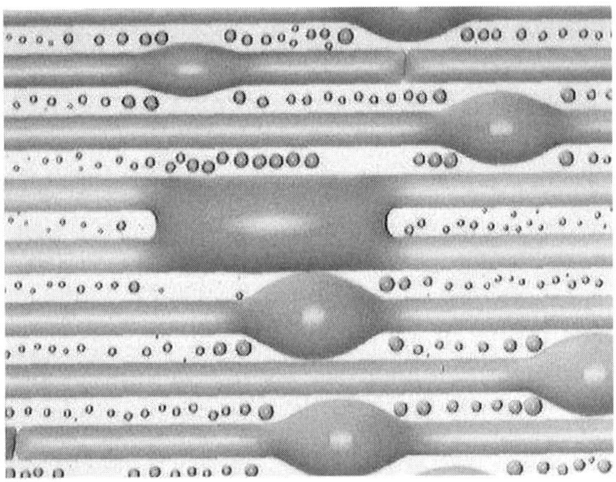

**Fig. 7.** Optical micrograph showing water morphologies that have been condensed on hydrophilic $MgF_2$ stripes on a hydrophobic PDMS substrate. The cylindrical homogeneous water filaments have become unstable and form one large bulge-like droplet per filament. In the center, two such bulge-like droplets have merged to a liquid bridge combining two initially separated stripes (image taken from Herminghaus et al. [16])

As pointed out by Gau et al. in [3], the interfacial instability of asymptotically long liquid filaments occurs when the Laplace pressure reaches its maximum value, i.e., when the contact angle $\theta$ of the cylindrical filament attains the value 90°. Here, $\theta$ denotes the lateral contact angle in a homogeneous part of the filament. "Overfilled" liquid filaments with a contact angle of $\theta > 90°$ may become unstable with respect to a longitudinal exchange of volume since the Laplace pressure is a monotonously decreasing function of the volume in this case. Hence, the liquid in such an overfilled homogeneous filament cannot be in a stable equilibrium. Any small deformation is amplified if the wavelength of the perturbation is sufficiently large, as it is the case in the Rayleigh-Plateau instability of a free standing liquid cylinder [67]. Corrugations of the liquid–vapor interface with a short wavelength are linked to an area increase and are therefore energetically costly. Hence, the instability is effectively suppressed if the wavelength $\lambda$ of a shape perturbation is smaller than a threshold wavelength

$$\lambda_{\mathrm{c}} = \frac{w\theta}{\sin\theta}\sqrt{\frac{\pi^2}{\theta^2 - \pi^2/4}}, \tag{4}$$

which is a function of the lateral contact angle $\theta$ and stripe width $w$, see Gau et al. [3]. According to Eq. (4), one can "overfill" liquid filaments up to a lateral contact angle $\theta$ larger than 90° on short stripes with a length $l < \lambda_{\mathrm{c}}(\theta)$. For "flat" cylindrical filaments with $\theta < 90°$, however, the Laplace pressure is a monotonously increasing function of the volume. Hence, flat filaments of *arbitrary* length are stable with respect to a longitudinal exchange of volume.

### 3.3.2. Partially Wettable Stripe

In contrast to perfectly wettable domains, one may expect that a hydrophilic stripe will not be fully covered by liquid if the material contact $\theta_{-}$ angle is finite. Partial wettability implies that sufficiently small liquid droplets are fully located inside the hydrophilic domain. Here, the droplets assume the shape of a spherical cap with the material contact angle $\theta_{-}$, see Fig. 8a, resp., d. As the liquid volume of the droplet increases, the radius of the contact line grows until it eventually touches the boundary of the hydrophilic stripe and becomes partially pinned. The evolution of the liquid shape during a further increase of volume depends strongly on the material contact angle $\theta_{-}$ on the hydrophilic stripe: For small contact angles $\theta_{-}$, the droplet grows into an elongated liquid filament, as depicted in Fig. 8b, resp., e. The liquid–vapor interface of a long filament is cylindrical and exhibits noticeable deviations from this ideal shape only close to its ends, only. In addition to the contact angle $\theta_{-}$ on the stripe, these homogeneous filaments are characterized by the lateral contact angle $\theta$ at the pinned contact line. If the material contact angle $\theta_{-}$ on the stripe is large, however, the shape of the liquid–vapor interface evolves into a three-dimensional bulge, cf. Fig. 8c, resp., f. In the asymptotic limit of very large volumes $V \gg w^3$, the bulge shape approaches a spherical cap with the contact angle $\theta_+$ on the hydrophobic matrix. Again, $w$ denotes the width of the stripe.

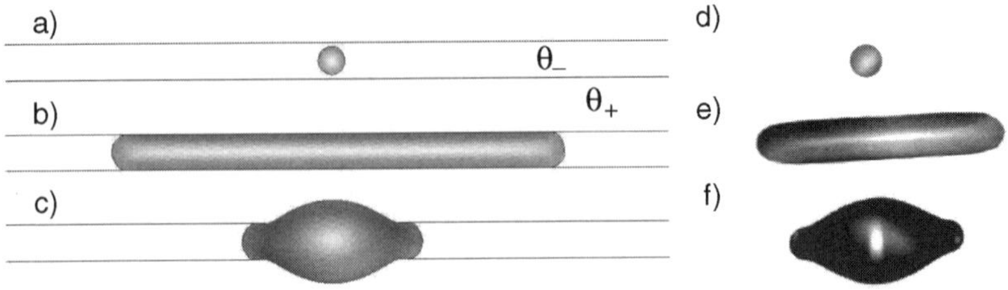


**Fig. 8.** The three principle droplet morphologies on a single, hydrophilic stripe with a variable contact angle $\theta_-$. (**a**) Small spherical droplet with a circular contact line at a contact angle of $\theta_- = 35°$ and volume $V = 0.05\,w^3$. (**b**) Elongated liquid filament at a contact angle $\theta_- = 35°$ and volume $V = 3\,w^3$. (**c**) bulge morphology at $\theta_- = 45°$ and $V = 10\,w^3$. The contact angle on the hydrophobic substrate is $\theta_+ = 180°$ in all cases. Micrographs in (**d–f**) show corresponding droplet morphologies that have been observed in experiments on hydrophobic substrates decorated with hydrophilic stripes of different wettability (taken from Darhuber et al. [4])

The above scenario describes, for instance, the condensation of liquid on stripes of fixed wettability. Alternatively, one may consider elongated liquid filaments with a fixed volume and vary the contact angle $\theta_-$ on the hydrophilic stripes. The liquid forms elongated filaments for small contact angles $\theta_-$, which shrink in length as the contact angle on the stripe is increased. At the same time, the lateral contact angle $\theta$ being always larger than $\theta_-$ increases until it reaches 90° at a contact angle $\theta_- = \theta_-^*$ on the stripe. In complete analogy to the filaments on perfectly wettable stripes, we find a discontinuous transition to a bulge at $\theta_- = \theta_-^*$. The contact angle $\theta_-^*$ where the filament decays into a bulge depends on the liquid volume for short liquid filaments but is virtually independent on the liquid volume for $V \gg w^3$. An analytical model shows, in full agreement with the results of numerical minimizations, that the limit of stability for long liquid filaments becomes

$$\theta_-^* = \arccos\left(\frac{\pi}{4}\right) \approx 38.24^\circ \, , \tag{5}$$

for large volumes, $V \gg w^3$ [5]. Note that expression (5) is only valid for hydrophobic contact angles $\theta_+ > 90°$ on the substrate.

The bulge-like morphology is stable for large volumes and large hydrophilic contact angles but decays into the filament morphology at a certain volume $V = \tilde{V}$ and a contact angle $\theta_- = \tilde{\theta}_-$ which, in general differs from $\theta_-^*$. This transition from a bulge to a filament can be understood from the Laplace pressure in the bulge. At a certain volume, the Laplace pressure in the bulge is high enough to extrude a liquid filament from the main body of the bulge onto the stripe. This process continues until the bulge has completely transformed into a filament since the Laplace pressure of a bulge increases for a decreasing volume while the pressure inside the filament is virtually independent on the filament length.

In contrast to the instability of liquid filaments, the angle $\tilde{\theta}_-$ corresponding to the instability of bulges does not become independent in the limit of large volumes. The points of instability of filaments and bulges can be mapped into the morphology diagram shown in Fig. 9, which displays regions of $\theta_-$ and $V$ where the filament F and the bulge B are found as locally or even globally stable morphologies. Filaments F become unstable on the horizontal dashed line in Fig. 9 according to the value given in expression (5). Note that the corresponding lines of instability $(\tilde{V}, \tilde{\theta}_-)$ and $(V^*, \theta_-^*)$, and the transition line, where the interfacial energy of both morphologies are equal, merge into a single point at small volume. The existence of such a bifurcation point of equilibrium configurations is not surprising as the liquid morphologies on the hydrophilic stripe need to become sufficiently large to allow a clear distinction between a bulge and a filament. The morphological transition between both morphologies has been experimentally demonstrated by Klinger et al. [17] by use of the electrowetting effect on a stripe electrode, see Fig. 10.

The analytical model describing the filament instability can be extended to "hydrophobic" contact angles $\theta_+ < 90°$. Again, we consider long liquid filaments of fixed volume and vary the wettability on the stripe. While increasing the contact angle $\theta_-$ on the stripe, the lateral contact angle $\theta$ of the filament at the pinned contact line increases. If, now, the contact angle $\theta_+$ on the substrate is smaller than $90°$, the contact line depins from the boundary of the stripe prior to the above-explained instability for $\theta_+ > 90°$. A cylindrical liquid filament with an unpinned contact lines is

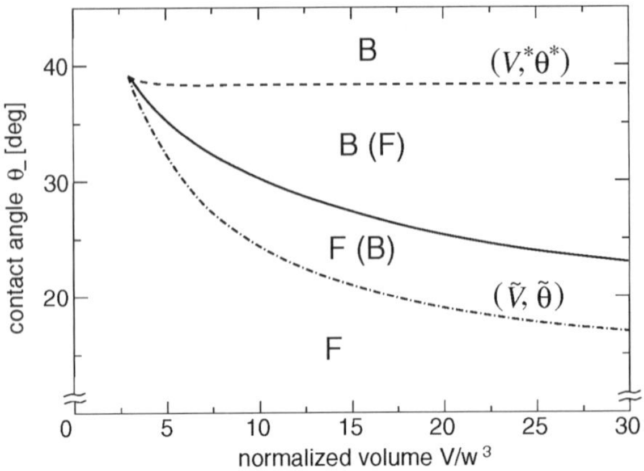


**Fig. 9.** Morphology diagram for liquid wetting a single hydrophilic stripe of finite contact angle $\theta_-$. A contact angle $\theta_+ = 180°$ is assumed on the hydrophobic matrix. The liquid volume $V$ is given in units of the stripe width, $w^3$. *Dashed lines* represent stability boundaries of a particular droplet morphology. Liquid filaments F are stable up to the almost horizontal line $(V^*, \theta_-^*)$, while the localized bulge morphology B decays into filaments for small volumes $\tilde{V}$ on the lower *dashed dotted line* $(\tilde{V}, \tilde{\theta})$. Brackets denote that the corresponding morphology is metastable in this region. The solid line displays the points of transition between both droplet morphologies, i.e., where the interfacial energies are identical

unstable and decays into a bulge [66, 68]. As a consequence, the threshold angle $\theta_-^*$ limiting the range of locally stable filament becomes a function of the contact angle $\theta_+$. According to Brinkmann and Lipowsky [5], the analytical model yields a relation

$$\theta_-^* = \arccos\left(\frac{\theta_+}{2\sin\theta_+} - \frac{\cos\theta_+}{2}\right). \tag{6}$$

The stability limit $\theta_-^*$ given by expression (5) is asymptotically valid for long filaments, i.e., for large volumes $V \gg w^3$.

### 3.3.3. Hydrophilic Rings

In the limit of large radii, $R \gg w$, annular domains can be considered as a special case of hydrophilic stripes. Accordingly, liquid morphologies on these "thin" annular domains should be comparable to the morphologies wetting hydrophilic stripes. The spectrum of liquid morphologies on such a

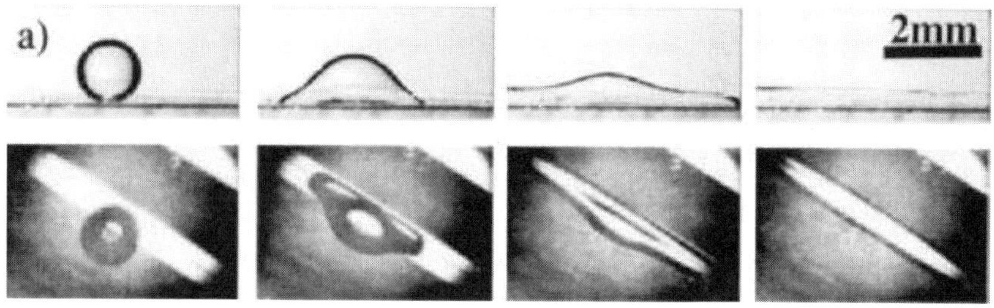


**Fig. 10.** Optical micrographs of electrowetting-induced transition between an elongated liquid filament and a droplet on a stripe electrode. *Top:* side view, *bottom:* top view. The evaporated stripe is visible as bright area in the top view. The electrode contacting the droplet consists of a thin gold wire aligned parallel to the stripe (images taken from Klingner and Mugele [17])

hydrophilic ring was theoretically and experimentally described by Lenz et al. [21] using solder as a wetting liquid. The molten solder was brought into contact with annular domains of highly wettable copper. As expected by the analogy to the stripe, axially symmetric morphologies that resemble an annular, closed liquid filament are found at small liquid volumes, cf. Fig. 11a. The shape of the liquid–vapor interface of this symmetric configuration can be explicitly parameterized by suitable segments of nodoids, a member of the family of Delaunay surfaces [21]. A second equilibrium morphology being analogous to the bulge morphology on the hydrophilic stripe occurs at larger liquid volume, breaking the exact symmetry of the annular domain. Numerical minimizations of the interfacial energy in the annular geometry are shown in Fig. 11b, c.

In a recent work, equilibrium wetting morphologies on ring-shaped hydrophilic domains were studied by Porcheron et al. [22] minimizing a density functional on a lattice including long-ranged interactions. They found a large spectrum of different morphologies, either partially or fully wetting the hydrophilic annular domain, similar to the morphologies found on hydrophilic stripes. At large volumes, the liquid forms a spherical cap that fully covers the hydrophobic part of the substrate enclosed by the hydrophilic ring, see also the shape minimizations shown in Fig. 11d. Bulges whose contact line has detached from the inner and outer boundary occur in the limit of very narrow annuli [22].

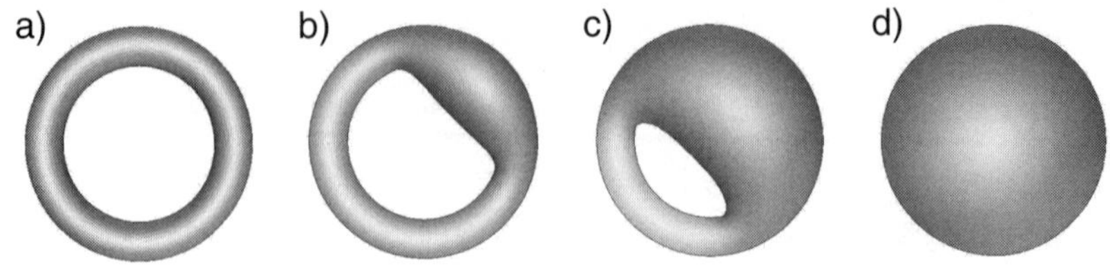


**Fig. 11.** A series of liquid shapes wetting an annular hydrophilic domain for different volumes, calculated with the software surface evolver. The contact angle on the annular domain is $\theta_- = 0°$, which ensures that the hydrophilic ring is always completely covered by liquid. A contact angle of $\theta_+ = 105°$ was assumed on the hydrophobic part of the substrate. The volume is (**a**) $V = 5.5w^3$ (**b**) $V = 7w^3$ (**c**) $V = 17w^3$, and (**d**) $V = 25w^3$

### 3.3.4. Liquid Wetting Several Stripes

The morphology of liquid in contact to more than one hydrophilic stripe of width $w$ and separation $r$ is controlled by the normalized distance $r/w$ in addition to the material contact angles $\theta_-$ and $\theta_+$, as well as the normalized liquid volume $V/w^3$.

In the limit of small separations $r \ll w$ between two hydrophilic stripes, one may regard the two stripes and the enclosed hydrophilic part of the substrate as a single hydrophilic stripe with a reduced effective wettability [5]. Hence, it is not surprising that the droplet morphologies found on two stripes resemble those on a single hydrophilic stripe. At small separation $r \ll w$ and large liquid volume $V \gg w^3$, two types of bridges exist in complete analogy with the droplet configurations on a single hydrophilic stripe: Homogeneous filament bridges with an almost cylindrical liquid-vapor interface as shown in Fig. 12a and bulge-like bridges, cf. Fig. 12b, c. The two types of bulge-like bridges are distinguished by the position of their contact line. The morphology in Fig. 12b exhibits a contact line being pinned at the outer domain boundaries of the pair of stripes; for Fig. 12c, it detaches from the domain boundary and makes an excursion across the hydrophobic matrix. Bulge-like bridges are localized along the stripes, whereas filament-like bridges are rather extended and try to maximize the contact with the hydrophilic stripes. Bulge-like bridges exhibit a lower Laplace pressure compared to the filament-like bridges if we compare both morphologies for the same liquid volume, contact angles, and stripe separation. In the range of large volumes $V \gg w^3$, both types of liquid bridges are symmetric with respect to the symmetry of the hydrophilic stripes.

A distinction between filament-like and bulge-like bridges cannot be made in the range of small liquid volumes. Here, the contact line of the bridging droplet is not necessarily simultaneously pinned to both outer boundaries of the hydrophilic stripes. In this range of volumes, the liquid prefers to form asymmetric shapes because it tries to avoid contact to the hydrophobic stripe enclosed by the two hydrophilic stripes.

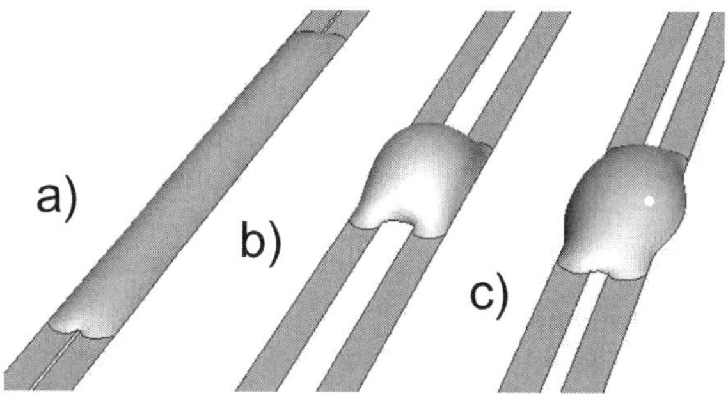


**Fig. 12.** Liquid morphologies wetting two hydrophilic stripes in the regime of large volumes. (**a**) The liquid forms elongated filament-like bridges on two highly wettable stripes at small separation between the stripes. Localized bulge-like droplets (**b**) and (**c**) appear on less hydrophilic stripes for lower wettabilities or for increased separation

Due to large the spectrum of possible equilibrium shapes for a given volume, the final equilibrium morphology of a liquid droplet placed onto a pattern of parallel hydrophilic stripes depends on its history and the dynamics of the wetting process. This issue was addressed theoretically and experimentally by Léopoldès et al. [69]. They studied the spreading of liquid droplets jetted onto a substrate of parallel hydrophilic stripes. Lattice Boltzmann simulation techniques were used to solve the Navier-Stokes equation governing the dynamics of droplet impact and subsequent spreading of the liquid on the substrate. It turned out that the droplet volume relative to the width of a hydrophilic stripe is the most relevant parameter with respect to the final liquid morphology. The droplet shapes are sensitive to the point of impact and the droplet velocity whenever the ratio of the initial droplet radius to the stripe width is around unity. The final droplet morphologies are identical to the equilibrium morphologies proposed by Brinkmann and Lipowsky [5] and by Darhuber et al. [4].

A periodic modulation of the wettability imposes a strong variation to the contact line curvature of the wetting droplet. This effect shown in the numerical minimization in Fig. 12b can be employed to measure the line tension of the three-phase contact line [70]. The line tension can be introduced as an excess of interfacial energies due to the presence of the three-phase contact line [26]. On chemically homogeneous substrates, this line energy is simple proportional to the length of the contact line. Contrary to interfacial tension, the line tension of the three-phase contact line can assume either negative or positive values [72, 73]. If the contact line is strongly curved, the equilibrium contact angle of the wetting liquid deviates from the equilibrium contact angle of a straight contact line given by the equation of Young-Dupré (3), see Swain and Lipowsky [71].

An experimental study of the systematic variations of the contact angle were performed by Pompe et al. [70]. A large droplet of the wetting liquid is placed onto a pattern of parallel stripes that have been created by micro contact printing of self assembly monolayer. The width of the stripes and the periodicity of the pattern is on the order of several 100 nm. The variations in the local wettability impose spacial oscillation onto the position of the contact line, both with the periodicity of the underlying stripe pattern. High curvatures lead to a modification of the equilibrium condition of Young-Dupré of the form

$$\cos\theta = \cos\theta_{\text{o}} - \frac{c\,\tau}{\gamma_{\text{lv}}}, \tag{7}$$

where $\theta_{\text{o}}$ is the contact angle given by the equation of Young-Dupré (3), $c$ is the curvature of the contact line in the substrate plane, and $\tau$ the line energy per length of the contact line. AFM measurements of liquid profiles (shown in Fig. 13) allow to compute the local contact angle $\theta$ and the local curvature $c$ of the contact line. The ratio of line tension $\tau$ to surface tension $\gamma_{\text{lv}}$ determines the slope of $\cos\theta$ plotted against the local curvature $c$ of the contact line. The measured values of the contact line tension $\tau$ range from $10^{-11}$ to $10^{-10}\,\text{Jm}^{-1}$ [28, 70] in accordance with theoretical predictions based on the effective interface model and experimental findings.

In the limit of large volumes, the length scale of the corrugation of the three-phase contact line due to the wettability pattern is small compared to the size of the droplet. In this instance, it is useful to define an apparent contact angle of the overall droplet shape, which averages between the hydrophilic and the hydrophobic contact angle of the stripes and substrate

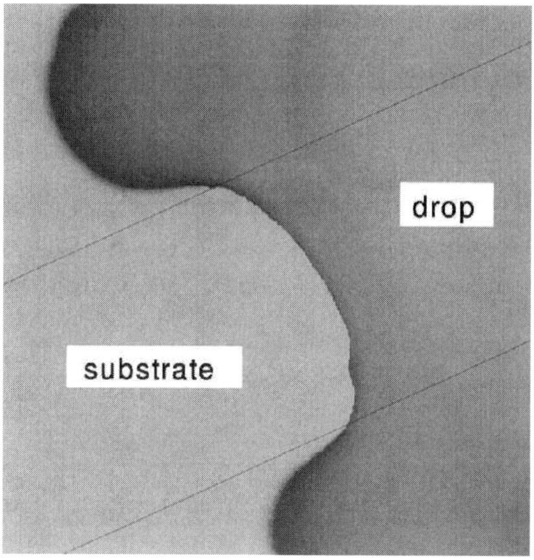


**Fig. 13.** High resolution AFM micrograph showing the contact line of a hexaethylene glycol droplet on a substrate decorated with a pattern of differently wettable stripes. The corrugation of the contact line is used to determine the dependence of the local contact angle on the curvature $c$ of the contact line. The boundaries of the hydrophilic and hydrophobic stripes are indicated as thin lines (Reprinted with permission from Pompe and Herminghaus [70]. Copyright (2000) by the American Physical Society)

matrix, respectively. Experiments [55, 74] and recent simulations [75] of these droplet morphologies confirm the well-known Cassie equation [76], which expresses the cosine of the apparent contact angle as an average of the cosine of the local material contact angles weighted by the fraction of the wetted area.

## 4. Wetting of Topographically Patterned Substrates

Liquid wetting morphologies similar to those found on plane substrates with chemical line patterns emerge on chemically homogeneous substrates with linear surface topographies such as steps, wedges, and grooves. Again, pinning of the three-phase contact line is the key to understand the rich variety of liquid equilibrium morphologies. Here, the appearance of different equilibrium morphologies is governed by the specific geometry of the topography, the equilibrium contact angle, and the liquid volume.

In this section, we will again start with an overview of the experimental techniques to prepare surface topographies and continue with the discussion of the liquid morphologies adsorbed to regular arrays of linear surface geometries. Particular attention is paid to the comparison of triangular [23] and rectangular [8] surface grooves, i.e., geometries for potential building blocks in open microfluidic systems.

Variations of the apparent contact angle can be utilized to trigger transitions between different liquid morphologies. Changes in the apparent contact angle allow, for instance, to toggle between dry and completely filled rectangular grooves [9], similar to the electrowetting-induced spreading of liquid on a stripe electrode shown in Fig. 10 [17].

### *4.1. Substrate Preparation*

Depending on –+ substrate material, surface grooves can be created by various techniques in a large range of lateral dimensions. We summarize the most common techniques used in fundamental research to fabricate topographically structured substrates: Comparably large grooves down to about 100 μm width can be directly micromachined in metal or polymeric material [12, 13, 77, 78]. Depending on the complexity of the structures, micromachining is a reasonably fast technique to produce single topographic substrates. The surface quality of such fabricated substrates, on the other hand, is often not sufficient for wetting experiments on small length scales.

On submicron length scales topographic structures can be imprinted into molten polymeric material, curable prepolymers, or even glass or silicone from a rigid master [79–86]. The surface quality of such fabricated grooves is good and is basically determined by the surface quality of the master. Similarly, topographic structures of the same length scales can be produced in polymeric materials by soft lithography where silicon rubber molds are used instead of rigid ones [56, 57, 87]. Imprinting techniques are advantageous in case larger quantities of the same kind of substrates are needed, but require more technical effort, e.g. to fabricate a suitable master.

In fundamental research, however, typically only a few substrates with the same topography are needed. Here, it is more important to explore many different surface geometries and wettabilities to map out a larger parameter space. Thus, standard photolithography techniques using glass or silicone substrates with subsequent etching procedures are a good choice. Well-defined small-scaled geometries with triangular and trapezoidal cross section can be fabricated using chemical etching procedures [88]. Anisotropic wet etching of <1, 0, 0>-silicon substrates with KOH leads to a wedge angle of $\psi = 54.7°$ given by the crystal lattice structure. Another advantage using glass or silicon substrates is that the wettability can be tailored in a wide range by self assembly monolayer as described in Sect. 3.1.

Grooves with rectangular cross section can be fabricated by isotropic reactive ion etching subsequent to the photo lithography. Using this technique, even deep grooves with large aspect ratio can be fabricated.

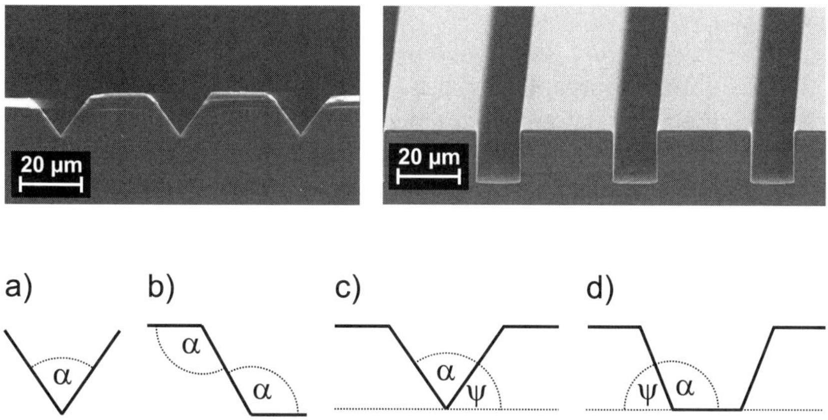


**Fig. 14.** *Top*: scanning electron micrographs of triangular and rectangular grooves on silicon-wafer fabricated by anisotropic wet etching and isotropic reactive ion etching, respectively. *Bottom*: Sketch of various cross sections (from left to right): infinite wedge (**a**), topographic step (**b**), finite wedge (**c**) and groove geometry (**d**) (top right image courtesy of J.-C. Baret, Philips Research Laboratories, Eindhoven)

Liquid equilibrium morphologies in surface grooves can be created using the same procedures as for chemically patterned substrates: dip coating a sample, dewetting liquid from a sample, or condensing liquid from the vapor phase. To analyze and distinguish the large variety of liquid shapes on topographic substrates, a three-dimensional analysis of the emerging wetting morphologies is usually required. Liquids that can be solidified are highly favorable because they allow to use imaging techniques, which might hardly be possible otherwise. The contour of a groove with small dimensions and the profile of the liquid–vapor interface can, e.g., be characterized by AFM optical or mechanical profiler, or scanning electron microscopy (SEM), as can be seen in Fig. 14.

## 4.2. Basic Topographies: Infinite Wedge and Step

### 4.2.1. Infinite Wedge

The most fundamental linear surface topography is an infinite wedge. Because of its invariance with respect to rescaling in all dimensions, a wedge has no intrinsic length scale and is solely characterized by a single angle. This angle can be the opening angle $\alpha$ or, as used in the sequel, the wedge angle $\psi = 90° - \alpha/2$, cf. Fig. 14. Thus, the contact angle $\theta$ and the wedge angle $\psi$ are the only control parameters for the appearance of liquid equilibrium morphologies in infinite wedges. In the particular case of Fig. 15, the triangular groove can be regarded as an infinite wedge since the degree of filling is sufficiently small and the liquid wets only the bottom part of the wedge.

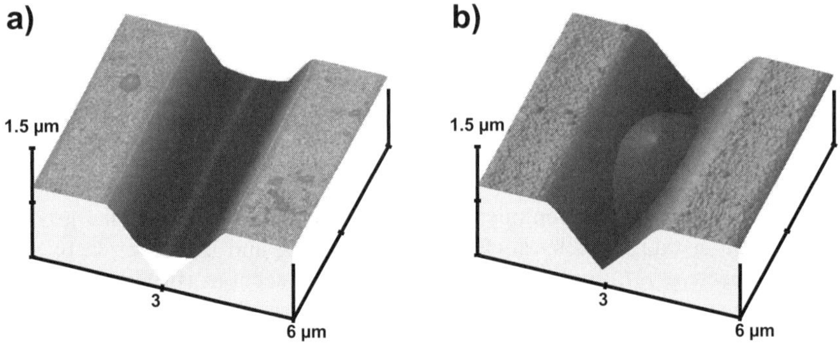


**Fig. 15.** Atomic force micrographs of polystyrene vapor deposited onto grooved substrates with a wedge angle of $\psi = 54.7°$. (**a**) A liquid wedge with negative mean curvature, $\theta \approx 33°$ and (**b**) droplet wetting morphology with positive mean curvature, $\theta \approx 64°$, cf. [23]

As shown in the AFM micrographs of Fig. 15, there are only two equilibrium wetting morphologies in infinite wedges: spherical droplets with positive mean curvature for $\theta > \psi$ and cylindrical liquid wedges with negative curvature extending along the entire length of the wedge for $\theta < \psi$. Increasing the contact angle $\theta$ to the filling angle $\theta = \psi$, the liquid–vapor interface of a liquid wedge becomes marginally stable with respect to a whole spectrum of long wavelength, perturbations. This peculiar behavior can be understood from the scaling invariance of the liquid configuration in a 2D- cross section of the wedge. Hence, even thermally excited shape fluctuations may become important in a statistical description of the filling transition in infinite wedges [89, 90].

As the contact angle $\theta$ approaches the wedge angle $\psi$ from above, the spherical droplets elongate while their tips become more pointed. At the same time, the Laplace pressure tends to zero. Eventually, in the limit $\theta = \psi$, the liquid extends to the entire length of the wedge and the liquid–vapor interface attains a plane configuration.

### 4.2.2. Tip Shape

It has been first noted by Shuttleworth and Bailey [91] that the tip of a liquid filament in wedges becomes more and more pointed as the contact angle $\theta$ approaches the critical filling angle $\psi$ from above, see Fig. 16. They assumed that the liquid–vapor interface at the tip of a filament becomes asymptotically plane. To characterize the geometry of the liquid tip, they introduced the tilt angle $\beta$ between the liquid–vapor interface and the bottom of the groove. By elementary geometry they derived a relation

$$\cos\beta \cos\psi = \cos\theta, \tag{8}$$

between the contact angle $\theta$, the tilt angle $\beta$, and the wedge angle $\psi$, which holds in the range $\theta > \psi$.

More than 20 years later, Concus and Finn showed that any minimal surface in an infinite wedge, i.e., a surface of zero mean curvature that forms a constant contact angle on the wedge walls, has to be a plane [92]. Since the shape of the wedge bottom is scaling invariant, one can argue that by "zooming in", the tip of a surface of finite constant mean curvature can be asymptotically regarded as a minimal surface on small length scales, i.e., as a surface with zero mean curvature. This implies that the terminal part of the liquid–vapor interface of any equilibrium morphology in the wedge will become asymptotically plane and that, in particular, the predictions of Shuttleworth and Bailey should hold. This is confirmed by experimental measurements of the tilt angle $\beta$ of the tip by AFM as a function of the contact angle $\theta$, cf. Fig. 16. As a consequence, the transition from filaments to liquid wedges does not happen in an abrupt manner. Instead, one observes a gradual elongation of the droplets involving pointed tips and a final divergence of the total droplet extension while lowering the contact angle $\theta$ down to a value below the filling angle $\theta = \psi$ of a wedge.

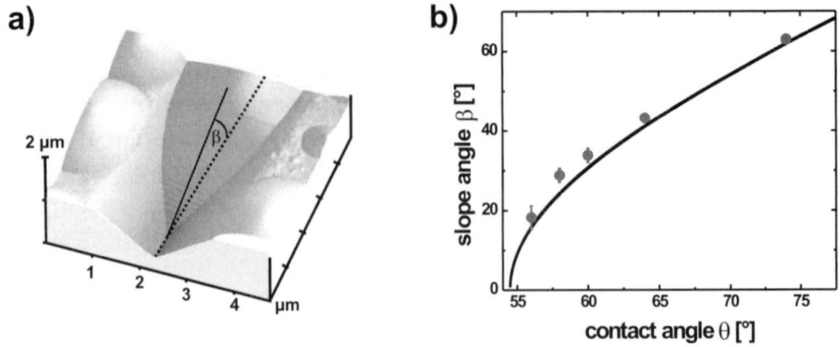


**Fig. 16.** (**a**) AFM micrograph showing the terminal part of the meniscus of a polystyrene droplet in a triangular groove. The tilt angle $\beta$ of the terminal meniscus is sketched. (**b**) Tilt angle $\beta$ for different contact angles $\theta$. Experimental data points were obtained from polystyrene droplets (PS) [23]. The solid line shows the expected relation between $\beta$ and $\theta$ for a wedge angle fixed to $\psi \simeq 54.7^{\circ}$ according to (8), see also [23]

### 4.2.3. Topographic Step

Starting from an infinite wedge, one can increase the complexity of the topography by adding corners and edges as sketched in Fig. 14. Consequently, new length scales are introduced in addition to the length scale $\propto V^{1/3}$ given by the volume of the liquid. For a topographic step of height $h$, one can expect that, besides the step angle $\alpha$ and the contact angle $\theta$, the rescaled volume $V/h^3$ is a relevant control parameter of liquid morphologies [7]. Provided that either the droplet volume is sufficiently low or the contact angle is smaller than the step angle, the morphologies are identical to those of the infinite wedge. At larger volumes and contact angles larger than the step angle $\theta > \alpha$ the contact line becomes partially pinned to the upper edge of the step. In this regime, elongated liquid structures with an almost homogeneous cross section ("liquid cigar") or localized droplets sitting on the lower semiplane of the sample ("step blobs") can be found, see Fig. 17. A common feature of both morphologies is that a part of the contact line is pinned to the upper edge of the step. For the particular value $\alpha = 90°$, the elongated liquid filament of finite length can be found as a metastable configuration in a range of contact angles $45^{\circ} < \theta \lesssim 54.5^{\circ}$, see Brinkmann and Blossey [6].

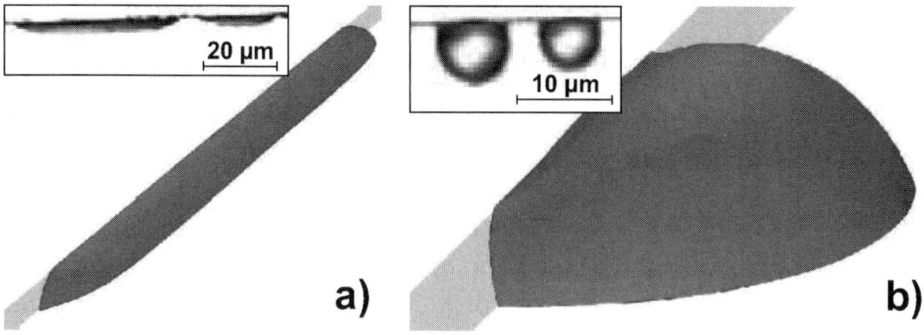


**Fig. 17.** (**a**) A liquid filament, or liquid "cigar", and a step-blob (**b**) at liquid volumes $V/h^3 = 25$ and step angle $\alpha = 110°$. The contact angle on the substrate is $\theta = 40°$ for the cigar and $\theta = 70°$ for the step-blob. *Insets:* optical micrographs of polystyrene wetting a rectangular step with (**a**) height $h \approx 780\,nm$ and contact angle $\theta \approx 41°$ (**b**) $h \approx 700\,nm$, $\theta \approx 76°$

Going beyond the geometry of an infinite wedge or step, one may consider linear substrate topographies involving more than two kinks. Other simple geometries are symmetric wedges of finite width (three kinks) as we will consider in the following.

### 4.3. Triangular Grooves

The geometry of a symmetric groove of triangular cross section, sketched in Fig. 14, is determined by a wedge angle $\psi$ and the distance $w$ between the two edges limiting the groove walls to the plane part of the substrate. As mentioned before, the liquid morphologies do not experience the finite extensions of a triangular groove at small liquid volumes $V \ll w^3$ or contact angles $\theta < \psi$. For larger volumes and for contact angles larger than the wedge angle, the finite width of the groove comes into play. The contact line of the wetting liquid may become partially pinned to the edges of the groove, as illustrated in Fig. 18. The spectrum of morphologies consists mainly of localized droplets D, liquid wedges W, and liquid filaments F. Droplets D of small volume are fully confined to the groove (as shown in Fig. 15.b) while larger droplets wet the ridges between two neighboring grooves. The contact line of a liquid filament F is pinned to the edges of the triangular grooves. In contrast to liquid wedges W, filaments extend over a finite length of the groove and exhibit an almost homogeneous cylindrical liquid–vapor interface with a positive mean curvature. For a fixed wedge angle, the magnitude of the mean curvature of sufficiently long and mechanically equilibrated liquid filaments depends solely on the contact angle and is independent on the liquid volume. Hence, we can approximate the complete shape of a liquid filament by a segment of a cylinder. In this way we obtain a simple analytical model of the liquid configuration, which allows to study the stability of the liquid–vapor interface. A detailed, three-dimensional approximation of the liquid–vapor interface can be reached in numerical minimizations using finite elements, see Sect. 2.3.

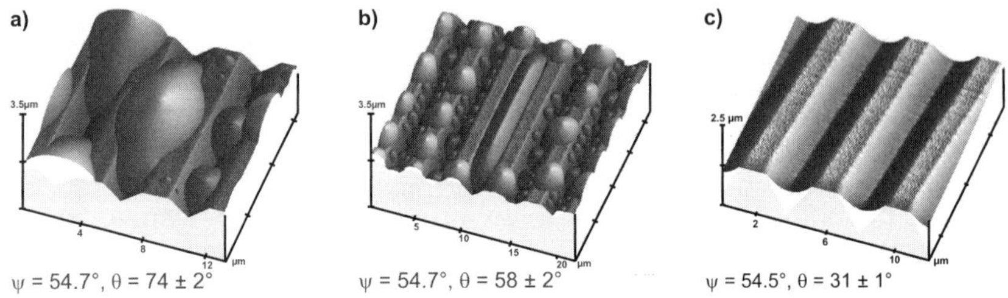


**Fig. 18.** AFM micrographs of vapor deposited polystyrene structures in grooves with triangular cross section with $\psi \simeq 54.7^\circ$. Localized droplets D form at high contact angles $\theta$, while extended filaments F with positive Laplace pressure are found for values of $\theta$ close above the wedge angle $\psi$. Liquid wedges W with a concave meniscus occur for contact angles $\theta < \psi$, cf. [23]

Basically, there are four independent control parameters that govern the appearance of different liquid morphologies in triangular grooves: the liquid volume $V$, the contact angle $\theta$, the wedge angle $\psi$, and the width $r$ of the ridges separating two neighboring grooves. It turned out that the contact angle $\theta$ and the wedge angle $\psi$ determine the equilibrium shape of the liquid at high volumes $V \gg w^3$.[1] Variations in $r$ influence solely overspilling droplets D in neighboring grooves, which are prone to coalesce and to the formation of larger localized droplet structures. The emergence of filaments F and wedges W, however, is independent on $r$ due to the full confinement of liquid in the grooves. Hence, the appearance of the different

[1]An additional condition $V < w^2 l$ on the volume ensures that the channel of length $l$ and width $w$ is not 'overfilled' by liquid.

droplet morphologies depending on wettability and groove geometry is best illustrated in the form of a morphology diagram in terms of the wedge angle $\psi$ and the contact angle $\theta$. The liquid volume of the droplets as a third control parameter has virtually no influence on the morphologies provided $V \gg w^3$.

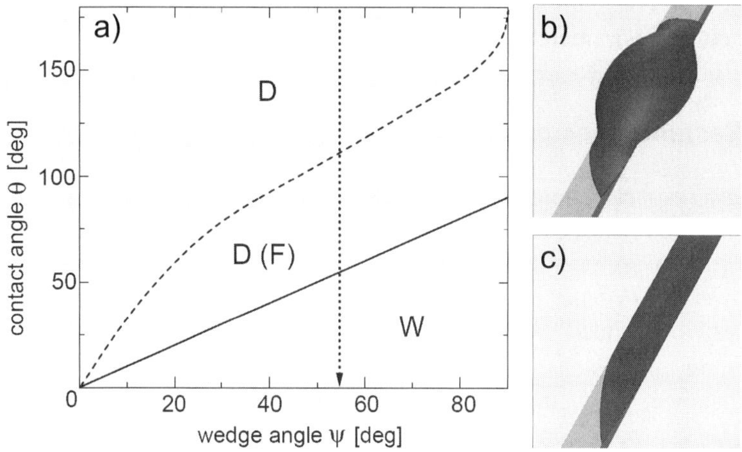


**Fig. 19.** Morphology diagram of liquid morphologies in triangular grooves in the asymptotic limit of large volumes $V \gg w^3$. The regions where localized droplets and liquid wedges represent the global minimum of the interfacial energy are denoted by D and W, respectively, while (F) indicates the region of metastable elongated filaments, see also [23]. The arrow highlights the special wedge angle $\psi \simeq 54.7^\circ$

A morphology diagram displays regions of relevant control parameters where certain equilibrium droplet shapes exist as locally or globally stable shapes. In the considered regime of large volumes, all liquid filaments F are metastable with respect to droplets D. Metastable configurations are given brackets. The solid line in Fig. 19 separates the regions where droplets D or liquid wedges W are global minima of the interfacial energy. The transition line is simply given by $\theta = \psi$. The dashed line bounds the region of control parameter where filaments F are found as metastable minima to large contact angles.

To illustrate the hysteretic behavior, let us make a gedanken experiment: we start with a lemon-shaped droplet morphology in regime D and decrease the contact angle. The droplet will completely disappear into the groove forming a liquid wedge W with negative Laplace pressure, if the contact angle is lowered below the solid line in Fig. 19, $\theta < \psi$. But increasing the contact angle back to its original value, the wedge will first undergo a continuous transition to an elongated filament F at $\theta = \psi$ followed by a discontinuous transition to a droplet morphology D when crossing the dashed line in Fig. 19. In regime D(F), liquid filaments and droplets can coexist on a sample, as their appearance depend on the history of the control parameter, liquid, volume or contact angle.

### 4.4. Rectangular Grooves

In many respects, the equilibrium shapes of liquid droplets found in rectangular grooves are comparable to the morphologies in triangular grooves, but differ in some essential details. The slope of the side walls and a second bottom corner is responsible for the main differences. Rectangular grooves (four kinks) can be regarded as an antisymmetric combination of two topographic steps, both with a step angle $\psi = 90^{\circ}$. For the wall to wall distance, $w$, being large compared to the depth of the groove, $h$, a rectangular groove can be regarded as two independent steps being wetted by two independent morphologies. As the distance $w$ between the opposed topographic steps is decreased such that $w \approx h$ it becomes more and more likely that liquid morphologies with a single meniscus being pinned to both groove edges appear. The combination of liquid structures wetting the entire cross section of the grooves and droplets wetting only a single corner creates a large spectrum of possible equilibrium morphologies in rectangular grooves as we will demonstrate in the following.

It is useful to define a dimensionless aspect ratio $X = w/h$ of the groove depth $h$ to the groove width $w$ to characterize the geometry of a rectangular groove. The most prominent liquid morphologies in rectangular grooves are shown in Fig. 20. Two dominant shapes appear in a wide range of contact angles $\theta$ and aspect ratios $X$: localized, droplet-like shapes D, and elongated liquid filaments F. But in contrast to triangular grooves, the liquid–vapor interface of the liquid filament can be bent either toward the liquid, F–, or toward the vapor phase, F+. The length of a liquid filament depends on the amount of adsorbed liquid while its cross section is independent on the volume, provided that the terminal part of the liquid–vapor interface does not experience the finite length of the groove.

A second class of liquid morphologies appears in combination with the localized droplets and elongated filaments As shown in Fig. 20, the liquid meniscus of the respective morphology may continue as fine liquid wedges W wetting the corners of the groove to either side of the droplet or filament [8, 23]. The liquid wedges may either be pinned to the upper edges of the groove pW or located in the lower corner of the grooves cW.

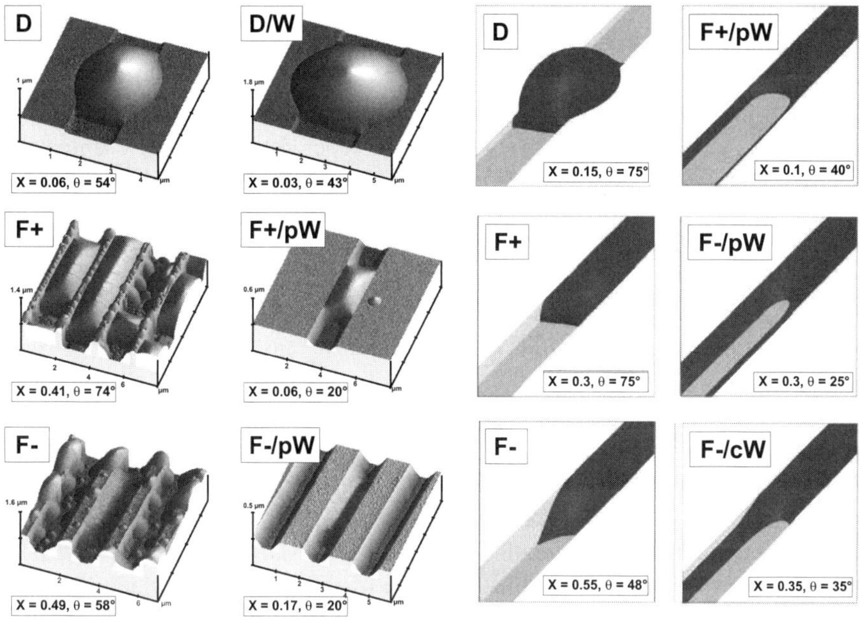


**Fig. 20.** AFM images (left) and numerically obtained shapes (right) of liquid structures in grooves with rectangular cross section. For contact angles $\theta > 45^{\circ}$, one observes overspilling droplets D that spread onto the ridges, extended filaments F+ with positive Laplace pressure, and extended filaments F– with negative Laplace pressure. For $\theta < 45^{\circ}$, one finds droplets D and filaments F that are connected to thin liquid wedges W in the groove corners. The Laplace pressure of filaments and wedges might be positive (+) or negative (–). For details, see [8, 23]

For large liquid volumes, $V \gg w^3$, the liquid shapes are essentially determined by the control parameter contact angle $\theta$ and aspect ratio $X$, similar to the triangular grooves. Here, it is particularly useful to present the various droplets shapes in the morphology diagram shown in Fig. 21 in terms of $\theta$ and $X$ as it provides a convenient way to compare the morphologies found in experiments and theory.

All transition lines between different mi were obtained in an analytical model of the liquid morphologies. This model allows to predict the respective regions of contact angles and aspect ratios where globally or locally stable morphologies are found. The morphology diagram Fig. 21 contains the wetting morphologies we identified in Fig. 20: Droplets D, liquid filaments F+ with positive Laplace pressure, and filaments F– with negative Laplace pressure. All these wetting morphologies can coexist with liquid wedges W for contact angles $\theta < 45°$. For small aspect ratios $X$ or large contact angles $\theta$, the liquid forms localized droplet morphologies D. For smaller contact angle or increasing aspect ratio, one may enter the region where filaments with positive Laplace pressure F+ exist as metastable configurations in addition to globally stable droplets D. The boundary of this region is shown as a dashed line in Fig. 21. Decreasing the contact angle even further or increasing the aspect ratio, the Laplace pressure of the filaments lowers and changes its sign when crossing the solid line in the morphology diagram, which denotes filaments of zero mean curvature.

The filament morphology F+ with a positive Laplace pressure is metastable with respect to localized droplets D. Hence, for large liquid volumes, the dashed line represents the largest contact angle for a given aspect ratio for which filaments with positive Laplace pressure can exist as locally stable shapes. Filaments in the regime F– of negative Laplace pressure are always globally stable. Because of the negative Laplace pressure in these filaments, liquid can be spontaneously drawn into rectangular grooves from a large feeding droplet or a liquid reservoir at zero pressure.

Similar to triangular grooves, we find a hysteretic behavior when switching the wettability, e.g., by electrowetting. A large droplet spreads into the grooves in the form of liquid filaments as long as $X > X_0$ with $X_0 = (\sqrt{2} - 1)/2 \simeq 0\ 21$ [8] when crossing the solid line between the regimes D(F+) and F–. Liquid wedges will grow out of the filament and coexist with them when lowering the contact angle further below $\theta = 45°$, provided the length of the groove is finite and the liquid volume is large enough. In case the grooves are sufficiently long the wedges will extract the entire liquid from the filament forming just two liquid wedges. The upper three-phase contact line of these liquid wedges will move closer to the upper edge of the groove and as the contact angle is lowered even further. Entering the regime F–/pW, the upper three-phase contact line of the liquid wedge becomes pinned at the edge of the groove. In general, the coexistence of liquid wedges and filaments depends on the length of the grooves and the amount of liquid in the groove.

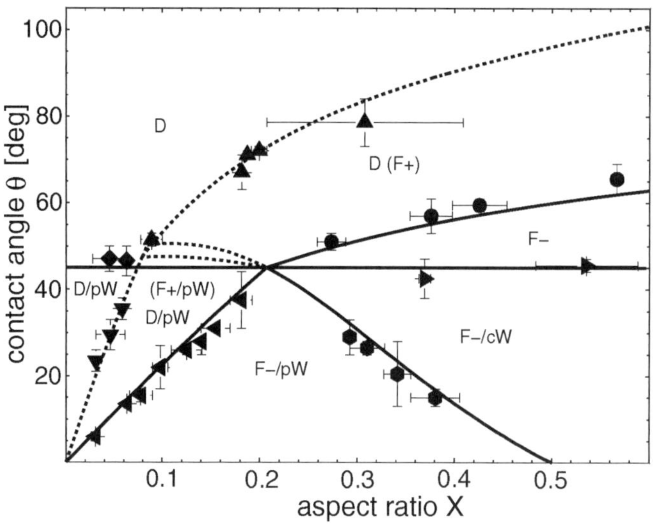


**Fig. 21.** Morphology diagram of liquid confined to rectangular grooves in the asymptotic limit of large volumes, $V \gg w^3$. The analytical results are in excellent agreement with numerical (not shown) and experimental results (full symbols), cf. Seemann et al. and Khare et al. [8, 23]. The different full symbols denote the respective transition between two regimes. All 185 experimentally determined shapes are fully consistent with the theoretical classification

Most of the morphology transitions are reversible and occur at the same contact angle. But increasing the contact angle above the line indicating filaments of zero mean curvature, the liquid forms metastable filaments with positive Laplace pressure rather than decaying into droplets. These filaments will become unstable when the contact angle enters the regime D above the dashed line. Analogous to the morphologies in triangular grooves, the emergence of liquid morphologies in the regime D(F+) depends on the history of each individual morphology and liquid filaments F+ with positive mean curvature and droplets D may coexist on the same sample.

In sufficiently shallow grooves with $X < X_0$, one may first observe the formation of small pinned liquid wedges growing out of a droplet D while decreasing the contact angle into the regime D/pW. Complete groove filling sets in if we enter the regime F–/pW. Here, the hysteretic behavior during an increase of the contact angle is much more complex and we refer to Seemann et al. [8] for details.

### 4.5. Switching Equilibrium Morphologies

Liquid filaments advance into triangular and rectangular grooves from droplets or large reservoirs upon decreasing the contact angle. The filling angle is given by the transition line between morphology D (F+) and W in case of triangular grooves, respectively the transition line between D and F– in case of rectangular grooves as shown in the morphology diagrams Figs. 19 and 21.

Hence, combining the topographically structured substrates with a technique, which allows to vary the apparent contact angle, induces morphological transitions which possibly allow for an on-demand imbibition of liquid into grooves, similar to what has been reported for substrates with wettable stripes [17]. But in contrast to planar substrates with wettability patterns, grooves can imbibe liquid even from a reservoir with zero Laplace pressure. Moreover, using grooved substrates may open the possibility to drain liquid completely from grooves by reversibly increasing the wettability. As shown by the morphology diagrams, filling and draining can be done at moderate contact angles. In case of chemically structured substrates, liquid filaments can only be fed by a reservoir with positive Laplace pressure, provided the contact angle of the wettable area is larger that, 0°. But in this regime, up to now, there is no technique available to switch the wettability reversibly. For surface grooves, on the other hand, electrowetting [24] is a versatile and easy-to-use tool to vary the apparent contact angle reversibly in the desired range [9].

The series of optical micrographs in Fig. 22 illustrate typical filling transition experiments varying the wettability by electrowetting on grooves with rectangular cross section [9] and on triangular grooves [23], respectively. The experiment shows a clear threshold behavior for the groove filling, in agreement with the morphology diagrams of Figs. 19 and 21. Because the liquid filaments are always in contact with a large reservoir or droplet, respectively, we do not find a hysteretic behavior. In contrast to the morphology transition of the entire liquid volume, the threshold contact angle for groove filling and draining is identical to the contact angle on the line of zero Laplace pressure.

If the groove filling was solely governed by the relative capillary pressure between the feeding droplet and the filaments, one would expect the filaments to grow indefinitely inside the grooves when the filling threshold is exceeded. However, one readily sees from Fig. 22 that the filaments extend to a finite length, which increases with applied voltage. This finite length is determined by the applied AC-voltage and the finite

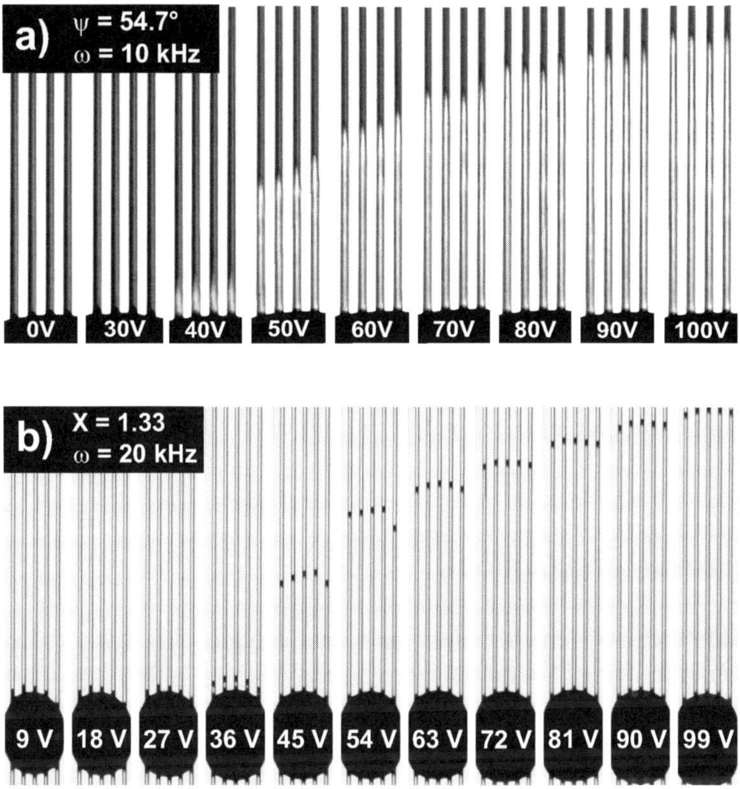


**Fig. 22.** Top view of a droplet advancing into grooves with (**a**) triangular and (**b**) rectangular cross section for different (equidistant) voltages. The apparent contact angle is lowered using the electrowetting effect applying an AC-voltage between an aqueous solution and a silicone substrate, where $\omega$ is the frequency of the applied voltage. Liquid filaments appear bright in the triangular grooves, whereas the tips of a liquid filament appear black in the optical reflection micrographs of rectangular grooves. The length of liquid filaments are in equilibrium at the respective voltage applied [9, 23] (bottom image reprinted with permission from Baret et al. [9]. Copyright (2005) American Chemical Society).

conductivity of the liquid, causing the voltage to drop along the liquid filament [9]. Hence, the apparent contact angle varies along the liquid filament. In equilibrium, the apparent contact angle at the tip of the filament equals the threshold contact angle for the groove filling, as can be determined from the morphology diagrams (Figs. 19 and 21). In case of rectangular grooves, the liquid can be drained reversibly from the grooves whereas the liquid filaments in triangular grooves decay into isolated droplets via a one-dimensional Rayleigh-Plateau type dewetting process rather than receding back to its feeding droplet [23, 93]. To understand this phenomenon, we recall the idea of Shuttleworth and Bailey [91], as discussed in Sect. 4.2. When approaching the filling angle from above, the tip of the liquid being pulled out of the feeding reservoir becomes more and more pointed while the slope angle of the liquid–vapor interface close to the tip decreases. A close inspection reveals the reason for the instability in triangular grooves during retraction: the contact line of the liquid protruding into the groove is not pinned at the upper edges of the grooves. Hence, during a sudden increase of the apparent contact angle, we generate a liquid filament with positive mean curvature whose contact lines are not pinned. These filaments are dynamically unstable and do not retract into the feeding droplet but decay into a regularly spaced train of isolated droplets.

## 5. Summary and Outlook

In a wide range of length scales, interfacial energies determine the equilibrium morphologies of nonvolatile liquids wetting substrates decorated with patterns of differently wettable domains. Provided the wettability on these surface domains is sufficiently high, it will be possible to confine the wetting liquid to various shapes if the volume falls into an appropriate range. Here, pinning of the three-phase contact line plays a central role for the mechanical stability of the liquid–vapor interface of the wetting liquid. At large volumes, the liquid–vapor interface of the wetting phase may undergo transitions to configurations characterized by a heterogeneous distribution of liquid on the wettable parts of the surface, as we have illustrated in various examples. These interfacial tension-driven instabilities could pose a serious problem in small scale printing techniques where an even distribution of liquid on the motives is highly desirable. Using the electrowetting effect to tune the wettability, it has been demonstrated for the particular geometry of a wettable stripe that the shape of a liquid droplet can be switched between an elongated and an almost spherical shape. An application of this morphological transition in microfluidics seems to be difficult because the contact angle of the transition to the elongated filament occurs is fairly small for large liquid volumes. The electrowetting effect, on the other hand, is restricted to larger contact angles by saturation effect and contact line instabilities.

Using surface topographies instead of chemically patterned, plane substrates offers a major advantage: The Laplace pressure of liquid morphologies in linear surface grooves of triangular and rectangular cross section can assume negative values in mechanical equilibrium. The sign of the pressure depends, besides on the groove geometry, on the wettability of the surface. An active control of the wettability opens the possibility to draw liquid into the grooves from a reservoir at zero Laplace pressure, e.g., from a large droplet. The contact angle required for imbibition into surface grooves is finite in contrast to plane wetting patterns even for large liquid volumes. This contact angle is determined by the geometry of the groove and becomes large for high aspect ratio grooves. This, in return, allows to adjust the filling angle to the range of contact angles accessible by electrowetting.

## Acknowledgment

The authors gratefully acknowledge stimulating discussions with Stephan Herminghaus and Reinhard Lipowsky. Furthermore, we thank Jean-Christophe Baret for fruitful collaboration. Financial support from the MPG and the DFG priority program SPP 1164, grant SE1118 is gratefully acknowledged.

## References

1. Lipowsky R (2001) Structured surfaces and morphological wetting transitions. Interface Sci 9:105–115
2. Herminghaus S, Pompe T, Fery A (2000) Scanning force microscopy investigation of liquid structures and its application to fundamental wetting research. J Adhesion Sci Technol 14:1767–1782
3. Gau H, Herminghaus S, Lenz P, Lipowsky R (1999) Liquid morphologies on structured surfaces: From microchannels to microchips. Science 283:46–49
4. Darhuber AA, Troian SM, Miller SM, Wagner S (2000) Morphology of liquid microstructures on chemically patterned surfaces. J Appl Phys 87:7768–7775
5. Brinkmann M, Lipowsky R (2002) Wetting morphologies on substrates with striped surface domains. J Appl Phys 92:4296–4306
6. Checco A, Gang O, Ocko BM (2006) Liquid nanostripes. Phys Rev Lett 96:056104
7. Brinkmann M, Blossey R (2004) Blobs, channels and "cigars": Morphologies of liquids at a step. Eur Phys J E 14:79–89

8. Seemann R, Brinkmann M, Lange FF, Kramer EJ, Lipowsky R (2005) Wetting morphologies at microstructured surfaces. Proc Natl Acad Sci USA 102:1848–1852
9. Baret JC, Decré M , Herminghaus S, Seemann R (2005) Electroactuation of fluid using topographical wetting transitions. Langmuir 21:12218–12221
10. Darhuber AA, Troian SM, Davis JM, Miller SM, Wagner S (2000) Selective dip-coating of chemically micropatterned surfaces. J Appl Phys 88:5119–5126
11. Darhuber AA, Troian SM (2005) Principles of microfluidic actuation by modulation of surface stresses. Annu Rev Fluid Mech 37:425–455
12. Mann JA, Romero L, Rye RR, Yost FG (1995) Flow of simple liquids down narrow-v grooves. Phys Rev E 52:3967–3972
13. Rye RR, Yost FG, Mann JA (1996) Wetting kinetics in surface capillary grooves. Langmuir 12:4625–4627
14. Yost FG, Rye RR, Mann JA (1997) Solder wetting kinetics in narrow V-grooves. Acta mater 45:5337–5345
15. Blossey R, Bosio A (2002) Contact line deposits on cDNA microarrays: A “twin-spot effect.” Langmuir 18:2952–2954
16. Herminghaus S, Gau H, Mönch W (1999) Artificial liquid microstructures. Adv Mater 11:1393–1395
17. Klingner A, Mugele F (2004) Electrowetting-induced morphological transitions of fluid microstructures. J Appl Phys 95:2918–2920
18. Lenz P, Lipowsky R (1998) Morphological transitions of wetting layers on structured surfaces. Phys Rev Lett 80:1920–1923
19. Bechinger C, Muffler H, Schäfle C, Sundberg O, Leiderer P (2000) Submicron metal oxide structures by a sol–gel process on patterned substrates. Thin Solid Films 366:135–138
20. Lenz P, Bechinger C, Schäfle C, Leiderer P, Lipowsky R (2001) Perforated wetting layers from periodic patterns of lyophobic surface domains. Langmuir 17:7814–7822
21. Lenz P, Fenzl W, Lipowsky R (2001) Wetting of ring-shaped surface domains. Europhys Lett 53:618–624
22. Porcheron F, Monson PA, Schoen M (2006) Wetting of rings on a nano-patterned surface: A lattice model study. Phys Rev E 73:041603
23. Khare K, Law BM, Herminghaus S, Brinkmann M, Seemann R (2006) Liquid morphologies on linear surface topographies. submitted.
24. Mugele F, Baret JC (2005) Electrowetting: From basics to applications. J Phys Condens Matter 17:R705–R774
25. Hu DL, Chan B, Bush JWM (2003) The hydrodynamics of water strider locomotion. Nature 424:663–666
26. Rowlinson JS, Widom B (1982) Molecular Theory of Capillarity. Clarendon, Oxford
27. Lipowsky R, Lenz P, Swain PS (2000) Wetting and dewetting of structured and imprinted surfaces. Colloids Surf A 161:3–22
28. Seemann R, Jacobs K, Blossey R (2001) Polystyrene nanodroplets. J Phys Condens Matter 13:4915–4923

29. Checco A, Cai YG, Gang O, Ocko BM (2006) High resolution non-contact AFM imaging of liquids condensed onto chemically nanopatterned surfaces. Ultramicroscopy 106:703–708
30. Do Carmo MP (1976) Differential Geometry of Curves and Surfaces. Prentice Hall, New Jersey
31. Brakke KA (1992) The surface evolver. Exp Math 1:141–165; (1996) The surface evolver and the stability of liquid surfaces. Philos Trans R Soc Lond A 354:2143–2157
32. Sagiv J (1980) Organized monolayers by adsorption. I. Formation and structure of oleophobic mixed monolayers on solid surfaces. J Am Chem Soc 102:92–98
33. Peters RD, Yang XM, Kim TK, Sohn BH, Nealey PF (2000) Using self-assembled monolayers exposed to X-rays to control the wetting behavior of thin films of diblock copolymers. Langmuir 16:4625–4631
34. Peters RD, Yang XM, Kim TK, Nealey PF (2000) Wetting behavior of block copolymers on self assembled films of alkylchlorosiloxanes: Effect of grafting density. Langmuir 16:9620–9626
35. Kim TK, Yang XM, Peters RD, Sohn BH, Nealey PF (2000) Chemical modification of self-assembled monolayers by exposure to soft X-rays in air. J Phys Chem B 104:7403–7410
36. Nissen J, Jacobs K, Radler JO (2001) Chemical modification of self-assembled monolayers by exposure to soft X-rays in air. Phys Rev Lett 86:1904–1907
37. Hoeppener S, Maoz R, Sagiv J (2006) Contact electrochemical replication of electrochemically printed monolayer patterns. Adv Matter 18:1286–1290
38. Bard AJ, Denuault G, Lee C, Mandler D, Wipf DO (1990) Scanning electrochemical microscopy – A new technique for the characterization and modification of surfaces. Acc Chem Res 23:357–363
39. Garcia R, Calleja M, Perez-Murano F (1998) Local oxidation of silicon surfaces by dynamic force microscopy: Nanofabrication and water bridge formation. Appl Phys Lett 72:2295–2297
40. Dai HJ, Franklin N, Han J (1998) Exploiting the properties of carbon nanotubes for nanolithography. Appl Phys Lett 73:1508–1510
41. Maoz R, Cohen SR, Sagiv J (1999) Nanoelectrochemical patterning of monolayer surfaces: Toward spatially defined self-assembly of nanostructures. Adv Mater 11:55–61
42. Geissler M, Xia YN (2004) Patterning: Principles and some new developments. Adv Mat 16:1249–1269
43. Gau H, Herminghaus S (2000) Ripening of ordered breath figures. Phys Rev Lett 84:4156–4159
44. Denkov ND, Velev OD, Kralchevsky PA, Ivanov IB, Yoshimura H, Nagayama K (1992) Mechanism of formation of 2-dimensional crystals from latex particles on substrates. Langmuir 8:3183–3190
45. Denkov ND, Velev OD, Kralchevsky PA, Ivanov IB, Yoshimura H, Nagayama K (1993) 2-dimensional crystallization. Nature 361:26–26

46. Burmeister F, Schäfle C, Matthes T, Bohmisch M, Boneberg J, Leiderer P (1997) Colloid monolayers as versatile lithographic masks. Langmuir 13:2983–2987
47. Widawski G, Rawiso M, Francois B (1994) Self-organized honeycomb morphology of star-polymer polystyrene films. Nature 369:387–389
48. Heier J, Kramer EJ, Walheim S, Krausch G (1997) Thin diblock copolymer films on chemically heterogeneous surfaces. Macromolecules 30:6610–6614
49. Wang R, Parikh AN, Beers JD, Shreve AP, Swanson B (1999) Nonequilibrium pattern formation in Langmuir-phase assisted assembly of alkylsiloxane monolayers. J Phys Chem B 103:10149–10157
50. Gleiche M, Chi LF, Fuchs H (2000) Nanoscopic channel lattices with controlled anisotropic wetting. Nature 403:173–175
51. Piner RD, Zhu J, Xu F, Hong SH, Mirkin CA (1999) "Dip-pen" nanolithography. Science 283:661–663
52. Ginger DS, Zhang H, Mirkin CA (2004) The evolution of dip-pen nanolithography. Angew Chem Int Ed 43:30–45
53. Wouters D, Schubert US (2004) Nanolithography and nanochemistry: Probe-related patterning techniques and chemical modification for nanometer-sized devices. Angew Chem Int Ed 43:2480–2495
54. López GP, Biebuyck HA, Frisbie CD, Whitesides GM (1993) Imaging of features on surfaces by condensation figures. Science 260:647–649
55. Drelich J, Miller JD, Kumar A, Whitesides GM (1994) Wetting characteristics of liquid drops at heterogeneous surfaces. Colloids Surf A 93:1–13
56. Xia YN, Whitesides GM (1998) Soft lithography. Annu Rev Mater Sci 28: 153–184
57. Gates BD, Xu QB, Stewart M, Ryan D, Willson CG, Whitesides GM (2005) New approaches to nanofabrication: Molding, printing, and other techniques. Chem Rev 105:1171–1196
58. Kumar A, Whitesides GM (1994) Patterned condensation figures as optional diffraction gratings. Science 263:60–62
59. Herminghaus S, Fery A, Reim D (1997) Imaging of droplets of aqueous solutions by tapping-mode scanning force microscopy. Ultramicroscopy 69:211–217
60. Rehse N, Knoll A, Konrad M, Magerle R, Krausch G (2001) Surface reconstruction of an ordered fluid: An analogy with crystal surfaces. Phys Rev Lett 87:035505
61. Luo CX, Xing RB, Zhang ZX, Fu J, Han YC (2004) Ordered droplet formation by thin polymer film dewetting on a stripe-patterned substrate. J Colloid Interface Sci 269:158–163
62. Kim E, Xia YN, Whitesides GM (1996) Micromolding in capillaries: Applications in materials science. J Am Chem Soc 118:5722–5731
63. Yang PD, Deng T, Zhao DY, Feng PY, Pine D, Chmelka BF, Whitesides GM, Stucky GD (1998) Hierarchically ordered oxides. Science 282:2244–2246
64. Donald AM (2003) The use of environmental scanning electron microscopy for imaging wet and insulating materials. Nat Mater 2:511–516

65. Lipowsky R, Brinkmann M, Dimova R, Franke T, Kierfeld J, Zhang XZ (2005) Droplets, bubbles, and vesicles at chemically structured surfaces. J Phys Condens Matter 17:S537–S558
66. Lenz P, Lipowsky R (2000) Stability of droplets and channels on homogeneous and structured surfaces. Eur Phys J E 1:249–262
67. Lord Rayleigh (1878) On the instability of jets. Proc London Math Soc 10: 4–13
68. Brinkmann M, Kierfeld J, Lipowsky R (2005) Stability of liquid channels or filaments in the presence of line tension. J Phys Condens Matter 17:2349–2364
69. Léopoldés J, Dupuis A, Bucknall DG, Yeomans JM (2003) Jetting micron-scale droplets onto chemically heterogeneous surfaces. Langmuir 19:9818–9822
70. Pompe T, Herminghaus S (2000) Three-phase contact line energetics from nanoscale liquid surface topographies. Phys Rev Lett 85:1930–1933
71. Swain PS, Lipowsky R (1998) Contact angles on heterogeneous surfaces: A new look at Cassie's and Wenzel's laws. Langmuir 14:6772–6780
72. Indekeu JO (1992) Line tension near the wetting transition: Results from an interface displacement model. Physica A 183:439–461
73. Indekeu JO (1994) Line tension at wetting. Int J Mod Phys B 8:309–345
74. Drelich J, Wilbur JL, Miller JD, Whitesides GM (1996) Contact angles for liquid drops at a model heterogeneous surface consisting of alternating and parallel hydrophobic/hydrophilic strips. Langmuir 12:1913–1922
75. Brandon S, Haimovich N, Yeger E, Marmur A (2003) Partial wetting of chemically patterned surfaces: The effect of drop size. J Colloid Interface Sci 263:237–243
76. Cassie ABD (1948) Contact angles. Discuss Faraday Soc 3:11–16
77. Rye RR, Mann JA, Yost FG (1996) The flow of liquids in surface grooves. Langmuir 12:555–565
78. Bruschi L, Carlin A, Mistura G (2002) Complete wetting on a linear wedge. Phys Rev Lett 89:166101
79. Chou SY, Krauss PR, Renstrom PJ (1995) Imprint of sub-25 nm vias and trenches in polymers. Appl Phys Lett 67:3114–3116
80. Chou SY, Krauss PR, Renstrom PJ (1996) Imprint lithography with 25-nanometer resolution. Science 272:85–87
81. Haisma J, Verheijen M, vandenHeuvel K, vandenBerg J (1996) Mold-assisted nanolithography: A process for reliable pattern replication. J Vac Sci Technol B 14:4124–4128
82. Heckele M, Bacher W, Muller KD (1998) Microsyst Technol 4:122–124
83. Bender M, Otto M, Hadam B, Vratzov B, Spangenberg B, Kurz H (2000) Fabrication of Nanostructures using a UV-based imprint technique. Microelectron Eng 53:233–236
84. Colburn M, Grot A, Choi BJ, Amistoso M, Bailey T, Sreenivasan SV, Ekerdt JG, Willson CG (2001) Patterning nonflat substrates with a low pressure, room temperature, imprint lithography process. J Vac Sci Technol B 19:2162–2172

85. Otto M, Bender M, Hadam B, Spangenberg B, Kurz H (2001) Characterization and application of a UV-based imprint technique. Microelectron Eng 57:361–366
86. Chou SY, Keimel C, Gu J (2002) Ultrafast and direct imprint of nanostructures in silicon. Nature 417:835–837
87. Seemann R, Kramer EJ, Lange FF (2004) Patterning of polymers: Precise channel stamping by optimizing wetting properties. New J Phys 6:111
88. Madou MJ (1997) Fundamentals of Microfabrication. CRC Press, Boca Raton, FL
89. Parry AO, Rascon C, Wood AJ (2000) Critical effects at 3D wedge wetting. Phys Rev Lett 85:345–348
90. Rascon C, Parry AO (2000) Geometry-dominated fluid adsorption on sculpted solid substrates. Nature 407:986–989
91. Shuttleworth R, Bailey GLJ (1948) The spreading of a liquid over a rough solid. Discuss Faraday Soc 3:16–22
92. Concus P, Finn R (1969) On behavior of a capillary surface in a wedge. Proc Natl Acad Sci USA 63:292–299
93. Khare K, Law BM, Gurevich E, Brinkmann M, Herminghaus S, Seemann R (2007) Dewetting of liquid filaments in wedge-shaped grooves. submitted

# Chapter 5
# Electrowetting: Thermodynamic Foundation and Application to Microdevices

Hyejin Moon and Chang-Jin "CJ" Kim

Mechanical and Aerospace Engineering Department
University of California at Los Angeles (UCLA)
Los Angeles, CA 90095, USA.

## 1. Introduction

Surface tension is a line force, directly proportional to the length. However, surface forces (e.g., pressure) are proportional to the square of the length, and body forces (e.g., inertia or gravity) are proportional to the cubic of the length. Therefore, surface tension becomes dominant over pressure or gravity as the dimension of interest gets smaller – in practice, below millimeter range. Many examples are found in our daily lives. An ant can carry a load that is several times heavier than its own weight, demonstrating its relative strength, but it also can be trapped in a water droplet showing its relative weakness. Small creatures climb up walls. Water striders hop on water surfaces. Note these examples are in millimeter scale where inertia forces are still comparable to surface tension. In micrometer scale, surface tension starts to dominate over not only the inertia forces but most other physical forces.

The importance of surface tension has been recognized early in the field of microelectromechanical systems (MEMS) through the problems it causes. Surface tension makes fabrication of slender elements challenging and hinders the proper operation of common microdevices. Especially for surface micromachining, which tends to produce slender micron-scale elements separated by only micron-scale gaps, stiction is a particularly severe and well-known problem [1, 2]. Naturally, as the first reaction, most effort on surface tension in the development of micromechanical devices has been made to avoid it. Intentional utilization of such a scale effect in designing microdevices from the conceptual level has been rare mainly because of the lack of precedents. Despite the challenges, there has been a new class of research to introduce surface tension as the main source of physical forces in microscale mechanical devices (i.e., micromachines) [3–7]. In these micromachines, liquid–gas menisci play a key role; often droplets and bubbles function as discrete mechanical components like the solid components in the regular machines with which we are familiar. For example, a vapor bubble functions as a virtual check valve against liquid flows in a microchannel with the effectiveness possible only in microscale [3, 8], and liquid–metal droplets function as micromechanical switches with the robustness (i.e., stable against physical disturbances) possible only in microscale [9, 10].

Aside from the above effort to develop micromachines, the interface between liquid and solid or between two immiscible liquids has attracted particular interest because of its biochemical application potential in microfluidics. Since surface effects dominate over volume effects, the gradient of surface tension may be a proper method to pump tiny amounts of liquid. There are numerous methods to change surface energy. In order for this surface effect to be practically employed to microfluidic devices, however, it is required that surface (interface) properties can be controlled in *real time* and *reversibly* via an easily manipulated external variable (such as incident illumination, electric potential, or local temperature gradient). In this chapter, we review the control of interfacial forces and energies by *electric potential* for its compatibility with MEMS devices. Although the discussion eventually leads to the emerging field of digital (or droplet) microfluidics and its implication for lab-on-a-chip, our two main goals in this chapter are to establish the theoretical foundation for electrical control of interfacial energies (forces) and to review how liquids are manipulated and routed using such interfacial control in microdevices. Although there exist alternative explanations (e.g., electrostatic [7, 11, 12] or dielectrophoretic [13, 14]) for the same or similar fluidic phenomena and operations, this chapter focuses on the theory of electrocapillarity and electrowetting and their thermodynamic foundation with an interest to help the designer of such droplet microdevice. A summary of the alternative explanations as well as a more comprehensive treatment of the electrowetting can be found in the excellent review by Mugele [15].

## 2. Theoretical Background

### 2.1. Surface Tension

The system shown in Fig. 1 is a liquid in contact with its vapor. Since the molecules in the liquid phase are packed more densely than those in the vapor phase, the intermolecular force between the molecules in the liquid phase is greater than the intermolecular force between molecules in the gas phase. To extend the area of the interface, it is necessary to bring molecules from the interior onto the surface, so that work is done against cohesive forces in the liquid. Consequently, molar free energy of the surface region of the liquid is higher than that of the bulk liquid.

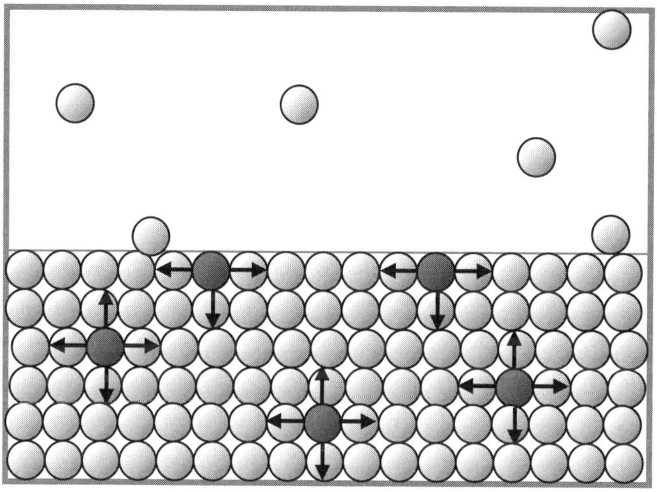

**Fig. 1.** Liquid–vapor system

In 1805, Thomas Young showed that the mechanical properties of the surface could be related to those of a hypothetical membrane stretched over the surface. The membrane was supposed to be in a state of tension. Surface tension acts parallel to the surface and pulls inward to oppose any attempt to extend the surface area. The unit of surface tension (equivalent to surface energy) in the SI system is Newton per meter ($N\ m^{-1}$) or Joule per meter squared ($J\ m^{-2}$).

### 2.2. Surface Thermodynamics

For a closed (i.e., no mass exchange with its environment) and pure chemical component system with no chemical reactions, the state functions volume $V$, internal energy $U$, and entropy $S$ form a complete basis, that is, we can describe all of thermodynamics by manipulating these three functions. Unfortunately, nature will not allow us to simply use $U$, $S$, and $V$ to describe it. The problem is that $S$ cannot be measured easily and so it is difficult to develop an equation of $U = U(S,V)$. For this reason, auxiliary energy functions such as enthalpy $H$, Helmholtz free energy $F$, and Gibbs free energy $G$ need to be introduced. We can choose the proper energy function to describe the system depending on the type of state functions of convenience. The definitions of state functions and their differential formulae are shown in Table 1.

**Table 1.** Types of state functions and their differential expression

| Name of function | Symbol and natural variables | Definition | Differential expression |
|---|---|---|---|
| Internal energy | $U(S,V)$ | $\Delta U = q + w$ | $dU = TdS - PdV$ |
| Enthalpy | $H(S,P)$ | $H = U + PV$ | $dH = TdS + VdP$ |
| Helmholtz free energy | $F(T,V)$ | $F = U - TS$ | $dF = -SdT - PdV$ |
| Gibbs free energy | $G(T,P)$ | $G = H - TS$ | $dG = -SdT + VdP$ |

Thermodynamic property at the surface or interfacial region needs to be treated differently from the property in the bulk phase. Figure 2 shows the interfacial region of two phases, $\alpha$ and $\beta$. Let us place the planes, $AA'$ and $BB'$, so that the properties of bulk phase $\alpha$ sustains up to the surface $AA'$, and those of the bulk phase $\beta$ up to the surface $BB'$. Thus, the concentration of component $i$ up to $AA'$ is bulk concentration $c_i^\alpha$, and of component $i$ up to $BB'$ uniformly $c_i^\beta$. Within the interfacial region (between $AA'$ and $BB'$), properties of the system vary *continuously* from pure $\alpha$ to pure $\beta$. In general, the thickness of the interfacial region is not more than about ten molecular diameters.

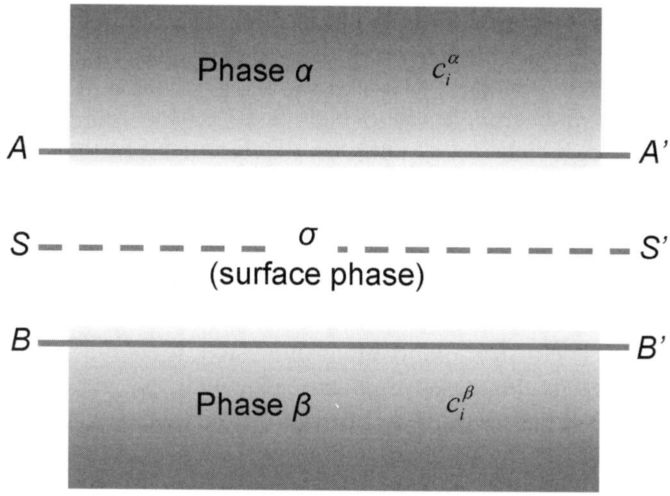


**Fig. 2.** Definition of surface phase

Let us draw another parallel surface $SS'$ to $AA'$ and $BB'$ within the interfacial region. Such a surface $SS'$ is called a surface phase and is designated by a superscript $\sigma$. From the conservation of mass, the total amount of component $i$ in the system will be

$$n_i = n_i^\alpha + n_i^\beta + n_i^\sigma .$$

In terms of the concentration $c$,

$$n_i^\alpha = c_i^\alpha V^\alpha \text{ and } n_i^\beta = c_i^\beta V^\beta .$$

The concept of surface phase can be extended to other extensive variables of the systems. For example,

Surface energy $U^\sigma = U - U^\alpha - U^\beta$,
Surface entropy $S^\sigma = S - S^\alpha - S^\beta$.

The choice of the dividing surface has so far been left arbitrary, but to make the mechanical properties of the model system conform to those of the real system, the division must be made at the surface of tension. In the case of a plane surface, however, the work done in extending the surface does not depend on the location of the surface of tension, and we can accordingly place the Gibbs surface $SS'$ wherever we please [16].

### 2.3. General Concept of Work [16]

In the mechanical systems, the work is always formulated as the product of two terms: an intensive factor, which is a generalized force, and an extensive factor, which is a generalized displacement. In the conventional thermodynamics that considers the work by the volume change, the generalized force is the pressure $P$ and the generalized displacement is the change of volume of the system $dV$, so that the mechanical work on the system equals $PdV$. The same formulation can apply to nonmechanical work as well.

In physical chemistry, one is often interested in charges carried out in electrical cells. In the case of electrical work, the generalized force becomes the electromotive force (EMF) $\Phi$ of the cell, and the generalized displacement becomes the charge $dQ$ transferred through the external circuit as the cell discharges ($dQ < 0$). The element of work done on the cell is $\Phi dQ$. Similarly, in magnetic systems, the intensive factor is the magnetic field strength $H$. If this acts on a substance to produce a magnetization $dM$ in the direction of the field, the work done on the substance is $HdM$. We can thus summarize the various examples of work in Table 2. Work done on a system is always taken as positive.

**Table 2.** Examples of work

| Generalized force | Displacement | Element of work ($dw$) |
|---|---|---|
| Tension $f$ | Distance $l$ | $fdl$ |
| Surface tension $\gamma$ | Area $A$ | $\gamma dA$ |
| Pressure $P$ | Volume $V$ | $-PdV$ |
| Electromotive force $\Phi$ | Charge $Q$ | $\Phi dQ$ |
| Magnetic field $H$ | Magnetization $M$ | $HdM$ |

### 2.4. Surface Tension in Thermodynamic Consideration

Consider a container filled with pure water and initially covered with its lid. Now we uncover the container very slowly so that the change of work

$dW$ is negligible. Assume that the process is being done at a constant temperature and pressure. Because of the constant pressure process, the mechanical work would be zero. The nonmechanical work we need to take into account in this process is the work to create the new surface of water $\gamma dA$. Here, $dA$ indicates the change area of the surface of water exposed to its vapor. In general, energy by a chemical potential $\mu_i$ also needs to be considered. Hence, the change of Gibb's energy is

$$\begin{aligned} dG &= -SdT + VdP + w_{\text{nonmechanical}} + \Sigma\mu_i dn_i \\ &= -SdT + VdP + \gamma dA + \Sigma\mu_i dn_i \quad . \end{aligned} \tag{1}$$

From the assumption that the process occurs with pure water at a constant temperature and pressure, the rearrangement of the (1) would be

$$\left(\frac{\partial G}{\partial A}\right)_{T,P,n_i} = \gamma \, . \tag{2}$$

Equation (2) states that the surface tension equals the partial derivative of the Gibbs free energy with respect to surface area at constant $T$ and $P$. Stated another way, surface tension (N $m^{-1}$ in SI unit) is also surface free energy per unit area (J $m^{-2}$ in SI unit).

### 2.5. Liquid–Liquid and Liquid–Solid Interfaces: Young's Equation [17]

The interfacial tension between two liquid phases, $\alpha$ and $\beta$, is designated by $\gamma_{\alpha\beta}$. Suppose that the interface has a unit area. Then if we pull the two phases apart we will form one unit of a surface of pure phase $\alpha$ with surface Gibbs energy $\gamma_{\alpha v}$ and one unit of a surface of pure phase $\beta$ with surface Gibbs energy $\gamma_{\beta v}$. The increase in Gibbs energy in this transformation is

$$\Delta G / A = w_A^{\alpha\beta} = \gamma_{\alpha v} + \gamma_{\beta v} - \gamma_{\alpha\beta} \, . \tag{3}$$

This increase in Gibbs energy per area is called the work of adhesion $w_A^{\alpha\beta}$ between the phases $\alpha$ and $\beta$. Note that since the pure phases $\alpha$ and $\beta$ are in contact with the vapor phase, we have written $\gamma_{\alpha v}$ for the interfacial tension between $\alpha$ and the vapor phase. Similarly, $\gamma_{\beta v}$ is the interfacial tension between phase $\beta$ and the equilibrium vapor phase.

If we pull apart a pure phase $\alpha$, two units of surface with surface Gibbs energy $\gamma_{\alpha v}$ are formed,

$$\Delta G / A = w_C^{\alpha} = 2\gamma_{\alpha v} \, .$$

This increase in Gibbs energy per area $w_C^\alpha$ is called the work of cohesion of $\alpha$. Similarly, $w_C^\beta = 2\gamma_{\beta v}$. Then

$$w_A^{\alpha\beta} = \frac{1}{2}w_C^\alpha + \frac{1}{2}w_C^\beta - \gamma_{\alpha\beta}$$

or

$$\gamma_{\alpha\beta} = \frac{1}{2}(w_C^\alpha + w_C^\beta) - w_A^{\alpha\beta}.$$

As the Gibbs energy of adhesion between the phases $\alpha$ and $\beta$ increases, $\gamma_{\alpha\beta}$ decreases. When $\gamma_{\alpha\beta} = 0$, there is no resistance to the extension of the interface between phase $\alpha$ and $\beta$; the two liquids mix spontaneously. In this case, the work of adhesion is the average of the work of cohesion of the two liquids

$$w_A^{\alpha\beta} = \frac{1}{2}(w_C^\alpha + w_C^\beta).$$

The same argument holds for the interfacial tension between a solid and a liquid. Thus, in analogy to (3), we have

$$w_A^{sl} = \gamma_{sv} + \gamma_{lv} - \gamma_{sl}. \tag{4}$$

Although $\gamma_{sv}$ and $\gamma_{sl}$ are not measurable, it is possible to obtain a relation between $\gamma_{sv} - \gamma_{sl}$ and the contact angle $\theta$ and $\gamma_{lv}$. To do this, we consider the liquid drop resting on a solid surface as in Fig. 3. If the liquid surface deforms slightly so that the area of the solid–liquid interface increases by $dA_{sl}$, then the Gibbs energy change is

$$dG = \gamma_{sl}\, dA_{sl} + \gamma_{sv}\, dA_{sv} + \gamma_{lv}\, dA_{lv}.$$

From Fig. 3, we have

$$dA_{sv} = -\, dA_{sl} \text{ and } dA_{lv} = dA_{sl}\cos\theta;$$

then,

$$dG = (\gamma_{sl} - \gamma_{sv} + \gamma_{lv}\cos\theta)\, dA_{sl}. \tag{5}$$

At the equilibrium state, $dG = 0$. Thus, the contact angle of the liquid to the solid surface can be defined with (6).

$$\cos\theta = \frac{\gamma_{sv} - \gamma_{sl}}{\gamma_{lv}}. \tag{6}$$

$$\text{or } \gamma_{sl} = \gamma_{sv} - \gamma_{lv} \cos\theta. \tag{7}$$

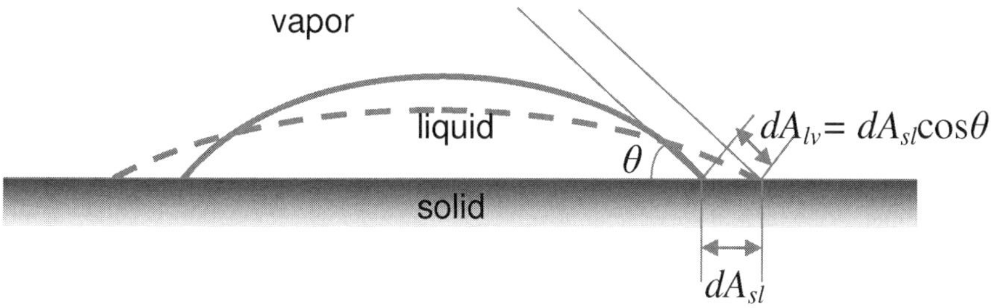


**Fig. 3.** Spreading of a liquid drop on a solid surface

### *2.6. Pressure Difference at the Curvilinear Surface*

The curvature of the surface can be defined with two principal radii at the given point $p$ on the surface. Principal radii are the maximum and the minimum radius of the curvature. They are on the planes that are perpendicular to each other and that are both perpendicular to the tangential plane to the surface at the point $p$. An example of principal radii is shown in Fig. 4.

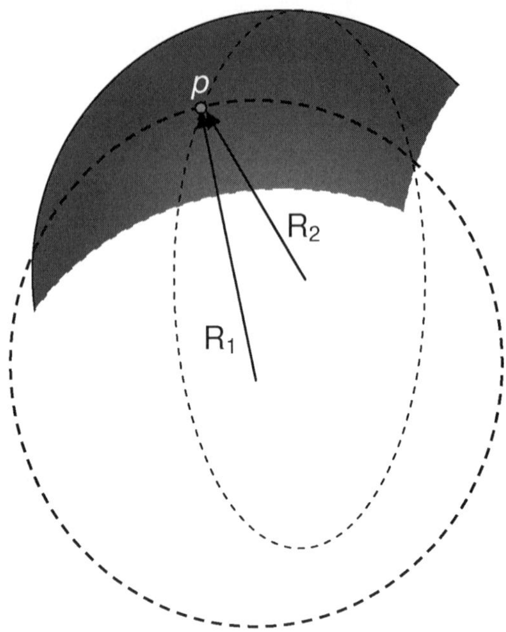


**Fig. 4.** Principal radii of a curvilinear surface

When an interface of two different phases forms a curvature, there is a pressure difference between the two phases

$$\Delta P = \gamma\left(\frac{1}{R_1} + \frac{1}{R_2}\right). \tag{8}$$

This is the well-known Laplace-Young equation. If a radius of curvature $R_1$ or $R_2$ is in the opposite direction to that shown in Fig. 4, it will have a negative sign. That is, a radius of curvature is positive if it is convex in the direction of the pressure and negative if it is concave in the direction of the pressure (Fig. 5). The Laplace-Young equation determines the shape of the interfacial surface, providing that the excess pressure $\Delta P$ is known at all points on the surface.

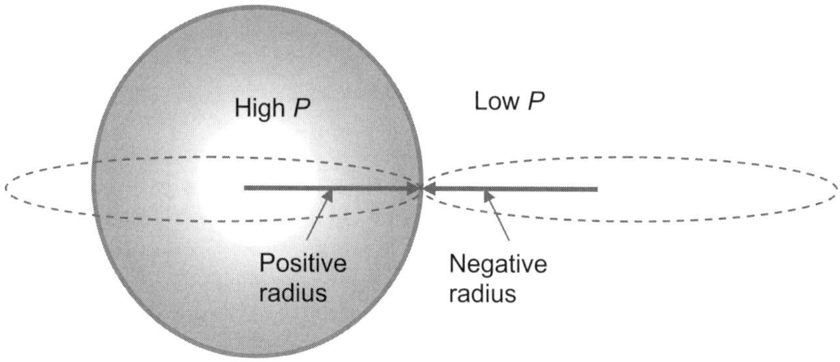


**Fig. 5.** A radius of curvature is positive if it is convex in the direction of the pressure and negative if it is concave in the direction of the pressure

### 2.6.1. Example: Application of the Laplace-Young Equation

The pressure difference between the inside and the outside of the liquid segment can be estimated by using the Laplace-Young equation. Consider a liquid segment that is placed between two parallel plates and its contact angle to the surface of plate is $\theta$. The radius of the liquid disk is $R$ at the interface of the liquid and the plate, and $r$ in the middle (Fig. 6). The spacing between two plates is $d$.

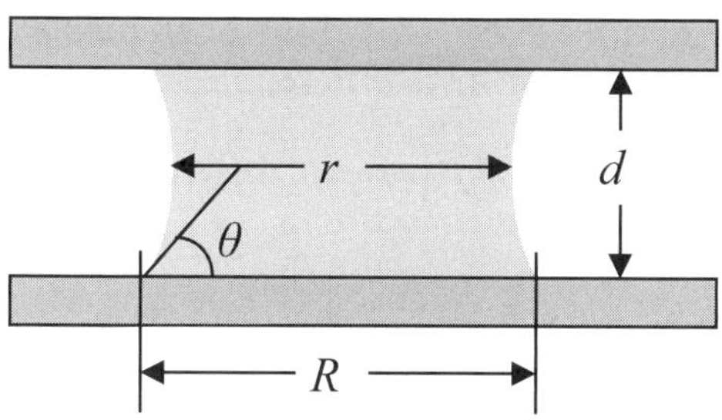


**Fig. 6.** Liquid segment between two parallel plates

Let us choose a star symbol (Fig. 7) as the point to find two principal radii. The two planes including principal radii should be perpendicular to the tangential plane of the curvature at the point indicated with the star.

Also the two planes should be perpendicular to each other. The tangential plane of the curvature is on the *y*-plane, $R_1$ is on the *x*-plane, and $R_2$ is on the *z*-plane.

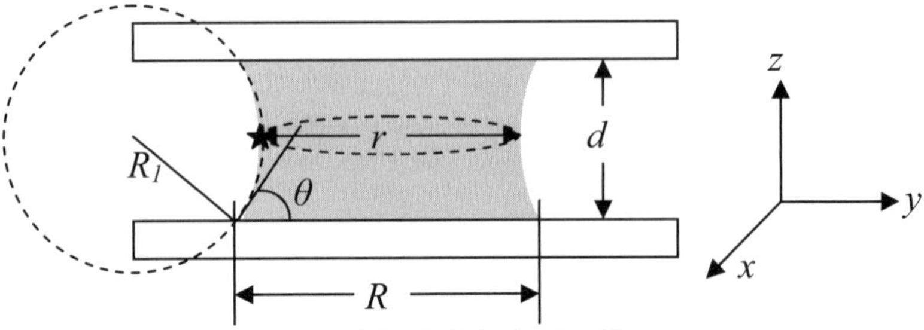


**Fig. 7.** Principal radii

Then the two principal radii will be $R_1 = -(d/2)/\cos\theta$ and $R_2 = r/2$. From the Laplace-Young equation, the pressure difference between the inside liquid and the outside liquid will be

$$\Delta P = P_{\text{in}} - P_{\text{out}} = \gamma(1/R_1 + 1/R_2) = \gamma\,(-2\cos\theta/d + 2/r).$$

## 2.7. Control of Surface Tension

As previously mentioned, the surface tension can be controlled by several ways. Before showing the examples of controlling the surface tension and the thermodynamic explanation on them, a mathematical tool needs to be introduced.

If $f(x_1, x_2, x_3, \ldots, x_n)$ is a function of a number $n$ of independent variables $x_1, x_2, \ldots, x_n$, the total differential $df$ is defined in terms of the partial derivatives $(\frac{\partial f}{\partial x_i})_{x_1, x_2, \ldots, x_n}$ and the differentials $dx_1, dx_2, \ldots, dx_n$ as

$$df = \sum (\frac{\partial f}{\partial x_i})_{x_1, x_2 \ldots} dx_i \ . \tag{9}$$

If there are only two independent variables, *x* and *y*, (9) becomes

$$df(x, y) = (\frac{\partial f}{\partial x})_y dx + (\frac{\partial f}{\partial y})_x dy. \tag{10}$$

Meanwhile, the integration of (10) at constant *x* and *y* results in

$$f(x,y) = \int df = \int (\frac{\partial f}{\partial x})_y dx + \int (\frac{\partial f}{\partial y})_x dy \\ = (\frac{\partial f}{\partial x})_y x + (\frac{\partial f}{\partial y})_x y. \tag{11}$$

Differentiate about $x$ and $y$, and (11) becomes

$$df(x,y) = d(\frac{\partial f}{\partial x})_y x + (\frac{\partial f}{\partial x})_y dx + d(\frac{\partial f}{\partial y})_x y + (\frac{\partial f}{\partial y})_x dy. \tag{12}$$

Subtract (12) from (10), and the rearrangement is

$$xd(\frac{\partial f}{\partial x})_y + yd(\frac{\partial f}{\partial y})_x = 0. \tag{13}$$

With the above mathematical tools in our hands, let us now consider real systems involving surface tension as an important component. Our interest is to consider possible means of controlling surface tension and understand theoretically how they affect the surface tension. Three exemplary systems are presented below, where the surface tension is affected by chemical variation, temperature change, or electric potential [4, 5, 18]. Each system has been implemented in reality, inducing physical motions by the surface tension, which is controlled by adding a surfactant, creating a temperature gradient, or applying an electric potential.

### 2.7.1. Example 1: Chemical Potential – Surface Tension System

Consider a system with a liquid solution. Now we add a chemical that tends to concentrate near the interface (e.g., surfactant). Figure 8 shows the process and the surface phase that we are examining. We can write for the surface Gibbs function,

$$dG^\sigma = -S^\sigma dT + VdP + w_{non-mechanical}. \tag{14}$$

The chemical adding process can be assumed constant $T$ and $P$ process, and the nonmechanical work in this example is the work by surface tension $\gamma dA$ and by chemical potential $\mu_i dn_i$. In this condition, (14) becomes

$$dG^\sigma = \gamma dA + \sum \mu_i dn_i^\sigma. \tag{15}$$

If we integrate this expression at constant $\gamma$ and $\mu_i$,

$$G^\sigma = \gamma A + \sum \mu_i n_i^\sigma.$$

When this is differentiated,

$$dG^{\sigma} = \gamma dA + A d\gamma + \sum n_i^{\sigma} d\mu_i + \sum \mu_i dn_i^{\sigma}. \quad (16)$$

On combining (15) with (16),

$$A d\gamma + \sum n_i^{\sigma} d\mu_i = 0. \quad (17)$$

The rearrangement of equation (17) becomes

$$\frac{d\gamma}{d\mu_i} = -\sum \frac{n_i^{\sigma}}{A}. \quad (18)$$

Equation (18) roughly states that the total variation of the con-concentration of $i$ components at the surface decreases the surface tension. Note that (18) is valid for planar interface only. For the curved interface, the degree of freedom increases one order higher because of the pressure difference of both sides of the curvature.

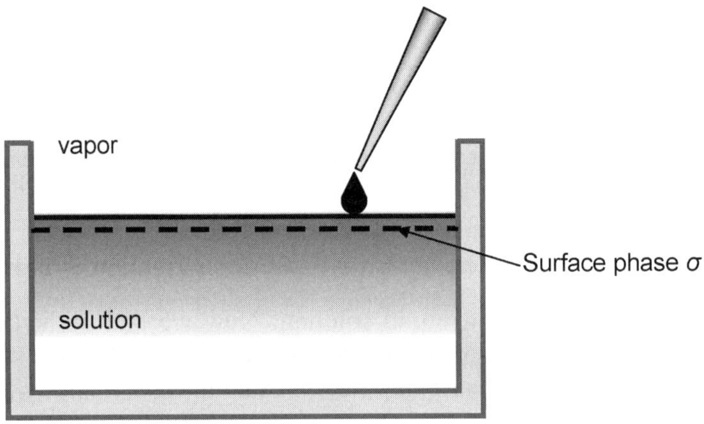


**Fig. 8.** Chemical adding process and choice of surface phase

An example of this system, which creates physical motion, can be found in a toy boat floating on water surface, as illustrated in Fig. 9. By adding a surfactant (e.g., soap) at the rear of the boat, the backward pull by the surface tension at the rear becomes smaller than the forward pull at the front, creating a net force and driving the boat forward. The pull by surface tension to side directions is symmetric and amounts to no net force.

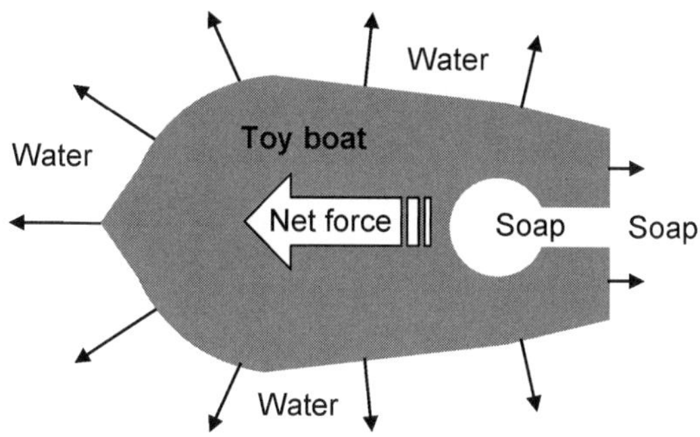


**Fig. 9.** Toy boat floating on water surface. The soap dropped in the key hole is released to the rear of the boat, inducing a forward motion of the boat

### 2.7.2. Example 2: Temperature – Surface Tension System

Consider heating the liquid segment. The liquid is pure and the volume remains the same during the heating. The pressure at the interface may vary and so internal energy $U = U\,(S, V)$ is the proper function to use for this example. Figure 10 shows the surface phase chosen at the liquid–vapor interface.

$$
\begin{aligned}
dU^{\sigma} &= TdS^{\sigma} - PdV + w_{non\text{-}mechanical} \\
&= TdS^{\sigma} - PdV + \gamma dA + \sum \mu_i dn_i^{\sigma}
\end{aligned}
$$

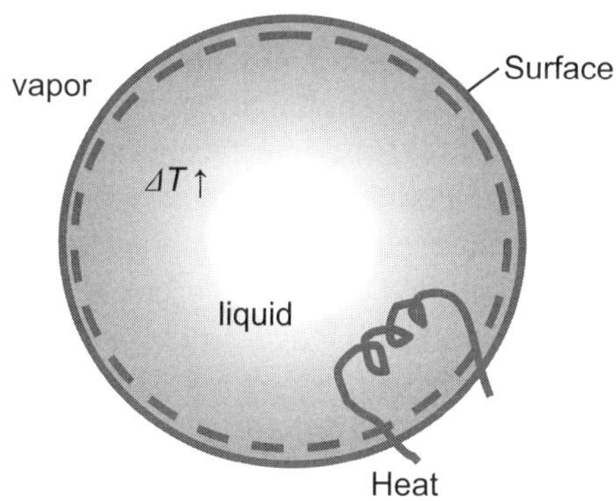


**Fig. 10.** Heating a liquid segment and the surface phase

Since the volume of the liquid remains the same through the process and the liquid is pure liquid, i.e., $dn_i^\sigma = 0$,

$$dU^\sigma = TdS^\sigma + \gamma dA . \tag{19}$$

Integrate (19) at constant $A$ and constant $S$ to obtain

$$U^\sigma = \gamma A + S^\sigma T . \tag{20}$$

Total differential of (20) brings

$$dU^\sigma = Ad\gamma + \gamma dA + S^\sigma dT + TdS^\sigma . \tag{21}$$

From 19 and 21, the Gibbs-Duhem equation is obtained as follows.

$$Ad\gamma + S^\sigma dT = 0$$

or (22)

$$\frac{d\gamma}{dT} = -\frac{S^\sigma}{A} .$$

Equation (22) states that as temperature increases, surface tension decreases (for a positive entropy process that is in the direction of a natural process). Early examples of this system being used to create physical motion can be found in several sources, through various experiments including microscale devices in recent years [3, 6].

### 2.7.3. Example 3: Electric Potential – Surface Tension System

Let us consider an interface between a liquid and an electrode across which electrical potential is applied, as shown in Fig. 11. With the electrical potential, charges may build up at the interface, and we choose to put the surface phase near the charge containing interface. If the electrode surface is flat enough and the surface phase is set up parallel to the electrode surface, we can assume that this charging process is performed under constant pressure and temperature. Hence, Gibbs energy may be a good candidate to describe the process

$$G = G\,(T, P, \text{others}).$$

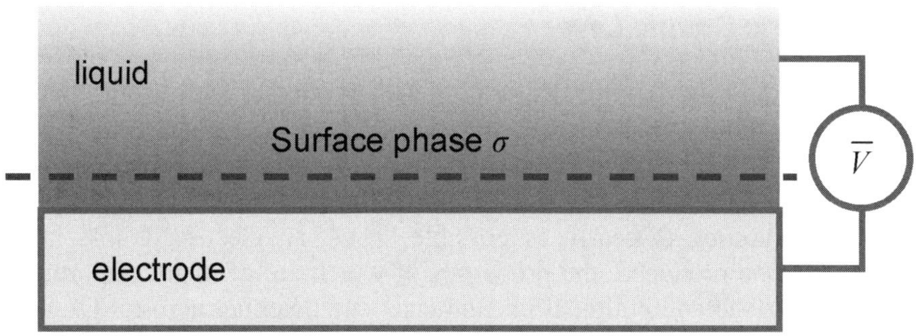


**Fig. 11.** Liquid contact with charged electrode

In this process, the nonmechanical work we should consider is work by surface tension $\gamma\, dA$ and by electric potential $\Phi\, dQ$.

$$dG = -S\,dT - V\,dP + w_{\text{nonmechanical}}, \tag{23}$$

where $$w_{\text{nonmechanical}} = \Phi\, dQ + \gamma\, dA.$$

Since it is a constant $T$ and $P$ process, the equation (23) becomes

$$dG = \Phi\, dQ + \gamma\, dA. \tag{24}$$

Integration at a constant charge and area gives

$$G = \Phi Q + \gamma A\,. \tag{25}$$

Differentiate equation (25) to obtain a form of the Gibbs-Duhem equation,

$$Q\, d\Phi + A\, d\gamma = 0,$$

$$\left(\frac{d\gamma}{d\Phi}\right)_{T,P,\mu_i} = -\frac{Q}{A}. \tag{26}$$

Equation (26) states that surface tension will decrease as electric potential is applied at the interface, and its rate is the charge per unit area. Note that the surface phase was set up near the liquid–electrode interface, which means $\gamma$ in this example is interfacial free energy at the liquid–electrode interface, or $\gamma = \gamma_{sl}$. Experiments of this system have been conducted by Lippmann over a century ago, and attempts to create physical motion can be found in various reports in recent years, as elaborated in the following sections.

## 3. Electrowetting and Its Recent Variations

### 3.1. Electric Double Layer

Consider the phase $\alpha$ (typically solid phase) in contact with another phase $\beta$ (typically liquid phase), as in Fig. 12. If $\alpha$ has a certain surface concentration of fixed charges, we may expect in general to find that opposite charges in $\beta$ will be distributed close to the interface so as to achieve local electric neutrality as nearly as possible. Thus, an electric double layer (EDL) will be produced, the properties of which are of major importance in many surface and colloidal phenomena. The potential across EDL will be $\Delta\Phi$, which cannot be directly measured.

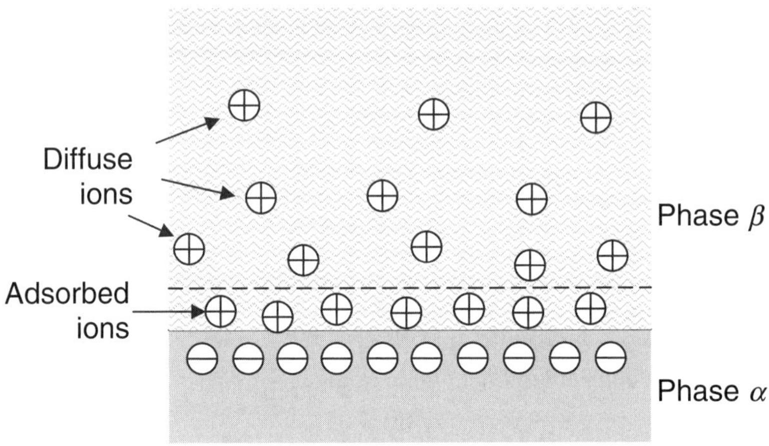


**Fig. 12.** Electric double layer (EDL)

The charging of the surface upon contact with the liquid can occur in two ways: (a) by the ionization or dissociation of surface groups (e.g., the dissociation of protons from surface carboxylic groups (–COOH → –$COO^-$ + $H^+$), which leaves behind a negatively charged surface), and (b) by the adsorption (binding) of ions from solutions onto a previously uncharged surface, e.g., the binding of $Ca^{2+}$ onto the zwitterionic headgroups of lipid bilayer surfaces, which charges the surfaces positively. Depending on the ionic conditions, even the air–water and hydrocarbon–water interfaces can become charged in this way. The adsorption of an ion from solutions can, of course, also occur onto oppositely charged surface site, e.g., the adsorption of cationic $Ca^{2+}$ to anionic $COO^-$ sites vacated by $H^+$ or $Na^+$. This is known as ion exchange [19].

The simplest model for EDL is the parallel capacitor model. The solid surface is charged by bounded ions, and counterions in a liquid phase form a parallel capacitor-like structure. The thickness of the layer is known to be in the range of about 1–10 nm. By applying opposite charges to the charged surface, it can be made to have a zero net charge at the solid–electrolyte interface. The applied potential corresponding to the zero net charge at the interface is called "potential of zero charge (PZC)."

### *3.2. Electrocapillarity: Lippmann's Experiment*

As we have seen previously, the presence of a net electric charge on a surface lowers the surface tension because repulsion between like charges reduces the work needed to extend the surface area. In 1875, Lippmann made the first quantitative measurements of this effect. Experimental system is schematically shown in Fig. 13. This set up consists of an electrochemical cell with a mercury electrode contained in a capillary tube and a nonpolarizable reference electrode (such as the calomel electrode). An external voltage source permits adjustment of the electric potential between the mercury capillary and electrode and the mercury lead of the calomel electrode.

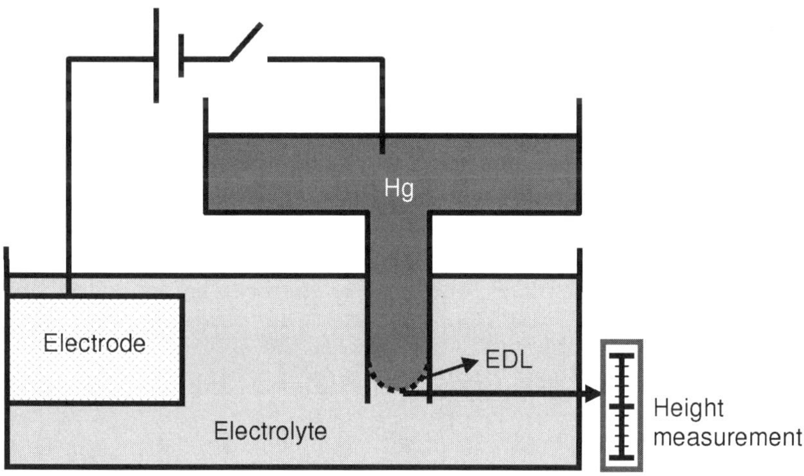


**Fig. 13.** Lippmann's experiment described in a simplified schematic. The height of mercury meniscus was read as a measurement of surface tension at the interface

The mercury electrode in contact with the nonreactive salt solution can be considered to behave as an ideal polarizable electrode. An applied electric potential gradient $\Delta\Phi$ changes the charge density at the mercury–

electrolyte interface, but does not induce any electron transfer through this interface. As $Q/A$ changes, the surface tension $\gamma$ of the mercury also changes, and the position of the mercury meniscus shifts in the capillary (observed with a low power microscope). The change in $\gamma$ is measured by the change in the height of the mercury reservoir necessary to restore the original position of the meniscus. One thus obtains a curve of $\gamma$ vs. $\Phi$, called the electrocapillary curve. An example is shown in Fig. 14.

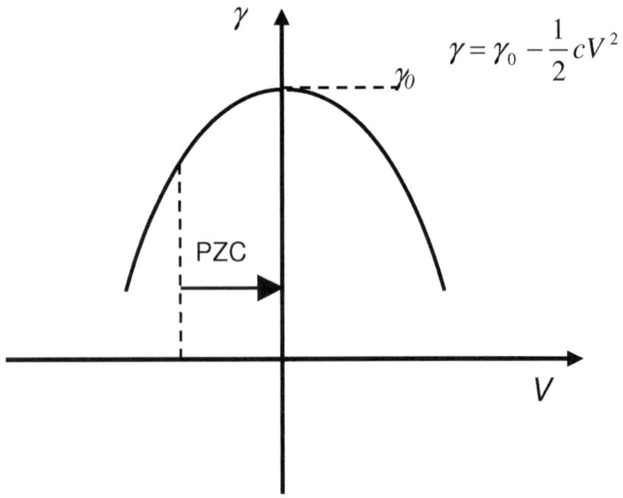


**Fig. 14.** Typical example of electrocapillary curve

The ideal nonpolarizable electrode and the ideal polarizable electrode are limiting cases that can be treated exactly by thermodynamics. With this assumption of ideal interfaces, one basic relation can be derived as shown previously:

$$\left(\frac{\partial \gamma}{\partial \Phi}\right)_{T,P,\mu} = -\frac{Q}{A}. \tag{27}$$

This equation is now referred as the Lippmann equation. According to equation (27), the slope of the electrocapillary curve yields the surface density of a charge on the electrode. From (27), we note that the maximum of the electrocapillary curve corresponds to a surface charge density of zero. For many electrolytes, PZC at the mercury surface is about 0.5 V (referred to a normal calomel electrode in KCl), but for some electrolytes it deviates considerably from that value.

### 3.3. Electrowetting: On Solid Electrode

Although Lippmann's electrocapillary experiment considered the interfacial energy (i.e., surface tension) between a liquid electrolyte and a *liquid* electrode (mercury), the same thermodynamic consideration can be applied for the interfacial energy between a liquid electrolyte and a *solid* electrode. As described in the discussion of EDL above, the charge density at the solid–liquid interface may vary with the application of an external electric field, and surface tension (energy) at the interface $\gamma_{sl}$ may change, which results in the spread of liquid on the solid electrode surface, as illustrated in Fig. 15. Since the electric field induces the spread of liquid (i.e., the liquid wets the surface) and no capillary structure is involved, this phenomenon is generally called "electrowetting" rather than electrocapillarity. Noting that capillary rise occurs when a liquid wets the inner surface of a capillary tube, one may interpret electrocapillary as a special case of electrowetting, i.e., electrowetting involving capillary structures.

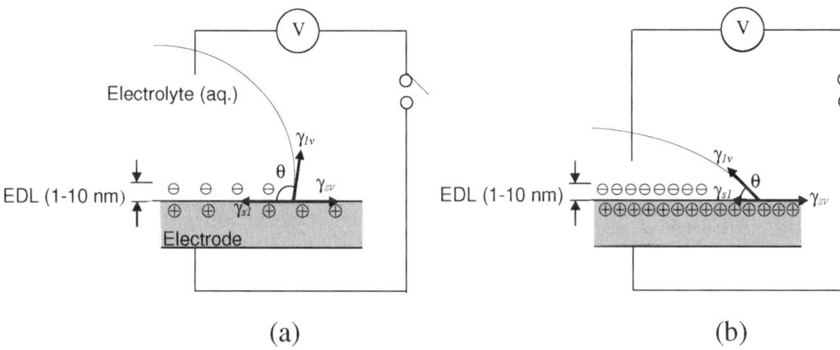


**Fig. 15.** A droplet on a solid electrode spreads when the double layer charge increases by an externally applied electric field [20, 21]. This phenomenon is called electrowetting, conventionally on an electrode.

Recall (27)

$$\left(\frac{\partial \gamma}{\partial V_{\mathrm{e}}}\right)_{\mathrm{T,P},\mu} = -\frac{Q}{A} = -q = -cV_{\mathrm{e}}, \qquad (28)$$

where $V_{\mathrm{e}}$ is the electric field across the EDL, $c$ is specific capacitance of EDL, and $q$ is specific charge density at EDL, which is typically

~0.05 C m$^{-2}$ for mercury and water interface. Integration of (28) results in the apparent form of the Lippmann equation

$$\gamma - \gamma_0 = -\frac{1}{2}c(V_e^2 - V_{e0}^2). \tag{29}$$

For the derivation to be valid, an ideal polarizable interface between the liquid and the electrode and an ideal nonpolarizable interface between the liquid and the counter electrode are required. Indeed, sometimes an EDL at the interface can be assumed as an ideal polarizable interface. However, formation of this kind of interface and its polarizability highly depend on the type of ions and the type of metals in contact.

Because only a small voltage drop can be sustained across the EDL, the contact-angle change $\Delta\theta$ that can be induced by this conventional electrowetting is relatively small, limited by an electron transfer from the electrode to a redox-active species in the liquid. Application of this electrowetting mechanism for micropumping has been introduced to the MEMS community in [22] and later reduced into working microdevices in [23–25].

### *3.4. Electrowetting: On Dielectric [26]*

Electrowetting may occur even on an insulator, if it is thin enough and covers an electrode, as the dielectric layer inserted between the liquid and the electrode can emulate the EDL in conventional electrowetting. However, significantly higher voltage needs to be applied across the dielectric to induce a change of charge density comparable to that in an EDL, since the specific capacitance (capacitance per area) that the dielectric layer forms between the liquid and the electrode is much lower than that of EDL. On the other hand, the dielectric, which prevents the electron transfer between the liquid and the electrode, can accommodate much higher voltages than EDL can in such a way that more contact-angle changes can be generated on it than was possible with conventional electrowetting on metal. To clearly indicate that wetting is on a dielectric material, in comparison with the conventional metal (i.e., conductive material), we name this wetting phenomenon electrowetting-on-dielectric (EWOD).

As early as 1980, Minnema [27] observed water spreading on a polymer sheet (200–900-μm thick polyethylene) when over 1,000 V of voltage was applied. Described as "water treeing" and studied with an interest different from today's electrowetting community, the observed phenomenon was nevertheless clearly of EWOD. Later, Berge [28] and Vallet [29] used a 53-μm thick PTFE and a 12-μm poly(ethylene terephthalate) (PET) film, respectively, to obtain reversible electrowetting with ~200 V. Welters [30] reported a great improvement of contact-angle reversibility by using Teflon AF1600 on a 6–35-μm thick high-dielectric-strength polymer. This configuration (hydrophobic coating on insulating polymer on electrode) has since been widely adopted by others, and improvement has continued. Verheijen [31] showed a better contact-angle reversibility by impregnating the surface with silicone oil, and Seyrat [32] further confirmed that using a nonpolar oil such as hexadecane or silicone can improve the reversibility and lifetime.

However, meaningful degree of changes in wetting, e.g., enough to overcome the pinning of the contact line, required typically over 100 V unless the surface is covered with oil, making instrumentation inconvenient and development of a handheld system impractical. Following the theoretical prediction of Lee [24, 25] and utilizing the integrated circuits (IC) thin-film technologies, Moon [21] has reported that a considerable reduction of the driving voltage is possible. For a given large contact-angle change (120° → 80°, large enough to drive a water droplet along a dry surface), driving voltages was reduced to as low as ~15 V by using a very thin (700 Å) dielectric of a high dielectric constant (~200). This early evolution of EWOD is summarized in Table 3 with the applied voltages, the materials of the dielectric layers, and the published years.

**Table 3.** Summary of early EWOD evolution [26]

| Ref. | Voltage | Dielectric layer(s) | Summary | Year |
|---|---|---|---|---|
| [27] | AC (50 or 1,000 Hz) | 200–900 μm PE | Irreversible | 1980 |
| [28] | AC (50–3,000 Hz) | 53 μm PTFE | Reversible<br>Saturation | 1993 |
| [29] | AC (50 Hz – a few kHz) | 12 μm PET | Contour line instabilities | 1996 |
| [30] | DC | 0.1μm Teflon on 6–35 μm Parylene C or polyimide | 8° reversibility<br>Capillary rise | 1998 |
| [31] | DC | Silicone oil on 30 nm Teflon on 10 μm Parylene N | 2° reversibility<br>Polarity, ion type, and molarity independent | 1999 |
| [21] | ~15 V DC | Teflon on 700 Å BST | 120° → 80°<br>Dry surface | 2002 |

Illustrated in Fig. 16, EWOD may be considered an electrowetting phenomenon under certain material configurations. The ideal dielectric blocks electron transfer, while sustaining the high electric field at the interface that results in charge redistribution when a potential is applied. By using a hydrophobic dielectric, the large initial contact angle provides room for a large contact-angle change $\Delta\theta$ upon electrowetting. Furthermore, by employing a dielectric layer between the liquid and electrode, virtually any kind of liquid can be used (barring the change of solid surface by adsorption from the liquid), regardless of the polarization of the interface.

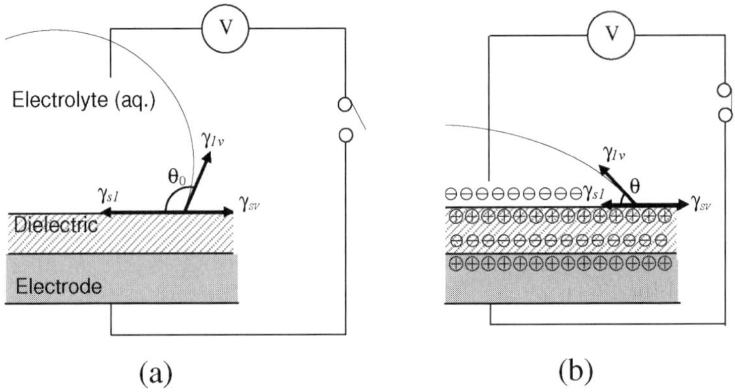


**Fig. 16.** A droplet on solid dielectric covering an electrode spreads when the charge is accumulated on the surface by an externally applied electric field between the liquid and the imbedded electrode. The electrowetting on this material configuration is called EWOD to clarify a deviation from the conventional electrowetting on bare electrode [21]

Lippmann's equation (29) can be expressed in terms of the contact angle $\theta$ by incorporating Young's equation (30). The resulting equation (32) is called the Lippmann-Young equation.

$$\gamma_{sl} = \gamma_{sv} - \gamma_{lv} \cos\theta, \tag{30}$$

$$\gamma - \gamma_0 = -\frac{1}{2} c(V_e^2 - V_{e0}^2), \tag{31}$$

$$\cos\theta = \cos\theta_0 + \frac{1}{\gamma_{lv}}\frac{1}{2}cV_e^2 . \tag{32}$$

Here, $\theta_0$ is the contact angle when the electric field across the interfacial layer is zero ($V_{e0} = 0$), $\gamma_{sl}$ is the solid–liquid surface tension, $\gamma_{lv}$ the liquid–gas surface tension, and $\gamma_{sv}$ the solid–gas surface tension. Note that $\gamma_{lv}$ and $\gamma_{sv}$ are assumed to be constant, independent of the applied potential. Further note that a change of the surface energy is at the solid–liquid interface, but visible to observation is movement of the liquid–gas interface.

## 4. Microfluidic Device Using Electrowetting

As it is convenient to control devices and systems electrically, electrowetting has become the most popular mechanism when it comes to microfluidic devices that drive droplets. Although many application examples have been reported to date, mostly biochemical, we will focus on the initial device development and the fundamental technologies rather than the applications.

### *4.1. Pumping by Electrowetting on Liquid Electrode: CEW*

First reported by Beni in 1981 [33], continuous electrowetting (CEW), illustrated in Fig. 17, is based on the same material configuration as the original Lippmann's experiment: liquid electrolyte and liquid electrode, or more specifically mercury in an electrolyte-filled glass capillary. However, there are discrepancies. In Lippmann's experiment, a column of mercury slightly grew from or shrunk back to a pool when a voltage is applied between the mercury pool and the electrolyte. For CEW, on the other hand, a droplet (or slug) of mercury is driven along the capillary by applying voltage between the electrolytes on the opposite sides of the mercury droplet.

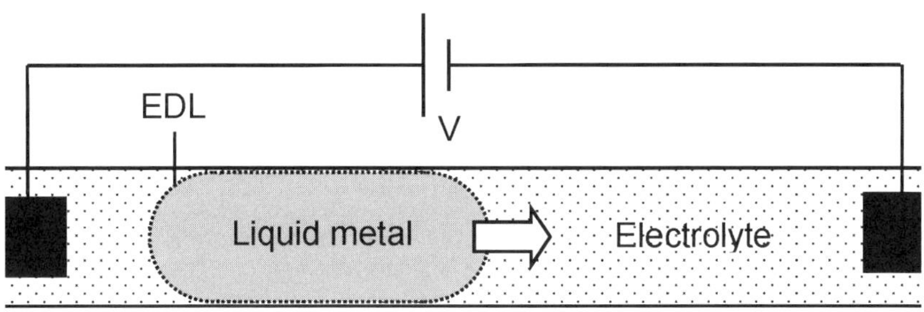


**Fig. 17.** A liquid–metal droplet moved by CEW in a capillary. Adapted from [26, 34]

More importantly, the actuation mechanism of CEW is different from that of the electrocapillarity in Lippmann's experiment. During the movement of mercury by CEW, a thin layer of electrolyte exists between the mercury and the capillary inner wall. When a voltage is applied between the electrolytes on the opposite sides of the mercury droplet, a potential gradient is established on the interface of the mercury and the thin electrolyte along the capillary length. As a result, the charge density in the EDL also varies along the interface, and so does the surface tension. This gradient in surface tension generates a net force and drives the mercury droplet along the capillary. The CEW mechanism has first been applied to optical switches [35, 36] and later successfully adopted as the basis of a liquid micromotor [37] and a micropump [38] fabricated by MEMS technologies.

### 4.2. Pumping by Electrowetting on Solid Electrode

While CEW describes electrical modulation of the liquid–liquid surface tension, general electrowetting includes modification of the solid–liquid surface tension as described in Sect. 3.3. When a voltage is applied between a solid electrode and electrolyte, the charge density of the EDL between them is changed, and so does the solid–liquid surface energy. If an electrolyte is introduced into a capillary whose inner surface is conductive, the initially nonwetting surface can be activated to become wetting and draw the liquid [22–25] as schematically described in Fig. 18.

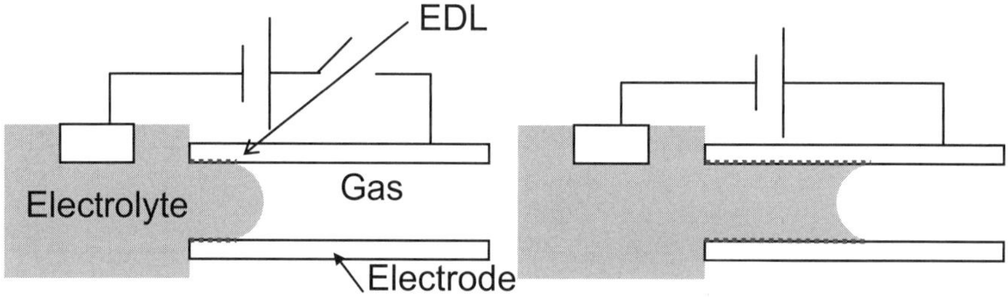


(a) Hydrophobic: retracting (b) Hydrophilic: advancing

**Fig. 18.** Aqueous liquid driven by electrowetting actuation into a capillary. Adapted from [24, 26]

To move a discrete droplet, the contact angle of the liquid segment changes locally by applying voltage to the electrode at one end of the droplet [24, 25]. One way to analyze the movement is through the pressure gradient generated inside the droplet. As illustrated in Fig. 19, when voltage is applied to the right side of the droplet, the contact angle of the droplet drops on the right side only. The curvature of the liquid segment on the right side decreases accordingly, and the liquid pressure in the wetting side (the right side) of the droplet becomes smaller than that of the left side. The pressure inside the droplet is no more at equilibrium, the left side having a larger pressure than the right end within the droplet. As a result, the liquid segment moves to the right.

$$\begin{aligned}\Delta P_1 &= \Delta P_{L1} - \Delta P_{R1} = \gamma_{lv}\left(\frac{1}{r_{L1}} - \frac{1}{r_{R1}}\right) \\ &= \gamma_{lv}\left(\frac{\cos\theta_{bL1} - \cos\theta_{tL1}}{d} - \frac{\cos\theta_{bR1} - \cos\theta_{tR}}{d}\right)\end{aligned} \quad . \tag{33}$$

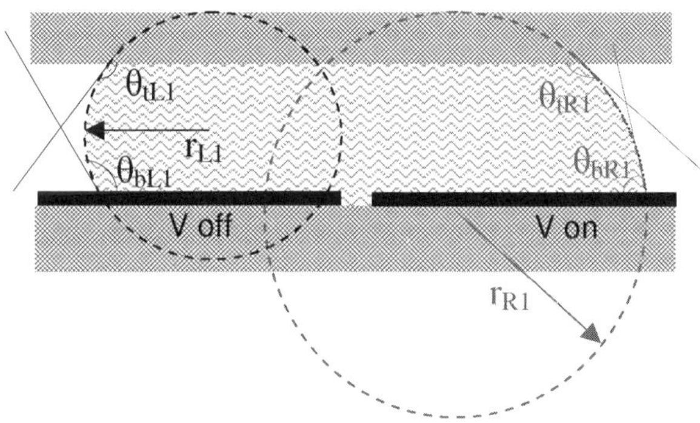


**Fig. 19.** A droplet under an internal pressure gradient induced by electrowetting actuation

### *4.3. Pumping by Electrowetting on Dielectric-Coated Solid Electrode (EWOD) [26, 39]*

Because of the large and reversible contact-angle change it allows, EWOD has become the most popular among electrowetting mechanisms. The solid–liquid surface tension is changed mainly because the applied voltage between the liquid and the embedded electrode changes the charge density of the dielectric layer. Just like the case with conventional electrowetting in the previous section, if an electrolyte is introduced into a capillary whose inner surface has the material configuration of EWOD, the initially nonwetting surface can be activated to become wetting and draw the liquid as in Fig. 20. Much improved reversibility over conventional electrowetting-based counterpart (i.e., electrowetting on metal) has been reported for the same device structures [24, 25].

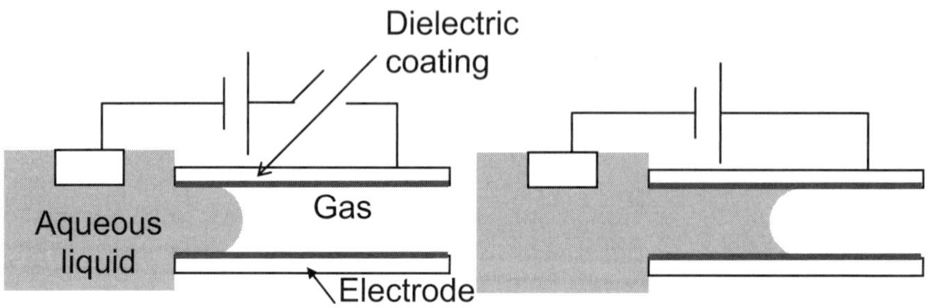


(a) Hydrophobic: retracting (b) Hydrophilic: advancing

**Fig. 20.** Aqueous liquid driven by EWOD into a capillary. Adapted from [24, 25]

Similar to the conventional electrowetting described in the previous section, discrete liquid droplets can be moved using a change of surface tension by EWOD as well [23, 24, 40]. One of the configurations that can be used is described in Fig. 21. When voltage *V* is applied across the droplet placed between the embedded electrodes in top and bottom plates, the liquid contact angles change according to the potential differences between the liquid and each electrode as shown in Fig. 21b. For simplicity, it is assumed that the contact angle changes the same if the applied voltages are the same. The bottom-left electrode is electrically floated and considered to have zero electrical potential, resulting in no contact-angle change. When the contact angles at the interfaces of the droplet change asymmetrically along the channel, the droplet moves in one direction (to the right side in Fig. 21b) due to the pressure gradient generated inside the droplet. The internal pressure gradient is calculated using the Laplace-Young equation, which expresses a pressure difference across an interface.

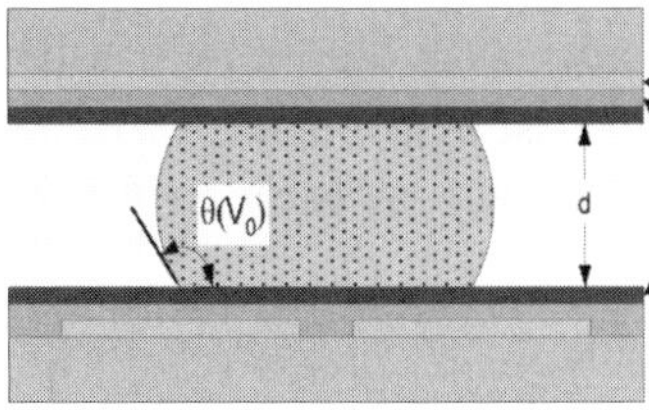


(a) Squeezed droplet between two plates before applying voltage $V$

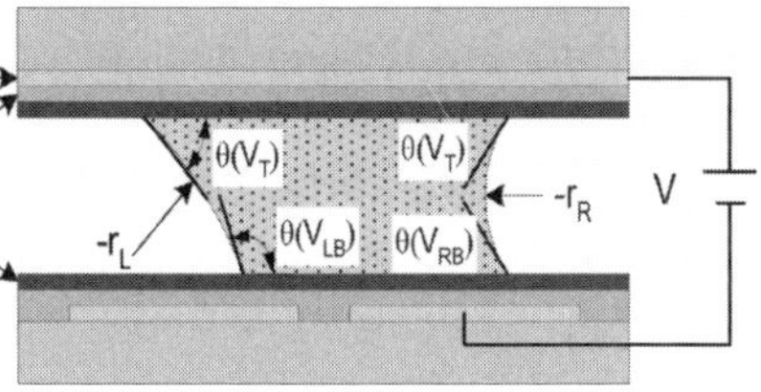


(b) Change in contact angles of the droplet after applying voltage $V$

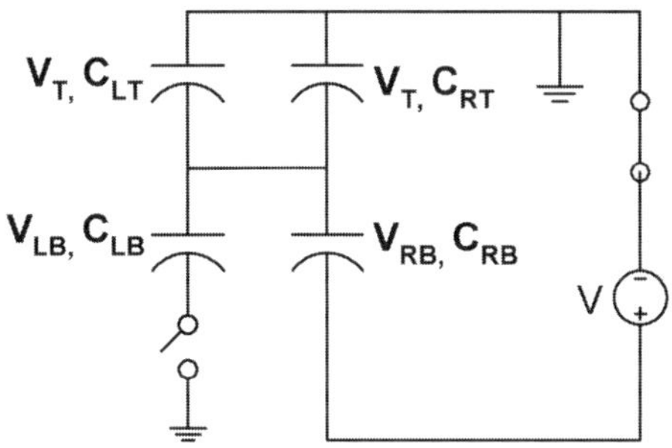


(c) Simple parallel-plate capacitor model

**Fig. 21.** Generating internal pressure imbalance by EWOD for moving a droplet. $\theta(V)$ = contact angle with applied voltage $V$; $r$ = radius of curvature; $d$ = distance between the parallel plates [39]

Driving pressure of the droplet is essentially the same as (33) but presented slightly differently and with more elaboration below. The pressure differences $\Delta P$ across the left liquid–air interface of the droplet and across the right interface (as illustrated in Fig. 21b) are

$$\Delta P_{\mathrm{L}} = P_{\mathrm{L}} - P_{\mathrm{a}} = \gamma_{lv}\left(\frac{1}{R_{\mathrm{L}}} + \frac{1}{r_{\mathrm{L}}}\right), \tag{34a}$$

$$\Delta P_{\mathrm{R}} = P_{\mathrm{R}} - P_{\mathrm{a}} = \gamma_{lv}\left(\frac{1}{R_{\mathrm{R}}} + \frac{1}{r_{\mathrm{R}}}\right), \tag{34b}$$

where $P_L$ and $P_R$ are local internal pressures inside the drop, on the left and right side, respectively, and $P_a$ is an atmospheric pressure outside the drop. $R_L$ and $R_R$ are the other major radii of the curvatures (in the plane parallel to the plates and not shown in the figure) and assumed equal. $\gamma_{lv}$ is the surface tension at the liquid–gas interface. The driving pressure $\Delta P$ of the drop is the difference between the internal pressures at the left side and right side of the drop and expressed as

$$\Delta P = P_L - P_R = \gamma_{lv}(\frac{1}{r_L} - \frac{1}{r_R}) \tag{35}$$

with radii of curvatures determined by the contact angles $\theta$. The radii of curvatures are geometrically determined by

$$\frac{1}{r_R} = -\frac{\cos\theta(V_T) + \cos\theta(V_{RB})}{d}, \tag{36a}$$

$$\frac{1}{r_L} = -\frac{\cos\theta(V_{LB}) + \cos\theta(V_T)}{d}. \tag{36b}$$

Substituting (36) into the driving pressure equation (35) gives

$$\begin{aligned}\Delta P &= \gamma_{lv}(\frac{1}{r_L} - \frac{1}{r_R}) \\ &= \frac{\gamma_{lv}}{d} \cdot \left[\cos(\theta(V_T)) + \cos(\theta(V_{RB})) - \cos(\theta(V_{LB})) - \cos(\theta(V_T))\right]\end{aligned} \tag{37}$$

The bottom-left electrode is floated and $V_{LB} = V_0 = 0$ while the top and the bottom plates have the same electrical properties, $C_{LT} = C_{LB} = C_{RT} = C_{RB} = C$. Combining with (32) and expressing the specific capacitance $c$ with the material and the geometry, (37) is simplified to

$$\Delta P = \frac{2}{9} \frac{\varepsilon_o \cdot \varepsilon}{dt} V^2, \tag{38}$$

where $\varepsilon$ is permittivity of vacuum, $\varepsilon_o$ and $t$ are the dielectric constant and the thickness of the dielectric layer, respectively, $\partial$ is the gap spacing between the top and bottom plate, and $V$ is the applied voltage.

Equation (38) shows that the driving pressure is always greater than zero when any voltage is applied. To translate a droplet in reality, however, the voltage has to be large enough to overcome the resistance against droplet movement. The maximum voltage one can apply is limited by an electrical failure through the dielectric. The resistance against droplet sliding is determined by the surface quality, such as smoothness and imperfections, which often is indicated by contact-angle hysteresis. With a given EWOD material configuration and surface quality, can we design more efficient EWOD devices? Understanding the mechanism of pumping and the resulting equations would serve us to explore more efficient driving of a droplet. The distance or the gap height $d$ is an important parameter for the droplet movement, and smaller gap tends to compensate the limitation in the maximum applicable voltage across the droplet.

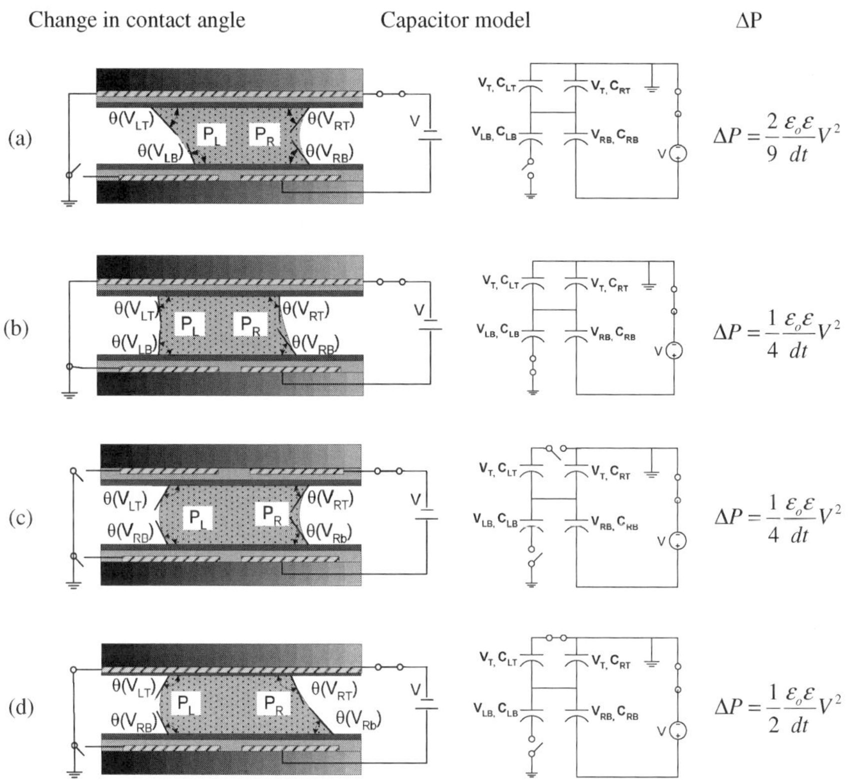


**Fig. 22.** Additional EWOD driving configurations and corresponding driving pressures [39]

Various configurations of electrical connection are also possible other than the one described above and summarized in Fig. 22. The configuration in Fig. 22b is identical to the one in Fig. 22a except for a grounded bottom left electrode. For the configuration shown in Fig. 22c, the top electrodes are also separated such that each electrode can have different potential levels. For the configuration shown in Fig. 22d, the top plate does not have any dielectric layer, but a very thin hydrophobic layer acts as a dielectric [41]. The driving pressure for each configuration is calculated by applying the same principles used to determine (38). According to the calculation, the most efficient way to generate the driving pressure is through the configuration of Fig. 22d, where the pressure is more than twice that of the configuration in Fig. 22a. Other two configurations have similar efficiencies to the first configuration. When considering efficiency only, the configuration shown in Fig. 22d would be best to use to drive the discrete droplets. However, the selection of the proper configuration requires other considerations such as materials, surface coating, device fabrication, device package, and applications.

### *4.4. Reconfigurable Digital (or Droplet) Microfluidics*

Electrowetting, including EWOD, applies most naturally to droplets, when it comes to handling liquids on a chip. Such microfluidics is often called droplet or digital microfluidics as opposed to the usual continuous microfluidics. An analogy can be made in electronics, where digital electronics processes packets of electronic signals (0 s and 1 s) while its analog counterpart processes the continuous flow of signals. To enable a lab-on-a-chip based on droplets operated by electrowetting actuation, four fundamental microfluidic operations may be identified as prerequisites for digital microfluidic circuits: (1) cutting, (2) merging, (3) creating, and (4) transporting liquid droplets [41, 42]. The most challenging among the four was creation of droplets (i.e., digitizing the sample) from a pool of liquid (reservoir) entirely by EWOD actuation (e.g., without any non-EWOD assistant such as pushing the sample in with an external pressure) or reducing the resistance (e.g., using oil).

Figure 23 shows the EWOD-based digital microfluidics envisioned by Lee [23]. The lab-on-a-chip creates droplets from the reservoirs and carries out most of the microfluidic operations with the discrete droplets. For example, Fig. 23 shows that in M+N, different samples or reagents are introduced and, after several droplet-based microfluidic operations, various kinds of products can be produced or analyses made on a single chip. Since the routing of the droplets is determined by electrical signals, rather than the predetermined physical features, the chip is reconfigurable. The reconfigurability is fully exercised if the EWOD chip can route droplets on a two-dimensional grid array [43, 44]. Since electric control of surface tension moves individual droplets locally, discrete pumps, valves, or moving micromechanical parts are not needed, making the system dramatically simple, versatile, and energy efficient. Applying voltages across a dielectric layer, EWOD actuation consumes little power – important for battery-operated systems. The main advantages of EWOD actuation are best demonstrated when a reconfigurable digital microfluidic chip is operated in a handheld system [43, 45].

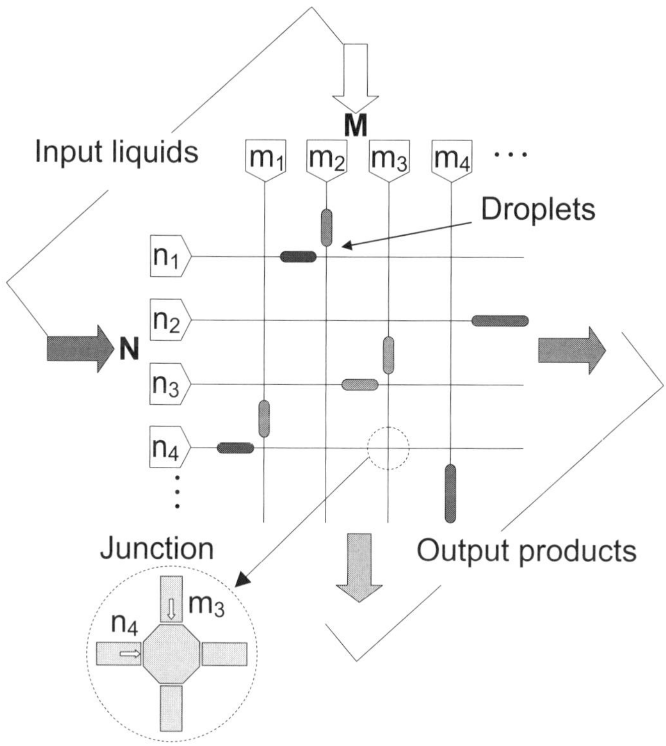


**Fig. 23.** Conceptual description of a reconfigurable digital microfluidic circuit [23]

## 5. Summary

Interfacial surface energy or tension can be controlled through various means including electric potential, the focus of this chapter. The relationship between the surface tension and the externally applied voltage was derived from the basic thermodynamic equations of the interfaces through the general concept of work, Young's equation, and the Young and Laplace equation. Starting from Lippmann's electrocapillarity, CEW was described as the first surface-tension-driven microactuation. Electrowetting was then described in its conventional configuration on solid electrode, leading to the more recent configuration of EWOD. A few examples of applications of these electrowetting mechanisms on microfluidic devices were described with an emphasis on EWOD.

## Acknowledgments

The work was supported by the NASA Institute of Cell Mimetic Space Exploration. The bulk of the materials used has been developed using Kim's NSF CAREER award, NSF XYZ-on-a-Chip program, and DARPA BioFlips program. The authors are indebted to Drs. Junghoon Lee, Shih-Kang "Scott" Fan, Ui-Chong Yi, and Sung K. Cho, who helped develop the understanding of the mechanism of electrowetting over the years. This review heavily relied on their dissertations. The help of Ms. Annie Lee is also appreciated in preparing the manuscript.

## References

1. Maboudian R and Howe RT (1997) Critical review: Adhesion in surface micromechanical structures. Journal of Vacuum Science and Technology B15: 1–20
2. Kim C-J, Kim JY, and Sridharan B (1998) Comparative evaluation of drying techniques for surface micromachining. Sensors and Actuators A: Physical A64: 17–26
3. Jun TK and Kim C-J (1998) Valveless pumping using traversing vapor bubbles in microchannels. Journal of Applied Physics 83: 5658–5664
4. Kim C-J (1999) Micromachines driven by surface tension. In: *Proceedings of 30th AIAA Fluid Dynamics Conference*, Norfolk, VA, pp. 1–6 (invited lecture)
5. Kim C-J (2001) Micropumping by electrowetting. In: *Proceedings of ASME Int Mechanical Engineering Congress and Exposition*, New York, NY, pp. IMECE2001/HTD-24200 (Invited)

6. Sammarco TA and Burns MA (1999) Thermocapillary pumping of discrete drops in microfabricated analysis devices. AIChE Journal 45: 350–366
7. Washizu M (1998) Electrostatic actuation of liquid droplets for microreactor applications. IEEE Transactions on Industry Applications 34: 732–737
8. Tseng F-G, Kim C-J, and Ho C-M (2002) A high resolution high frequency monolithic top-shooting microinjector free of satellite drops: Part I, Concept, design and model. Journal of Microelectromechanical Systems 11: 427–436
9. Latorre L, Kim J, Lee J, de Guzman P-P, Lee HJ, Nouet P, and Kim C-J (2002) Electrostatic actuation of microscale liquid-metal droplets. Journal of Microelectromechanical Systems 11: 302–308
10. Shen W, Edwards RT, and Kim C-J (2006) Electrostatically-actuated metal-droplet microswitches integrated on CMOS chip. Journal of Microelectromechanical Systems 15: 879–889
11. Kang KH (2002) How electrostatic fields change contact angle in electrowetting. Langmuir 18: 10318–10322
12. Kang KH, Kang IS, and Lee CM (2003) Wetting tension due to coulombic interaction in charge-related wetting phenomena. Langmuir 19: 5407–5412
13. Jones TB, Gunji M, Washizu M, and Feldman MJ (2001) Dielectrophoretic liquid actuation and nanodroplet formation. Journal of Applied Physics 89: 1441–1448
14. Vykoukal J, Schwartz J, Becker F, and Gascoyne P (2001) A programmable dielectrophoretic fluid processor for droplet-based chemistry. In: *Proceedings Micro Total Analysis Systems*, Netherlands, pp. 72–74
15. Mugele F (2005) Electrowetting: From basics to applications. Journal of Physics: Condensed Matter 17: R705–R774
16. Moore WJ (1972) *Physical Chemistry*, 2nd edn. Prentice-Hall, Englewood Cliffs, NJ
17. Castellan GW (1983) *Physical Chemistry*, 4th edn. Addison-Wesley Reading, MA
18. Kim C-J, *Class Note: UCLA MAE 281 Microsciences,* 1998.
19. Israelachvili J (1992) *Intermolecular and Surface Forces*, 2nd edn. Academic Press, London
20. Moon H (2005) Electrowetting-on-dielectric microfluidics: Modeling, physics, and MALDI application. Ph.D. Dissertation, University of California at Los Angeles (UCLA)
21. Moon H, Cho SK, Garrell RL, and Kim C-J (2002) Low voltage electrowetting-on-dielectric. Journal of Applied Physics 92: 4080–4087
22. Matsumoto H and Colgate JE (1990) Preliminary investigation of micro-pumping based on electrical control of interfacial tension. In: *Proceedings of IEEE MEMS Workshop*, Napa Valley, CA, USA, pp. 105–110
23. Lee J, Moon H, Fowler J, Kim C-J, and Schoellhammer T (2001) Addressable micro liquid handling by electric control of surface tension. In: *Proceedings of IEEE International Conference on MEMS*, Interlaken, Switzerland, pp. 499–502

24. Lee J-H (2000) Microactuation by continuous electrowetting and electrowetting: Theory, fabrication, and demonstration. Ph.D. Dissertation, University of California at Los Angeles (UCLA)
25. Lee J, Moon H, Fowler J, Schoellhammer T, and Kim C-J (2002) Electrowetting and electrowetting-on-dielectric for microscale liquid handling. Sensors and Actuators A 95: 259–268
26. Fan S-K (2003) Digital microfluidics by cross-reference EWOD actuation: Principle, device & system. Ph.D. Dissertation, University of California at Los Angeles (UCLA)
27. Minnema L, Barneveld HA, and Rinkel PD (1980) An investigation into the mechanism of water treeing in polyethylene high-voltage cables. IEEE Transactions on Electrical Insulation EI-15: 461–472
28. Berge B (1993) Electrocapillarity and wetting of insulator films by water. Comptes Rendus de l'Academie des Sciences Series II 317: 157–163
29. Vallet M, Berge B, and Vovelle L (1996) Electrowetting of water and aqueous solutions on poly(ethylene terephthalate) insulating films. Polymer 37: 2456–2470
30. Welters WJJ and Fokkink LGJ (1998) Fast electrically switchable capillary effects. Langmuir 14: 1535–1538
31. Verheijen HJI and Prins MWJ (1999) Reversible electrowetting and trapping of charge: model and experiments. Langmuir 15: 6616–6620
32. Seyrat E and Hayes RA (2001) Amorphous fluoropolymers as insulators for reversible low-voltage electrowetting. Journal of Applied Physics 90: 1383–1386
33. Beni G, Hackwood S, and Jackel JL (1982) Continuous electrowetting effect. Applied Physics Letters 40: 912–914
34. Lee J and Kim C-J (1998) Liquid micromotor driven by continuous electrowetting. In: *Proceedings of IEEE MEMS Workshop*, Heidelberg, Germany, pp. 538–543
35. Jackel JL, Hackwood S, and Beni G (1982) Electrowetting optical switch. Applied Physics Letters 40: 4–5
36. Jackel JL, Hackwood S, Veselka JJ, and Beni G (1983) Electrowetting switch for multimode optical fibers. Applied Optics 22: 1765–1770
37. Lee J and Kim C-J (2000) Surface-tension-driven microactuation based on continuous electrowetting. Journal of Microelectromechanical System 9: 171–180
38. Yun K-S, Cho I-J, Bu J-U, Kim C-J, and Yoon E (2002) A surface-tension driven micropump for low-voltage and low-power operations. Journal of Microelectromechanical System 11: 454–461
39. Yi U-C (2004) Soft printing of biological liquids for microarrays: Concept, principle, fabrication, and demonstration. Ph.D. Dissertation, University of California at Los Angeles (UCLA)
40. Pollack MG, Fair RB, and Shenderov AD (2000) Electrowetting-based actuation of liquid droplets for microfluidic applications. Applied Physics Letters 77: 1725–1726

41. Cho SK, Moon H, and Kim C-J (2003) Creating, transporting, cutting, and merging liquid droplets by electrowetting-based actuation for digital microfluidic circuits. Journal of Microelectromechanical Systems 12: 70–80
42. Cho SK, Moon H, Fowler J, and Kim C-J (2001) Splitting a liquid droplet for electrowetting-based microfluidics. In: *Proceedings of International Mechanical Engineering Congress and Exposition*, New York, NY, USA, pp. IMECE2001/MEMS-23831
43. Fan S-K, Hashi C, and Kim C-J (2003) Manipulation of multiple droplets on NxM grid by cross-reference EWOD driving scheme and pressure-contact packaging. In: *Proceedings of IEEE International Conference on MEMS*, Kyoto, Japan, pp. 694–697
44. Gong J and Kim C-J (2005) Two-dimensional digital microfluidic system by multi-layer printed circuit board. In: *Proceedings of IEEE Conference on MEMS*, Orlando, FL, pp. 726–729
45. Gong J, Fan S-K, and Kim C-J (2004) Portable digital microfluidics platform with active but disposable lab-on-chip. In: *Proceedings of IEEE Conference on MEMS*, Maastricht, the Netherlands, pp. 355–358

# Chapter 6
# Magnetic Beads in Microfluidic Systems – Towards New Analytical Applications

Martin A.M. Gijs

Ecole Polytechnique Fédérale de Lausanne (EPFL), Institute of Microelectronics and Microsystems, CH-1015 Lausanne, Switzerland.

## 1. Introduction

Key advantages of using microfluidic systems of reduced dimension for analytical applications are known to be (a) the possibility of using minute quantities of sample and reagents (down to picoliters), (b) relatively fast reaction times, when molecular diffusion lengths are of the order of the microchannel dimension, and (c) a large surface-to-volume ratio offering an intrinsic compatibility between the use of a microfluidic system and surface-based assays. Reviews of various fluidic operations in microfluidic systems, like sample preparation, sample injection, sample manipulation, reaction, separation and detection, published in the period 1998–2004, were presented by Auroux et al. [1] and Vilkner et al. [2].

Also nanomaterials and nanoparticles have become a hot topic in research. Functional nano- and microparticles ('beads') offer a large specific surface for chemical binding and a polymer colloid or microsphere solution has a low viscosity compared to solutions having the same amount of solid, giving it special properties. Such small particles can be advantageously used as a 'mobile substrate' for bio-assays or even for in vivo applications; they can be easily recovered from a dispersion, reversibly re-dispersed etc. Several reviews on the preparation and use of polymer particles and polymer colloids for medical, biological and optical applications exist [3, 4].

*Magnetic* nano- and microparticles offer still an additional advantage: having embedded magnetic entities, they can be magnetically manipulated using permanent magnets or electromagnets, independent of normal microfluidic or biological processes. This extra degree of freedom is at the basis of a still improved exposure of the functionalized bead surface to the surrounding liquid and of higher sample pre-concentration efficiencies, due to the increased relative motion of the bead with respect to the fluid. Pankhurst et al. [5] reviewed the applications of magnetic nanoparticles in biomedicine with focus on the underlying physics. Gijs [6] reviewed the handling and manipulation of magnetic particles in microfluidic systems, pointing out the advantages and prospects of these concepts for future analysis applications. The review of Pamme [7] has a somewhat wider scope and discusses various developments within the field of magnetism and microfluidics.

## 2. Types of Magnetic Beads

The synthesis of (non-magnetic) polymer micoparticles were reviewed by Kawaguchi [3]. The articles of Landfester and Ramírez [8], Bergemann et al. [9] and Grüttner et al. [10] present short review sections on the synthesis and chemical modifications of magnetic beads. In early publications, magnetic bead solutions were produced by grinding magnetite ($Fe_3O_4$) with long-chain hydrocarbons and a grinding agent [11]. Later magnetic fluids were produced by precipitating an aqueous $Fe^{3+}/Fe^{2+}$ solution with a base, coating these particles with an adsorbed layer of oleic acid and then dispersing them in a non-aqueous fluid [12]. Both type of processes result in very small magnetic particles with a surfactant coating in a non-aqueous liquid carrier in which the hydrophobic magnetite particles are dispersed. However, a lot of applications of magnetic beads rely on water as the continuous phase.

A water-based magnetic fluid was realized by conversion of iron products to magnetic iron oxide in an aqueous medium under controlled pH conditions [13]. Polymer-coated magnetic beads can be produced by in situ precipitation of magnetic materials in the presence of a polymer. In this way, magnetic beads surrounded by a hydrophilic polymer shell have been made, choosing for the polymer the water-soluble dextran [14], poly(ethylene imine) [15], poly(vinyl alcohol) [16], poly(ethylene glycol) [17] etc.

Different types of magnetic beads have their individual advantages and disadvantages. Polystyrene-coated magnetic particles are known for their excellent size distribution and spherical shape [18, 19]. However, their hydrophobic surface results in a high amount of unspecific protein and antibody binding on the particle surface, so that it needs to be modified chemically. Magnetic silica particles are very efficient in adsorbing proteins and DNA on their surface, but are hardly available with a small size distribution and an ideal spherical shape [20, 21]. Magnetic polysaccharide particles are important for many in vivo applications. They combine biocompatibility with availability in a size range below 300 nm [22, 23], but the particles are irregular in shape and the soft particle matrix causes them to be sensitive to mechanical stress. Also magnetic poly(lactic acid) particles play an important role in in vivo applications [24, 25]: they are biodegradable and their degradation time in the blood can be adjusted by their molecular weight and exact chemical composition.

From a physics point of view, the magnetic content of a particle is of particular importance. Small nanoparticles consist of a single-domain, because they have a dimension that is typically of the order or smaller than the typical thickness of a magnetic domain wall. Mono-domain magnetic particles become superparamagnetic (i.e. their time-averaged magnetization without a magnetic field is zero), when their magnetic energy is lower than about 10 times the thermal energy $k_B T$, with $r$ the particle radius of a supposed spherical particle and $k_B$ the Boltzmann constant [26]. At room temperature, $k_B T = 4.0 \times 10^{-21}\,\mathrm{J}$, and one finds a maximum radius $r$ = 6 nm for a superparamagnetic spherical particle of iron. Iron oxides, such as magnetite ($Fe_3O_4$) or maghemite ($\gamma$-$Fe_2O_3$) are more stable against oxidation and are preferentially used as core material, instead of iron. Typically, superparamagnetic particles of $Fe_3O_4$ with diameters in the 5–100 nm range are used. The magnetization curve of an ensemble of such superparamagnetic particles is hysteresis-free. As a consequence, suspended superparamagnetic particles tagged to the biomaterial of interest can be removed from a matrix using a magnetic field, but they do not agglomerate after removal of the field (i.e. they stay suspended). Hence, it is very easy to switch-on and -off the magnetic interaction. Larger magnetic particles (typically 0.5–5 μm in diameter) can have a single magnetic core or have a core composed of multiple more or less magnetically interacting nanoparticles in a non-magnetic matrix. Such mostly have a multi-domain structure and are characterized by a hysteretic magnetization characteristic. Two of such particles will have a maximum magnetic attraction energy much larger than the thermal energy, resulting in strong dipolar forces between the particles [27]. As a consequence, when exposed to an external magnetic induction, the magnetic microparticles acquire a magnetic dipole moment and coalesce, under influence of the magnetic dipole interaction into a supra-particle structure (SPS) consisting of chain-like 'columnar' structures along the field direction.

## 3. Forces on Magnetic Beads

It is important to recognize that a magnetic field *gradient* is required to exert a translation force, a uniform field giving rise only to a torque. The magnetic force acting on a point-like magnetic dipole or 'magnetic moment' **m** can be written as the gradient of the magnetic energy [11, 28]

$$\mathbf{F}_{\mathrm{m}} = \frac{1}{\mu_0} \nabla(\mathbf{m} \cdot \mathbf{B}) \approx \frac{1}{\mu_0} (\mathbf{m} \cdot \nabla)\mathbf{B} \tag{1}$$

The second part of the equation holds when the magnetic moment of the particle is not varying in space ($\nabla \cdot \mathbf{m} = 0$). Note that this is only a correct assumption when the moment is permanent or the magnetic particle is in such large field that its magnetization is completely saturated. In the case of superparamagnetic nanoparticles in a liquid medium, one can write for the moment $\mathbf{m} = V\mu_0 \mathbf{M} = V\mu_0 \chi \mathbf{H}$, with $\mathbf{M}$ the magnetization, $V$ the volume of the particle and $\chi$ the difference in magnetic susceptibility between the particle and the medium. Using the relation $\mathbf{B} = \mu_0 \mathbf{H}$, (1) becomes

$$\mathbf{F}_{\mathrm{m}} = \frac{V\chi}{\mu_0} (\mathbf{B} \cdot \nabla) \mathbf{B} \tag{2}$$

A permanent magnet typically is characterized by a magnetic induction $B_{\mathrm{m}} = 0.5 - 1\mathrm{T}$ and a field gradient $\nabla B \approx B_{\mathrm{m}}/w$, with $w$ the geometrical dimension of the permanent magnet. For a cylindrical permanent magnet with a diameter Ø = 5 mm, one induces on a spherical particle with radius $r = 500$ nm and $\chi = 1$ a magnetic moment $m = 2.6 \times 10^{-19}$ Tm$^3$, resulting in a magnetic force of about 40 pN. For a current-fed coil, the generated field is much smaller: a flat millimeter-size coil with 10 windings and a current of 0.5–1 A generates typically a magnetic induction of 1–10 mT, at least 100 times smaller than the permanent magnet. Consequently, also the gradient is a factor 100 lower, so that the force of (2) is a factor $10^4$ larger, explaining the frequent use of permanent magnets for actuation of magnetic beads in literature.

In many applications, a magnetically labeled material is separated from a liquid solution by passing the fluid mixture through a region where there is a magnetic field gradient that can immobilize the tagged material via magnetic forces. In all examples, the magnetic force needs to be appreciably larger than the hydrodynamic drag force acting on the magnetic particle. The hydrodynamic drag force is a consequence of the velocity difference between the magnetic particle and the liquid $\Delta\mathbf{v}$ and, for a spherical particle with radius $r$, is given by [29]

$$\mathbf{F}_{\mathrm{d}} = 6\pi\eta r\Delta\mathbf{v} f_{\mathrm{D}} \tag{3}$$

where $\eta$ is the viscosity of the medium surrounding the particle (for water, $\eta = 8.9 \times 10^{-4}\,\mathrm{N\,s\,m^{-2}}$) and $f_{\mathrm{D}}$ the drag coefficient that takes into account the influence of an eventual solid wall in the vicinity of the moving particle. Far away from the wall, $f_{\mathrm{D}} \approx 1$, while, when approaching the wall to a distance that is comparable with the particle size, $f_{\mathrm{D}}$ can increase up to a factor 3 [30]. Equalizing (2) and (3) and assuming $f_{\mathrm{D}} = 1$ permits to determine the maximum flow rate a particle can withstand when exposed to a magnetic immobilization force, or the maximum particle flow rate that can be generated by a magnetic force in a surrounding static liquid:

$$\Delta\mathbf{v} = \frac{2r^2\chi(\mathbf{B}\cdot\nabla)\mathbf{B}}{9\mu_0\eta} \equiv \frac{1}{\mu_0}\xi(\mathbf{B}\cdot\nabla)\mathbf{B} \tag{4}$$

with

$$\xi \equiv \frac{2r^2\chi}{9\eta} = \frac{V\chi}{6\pi r\eta} \tag{5}$$

the 'magnetophoretic mobility' of the particle, a parameter describing how magnetically manipulable the particle is. In general, the quantity $(\mathbf{B}\cdot\nabla)\mathbf{B}$ can be a strongly varying function in space.

Besides the magnetic and viscous forces that act in the horizontal plane of the chip, there exist other types of forces that are aligned perpendicularly to the channel wall, like the electrostatic force $F_{el}$ and the Van der Waals force $F_{\mathrm{VdW}}$ [31]. The Van der Waals force between two atoms is in most cases an attractive force that originates from electromagnetic interactions between dipoles and/or induced dipoles and that is acting on a distance smaller than 0.1–1 µm [32].

When a substrate or a particle is immersed into an aqueous solution, a surface charge can be induced through adsorption of ions present in the liquid and/or dissociation of surface groups. This surface charge is neutralized by oppositely charged mobile charges in the liquid, forming the so-called double layer. When the double layer of two surfaces overlap, an electrostatic interaction is induced resulting in either a repulsive or an attractive force $F_{el}$. This force can be derived from the electrostatic interaction energy and is written as follows [31, 33]:

$$F_{el} = \frac{2\pi\varepsilon\kappa r}{1-\exp(-2\kappa z)}\left[2\Psi_{0,s}\Psi_{0,p}\exp(-\kappa z)+\left(\Psi_{0,s}^{2}+\Psi_{0,p}^{2}\right)\exp(-2\kappa z)\right] \tag{6}$$

where $\varepsilon$ is the permittivity of the medium, $\Psi_{0,x}$ the surface potential of the particle or the substrate, $z$ the vertical distance between the substrate and the particle and $\kappa$ is the Debye-Hückel inverse double-layer thickness, which is defined as

$$\kappa = \left(\sqrt{\frac{\varepsilon k_B T}{2000\, N_A\, e^2 I}}\right)^{-1} \tag{7}$$

with $e$ the elementary charge, $N_A$ Avogadro's constant and $I$ the ionic strength of the buffer medium (Mole/l). An interesting basic study of electrostatic interactions between a $SiO_2$ surface and acrylate-coated particles showed that sticking of the particles to the microfluidic chip can be avoided by working at sufficiently high pH ($pH > 6$) and at not too high ionic strength of the medium ($I < 10\,mM$) [31].

## 4. Magnetic Bead Separation

Magnetic separator design can be as simple as the application and removal of a permanent magnet to the wall of a test tube to cause aggregation, followed by removal of the supernatant. However, it is preferable to increase the separator efficiency by producing regions with a high magnetic field gradient to capture the magnetic nanoparticles as they flow by in the carrier medium. For example, the separation can take place within a laminar flow of a carrier fluid along a thin microchannel. A field gradient is imposed across the thin dimension of the channel, perpendicular to the direction of the flow. A division of the flow at the channel outlet using a stream splitter completes the separation into fractions. The perpendicular particle separation velocity is induced by the field gradient and can easily be calculated using (4) (Fig. 1).

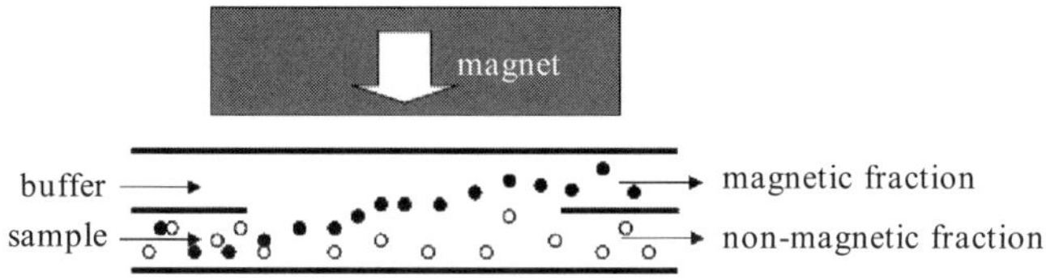


**Fig. 1.** Separation using a magnetic field gradient of a sample containing non-magnetic and magnetically labeled molecules into a magnetic and non-magnetic fraction

Microfluidic channels made in silicon wafers have been combined with closely positioned permanent magnets to separate magnetic from non-magnetic particles into different output channels [34]. In many of the 'macroscopic' applications, high magnetic field gradients were achieved by using the high background field gradient of a permanent magnet or an electromagnet in combination with a filter of closely packed steel wool. Most of the latter approaches, although successful, have the drawback that close monitoring of the separation process is difficult by the nature of the filter geometry.

A truly integrated magnetic bead sorter combines micromagnetics with microfluidics and is characterized by a magnetic environment where particles with distinct magnetic properties may be manipulated differently than other particles on the length scale of the microfluidic circuit. The group of C.H. Ahn, as one of the first, reported such integrated bead separation systems [35–37]. Such devices can be used, for example, for miniaturized cell sorting, where the magnetic beads are used as magnetic 'label' for actuation, or in a miniaturized bio-reaction system, where the magnetic beads play the role of carrier substrate in an assay [35, 38]. Figure 2 is an experimental separation of 1 μm diameter magnetic beads from a liquid flow. The separation is realized within a microfluidic channel in a silicon wafer that is integrated on top of a set of two planar electromagnets. Each of the electromagnets has a width of 50 μm and consists of a spiral Cu winding, electroplated in a photoresist mold and integrated with a soft magnetic permalloy yoke and backing plate. The latter serves to enhance the generated field by about a factor of two with respect to the field generated by the Cu winding only. The coil on the right is actuated with a current of 0.3 A and separates the beads from the flowing solution, while the coil on the left is without a current and attracts no magnetic beads from the flow. Comprehensive theoretical and experimental studies of various types of microelectromagnets on-chip were presented more recently [39–44].

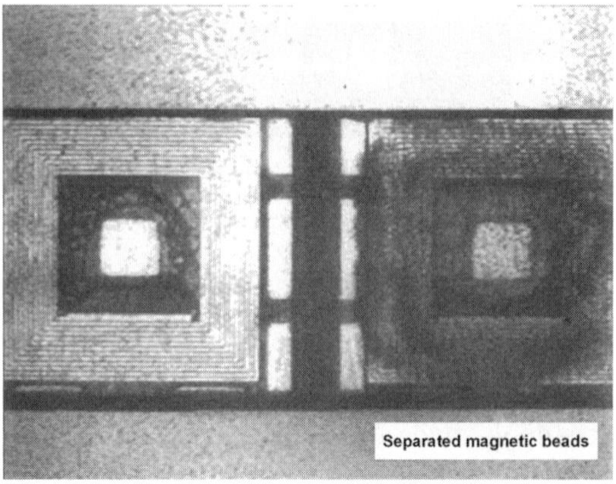


**Fig. 2.** Separation of 1 μm diameter magnetic beads from a fluid flow. The coil on the right is a planar electromagnet actuated with a current of 0.3 A, while the coil on the left is without a current. The width of the coil is 50 μm [reprinted from Choi et al. [36], with permission © 2001 Elsevier]

An important difference between the use of a permanent magnet and an electromagnet is the much lower generated magnetic induction of the latter. A permanent magnet easily generates a magnetic induction of 0.5–1 T, while the magnetic induction of a simple planer coil is in the mTesla range. Following the discussion of (2) and (3), one understands that an integrated electromagnet will produce a much smaller magnetic force, so that fluid flow in the microchannels needs to be strongly limited or that magnetic beads need to pass at a very close distance with the planar coil. Therefore, an interesting idea was to combine a uniform external magnetic field, to impose a large magnetic moment to the beads, with integrated linear conductors for the generation of a local magnetic field gradient [45]. Such setup allowed for the application of relatively strong and very local applied forces; up to 0.5 pN for 2.8 μm diameter polystyrene-coated particles.

In other work, a magnetic separation system was realized using an array of 15 μm wide electroplated Ni posts as filtering elements placed inside microfluidic channels made by soft lithography in PDMS [46]. Once magnetized by a magnetic field from an external NdFeB permanent magnet, these Ni posts attracted the magnetic field lines and thereby generated strong magnetic field gradients that efficiently could trap superparamagnetic beads passing them in a flowing stream of water. These Ni post arrays were also used to separate magnetic beads from non-magnetic beads. Similar work was published on the trapping of magnetic particles flowing through a microfluidic channel realized in PDMS [47]. Here, one has used magnetic 'anchors' formed by 4 μm diameter magnetic beads embedded in the sidewall of the microfluidic channel during the PDMS molding process.

The idea of integrating an array of integrated soft-magnetic elements embedded beneath or at the side of a microfluidic channel was reported on several occasions in literature [48, 49]. Such elements, which are polarized by a moderate bias field provided by an external magnet, produce a non-uniform field distribution characterized by a considerable field gradient. Functionalization of a microchannel with two types of magnetic beads using hydrodynamic focusing has also been demonstrated (see Fig. 3) [50]. The soft magnetic elements placed on both sides of the channel are magnetized by a relatively weak applied external magnetic field and provide magnetic field gradients attracting magnetic beads. Flows with two differently functionalized magnetic beads and a separating barrier flow are introduced simultaneously at the two channel sides and the centre of the microfluidic channel, respectively. On-chip experiments with fluorescence labeled beads demonstrate that the two types of beads are captured at each of the channel sidewalls. On-chip hybridization experiments show that the microfluidic systems can be functionalized with two sets of beads carrying different probes that selectively recognize a single base pair mismatch in target DNA. By switching the places of the two types of beads it was shown that the microsystem can be cleaned and functionalized repeatedly with different beads with no cross-talk between experiments.

A continuous diamagnetic capture mode magnetophoretic microseparator for separating red and white blood cells from diluted whole blood based on their native magnetic properties has been proposed [51–53]. The device was realised using microfabrication technology enabling the integration of microscale magnetic flux concentrators in an aqueous microfluidic environment. The micro-separator separated the blood cells using a high-gradient magnetic separation method without the use of magnetic bead labels. In other work [54–56], a continuous-flow microfluidic device that enabled cell by cell separation of cells selectively tagged with magnetic nanoparticles was demonstrated. The cells flew over an array of microfabricated magnetic stripes, which create a series of high magnetic field gradients that trap the magnetically labeled cells and alter their flow direction. The device demonstrated the separation of leukocytes form whole human blood. Magnetic field flow fractionation and analysis of different kind of magnetic nanoparticles was also demonstrated in a capillary using an external permanent magnet oriented orthogonal to a capillary flow induced by an applied pressure [57]. Retention and elution of the magnetic particles is a strong function of their magnetic moments, as stronger magnetic particles form thin clusters near the capillary wall with less interaction with the parabolic flow profile induced by a pressure gradient, while magnetically weaker particles feel the full viscous drag from the centre of the flow, resulting in an effective separation.

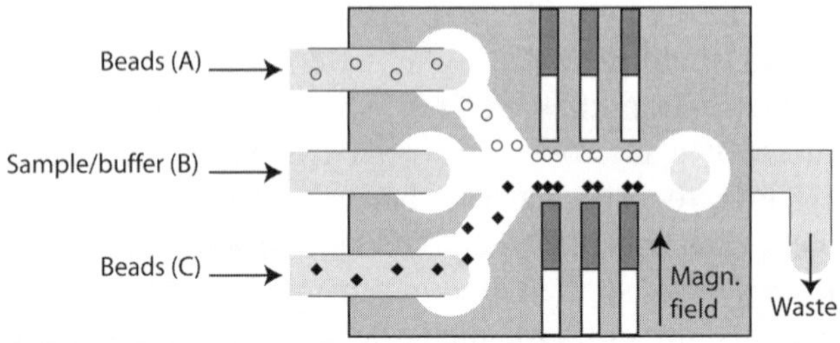


**Fig. 3.** Microsystem with soft magnetic elements and using hydrodynamic focusing. Two types of beads are applied through inlets A and C, respectively, and separated by a buffer flow applied through inlet B. The applied magnetic field and the magnetized magnetic elements with captured beads are indicated [reprinted from Smistrup et al. [50], with permission © 2005 The Royal Society of Chemistry]

The separation of magnetic microparticles was also achieved on-chip using free-flow magnetophoresis [58]. In this case, the magnetic particle sample solution and a buffer liquid are introduced in a microfluidic chip comprising a square separation chamber. A magnetic field gradient is applied transverse to the flow direction but in the plane of the substrate, rather than perpendicular to the plane of the chip. In this case, the complete width of the chamber can be used to separate non-magnetic and different kind of magnetic particles, rather than using the gradient over the much smaller vertical dimension of a microfluidic channel.

## 5. Magnetic Bead Transport

In magnetic transport, magnetic forces effectively transport the particles: it requires magnetic fields and magnetic forces that act on a larger range than necessary for separation, where magnetic beads approach very closely the magnetic actuation region by the fluid motion. Manipulation of magnetic beads in general and transport in particular is a difficult task, as the magnetic susceptibility $\chi$ of the magnetic beads is rather weak (typically $\chi \leq 1$), due to the small magnetic volumes of the particles (see (2)). This explains why mostly the large field of (mechanically moving) permanent magnets has been used for the separation, transport and positioning of magnetic microbeads [22]. In an approach towards miniaturization and automation of analytical applications, a system has been proposed in which liquid movement is substituted with magnetically induced movement of magnetic particles [59]. Fluidic channels were realised on a plastic cartridge of centimeter size and magnetic transport was induced by mechanically moving external permanent magnets. In another approach, magnetic particles have been transported over millimeter distances in a microfluidic channel using an array of electromagnets actuated in a four-phase scheme [60]. Each electromagnet consisted of a 0.3 mm diameter magnetic needle core with a wire-wound coil of 300 turns. For coil currents of the order of 0.5 A, forces of 0.1 pN were demonstrated.

Also miniaturized magnetic solutions have been proposed for bead transport, thereby taking full profit from batch microfabrication technologies. Typically, the size of the micropatterned magnets determines the spatial range, where appreciable magnetic forces acting on the microbeads exist. Serpentine gold wires micropatterned on silicon substrates have been combined with microfluidic structures realized in PDMS to transport 4.5 μm polystyrene-coated magnetic beads [61]. By engineering the magnetic field produced by different current-carrying wires, a microsystem was realized that could generate local magnetic field maxima that trap the magnetic beads. When the field maxima change locations, the microbeads follow those maxima. The device allowed precise positioning and transport over 100 μm distances in a single actuation event, which is partly due to the presence of a permanent magnet placed in proximity of the microfluidic chip, the role of which principally is to enhance the magnetic force by inducing a magnetic moment in the magnetic beads. A micro-electromagnet wire matrix, based on two layers of mutually orthogonal arrays of linear wires, has demonstrated magnetic transport of 1–2 μm size magnetic particles over 20 μm distances in a single actuation event [62]. This is a typical working range for the magnetic force, generated by a current-carrying conductor, when no external permanent magnet is used to induce a magnetic moment to the beads. Transport using current carrying wires or microcoils in combination with a magnetic substrate or film to enhance the magnetic forces has been demonstrated more recently [41, 63].

In another approach, a simple planar coil array-based magnetic transport system has been proposed, in which an individual coil is capable of displacing beads over millimeter distances in a liquid-containing capillary [64]. A drastic increase of the magnetic energy and magnetic forces acting on the beads was obtained by placing the complete coil array in a uniform static magnetic field that imposes a permanent magnetic moment to the microbeads. The very small magnetic field gradient of a simple planar coil proved then to be sufficient to displace 1 μm diameter beads over a distance of the order of the coil size. The coils were realized using simple Printed Circuit Board (PCB) technology (100 μm Cu winding width, 35 μm winding height, 200 μm winding pitch) and had a small number of windings ($N = 4$–10). A single coil typically generates a magnetic field gradient of about 5 mT $(mm)^{-1}$ for a maximum allowed current density of 400 A $(mm)^{-2}$. Arranging adjacent coils with spatial overlap over two layers of the PCB circuit and actuating them in a specific three-phase scheme assured the long-range displacement of the microbeads. Moreover, it was found that these polarized beads formed columns with a length of the order of the microfluidic channel size, due to magnetic dipole interactions. This column formation was partly at the basis of the strong magnetic force due to the larger magnetic volume and lower demagnetization factor of the structure.

An original transporting device consisting of a set of two tapered conductors, shifted linearly over half a period was demonstrated by Wirix-Speetjens et al. [31, 65]. This device was fabricated using standard semiconductor fabrication and photolithography techniques. A 300 nm thick thermal $SiO_2$ layer was grown on a Si wafer. The current conductors (TiW 10 nm/Au 150 nm/TiW 10 nm) were evaporated and patterned using a lift-off process. A 250 nm thick $SiO_2$ layer was then sputtered onto the wafer to form a passivation layer. Figure 4 shows a sequence of images demonstrating the step-wise movement of the magnetic particles. By sending a dc current through one of the current conductors, a magnetic field is generated that magnetizes the particles. Once the particles are magnetized, they move toward the edge of the conductor, after which they are attracted to the nearest narrow cross section, driven by the field gradient that is created by tapering the conductor. This way, one-dimensional step-wise transport is realized by applying low frequency non-overlapping clock pulses to both conductors alternatively. Figure 4a shows a series of magnetic particles positioned at the narrow sections of the bottom conductor while a dc current is flowing through, while Figure 4b shows the new position of the particles after a subsequent current pulse through the upper conductor. On Figure 4c, one can clearly see that the magnetic particles have moved forward over a whole period once the current is flowing through the bottom conductor again. The average speed of a magnetic particle is defined as the distance over which the particle is transported parallel to the conductor edge divided by the minimal time needed to reach the next minimal cross-section and is of the order of several 10 μm $s^{-1}$.

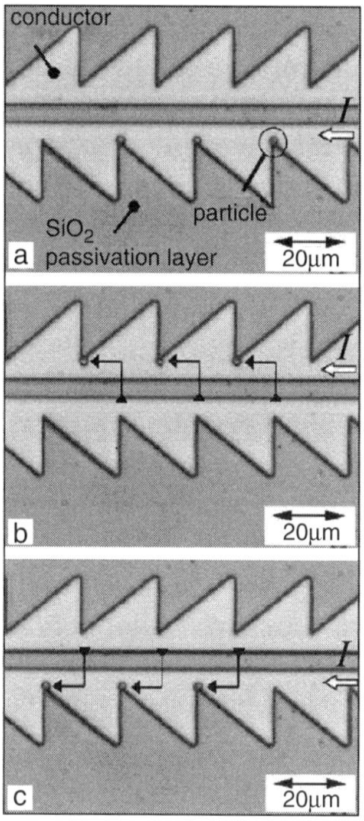


**Fig. 4.** Sequence of microscopic images showing the step-wise movement of single magnetic particles on a transporting device based on tapered conductors. I indicates through which conductor the current is flowing [reprinted from Wirix-Spectjens et al. [31], with permission © 2005 IEEE]

## 6. Magnetic Beads as Labels for Detection

Detection of biological molecules is usually accomplished using biomolecular recognition between the target molecule and a specific receptor (e.g. an antibody) that is tagged with a label. The label may be a radioisotope, enzyme, fluorescent molecule or charged molecule, for example. Recently, also magnetic beads have been used as labels for biosensing. Magnetic labels have several advantages over other labels. The magnetic properties of the beads are stable over time, in particular, because the magnetism is not affected by reagent chemistry or subject to photo-bleaching (a problem with fluorescent labels). There is also no significant magnetic background present in a biological sample and magnetic fields are not screened by aqueous reagents or biomaterials. In addition, magnetism may be used to remotely manipulate the magnetic particles. A number of very sensitive magnetic field detection devices have been developed during recent years, like giant magnetoresistance (GMR) [66] and spin-valve [67, 68] magnetic sensors that enable the measurement of extremely weak magnetic fields, for example the magnetic field generated by the magnetization of a magnetic particle. A basic GMR unit element or spin-valve device consists of a pair of magnetic films separated by a non-magnetic conducting layer [69]. When an external magnetic field rotates the magnetizations of the magnetic layers towards alignment, spin-dependent electron scattering is reduced at the interfaces within the device, decreasing its electrical resistance. GMR sensors can be microscopic in size and sensitive to the presence of micron and smaller size magnetic particles, when in close proximity. Besides GMR sensors, measurements of single magnetic beads have been demonstrated using miniaturized silicon Hall sensors [70] and planar Hall effect sensors based on permalloy thin films [71, 72].

Researchers at the Naval Research Laboratory, Washington DC [73–76], followed by others [77–90], have developed microsystems for the capture and detection of micron-sized, paramagnetic beads on a chip containing GMR or spin-valve sensors. A review of research in this specific area was presented by Graham et al. [91]. The technology is mostly based on a sandwich assay, where the target molecule is bound to an immobilized probe on the GMR sensor, after which the magnetic label is bound to the target using specific ligand–receptor interactions. A magnetic field gradient can be applied to attract magnetic beads to the chip or to selectively pull off only those beads not bound to the surface by specific binding. The field can be generated by external permanent magnets or electromagnets [92], electromagnets integrated with the chip [76] or by current carrying lines positioned close to the chip [93]. Bound beads are detected by the GMR sensors by applying a uniform magnetic field perpendicular to the substrate, imposing a magnetic moment to the superparamagnetic beads. This induced moment generates an in-plane magnetic field that is measured by the GMR sensor. Applying the uniform field normal to the plane of the GMR sensor rather than in-plane has the advantage that, due to demagnetization effects, a much larger magnetizing field can be applied to the beads without saturating the sensor [94].

Figure 5 is an example of the basics of an immunosorbent assay using magnetic nanoparticles [89]. The sensor requires the creation of (1) a capture antibody surface directly above the GMR sensing area and (2) a labeling step in which a magnetic nanoparticle is coupled to the target antigen. The presence of the particle-labeled antigen is then determined by the change in resistance before and after exposing the capture antibody surface to the labeled sample. Using a capture antibody surface composed of mouse IgG and 60 nm magnetic nanoparticles modified with a layer of goat α-mouse IgG, different surface concentrations of the magnetic nanoparticles solution could be monitored (down to 1 nanoparticle $(\mu m)^{-2}$).

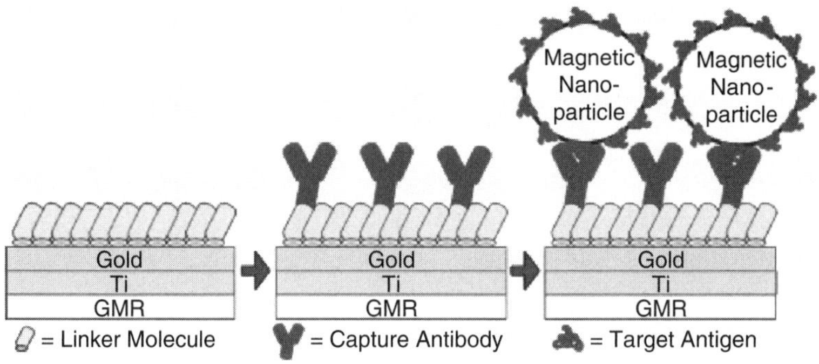


**Fig. 5.** Immunosorbent assay using magnetic beads. A capture antibody just above the GMR sensor probes the target antigen that is coupled to a magnetic nanoparticle. Detection is by measurement of the particle-generated magnetic field by the GMR sensor [reprinted from Millen et al. [89], with permission from © 2005 American Chemical Society]

Figure 6 is a photomicrograph of the GMR sensor platform before creation of the capture antibody surface. Each sensor consists of four GMR sensors that are laid out in a serpentine pattern and electrically wired to function as the four resistors in a Wheatstone bridge. Two of the GMR sensors act as the sense resistors ($R_{S1}$ and $R_{S2}$) and are set in proximity to form an interdigitated electrode 'sense pad' that is positioned within a microfluidic channel. Each sense GMR consists of a 2 μm wide strip that has a total length of over 11,000 μm and a spacing between neighboring strips of 2 μm. The two remaining GMR sensors function as reference resistors ($R_{R1}$ and $R_{R2}$) and are employed to reduce complications associated with the sensitivity of GMR sensors to temperature and stray magnetic fields.

Reiss et al. [83, 84, 86, 87, 95] developed similar sensors using 0.35 μm magnetic particles. Also a dose–response curve for a real biochemical assay was given and compared with fluorescence detection. The format of the test is rather peculiar, in that the concentration of probe DNA on the surface was varied rather than the concentration of DNA in the incubation step. It is thus more an assay of the likelihood of DNA attachment than of concentration in a test sample.

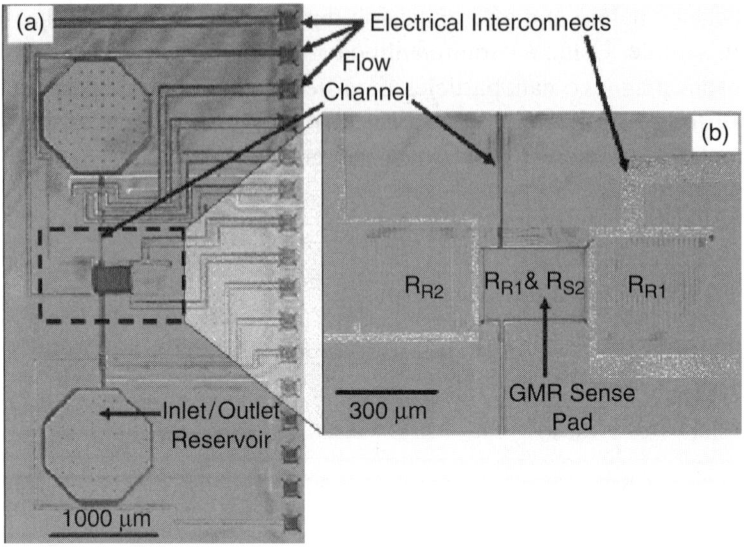


**Fig. 6.** GMR sensor and electrical and fluidic interconnects: (**a**) GMR sensor, 6 µm microfluidic channels, 1,000 µm inlet/outlet reservoirs, electrical interconnects and wire bonding pads. (**b**) Expanded view of the GMR sensor configured as a Wheatstone bridge. Two sensing resistors, $R_{S1}$ and $R_{S2}$ are interdigitated and positioned as a sense pad below a fluid reservoir, whereas the two reference resistors, $R_{R1}$ and $R_{R2}$, are laid on either side of the sense pad [reprinted from Millen et al. [89], with permission © 2005 American Chemical Society]

The GMR sensor sensitivity increases with decreasing surface area of the sensor; however, the chemical sensitivity, or number of analyte molecules that can hybridize to the surface increases with increasing surface area. Theoretical modeling showed that a GMR sensor can detect a single superparamagnetic bead of any size [88, 94, 96], as long as all system dimensions (bead size, sensor size, distance between bead and sensor) scale down proportionally. When the sensor size can be of the size of the bead, it should be possible to detect single beads with a radius smaller than 100 nm.

Most bead-sensing experiments today have been done with 2.8 µm size beads composed of magnetic $\gamma$-$Fe_2O_3$ and $Fe_3O_4$ nanoparticles dispersed in a polystyrene matrix with an average magnetic content of 17 weight % [19]. To maximize the sensor response, the magnetic beads should have a magnetization as high as possible and yet remain non-remnant to avoid clustering when suspended in solution. With the goal of achieving larger signals from the magnetic labels, one has developed soft ferromagnetic beads with 100 % magnetic content and much higher saturation magnetization

[74]. NiFe beads of 1 μm diameter showed a susceptibility of ~3, the maximum obtainable value for a uniformly magnetized sphere. Because of this property, smaller solid ferromagnetic beads could effectively be used as biomagnetic labels, which would also increase the dynamic range of biosensor assays by allowing more labels/unit area.

## 7. Separation and Mixing Using Magnetic Supraparticle Structures

When placed in a magnetic field, superparamagnetic particles will acquire a magnetic moment and they will start interacting by the magnetic dipole interaction. This interaction induces a spontaneous clustering of the particles into larger, often complicated, structures, sometimes called magnetic supraparticle structures (SPS). The shape of an SPS depends on parameters, like the size of the magnetic moment of the microbeads and the magnetic dipolar interaction between different beads. These properties are dependent on the amplitude and frequency of the applied magnetic field, the shape and magnetic content of the beads, the concentration of the magnetic particles in the fluid, the temperature etc. Despite the complexity of the aggregation process of a magnetic fluid into a SPS, the physical effects of a magnetic field on such structure are now very well understood [97, 98].

Self-assembled magnetic SPS have been used for long DNA separation in microfluidic channels. This represents a convenient solution, since no microlithography is required to define geometrical constrictions that are simply defined by the porous magnetic matrix [99]. Experimental separations using the SPS matrix have been combined with theoretical modeling [100] and, recently, computer-piloted flow control and injection of the experiment resulted in a quantitative and reproducible separation of long DNA by a SPS matrix [101].

Besides the use of magnetic SPS as a static stationary phase for separation in microfluidic channels, researchers have also been investigating the potential of manipulating *dynamically* magnetic particle aggregates for microfluidic applications. Under the influence of a varying magnetic field generated by a mechanically moving permanent magnet, magnetic SPS composed of 1–2 μm diameter superparamagnetic beads could be rotated through all axes in a microfluidic channel, without loss of structrual form allowing dynamic micron-scale movement without direct mechanical, electrical or photonic interactions [102]. A number of potential applications of this phenomenon were mentioned in the cited paper, like binding biomolecules on the magnetic particles for immuno-assays, studying subcellular biomechanics and microfluidic mixing in pico- and femtoliter volumes.

Active fluid mixing was demonstrated in microchannels made in a micromachined microfluidic chip of PMMA: mixing was based on the manipulation by a local alternating magnetic field of self-assembled porous structures of magnetic microbeads that are placed over the section of the channel [103, 104]. Using a sinusoidally varying magnetic field (1 Hz $< f <$ 100 Hz), a rotational motion of the SPS was induced, thereby strongly enhancing the fluid perfusion through the SPS that behaved as a dynamic random porous medium. The localized time-dependent magnetic field was generated using a current-fed coil and a soft magnetic yoke structure, which is partly integrated within a plastic microfluidic chip. The mixing is the result of the chaotic splitting of fluid streams through the dynamic and randomly porous structure of the SPS and the relative motion of magnetic entities with respect to the fluid flow. It was quantified by monitoring the fluorescent intensity of initially parallel fluorescent and non-fluorescent laminar streams. A 95% mixing efficiency over a channel length as small as the channel width (200 μm) and at flow rates of 0.5 cm $s^{-1}$ was obtained, demonstrating the large lateral mass transfer induced by the SPS as a consequence of the highly heterogeneous and dynamic nature of the SPS.

## 8. Magnetic Beads as Substrates for Bio-assays

Immuno-assays are probably the most important analysis method for biological molecules, as the molecular recognition reaction provides high selectivity and chemical sensitivity. Although conventional microtiter plate assays continue to play an important role, enzyme assays, DNA binding and competitive immuno-assays have been performed on microdevices [105–107]. Heterogenous assays, where reactions occur both in solution and in a solid phase, offer the advantage of easy separation of chemical complexes from reactants. Biomolecule immobilization on a solid phase, formed by the surface of micro- or nano-particles, evidently results in a small-volume and localized assay [108, 109]. Such solid phase provides a high surface-to-volume ratio, reducing diffusion times during the microfluidic procedures and increasing the density of binding sites, which is beneficial for a high detection signal and sensitivity. In addition, such assay allows for a rapid regeneration and exchange of the solid support when needed. For example, a small quantity of polystyrene magnetic microbeads containing immobilized biomolecules was injected into a neutral hydrophilic-coated fused-silica capillary [109]. The short plug (2–3 mm) of beads was held fixed by a magnet placed in the cartridge of a Capillary Electrophoresis (CE) system. The microfluidic procedure involved antigen capture from the sample solution to the beads, rinsing, elution and electrophoretic separation of the sample. The beads could be replaced after each run, eliminating the need to regenerate the solid support.

In another study, a small-volume heterogeneous immuno-assay was demonstrated using both classical fused silica capillaries and glass microfluidic chips [110]. The system demonstrated the presence of parathyroid hormone at concentrations of 1 $\mu g\ L^{-1}$, using a sandwich assay, which is based on the trapping or capture of the analyte by one antibody and the detection by another (fluorescent) antibody. Also a calibration curve was generated at clinically relevant levels (2–50 $\mu g\ L^{-1}$) for the detection of interleukin-5, an important cytokine for asthma, immune response cascades and cancer.

Dynamic DNA hybridization was demonstrated by immobilizing DNA targets on magnetic beads via streptavidin–biotin conjugation or base pairing between oligonucleotide residues [111]. The DNA/bead complex was introduced into the device, after which hybridization took place with a complimentary probe. The hybridized probe was then removed by heat denaturation to allow the DNA sample to be interrogated again by another probe with a different sequence of interest. Demonstration of specific hybridization reactions in an array format was achieved using four synthesized DNA targets and five probes in sequence, indicating the potential of this approach for gene expression analysis.

Figure 7 illustrates the principle of another study, where mRNA was purified from total RNA in a flowing stream within a microchip using polystyrene magnetic microbeads, labeled with a polythymine(T) tail for mRNA capture [112]. A few nanogram of mRNA could be captured per microgram of total RNA. This suggests it will be possible to perform cDNA library construction on-chip with only small amounts of mRNA.

Choi et al. reported on the development and characterization of an integrated microfluidic biochemical detection system for fast and low-volume immuno-assays using magnetic beads [38]. First, antibody-labeled polystyrene magnetic beads were separated from a liquid solution by a set of planar coils. While holding the antibody-coated beads, antigens were injected into the channel. Only target antigens were immobilized, and thus separated onto the magnetic bead surface due to the specific antibody/antigen reaction. Detection of the target antigen was based on an electrochemical enzymatic reaction in a sandwich immuno-assay.

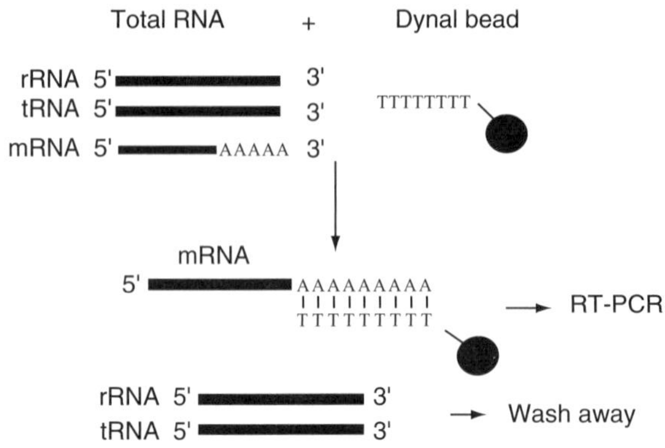


**Fig. 7.** Affinity purification of mRNA from total RNA using paramagnetic, poly-T coated magnetic beads. Reverse transcription-polymerase chain reaction (RT-PCR) may then be used to amplify the presence of a target gene [reprinted from Jiang and Harrison [112], with permission © 2000 The Royal Society of Chemistry]

In another study, a fully integrated biochip cartridge that consists of microfluidic mixers, valves, pumps, channels, chambers, heaters and DNA micro-array sensors was developed to perform DNA analysis of complex biological sample solutions [113]. Sample preparation (including magnetic bead-based cell capture, cell pre-concentration and purification, and cell lysis), polymerase chain reaction (PCR), DNA hybridization and electrochemical detection were performed in this fully automated miniature device. Cavitation microstreaming was implemented to enhance target cell capture from whole blood samples using immunomagnetic beads and accelerate DNA hybridization reaction. Thermally actuated paraffin-based microvalves were developed to regulate flows. Electrochemical pumps and thermopneumatic pumps were integrated on the chip to provide pumping of liquid solutions. Figure 8a and b are a schematic diagram and a photograph of the fluidic chip, respectively. The device is completely self-contained: no external pressure sources, fluid storage, mechanical pumps, or valves are necessary for fluid manipulation, thus eliminating possible sample contamination and simplifying device operation. Pathogenic bacteria detection from milliliter whole blood samples and single-nucleotide polymorphism analysis was demonstrated directly from diluted blood.

A GMR biosensor consisting of GMR type $Cu/Ni_{80}Fe_{20}$ multilayers, patterned into a spiral-shaped line to cover the area of a typical DNA spot (70 μm diameter) was presented by Schotter et al. [86]. The probe DNA was assembled on top of the sensor elements in different concentrations ranging from 16 pg $\mu l^{-1}$ up to 10 ng $\mu l^{-1}$. Complementary biotin-labeled analyte DNA is hybridized to the probe DNA at a concentration of 10 ng $\mu l^{-1}$. A number of different commercially available streptavidin-labeled magnetic microparticles were investigated to determine the most appropriate markers. The experimental comparison showed that the relative sensitivity of the magnetoresistive biosensor is superior to the fluorescent detection at low probe DNA concentrations (below 10 ng $\mu l^{-1}$). In other work, the real-time spin valve sensor response of 250 nm magnetic nanoparticles functionalized with a DNA target when hybridizing onto a complementary DNA probe was monitored [114]. The experiments demonstrated how a combination of magnetic-field-assisted hybridization, concentration of beads near the sensor and detection with an optimized spin-valve sensor can lead to accelerated biosensing with considerable sensitivity (detection limit of 16 pg $\mu l^{-1}$). The ability to manipulate magnetic beads on a chip surface using different test structures further revealed the potential of the dynamic magnetic biosensor approach for a broad spectrum of lab-on-a-chip applications, where single or a small amount of biomolecules attached to magnetic particles are to be guided to specific places on the chip. An interesting conclusion of this paper was that magnetic biosensors will not compete in sensitivity with the (fluorescence-based) high-density arrays in

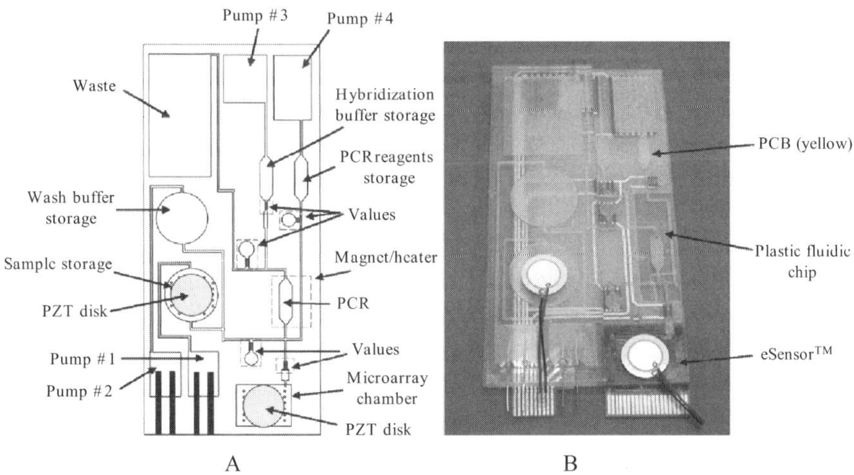


**Fig. 8.** (**a**) Plastic self-contained, fully integrated fluidic chip for sample preparation, PCR and DNA micro-array detection. Pumps 1–3 are electrochemical pumps, and pump 4 is a thermopneumatic pump. (**b**) The integrated device that consists of a plastic fluidic chip, a PCB, and a Motorola eSensor micro-array chip [reprinted from Liu et al. [113], with permission © 2004 American Chemical Society]

plate, which is made hydrophobic by applying a surface treatment. The droplets that form in the solution are suspended in the oil because of the differences in surface tension between the two liquids and the superior wettability of the oil to the plate. The drops contain magnetic microparticles that serve both as force mediators for the magnetic actuation and as a mobile substrate for the molecules. Hydrophilic magnetic beads were used to maintain them in a droplet during manipulation and to avoid sticking to the glass surfaces. Such droplet handling can be at the basis of a new type of micro-chemical analysis system, where droplets can be magnetically transported thanks to their magnetic bead content, can be fused to trigger chemical reactions, or can be split up in separate droplets. All of these basic actions have been demonstrated by magnetic actuation with a mechanically moving permanent magnet and using 30 nL droplets containing 30 μm diameter magnetic beads.

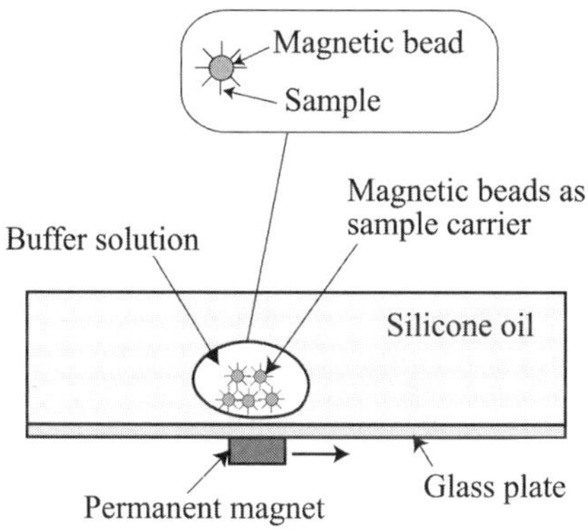


**Fig. 9.** Magnetic handling of a bead droplet in a silicone oil environment utilizing interfacial tension and wettability [reprinted from Shikida et al. [118], with permission © 2006 Elsevier]

Also a magnetic droplet manipulation system in which aqueous droplets are actuated through a multilayer set of coils on a PCB has been presented [119]. The PCB generates a changeable magnetic-field topology over the chip surface that allows variation in the force that acts on the magnetic particles inside the droplets. On top of the PCB, a reservoir containing silicone oil was placed. The density of the oil is adjusted such that an aqueous droplet can sink to the bottom of the reservoir where its spherical shape is approximately retained. The submerged and nearly spherical droplets slide, driven by the field gradient, over the reservoir bottom, which consists of a very thin hydrophobic Teflon layer (12.5 μm). This layer keeps the droplet and the coil multilayer a small distance apart and prevents adhesion-related problems. If a fixed position of the droplet is desired, the friction between the droplet and the reservoir bottom can be increased by locally modifying the surface of the Teflon layer with oxygen plasma. The resultant hydrophilic pattern serves as a droplet immobilization site for droplet merging and splitting operations. By combining droplet manipulation steps, a complete DNA-purification procedure was demonstrated. The latter employs six stages of droplet merging, mixing and splitting. The buffer solutions for binding, washing and elution are present on the chip in the form of immobilized 10 μL droplets. Figure 10 is a schematic summary of the complete procedure, which involves DNA-to-particle binding (a), washing (b) and elution of the DNA from the particles (c).

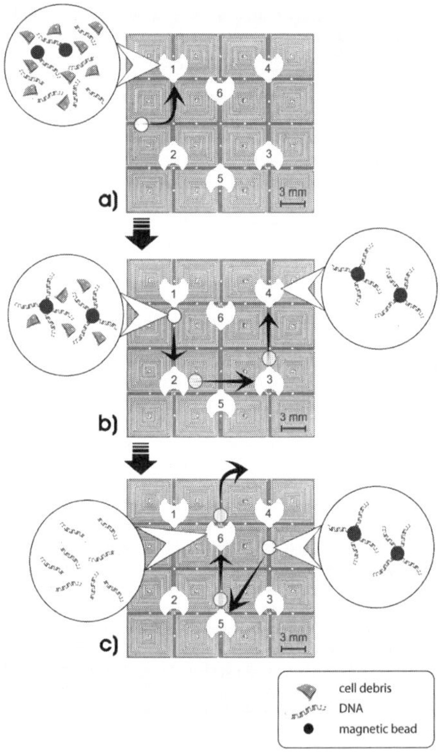


**Fig. 10.** DNA purification from a lysed cell sample solution: (**a**) the binding of the DNA to the magnetic particles in the binding/lysis buffer (position 1); (**b**) a series of three washing steps to remove any transferred cell debris, proteins and lipids (positions 2–4) and (**c**) removal of DNA from the magnetic beads through magnetic agitation in two immobilized droplets of elution buffer solution (positions 5–6) and, subsequently, removal of the magnetic particle from the chip [reprinted from Lehmann et al. [119], with permission © 2006 Wiley-VCH]

The bio-analytical performance characteristics of this system was evaluated by using three different types of genomic material of increasing complexity, HinDIII-digested λ-phage DNA, human genomic DNA fragments (>50 kb), and mammalian Jurkat ATCC TIB-152 cells. On-chip DNA purification from whole cell samples down to 10 cells per 10 μL, could be demonstrated. This combination of good DNA recovery and high sensitivity makes this system a very promising platform for DNA analysis in a miniaturized format, including purification, PCR amplification and detection.

## 10. Conclusion

An update on recent advances in the handling, manipulation and detection of magnetic beads in microfluidic systems was presented. Probably the most prominent advantage of magnetic beads over other solid phases lies in the fact that the particles can be magnetically probed and manipulated using permanent magnets or electromagnets, independent of normal chemical or biological processes. Some of the classical applications, like magnetic separation, have already found their way down to miniaturized fluidic or 'lab-on-a-chip' systems that strongly limit the consumption of samples and reagents; in such systems, magnetic beads effectively provided a chemically active substrate with a large surface-to-volume ratio.

The booming area of miniaturized applications of magnetic beads offers many exciting possibilities for future developments. It is a highly multidisciplinary area, requiring contributions from inorganic chemistry involved in the preparation of the magnetic beads, through biochemistry and medical science to allow for their functionalization and, of course, the basic physics of magnetism and magnetic materials. Nearly all important functions in a bio-assay can be realized using magnetic beads: sample purification, providing a solid substrate to the sample, mixing, labeling, manipulation and transport and, finally, separation. Combining such particles and their detection system with the recently developed magnetic manipulation techniques already resulted in highly integrated bio-assay systems, in which all functions (from sample treatment to read-out) could be extremely miniaturized. Droplet-based systems in which aqueous magnetic particle containing droplets are transported in an oil medium are particularly attractive for their simplicity, reconfigurability, self-containment and the reduction of sample evaporation effects. It is now generally recognized that nanotechnologies and biosciences will be one of the leading and most promising areas of research and development in the twenty-first century. Magnetic bead microsystems will play an important role in these developments.

## Acknowledgements

We thank the Swiss National Science Foundation (Grant 200020-107372) for financial support of our research. The author gratefully acknowledges Amar Rida, Victor Fernandez, Hicham Majd, Smail Hadjidj, Ulrike Lehmann, Virendra Parashar and Caroline Vandevyver for useful discussions and collaboration.

## References

1. Auroux PA, Iossifidis D, Reyes DR and Manz A (2002) Micro total analysis systems. 2. Analytical standard operations and applications. Analytical Chemistry 74 (12):2637–2652
2. Vilkner T, Janasek D and Manz A (2004) Micro total analysis systems. Recent developments. Analytical Chemistry 76 (12):3373–3385
3. Kawaguchi H (2000) Functional polymer microspheres. Progress in Polymer Science 25 (8):1171–1210
4. Kruis FE, Fissan H and Peled A (1998) Synthesis of nanoparticles in the gas phase for electronic, optical and magnetic applications – A review. Journal of Aerosol Science 29 (5–6):511–535
5. Pankhurst QA, Connolly J, Jones SK and Dobson J (2003) Applications of magnetic nanoparticles in biomedicine. Journal of Physics D: Applied Physics 36 (13):R167–R181
6. Gijs MAM (2004) Magnetic bead handling on-chip: New opportunities for analytical applications (review). Microfluidics and Nanofluidics 1:22–40
7. Pamme N (2006) Magnetism and microfluidics. Lab on a Chip 6 (1):24–38
8. Landfester K and Ramirez LP (2003) Encapsulated magnetite particles for biomedical application. Journal of Physics: Condensed Matter 15 (15):S1345–S1361
9. Bergemann C, Muller-Schulte D, Oster J, a Brassard L and Lubbe AS (1999) Magnetic ion-exchange nano- and microparticles for medical, biochemical and molecular biological applications. Journal of Magnetism and Magnetic Materials 194 (1–3):45–52
10. Gruttner C, Rudershausen S and Teller J (2001) Improved properties of magnetic particles by combination of different polymer materials as particle matrix. Journal of Magnetism and Magnetic Materials 225 (1–2):1–7
11. Papell SS (1965) US patent 3215572
12. Reimers GW and Khalfalla SE (1974) US patent 3843540
13. Kelley JR (1977) US patent 4019994
14. Molday RS (1984) US patent 4452773
15. Rembaum A (1981) US patent 4267234

16. Lee JW, Isobe T and Senna M (1996) Magnetic properties of ultrafine magnetite particles and their slurries prepared via in-situ precipitation. Colloids and Surfaces a-Physicochemical and Engineering Aspects 109:121–127
17. Suzuki M, Shinkai M, Kamihira M and Kobayashi T (1995) Preparation and characteristics of magnetite-labeled antibody with the use of poly(ethylene glycol) derivatives. Biotechnology and Applied Biochemistry 21:335–345
18. Singer JM (1987) Future directions in polymer colloids, in: NATO ASI Series. Series E, Applied Sciences, 138, edited by M.S. El-Asser and R.M. Fitch (Published in cooperation with NATO Scientific Affairs Division by Nijhoff, Dordrecht, The Netherlands). pp. x, 402.
19. Dynal. p. Dynal S.A., Oslo, Norway, product catalogue, http://www.dynal.no.
20. Kleiber J, Walter T, Harttig H, Lesniak C, Mennig M, Riedling M and Schmidt H (2001) US patent 6255477B1
21. Undisclosed_inventors (2001) European patent 1154443A1
22. Miltenyi S, Muller W, Weichel W and Radbruch A (1990) High-gradient magnetic cell-separation with macs. Cytometry 11 (2):231–238
23. Gruttner C and Teller J (1999) New types of silica-fortified magnetic nanoparticles as tools for molecular biology applications. Journal of Magnetism and Magnetic Materials 194 (1–3):8–15
24. Hafeli UO, Sweeney SM, Beresford BA, Sim EH and Macklis RM (1994) Magnetically directed poly(lactic acid) Y-90 microspheres – Novel agents for targeted intracavitary radiotherapy. Journal of Biomedical Materials Research 28 (8):901–908
25. Hafeli UO and Pauer GJ (1999) In vitro and in vivo toxicity of magnetic microspheres. Journal of Magnetism and Magnetic Materials 194 (1–3):76–82
26. LesliePelecky DL and Rieke RD (1996) Magnetic properties of nanostructured materials. Chemistry of Materials 8 (8):1770–1783
27. Chikazumi S (1964) Physics of Magnetism. Robert E. Krieger Publishing Company, Malabar, Florida, USA
28. Zborowski M, Sun LP, Moore LR, Williams PS and Chalmers JJ (1999) Continuous cell separation using novel magnetic quadrupole flow sorter. Journal of Magnetism and Magnetic Materials 194 (1–3):224–230
29. White FM (1999) Fluid Mechanics. McGraw-Hill, New York
30. Faxen H (1922) The resistance against the movement of a rigour sphere in viscous fluids, which is embedded between two parallel layered barriers. Annalen Der Physik 68 (10):89–119
31. Wirix-Speetjens R, Fyen W, Xu KD, De Boeck J and Borghs G (2005) A force study of on-chip magnetic particle transport based on tapered conductors. IEEE Transactions on Magnetics 41 (10):4128–4133
32. Gregory J (1981) Approximate expressions for retarded Vanderwaals interaction. Journal of Colloid and Interface Science 83 (1):138–145
33. Kar G, Chander S and Mika TS (1973) Potential-energy of interaction between dissimilar electrical double-layers. Journal of Colloid and Interface Science 44 (2):347–355

34. Blankenstein G and Larsen UD (1998) Modular concept of a laboratory on a chip for chemical and biochemical analysis. Biosensors & Bioelectronics 13 (3–4):427–438
35. Choi JW, Ahn CH, Bhansali S and Henderson HT (2000) A new magnetic bead-based, filterless bio-separator with planar electromagnet surfaces for integrated bio-detection systems. Sensors and Actuators B: Chemical 68 (1–3):34–39
36. Choi JW, Liakopoulos TM and Ahn CH (2001) An on-chip magnetic bead separator using spiral electromagnets with semi-encapsulated permalloy. Biosensors & Bioelectronics 16 (6):409–416
37. Ahn CH, Allen MG, Trimmer W, Jun YN and Erramilli S (1996) A fully integrated micromachined magnetic particle separator. Journal of Microelectromechanical Systems 5 (3):151–158
38. Choi JW, Oh KW, Thomas JH, Heineman WR, Halsall HB, Nevin JH, Helmicki AJ, Henderson HT and Ahn CH (2002) An integrated microfluidic biochemical detection system for protein analysis with magnetic bead-based sampling capabilities. Lab on a Chip 2 (1):27–30
39. Ramadan Q, Samper V, Poenar DP and Yu C (2006) An integrated microfluidic platform for magnetic microbeads separation and confinement. Biosensors & Bioelectronics 21 (9):1693–1702
40. Ramadan Q, Samper V, Poenar D, Liang Z, Yu C and Lim TM (2006) Simultaneous cell lysis and bead trapping in a continuous flow microfluidic device. Sensors and Actuators B: Chemical 113 (2):944–955
41. Ramadan Q, Yu C, Samper V and Poenar DP (2006) Microcoils for transport of magnetic beads. Applied Physics Letters 88 (3):032501
42. Ramadan Q, Samper V, Poenar D and Yu C (2004) On-chip micro-electromagnets for magnetic-based bio-molecules separation. Journal of Magnetism and Magnetic Materials 281 (2–3):150–172
43. Smistrup K, Tang PT, Hansen O and Hansen MF (2006) Micro electromagnet for magnetic manipulation in lab-on-a-chip systems. Journal of Magnetism and Magnetic Materials 300 (2):418–426
44. Smistrup K, Hansen O, Bruus H and Hansen MF (2005) Magnetic separation in microfluidic systems using microfabricated electromagnets-experiments and simulations. Journal of Magnetism and Magnetic Materials 293 (1):597–604
45. Tondra M, Granger M, Fuerst R, Porter M, Nordman C, Taylor J and Akou S (2001) Design of integrated microfluidic device for sorting magnetic beads in biological assays. IEEE Transactions on Magnetics 37 (4):2621–2623
46. Deng T, Prentiss M and Whitesides GM (2002) Fabrication of magnetic microfiltration systems using soft lithography. Applied Physics Letters 80 (3):461–463
47. Forbes ZG, Yellen BB, Barbee KA and Friedman G (2003) An approach to targeted drug delivery based on uniform magnetic fields. IEEE Transactions on Magnetics 39 (5):3372–3377

the short term. They will rather be complementary in being a fast, inexpensive and sensitive real-time indicator of the up- or down-regulation of genes or markers of certain diseases that can be used for routine diagnostics at the point of care.

The development of a generic semi-disposable microfluidic biosensor for the highly sensitive detection of pathogens via their nucleic acid sequences was presented by Kwakye et al. [115]. Disposable microchannels with defined areas for capture and detection of target pathogen RNA sequence were created in PDMS and mounted onto a reusable PMMA stand. Two different DNA probes complementary to unique sequences on the target pathogen RNA serve as the biorecognition elements. For signal generation and amplification, one probe is coupled to dye encapsulating liposomes while the second probe is coupled to superparamagnetic beads for target immobilization. The probes hybridize to target RNA and the liposome-target-bead complex is subsequently captured on a magnet. The amount of liposomes captured correlates directly to the concentration of target sequence and was quantified using a fluorescence microscope. Dengue fever virus serotype 3 sequences and probes were used as a model analyte system to test the sensor. Probe binding and target capture conditions were optimized for sensitivity resulting in a detection limit of as little as 10 amol $\mu L^{-1}$. Also an electrochemical microfluidic biosensor with an integrated mini-potentiostat for the quantification of RNA was developed by the same group and was based on nucleic acid hybridization and liposome signal amplification [116]. Specificity of the biosensor was ensured by short DNA probes that hybridize with the target RNA or DNA sequence. The reporter probe was coupled to liposomes entrapping the electrochemically active redox couple potassium ferri/ferrohexacyanide. The capture probes were coupled to superparamagnetic beads that were isolated on a magnet in the biosensor. Upon capture, the liposomes were lysed to release the electrochemical markers that were detected on an interdigitated microelectrode array in the biosensor just downstream of the magnet. The device achieved a detection limit of 0.01 μM of the combined ferri/ferrohexacyanide concentration, which was 10 times lower than that of a standard lab-bench system.

## 9. Magnetic Beads in Droplets

An interesting study reported on the magnetic actuation and transport of magnetic beads contained in water droplets that were suspended in a silicone oil solution [117, 118]. As illustrated in Fig. 9, an aqueous buffer solution containing magnetic beads is dropped into silicone oil on a glass

48. Furlani EP (2006) Analysis of particle transport in a magnetophoretic microsystem. Journal of Applied Physics 99 (2):024912
49. Smistrup K, Lund-Olesen T, Hansen MF and Tang PT (2006) Microfluidic magnetic separator using an array of soft magnetic elements. Journal of Applied Physics 99 (8):08P102
50. Smistrup K, Kjeldsen BG, Reimers JL, Dufva M, Petersen J and Hansen MF (2005) On-chip magnetic bead microarray using hydrodynamic focusing in a passive magnetic separator. Lab on a Chip 5 (11):1315–1319
51. Han KH and Frazier AB (2006) Paramagnetic capture mode magnetophoretic microseparator for high efficiency blood cell separations. Lab on a Chip 6 (2):265–273
52. Han KH and Frazier AB (2005) Diamagnetic capture mode magnetophoretic microseparator for blood cells. Journal of Microelectromechanical Systems 14 (6):1422–1431
53. Han KH and Frazier AB (2004) Continuous magnetophoretic separation of blood cells in microdevice format. Journal of Applied Physics 96 (10):5797–5802
54. Inglis DW, Riehn R, Sturm JC and Austin RH (2006) Microfluidic high gradient magnetic cell separation. Journal of Applied Physics 99 (8):08K101
55. Inglis DW, Davis JA, Austin RH and Sturm JC (2006) Critical particle size for fractionation by deterministic lateral displacement. Lab on a Chip 6 (5):655–658
56. Inglis DW, Riehn R, Austin RH and Sturm JC (2004) Continuous microfluidic immunomagnetic cell separation. Applied Physics Letters 85 (21): 5093–5095
57. Latham AH, Freitas RS, Schiffer P and Williams ME (2005) Capillary magnetic field flow Fractionation and analysis of magnetic nanoparticles. Analytical Chemistry 77 (15):5055–5062
58. Pamme N and Manz A (2004) On-chip free-flow magnetophoresis: Continuous flow separation of magnetic particles and agglomerates. Analytical Chemistry 76 (24):7250–7256
59. Ostergaard S, Blankenstein G, Dirac H and Leistiko O (1999) A novel approach to the automation of clinical chemistry by controlled manipulation of magnetic particles. Journal of Magnetism and Magnetic Materials 194 (1–3):156–162
60. Joung J, Shen J and Grodzinski P (2000) Micropumps based on alternating high-gradient magnetic fields. IEEE Transactions on Magnetics 36 (4):2012–2014
61. Deng T, Whitesides GM, Radhakrishnan M, Zabow G and Prentiss M (2001) Manipulation of magnetic microbeads in suspension using micromagnetic systems fabricated with soft lithography. Applied Physics Letters 78 (12):1775–1777
62. Lee CS, Lee H and Westervelt RM (2001) Microelectromagnets for the control of magnetic nanoparticles. Applied Physics Letters 79 (20):3308–3310

63. Wang ZH, Lew WS and Bland JAC (2006) Manipulation of superparamagnetic beads using on-chip current lines placed on a ferrite magnet. Journal of Applied Physics 99 (8):08P104
64. Rida A, Fernandez V and Gijs MAM (2003) Long-range transport of magnetic microbeads using simple planar coils placed in a uniform magnetostatic field. Applied Physics Letters 83 (12):2396–2398
65. Wirix-Speetjens R, Fyen W, De Boeck J and Borghs G (2006) Enhanced magnetic particle transport by integration of a magnetic flux guide: Experimental verification of simulated behavior. Journal of Applied Physics 99 (8):08P101
66. Baibich MN, Broto JM, Fert A, Vandau FN, Petroff F, Eitenne P, Creuzet G, Friederich A and Chazelas J (1988) Giant Magnetoresistance of (001)Fe/(001) Cr magnetic superlattices. Physical Review Letters 61 (21):2472–2475
67. Dieny B, Speriosu VS, Metin S, Parkin SSP, Gurney BA, Baumgart P and Wilhoit DR (1991) Magnetotransport properties of magnetically soft spin-valve structures. Journal of Applied Physics 69 (8):4774–4779
68. Freitas PP, Silva F, Oliveira NJ, Melo LV, Costa L and Almeida N (2000) Spin valve sensors. Sensors and Actuators A: Physical 81 (1–3):2–8
69. Hartmann U, ed. (1999) Magnetic multilayers and giant magnetoresistance. Springer Verlag, Berlin-Heidelberg.
70. Besse PA, Boero G, Demierre M, Pott V and Popovic R (2002) Detection of a single magnetic microbead using a miniaturized silicon Hall sensor. Applied Physics Letters 80 (22):4199–4201
71. Ejsing L, Hansen MF, Menon AK, Ferreira HA, Graham DL and Freitas PP (2004) Planar Hall effect sensor for magnetic micro- and nanobead detection. Applied Physics Letters 84 (23):4729–4731
72. Ejsing L, Hansen MF, Menon AK, Ferreira HA, Graham DL and Freitas PP (2005) Magnetic microbead detection using the planar Hall effect. Journal of Magnetism and Magnetic Materials 293 (1):677–684
73. Baselt DR, Lee GU, Natesan M, Metzger SW, Sheehan PE and Colton RJ (1998) A biosensor based on magnetoresistance technology. Biosensors & Bioelectronics 13 (7–8):731–739
74. Miller MM, Sheehan PE, Edelstein RL, Tamanaha CR, Zhong L, Bounnak S, Whitman LJ and Colton RJ (2001) A DNA array sensor utilizing magnetic microbeads and magnetoelectronic detection. Journal of Magnetism and Magnetic Materials 225 (1–2):138–144
75. Rife JC, Miller MM, Sheehan PE, Tamanaha CR, Tondra M and Whitman LJ (2003) Design and performance of GMR sensors for the detection of magnetic microbeads in biosensors. Sensors and Actuators A: Physical 107 (3):209–218
76. Edelstein RL, Tamanaha CR, Sheehan PE, Miller MM, Baselt DR, Whitman LJ and Colton RJ (2000) The BARC biosensor applied to the detection of biological warfare agents. Biosensors & Bioelectronics 14 (10–11):805–813
77. Coehoorn R and Prins MWJ (2003) Patent application WO 03/054523 A2

78. Graham DL, Ferreira HA, Freitas PP and Cabral JMS (2003) High sensitivity detection of molecular recognition using magnetically labelled biomolecules and magnetoresistive sensors. Biosensors & Bioelectronics 18 (4):483–488
79. Ferreira HA, Graham DL, Feliciano N, Clarke LA, Amaral MD and Freitas PP (2005) Detection of cystic fibrosis related DNA targets using AC field focusing of magnetic labels and spin-valve sensors. IEEE Transactions on Magnetics 41 (10):4140–4142
80. Ferreira HA, Feliciano N, Graham DL and Freitas PP (2005) Effect of spin-valve sensor magnetostatic fields on nanobead detection for biochip applications. Journal of Applied Physics 97 (10):10Q904
81. Ferreira HA, Feliciano N, Graham DL, Clarke LA, Amaral MD and Freitas PP (2005) Rapid DNA hybridization based on ac field focusing of magnetically labeled target DNA. Applied Physics Letters 87 (1):013901
82. Graham DL, Ferreira HA, Feliciano N, Freitas PP, Clarke LA and Amaral MD (2005) Magnetic field-assisted DNA hybridisation and simultaneous detection using micron-sized spin-valve sensors and magnetic nanoparticles. Sensors and Actuators B: Chemical 107 (2):936–944
83. Reiss G, Brueckl H, Huetten A, Schotter J, Brzeska M, Panhorst M, Sudfeld D, Becker A, Kamp PB, Puehler A, Wojczykowski K and Jutzi P (2005) Magnetoresistive sensors and magnetic nanoparticles for biotechnology. Journal of Materials Research 20 (12):3294–3302
84. Bruckl H, Brzeska M, Brinkmann D, Schotter J, Reiss G, Schepper W, Kamp PB and Becker A (2004) Magnetoresistive logic and biochip. Journal of Magnetism and Magnetic Materials 282:219–224
85. Brzeska M, Panhorst M, Kamp PB, Schotter J, Reiss G, Puhler A, Becker A and Bruckl H (2004) Detection and manipulation of biomolecules by magnetic carriers. Journal of Biotechnology 112 (1–2):25–33
86. Schotter J, Kamp PB, Becker A, Puhler A, Reiss G and Bruckl H (2004) Comparison of a prototype magnetoresistive biosensor to standard fluorescent DNA detection. Biosensors & Bioelectronics 19 (10):1149–1156
87. Schotter J, Kamp PB, Becker A, Puhler A, Brinkmann D, Schepper W, Bruckl H and Reiss G (2002) A biochip based on magnetoresistive sensors. IEEE Transactions on Magnetics 38 (5):3365–3367
88. Li GX, Joshi V, White RL, Wang SX, Kemp JT, Webb C, Davis RW and Sun SH (2003) Detection of single micron-sized magnetic bead and magnetic nanoparticles using spin valve sensors for biological applications. Journal of Applied Physics 93 (10):7557–7559
89. Millen RL, Kawaguchi T, Granger MC, Porter MD and Tondra M (2005) Giant magnetoresistive sensors and superparamagnetic nanoparticles: A chip-scale detection strategy for immunosorbent assays. Analytical Chemistry 77 (20):6581–6587
90. Tondra M, Popple A, Jander A, Millen RL, Pekas N and Porter MD (2005) Microfabricated tools for manipulation and analysis of magnetic microcarriers. Journal of Magnetism and Magnetic Materials 293 (1):725–730

91. Graham DL, Ferreira HA and Freitas PP (2004) Magnetoresistive-based biosensors and biochips. Trends in Biotechnology 22 (9):455–462
92. Lee GU, Metzger S, Natesan M, Yanavich C and Dufrene YF (2000) Implementation of force differentiation in the immunoassay. Analytical Biochemistry 287 (2):261–271
93. Lagae L, Wirix-Speetjens R, Das J, Graham D, Ferreira H, Freitas PPF, Borghs G and De Boeck J (2002) On-chip manipulation and magnetization assessment of magnetic bead ensembles by integrated spin-valve sensors. Journal of Applied Physics 91 (10):7445–7447
94. Tondra M, Porter M and Lipert RJ (2000) Model for detection of immobilized superparamagnetic nanosphere assay labels using giant magnetoresistive sensors. Journal of Vacuum Science & Technology a-Vacuum Surfaces and Films 18 (4):1125–1129
95. Schepper W, Schotter J, Bruckl H and Reiss G (2006) A magnetic molecule detection system – A comparison of different setups by computer simulation. Physica B: Condensed Matter 372 (1–2):337–340
96. Li GX and Wang SX (2003) Analytical and micromagnetic modeling for detection of a single magnetic microbead or nanobead by spin valve sensors. IEEE Transactions on Magnetics 39 (5):3313–3315
97. Liu J, Lawrence M, Wu A, Ivey ML, Flores GA, Javier K, Bibette J and Richard J (1995) Field-Induced Structures in Ferrofluid Emulsions. Physical Review Letters 74 (14):2828–2831
98. Flores GA, Liu J, Mohebi M and Jamasbi N (1999) Field-induced columnar and bent-wall-like patterns in a ferrofluid emulsion. International Journal of Modern Physics B 13 (14–16):2093–2100
99. Doyle PS, Bibette J, Bancaud A and Viovy JL (2002) Self-assembled magnetic matrices for DNA separation chips. Science 295 (5563):2237–2237
100. Dorfman KD and Viovy JL (2004) Semiphenomenological model for the dispersion of DNA during electrophoresis in a microfluidic array of posts. Physical Review E 69 (1):art. no. 011901
101. Minc N, Futterer C, Dorfman KD, Bancaud A, Gosse C, Goubault C and Viovy JL (2004) Quantitative microfluidic separation of DNA in self-assembled magnetic matrixes. Analytical Chemistry 76 (13):3770–3776
102. Hayes MA, Polson NA and Garcia AA (2001) Active control of dynamic supraparticle structures in microchannels. Langmuir 17 (9):2866–2871
103. Rida A, Lehnert T and Gijs MAM (2003) Microfluidic mixer using magnetic beads, in: 7th International Conference on Miniaturized Chemical and Biochemical Analysis Systems, October 5–9, 2003, Squaw Valley, California USA. pp. 579–582
104. Rida A and Gijs MAM (2004) Dynamics of magnetically retained supraparticle structures in a liquid flow. Applied Physics Letters 85:4986–4988

105. Koutny LB, Schmalzing D, Taylor TA and Fuchs M (1996) Microchip electrophoretic immunoassay for serum cortisol. Analytical Chemistry 68 (1):18–22
106. Chiem N and Harrison DJ (1997) Microchip-based capillary electrophoresis for immunoassays: Analysis of monoclonal antibodies and theophylline. Analytical Chemistry 69 (3):373–378
107. Hadd AG, Raymond DE, Halliwell JW, Jacobson SC and Ramsey JM (1997) Microchip device for performing enzyme assays. Analytical Chemistry 69 (17):3407–3412
108. Sato K, Tokeshi M, Odake T, Kimura H, Ooi T, Nakao M and Kitamori T (2000) Integration of an immunosorbent assay system: Analysis of secretory human immunoglobulin A on polystyrene beads in a microchip. Analytical Chemistry 72 (6):1144–1147
109. Rashkovetsky LG, Lyubarskaya YV, Foret F, Hughes DE and Karger BL (1997) Automated microanalysis using magnetic beads with commercial capillary electrophoretic instrumentation. Journal of Chromatography A 781 (1–2):197–204
110. Hayes MA, Polson NA, Phayre AN and Garcia AA (2001) Flow-based microimmunoassay. Analytical Chemistry 73 (24):5896–5902
111. Fan ZH, Mangru S, Granzow R, Heaney P, Ho W, Dong QP and Kumar R (1999) Dynamic DNA hybridization on a chip using paramagnetic beads. Analytical Chemistry 71 (21):4851–4859
112. Jiang GF and Harrison DJ (2000) mRNA isolation in a microfluidic device for eventual integration of cDNA library construction. Analyst 125 (12):2176–2179
113. Liu RH, Yang JN, Lenigk R, Bonanno J and Grodzinski P (2004) Self-contained, fully integrated biochip for sample preparation, polymerase chain reaction amplification, and DNA microarray detection. Analytical Chemistry 76 (7):1824–1831
114. Lagae L, Wirix-Speetjens R, Liu CX, Laureyn W, Borghs G, Harvey S, Galvin P, Ferreira HA, Graham DL, Freitas PP, Clarke LA and Amaral MD (2005) Magnetic biosensors for genetic screening of cystic fibrosis. IEE Proceedings – Circuits Devices and Systems 152 (4):393–400
115. Kwakye S and Baeumner A (2003) A microfluidic biosensor based on nucleic acid sequence recognition. Analytical and Bioanalytical Chemistry 376 (7):1062–1068
116. Kwakye S, Goral VN and Baeumner AJ (2006) Electrochemical microfluidic biosensor for nucleic acid detection with integrated minipotentiostat. Biosensors & Bioelectronics 21 (12):2217–2223
117. Shikida M, Inouchi K, Honda H and Sato K (2004) Magnetic handling of droplet in micro chemical analysis system utilizing surface tension and wettability, in: 17th IEEE International Conference on Micro Electro Mechanical Systems (MEMS 2004). Maastricht, The Netherlands, pp. 359–362.

118. Shikida M, Takayanagi K, Inouchi K, Honda H and Sato K (2006) Using wettability and interfacial tension to handle droplets of magnetic beads in a micro-chemical-analysis system. Sensors and Actuators B: Chemical 113 (1):563–569
119. Lehmann U, Vandevyver C, Parashar VK and Gijs MAM (2006) Droplet-based DNA purification in a magnetic lab-on-a-chip. Angewandte Chemie International Edition 45 (19):3062–3067

# Chapter 7
# Manipulation of Microobjects by Optical Tweezers

Shoji Maruo[1,2]

[1] Dept. of Mechanical Engineering, Yokohama National University, Yokohama, 240-8501, Japan
[2] PRESTO, Japan Science and Technology Agency

## 1. Introduction

Radiation pressure from a tightly focused laser beam can be used as optical tweezers to confine, position, and transport microparticles. Ashkin's group first demonstrated this technique in 1986 [1]. Optical tweezers provide unique features such as remote manipulation of micro/nano particles in liquid, noninvasive manipulation of biological samples, precise manipulation in sealed environment, and extremely small torque of the order of $10^{-12}$ Nm. For these reasons, optical tweezers and its related techniques have been widely applied to studies on biological cells and DNA molecules, microchemistry with microdroplets and beads. The history and previous works on optical tweezers were introduced in some review reports [2–5].

Although optical tweezers using a single focused laser beam have already used in many applications in various research fields, advanced technologies using optical trapping have increasingly developed such as multiple trapping based on holographic methods and special laser beams including Laguerre Gaussian beam and Bessel beam [3]. The advanced methods play an important role to expand the ability for the manipulation of microparticles.

In most applications of optical manipulation techniques, target samples such as microparticles, droplets, and cells are merely trapped or transferred to the desired position. By contrast, in some special cases, microparticles trapped in the focus of a laser beam can be also rotated by optical torque. Unlike other microfabricated electrostatic motors, the optical rotators offer unique advantages such as remote drive in liquid and no friction without mechanical contact. These features are useful for driving both microparticles and microfabricated microobjects as a tool for microfluidic devices such as micropumps and micromixers. In particular, sophisticated microstructures produced by threedimensional (3D) microfabrication offer microfluidic devices driven by light. Accordingly, the optical rotators have received a lot of attention. In this chapter, an overview of optical manipulation techniques related to microfluidics and biological applications is given.

## 2. Single-Particle Manipulation with a Focused Laser Beam

### *2.1. Trapping of a Micro/nano Particle with a Focused Laser Beam*

A photon not only has energy but also momentum. If the direction of light propagation is changed due to the reflection and refraction, a momentum change of photon is generated. According to Newton's third law, a radiation pressure is exerted on the interface between two media as a reaction to the momentum change. In electromagnetic theory of light, the photon force is well-known as optical radiation pressure. In case of the size of a microparticle is larger than the wavelength of incident light, the radiation pressure exerted on the microparticle can be analyzed by ray optics [6]. As shown in Fig. 1, when a laser beam is focused on a transparent microparticle whose refractive index is larger than that of the surrounding liquid, the laser beam is reflected and refracted at the surface of the microparticle. The laser beams A and B incident to the particle are mainly refracted; thereby radiation pressure is exerted at a vertical direction on the top of the microparticle. When the refracted light exits from the microparticle, radiation pressure is also exerted on the bottom of the particle. The net radiation pressure surrounding the surface of the microparticle is directed to the focus of the laser beam. Since a part of the laser beam whose incident angle is large play an important role to levitate the microparticle, an objective lens of high numerical aperture is normally used for the manipulation of microparticles. The force of optical trapping is proportional to the intensity of the trapping laser beam.

For the last two decades, optical trapping of a dielectric microparticle using a single focused laser beam has been widely applied to many application fields, including single molecule measurement of DNA and motor proteins [7–9], near-field scanning optical microscopes (NSOM) [10–12], and lasing in a microparticle [13]. For the measurement of the interaction between a microparticle and biological samples, high-precision measurements of both the position of the microparticle and the force between the particle and target molecules or surface are needed. In general, to evaluate the optical trapping force interacting with biological samples on the order of pN, the position of the microparticle trapped at a focus is measured with a quadratic photodiode [14–16]. Figure 2 shows a typical schematic diagram of pN force measurement system using a quadratic photodiode [15]. The quadrant diode is placed in a plane conjugate to the back focal plane of the objective lens. The change of the position of the microbead can be detected by monitoring the difference of signals of four detectors of the quadratic photodiode. Since the image of a microbead interacting with biomolecules is also formed on the photodiode with an objective lens of high numerical aperture, single-molecule analysis of motor proteins and DNA is achieved.

By contrast, applying optical trapping to a near-field optical microscope (NSOM), a trapped microparticle is just employed as a microprobe to scattering optical fields around the sample surface [10–12]. The intensity of scattering light is measured with a photomultiplier tube. Since the laser-trapped microprobe is scanned along the sample surface while a sample stage is moved in two dimensions, the near-field optical image is obtained by measuring the intensity of scattering light. In addition, the surface force between a microparticle and a substrate in solution can also be analyzed by measuring the displacement of the laser-trapped probe. As an example of NSOM with a laser-trapped microprobe, the experimental setup is shown in Fig. 3 [17]. In this setup, the amount of displacement is measured by detecting the intensity of light scattering at the probe illuminated by an evanescent field. A sensitivity of 27 fN and a range of 5.2 pN have been achieved. Such a surface force measurement using a laser-trapped microparticle can be useful for the analysis of surface conditions and fluidic properties inside a microchannel.

Not only spherical but also ellipsoidal and cylindrical microobjects can be stably trapped at a focus [18]. Most of the nonspherical particles tend to direct their major axis in the direction of the optical axis according to the net radiation pressure. As a result, nonspherical particles such as ellipsoids and cylinders are stably trapped at the focus as their major axis aligns with the optical axis. For example, when a laser beam is focused into a liquid containing multiple microparticles, several microparticles are simultaneously trapped at the focus and stand in line along the optical axis. This phenol-menon is utilized to make a rod with multiple microbeads functioning as a microvalve [19]. On the other hand, special particles with anisotropic shape or microparticles made from birefringent materials are trapped and rotated at the focus. The rotation of microparticles is described in the following Sects. 2.3 and 2.4.

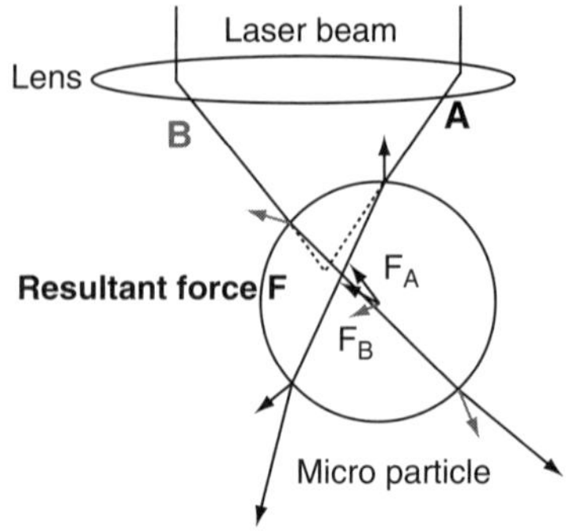


**Fig. 1.** Optical trapping of a dielectric microparticle with a focused laser beam

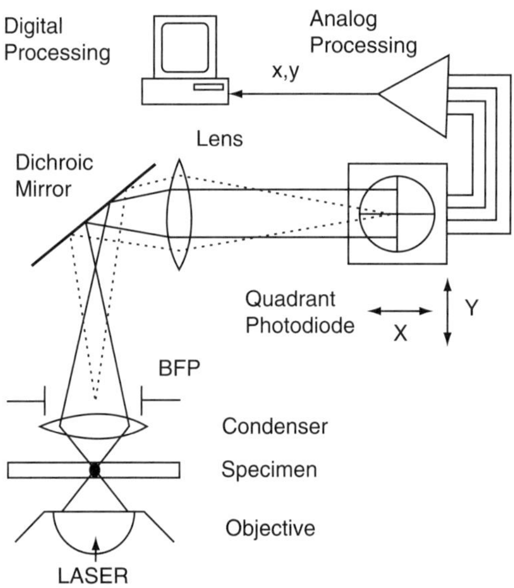


**Fig. 2.** Optical system for pN-order force measurement using a quadratic photodiode [15]

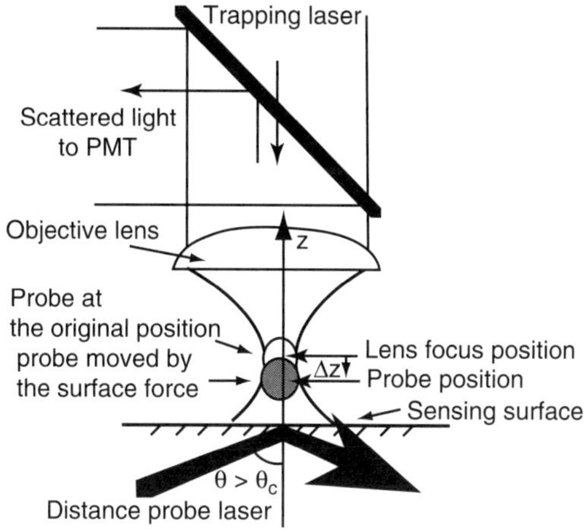


**Fig. 3.** Near-field microscope using a laser-trapped microparticle for surface force measurement [17]

## *2.2. Trapping of a Metallic Particle*

In case the laser beam is incident on a metal surface, since it is reflected in the opposite direction, the radiation pressure is exerted orthogonal to the metal surface. In general, therefore, metallic Mie particles, whose diameters are larger than the wavelength of light, are pushed away with a repulsive force owing to the momentum transfer of light. For this reason, circular scanning of a laser beam [20] or a doughnut beam of $TEM_{01}$ mode [21] were employed for the trapping of a metallic microparticle in three-dimensions by making use of this repulsive force.

By contrast, for two-dimensional trapping of a metallic Mie particle, a tightly focused laser beam can be used. In this case, rays with large inclination about the beam axis generate an effective force for lateral trapping [22]. In addition, Furukawa demonstrated that a softly focused Gaussian beam is also useful for trapping of metallic particles. When a Gaussian beam is focused on a metallic particle, a creeping wave is induced by a kind of surface plasmon surrounding the particle. This causes an attractive force as shown in Fig. 4b [23]. Figure 4 classifies the radiation pressure exerted on a Mie particle. Since a creeping wave deflects the incident laser beam, an attractive force is generated owing to momentum transfer of light. The numerical analysis of the above two methods was reported by Pu Chun Ke and Min Gu [24].

Metallic Rayleigh particles, whose diameters are smaller than the wavelength of light, can be also trapped by a tightly focused laser beam [25, 26]. In these demonstrations, the radiation pressure force exerted on Rayleigh particle is not due to scattering and absorption forces but gradient force is utilized for stable trapping. Although scattering and absorption forces are proportional to the light intensity and point along the direction of the incident laser beam, the gradient force is proportional to the gradient of light intensity and points toward to the focus. Therefore metallic Rayleigh particles are trapped at the focus.

An optically trapped metallic particle was utilized as a probe of NSOM [12]. Since the trapped metallic particle enhances the intensity of scattering light, the image contrast of NSOM can be improved compared with an image obtained by using a dielectric microparticle as a probe.

Fixation and pattering of metallic nanoparticles were also demonstrated by Masuhara's group [27]. In their method, a single gold nanoparticle of 80 nm diamater was optically trapped and manipulated to a surface of a glass substrate, and then fixed on it by irradiation with ultraviolet laser beam. Figure 5 shows an atomic force microscope (AFM) image of fixed and patterned gold nanoparticles. These nanoscale manipulation and fixation techniques will be useful for further studies and applications on nanoscience and nanotechnology, for example, spatial patterning of catalytic nanoparticles in a microcapillary of a micrototal analysis system.

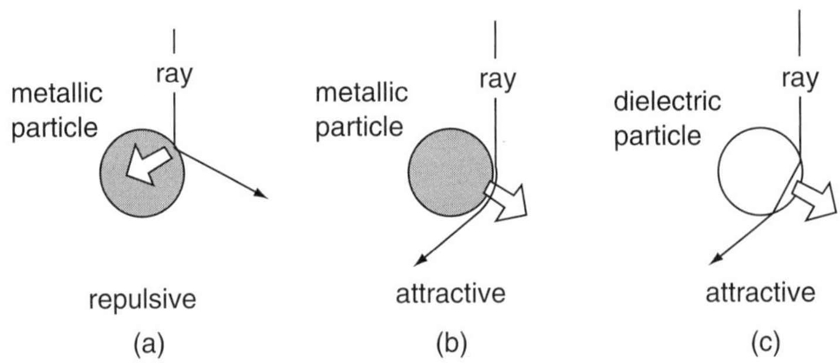


**Fig. 4.** Classification of radiation pressure exerted on metallic and dielectric Mie particles [23]. (**a**) Repulsive force arising from surface reflection, (**b**) attractive force of a creeping wave, and (**c**) attractive force arising from refraction from a dielectric particle

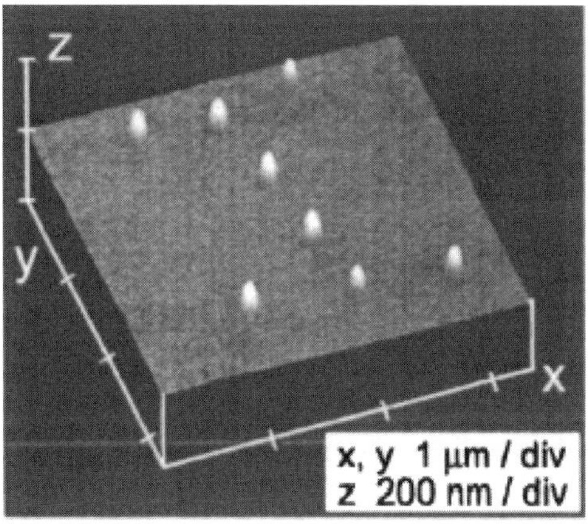


**Fig. 5.** An AFM image of fixed and patterned gold nanoparticles [27]

### *2.3. Rotation of a Birefringent Microparticle*

It is well known that circularly polarized light has angular momentum [28, 29]. When a circularly polarized laser beam is focused on a birefringent plate, the angular momentum of a circularly polarized light beam exerts a torque on the plate [30]. The amount of the momentum change depends on the phase retardation of the birefringent plate. If the retardation of the birefringent plate is $\lambda/2$, right-handed circularly polarized light is transformed into left-handed one. In this case, the change of spin angular momentum is maximal, so the torque exerted on the plate also attains to maximum value $h/\pi p$, where $h$ is Planck's constant. Friese et al. demonstrated that a calcite particle 1 µm thick could be trapped and rotated at a rotation frequency of 357 Hz with a circularly polarized laser beam of 300 mW [30].

By contrast, when a linearly polarized laser beam is focused on a birefringent microparticle, the microparticle can be trapped and angularly aligned [31]. The principle of the angular alignment of a birefringent microparticle is shown in Fig. 6. As shown in this figure, when the laser beam is focused on a birefringent microparticle, the fast axis of the birefringent microparticle initially makes an angle $\theta$ with respect to the oscillation plane of the incident electric field vector. And then the microparticle rotates so that the fast axis becomes coincident with the vibration plane as shown in Fig. 6b. Higurashi et al. demonstrated that a birefringent microparticle made from fluorinated polyimide is angularly aligned by rotating the direction of linear polarization [31]. Figure 7 shows the experiments on angular alignment of a birefringent microobject produced by a photolithographic technique. The fluorinated polyimide has a large birefringence ($\Delta n = n_2 - n_1 = 1.49 - 1.62 = 0.1$ for a wavelength of 1,064 nm). In the future, the angular alignment of microparticles will be useful for microoptomechanical systems, such as the assembly of microoptical components. In addition, an optically trapped birefringent microobject can be used as a torsion spring of a sensing probe for detecting small forces.

Not only natural birefringent materials but also artificial birefringent microstructures can be trapped and rotated in a focused laser beam. Artificial birefringent is well known as form birefringence [32]. Neale et al. demonstrated that a microgear that has a grating-like structure with submicron period could be trapped and rotated [33]. When a circularly polarized laser beam is incident on the gear with the grating as shown in Fig. 8, since the effective refractive indices for polarized laser beams perpendicular and parallel to the grating are different, an angular momentum transfer occurs. As a result, just focusing a laser beam could rotate the form birefringent microgear. Microfabricated microstructures with form birefringence are easily applied to various applications such as light-driven MEMS and microfluidic control devices such as micropumps and microstirrers.

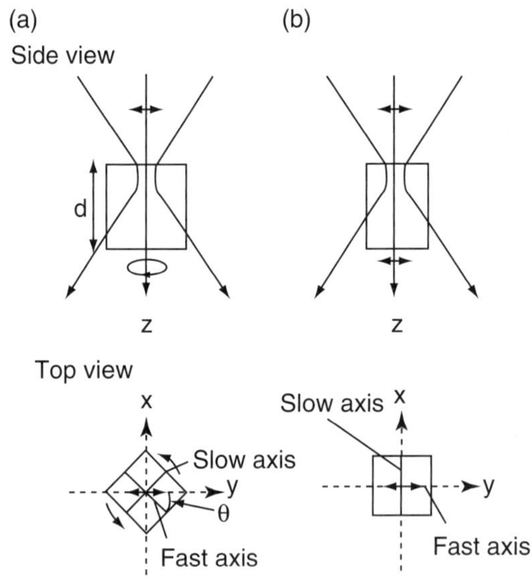


**Fig. 6.** Schematic diagram showing the optically induced angular alignment of a birefringent microobject [31]. (**a**) The fast axis of the microobject makes an angle $\theta$ with respect to the oscillation plane of the incident electric field vector. (**b**) The fast axis becomes coincident with the vibration plane

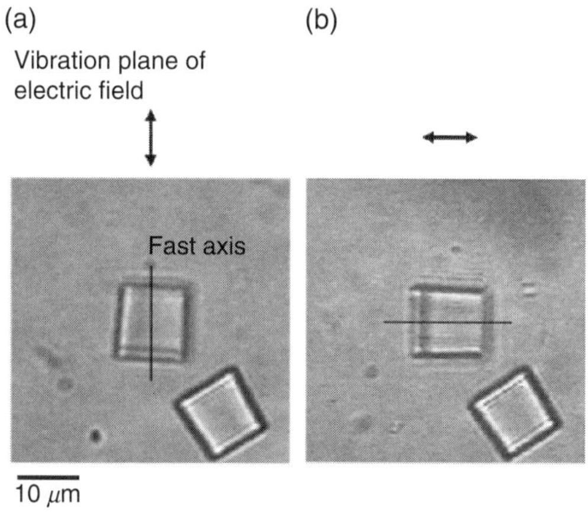


**Fig. 7.** Optically induced angular alignment of a birefringent microobject (thickness: 15 μm). (**a**) and (**b**) The direction of linear polarization was rotated by 90°. The trapping laser beam was irradiated perpendicular to the plane of the photograph and focused around the top surface of the microobject

### *2.4. Manipulation of a Micromachined Object*

Higurashi et al. have first demonstrated that microfabricated microobjects can be trapped and rotated by focusing a laser beam independently of polarization [34]. Giving the microobjects an anisotropic geometry produces a torque as a result of the net radiation pressure on their surfaces from a symmetrical input intensity profile. Unlike birefringent microobjects, the microfabricated microobject is not driven by the intrinsic angular momentum of a photon but by the momentum transfer caused by refraction or scattering. They used microobjects that were flat on the top and bottom, but had a shape assymetry on the side. Figure 9 shows the origin of the optically induced rotation for the designed microobject. In this case, the microobjects have a higher refractive index than the surroundings medium in order to trap the microobject at a focus. Therefore, the radiation-pressure force $F$ arising from the momentum change $\Delta P$ due to ray refraction is directed to the surrounding medium perpendicular to the side. The net forces on the side faces of the designed microobject lead to clockwise rotation. Higurashi et al. fabricated microobjects with fluorinated polyimide whose refractive index is 1.53 at a wavelength of 1064 nm and rotated the dissymmetry microobject in ethanol (refractive index: 1.36) [35]. Figure 10 shows the laser-power dependence of the rotation speed for $TEM_{00}$- and

$TEM_{01}$- mode laser beams. Since a $TEM_{01}$-mode Gaussian laser beam has a doughnut-like intensity profile, the torque exerted on the side is improved. As a result, the rotation speed is increased.

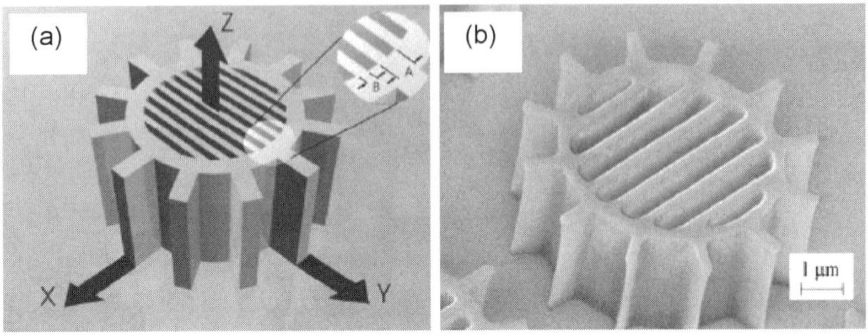


**Fig. 8.** Form birefringence microgears. (**a**) Diagram of a form birefringent microgear. A is the pitch of the photonic lattice and B is the rib width. (**b**) SEM micrograph of an actual microgear before release from the substrate [33]

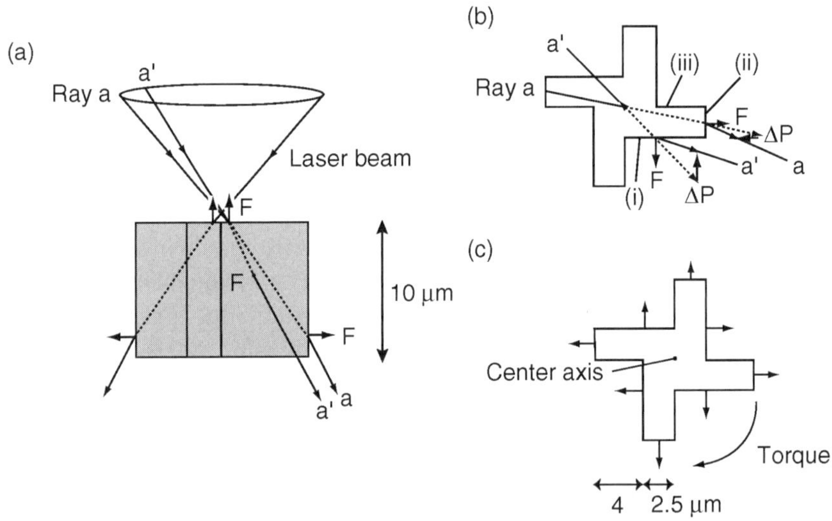


**Fig. 9.** Origin of the optically induced rotation for the designed microobject [34]. (**a**) Side view. (**b**) Top view. The radiation pressure force $F$ arises from the momentum change $\Delta P$ due to ray refraction. (**c**) Forces on the side faces of the designed microobject, which lead to clockwise rotation

On the other hand, they also fabricated other types of asymmetrically shaped microobjects with a lower refractive index relative to that of surrounding medium. Figure 11 shows the schematic design of optical trapping and optically induced rotation of a low relative-refractive index microobject [35]. While a focused laser beam passes through the microobject from the inside to the outside, radiation pressure is exerted on its sides. The net radiation pressure causes a rotation of a low relative-index microobject. Figure 12 shows photographs of such microobjects. Since the horizontal cross sections of the two microobjects are mirror images, they are rotated in opposite directions by focusing a laser beam. A microobject having an opening and shape anisotropy in its interior can also be trapped and rotated about an axis perpendicular to the trapping laser beam axis [36]. The microobject (12 μm in diameter) is produced by oxygen reactive ion etching of a fluorinated polyimide film. Figure 13 shows a series of four frames from a video recording demonstrating the trapping and rotation of the microobject about an axis perpendicular to the laser beam axis in ethanol. It is clear that this rotation is caused by the optical torque exerted on the inner walls of the microobject, because radiation pressure forces exerted on its outside act radially and do not contribute to the optical torque. Such an optically induced rotation has potential applicability for optically driven micromotors and sensitive microprobes for measuring the properties of microfluid dynamics.

Although most of microobjects designed for optically induced rotation were made by using lithographic techniques, 3D two-photon micro-fabrication techniques have also been employed for producing real objects with asymmetric shapes in recent years. The 3D two-photon microfabrication that was first demonstrated by Maruo et al. in 1997 [37, 38] makes it possible to produce arbitrary 3D polymeric micro-microstructures with 100 nm resolution, so optically induced microrotators are easily fabriccated according to their desired shapes. Galajda and Ormos demonstrated some types of special microrotors that generate optical torque by focusing a laser beam [39, 40]. Figure 14 shows the schematic diagram of an arrangement of a light-driven rotor able to rotate in both directions. When a laser beam is focused above the rotor, it is reflected on the outer side of the rotor. On the other hand, when the laser beam is focused below the rotor, the laser beam is reflected on the inner side of the rotor. Since the rotation directions caused by the momentum change of the laser beams are opposite, the rotation direction can be reversed just by changing the focus position along the optical axis.

One of the promising applications of optically induced microrotors is mixing in a microchannel or microchamber in biochip technologies. Ukita et al. fabricated a micromixer in which a specially designed microrotor was optically trapped and rotated at the junction of a Y-branch channel [41]. The microrotor was rotated at a speed of 90 rpm at a laser power of 160 mW.

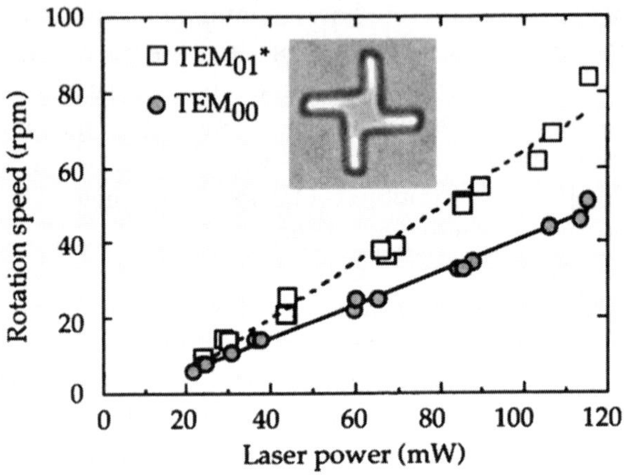


**Fig. 10.** Laser-power dependence of the rotation speed for $TEM_{00}$- and $TEM_{01}$* - mode laser beams [35]

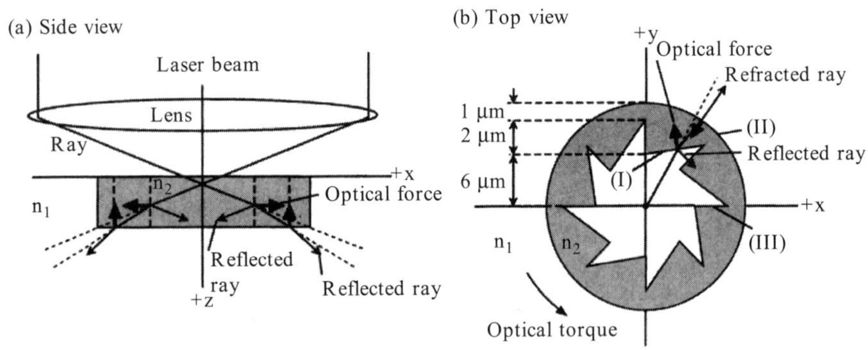


**Fig. 11.** Schematic illustrating the origin of optical trapping and optically induced rotation of a low relative-refractive index microobject ($n_1 > n_2$) [35]

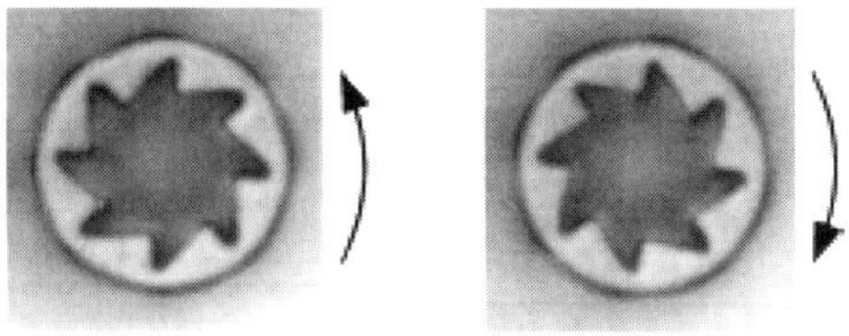

Fig. 12. Photos and direction of rotation for low relative-refractive index microobjects whose horizontal cross sections are mirror images [35]

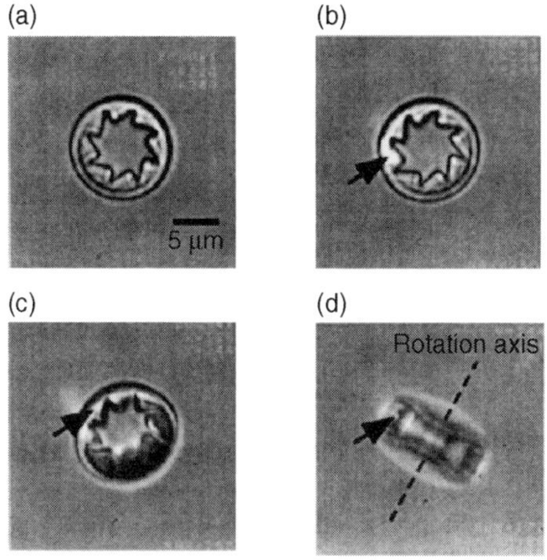


Fig. 13. A series of four frames from a video record demonstrating the trapping and rotation of the microobject about an axis perpendicular to the laser beam axis in ethanol [36]

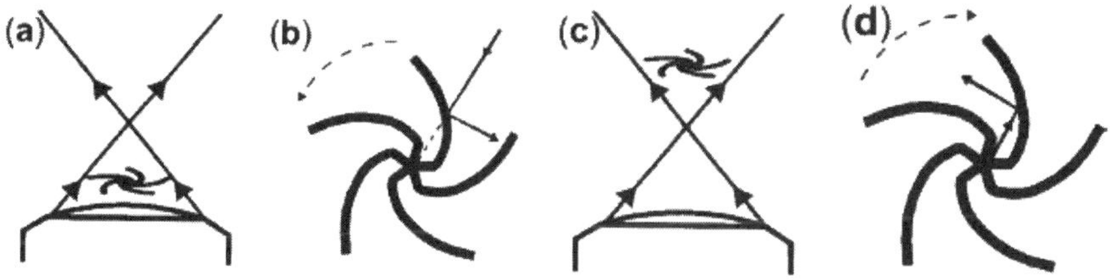


Fig. 14. Schematic of the arrangement of a light-driven rotor able to rotate in both directions [40]. (**a**) Position of the rotor in the focused laser beam and (**b**) characteristic reflections to drive rotation in one direction, (**c**) and (**d**) respective equivalent figures for the rotation in the opposite direction

# 3. Multiparticle Manipulation Techniques

## 3.1. Single Beam Based Manipulation

### 3.1.1. Time-Divided Laser Scanning for the Manipulation of Multiple Microparticles

High speed scanning of a single laser beam makes possible to trap many microparticles simultaneously. Sasaki et al. demonstrated that polystyrene microparticles (diameter: 2 μm) were trapped along desired trajectories by scanning a focused laser beam using galvano scanners [42]. Figure 15 shows microparticles arranged as a Chinese character that means light. In this case, time-averaged optical intensity distribution, which is generated by repeatedly scanning a trapping laser beam at 13–50 Hz, is utilized to trap multiple microparticles along the trajectory of the laser beam in ethylene glycol. In addition, if the scanning speed and laser power is optimized, an asymmetric optical potential field is generated on a moving microparticle as shown in Fig. 16 [42]. The asymmetric optical field causes to move the microparticle along the trajectory of the laser beam. As a result, multiple microparticles can be transported along the circular trajectory of the laser beam as shown in Fig. 16. In this experiment, the particles were not trapped at fixed positions, but moved along the produced pattern. Figure 16 shows sequential images recorded at intervals of 0.6 s. The circle with a diameter of 13.5 μm was created by repetitive scanning of the trapping beam at a repetition rate of 15 Hz in the right-handed rotation.

Recently Arai et al. proposed a synchronized laser scanning method to trap multiple microparticles by use of a galvano scanning system [43]. In their method, a single laser beam is scanned not continuously but discretely. The discrete scanning of a single laser beam requires the optimization of the scanning patterns, because both the position and the velocity stability of the optically trapped microparticles are strongly affected by Brownian motion while each microparticle is not irradiated by the laser beam. Therefore, the laser beam should be irradiated on each microparticle intermittently to trap the desired position stably at every moment.

To allow capture of much more microparticles, Mio et al. developed a laser-scanning micromanipulation system that used piezoelectric mirrors instead of galvano mirrors to steer the laser beam [44]. Piezoelectric systems have the advantage that they can scan much faster than galvano systems, reaching frequencies of 106 Hz. To trap multiple particles in any pattern using a single scanning laser beam, the repetition rate of the

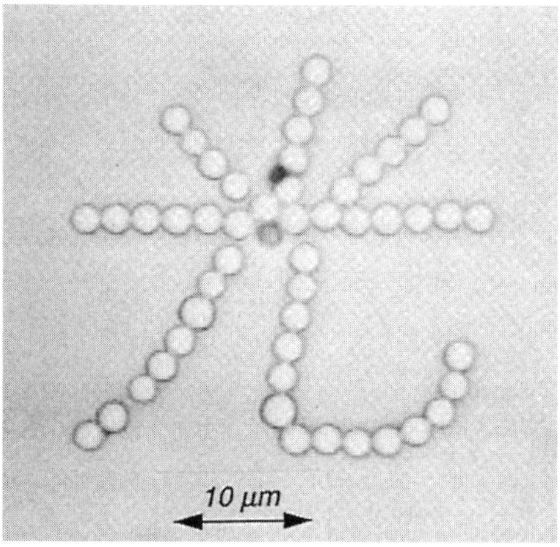


**Fig. 15.** Spatial patterning of polystyrene latex particles (diameter: 2μm) in ethylene glycol [42]. The Chinese character for light is formed

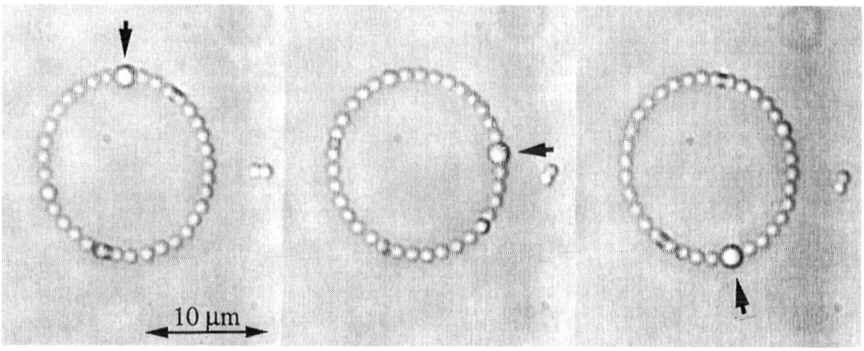


**Fig. 16.** Sequential images of the particle flow observed at intervals of 0.6 s [42]. The sample is polystyrene latex particles (diameter: 1 μm) in 1-pentanol

scanning beam must be faster than the time scale of particle Brownian motion. For this reason, high-speed scanning with piezoelectric systems is suitable for numerous microparticles simultaneously. Mio et al. mentioned that their system allows to trap 50 particles in water at a laser output power of less than 0.8 W and a piezoelectric mirror frequency around 1,200 Hz.

### 3.1.2 Continuous Transportation of Multiple Particles

Since Brownian motion cannot create a steady flux in a system at equilibrium, micro or nanoparticles cannot be transported in one direction in a symmetric and static optical potential field. By contrast, microparticles subjected to an asymmetric periodic potential can display net directional

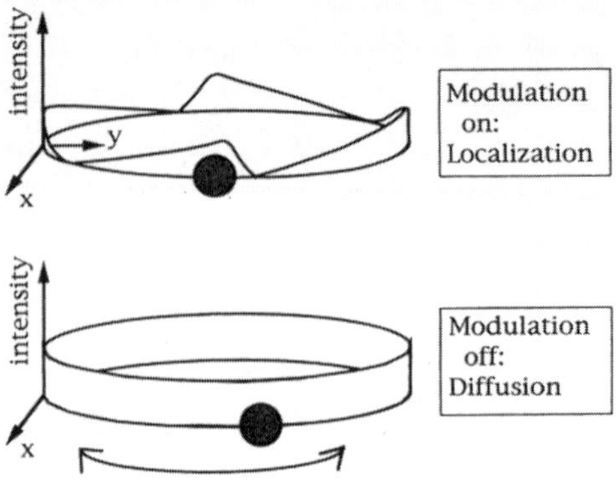


**Fig. 17.** Spatially asymmetric modulation of the beam intensity along the circle described by the optical trap [45]

motion even if the space-averaged force is zero. Faucheux et al. demonstrated such a behavior using a kind of optical trapping technique [45]. They generated an asymmetric potential whose intensity is modulated by scanning a laser beam circularly as shown in Fig. 17. The microparticle exhibit net motion with a velocity that depends on the modulation time. This demonstrated that the broken spatial symmetry and time modulation are indeed enough to induce directed motion from random noise. Recently, Lee et al. also demonstrated that colloidal silica spheres could be moved by using a spatially symmetric three-state ratchet potential comprised of discrete potential wells [46]. Particle sorting by size or refractive index is also achieved by controlling the spatial fringe periodicity of an asymmetric time-modulated light pattern with a sawtooth time modulation [47].

### 3.1.3. Bessel Beam for the Manipulation of Multiple Particles

In normal optical trapping, a focused laser beam is used to trap and manipulate single or multiple microparticles. The optical trapping technique allows to confine the particle in three-dimensions with an objective lens of high-numerical aperture. However, the region to trap a microparticle is limited by the working distance (about 100 μm at numerical aperture in the range of 1.2–1.4) of the objective lens. On the other hand, if a zeroth order Bessel light beam [48] that is a solution of scalar Helmholtz equation, is used as a trapping laser beam, the trapping region is extremely extended compared to that of the focused laser beam.

The electric field amplitude of a Bessel light beam is given by

$$E(r,z) = A\exp(ik_z z)J_0(k_r r) \tag{1}$$

Here $J_0$ is the zeroth-order Bessel function and $k_r$ and $k_z$ are the radial and longitudinal components of the free-space wave vector $k$ [49]. As inherent in (1), a Bessel beam theoretically requires an infinite energy. Practically, as a finite approximation of the Bessel beam, a conical glass prism known as an axicon is used [50]. The focal depth of a Bessel beam generated by an axicon is estimated by

$$z_{max} \approx \frac{w_0}{(n-1)\gamma} \tag{2}$$

where $\gamma$ and $n$ are the opening angle and refractive index of the axicon. $w_0$ is the waist of the illuminated Gaussian light beam. The focal depth is much larger than that of a high-numerical aperture objective lens. Therefore the use of a Bessel beam offers unique advantages as follows. For example, as shown in Fig. 18, Dholakia's group asserted that the central maximum of a Bessel beam provides a good localization of microparticles throughout a sample cell that are too large to be manipulated with a normal focused laser beam [49]. Dholakia's group also demonstrated that multiple microparticles could be trapped simultaneously along the central maximum of a Bessel beam [51]. Figure 19 shows the multiple optical trapping of microparticles using a Bessel beam. In this case, microparticles stored in two cells that are 3 mm apart are simultaneously trapped at the central maximum of the Bessel beam. Thus, a Bessel light beam will be useful for manipulating microparticles inside a thick chamber or a 3D microchannel owing to its long focal depth. This technique is also useful for particle manipulation inside a sophisticated 3D microchannel.

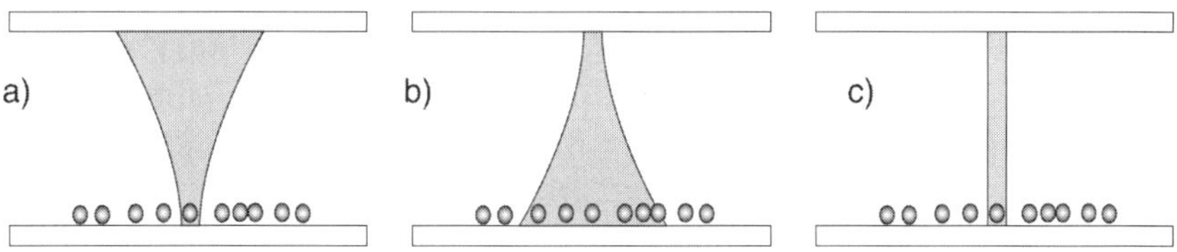


**Fig. 18.** Comparison of optical trapping methods using a Gaussian beam and a Bessel beam [49]. (**a**) and (**b**) final destination and initial condition with a Gaussian beam. (**c**) Optical trapping in deep area using a Bessel beam

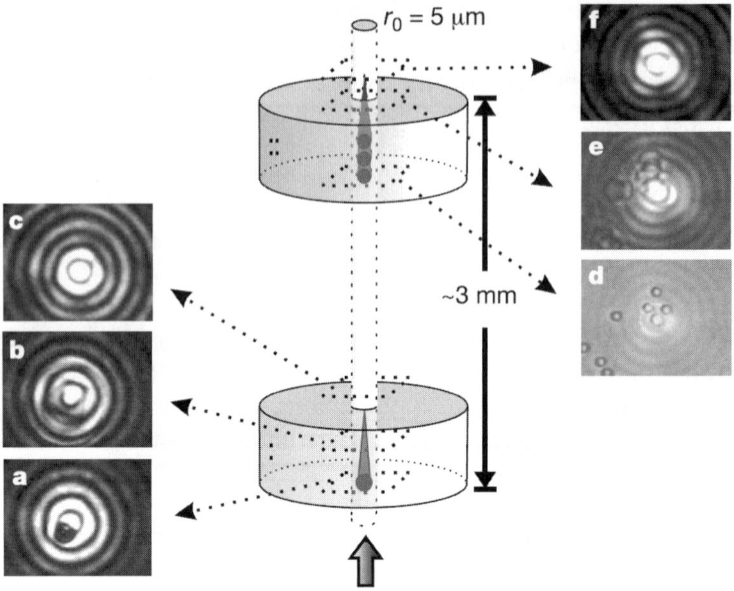


**Fig. 19.** Inverted optical tweezers experimental set-up using a Bessel beam [51]. The cells are 3 mm apart and 100 μm deep. (**a**–**f**), Frames from a video taken of objects captured by the Bessel beam

### *3.2. Holographic Optical Tweezers*

The use of a diffractive optical element makes it possible to statically transform a single laser beam into an array of beams [52]. By contrast, a computer-controlled spatial light modulator (SLM) allows the generation of multiple traps that can be reconfigured at video frame rates or higher [53, 54]. Figure 20 illustrates an optical setup for holographic optical tweezers using a SLM. The plane of the SLM, a, is imaged into the pupil of the microscope objective plane a* [54]. The planes b and c are imaged into the focal region of the microscope. By changing an input hologram, the creation of multiple foci and a 3D independent movement of the foci are easily realized. For example, holographic optical tweezers have been used to create 2D arrays of trapped microparticles [53], sorting of micro-sphere mixtures of distinct sizes and colors, simultaneous trapping of a colloidal mixture consisting of particles with both lower and higher refractive indices than that of the suspending medium and simultaneous independent manipulation of multiple semiconductor nanowires [55–57].

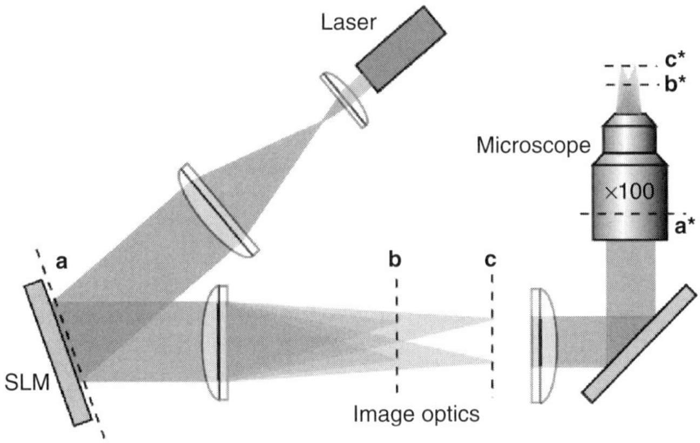


**Fig. 20.** Schematic diagram an optical tweezer using a SLM. The plane of the SLM, a, is imaged into the pupil of the microscope objective plane a*. The planes b and c are imaged into the focal region of the microscope [54]

In addition to lateral displacements, SLMs can also shift the axial position of the trap away from the focal plane of an objective lens. Using this technique it has been possible to manipulate microparticles both laterally and axially, and arrange them into 3D configurations [54, 58]. Figure 21 shows glass beads (2 μm diameter) trapped in the corners of an imaginary rotating tetrahedron [54]. The upper-right column shows the corresponding video frames. The phase-hologram patterns used for the generation of the corresponding optical traps are shown in the bottom-right column. Controlled 3D translation and rotation of an asymmetric microobject has also been demonstrated by using a pair of closely separated optical traps [59].

SLMs can create beams of light with helical wave fronts [3, 60, 61]. The helical modes focus to rings, rather than points, and also carry angular momentum. The doughnut beams form toroidal optical traps known as optical vortices, which are capable of localizing and applying torques to small volumes of matter. Figure 22a shows a schematic diagram of dynamic holographic optical tweezers creating an optical vortex [60]. When the phase mask shown in Figure 22a is imposed on the incident $TEM_{00}$ beam, the beam is converted into a helical beam that is focused into an optical vortex. Figure 22b shows an image of the resulting optical vortex obtained by placing a mirror in the focal plane. The central spot is the diffraction-limited focus of a separate coaxial $TEM_{00}$ beam and coincides with the optical axis. Figure 22c shows time-lapse image of a single colloidal sphere traveling around the optical vortex. The above techniques using SLMs can be used for multipurpose biochips and biomanipulation tools driven by light.

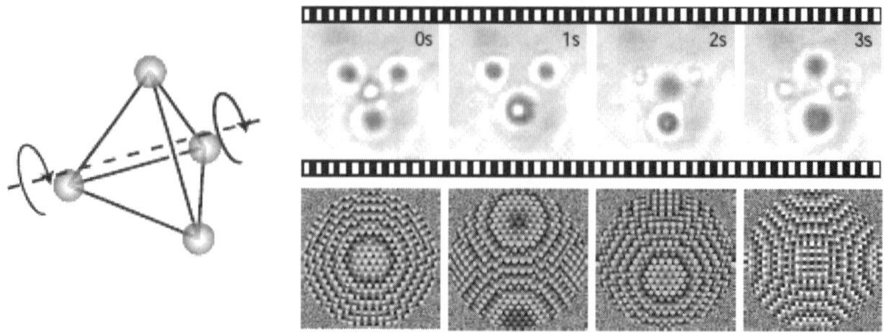


**Fig. 21.** Glass beads (2 μm diameter) trapped in the corners of an imaginary rotating tetrahedron [54]. The right columns show the corresponding video frames and the corresponding phase-hologram patterns

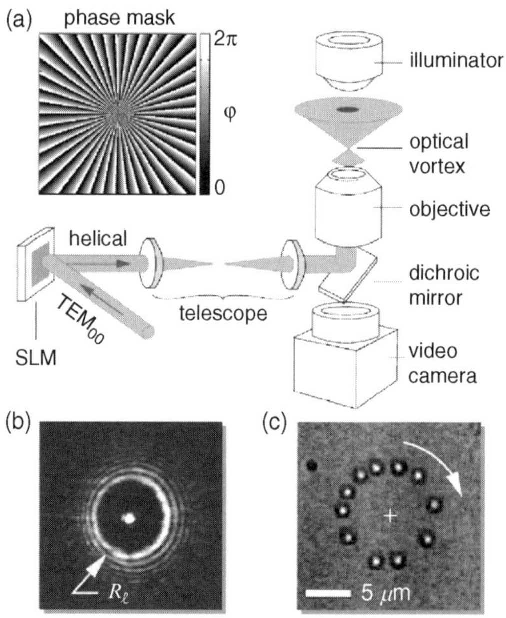


**Fig. 22.** Movement of microparticles along a circular trajectory using holographic optical tweezers creating an optical vortex [60]. (**a**) Dynamic holographic optical tweezers creating an optical vortex. (**b**) Image of the resulting optical vortex obtained by placing a mirror in the focal plane. (**c**) Time-lapse image of a single colloidal sphere traveling around the optical vortex

### 3.3. Evanescent Waves for the Propulsion of Microparticles

Evanescent waves, which are generated at the surface of a prism or waveguide owing to total internal reflection of light, have also been used for the propulsion of microparticles. In 1992, Kawata and Sugiura first demonstrated that small particles with diameters of 1–27 μm can be moved in the evanescent fields produced by a laser beam [62]. In this case, the evanescent field is scattered at a microparticle, so that the momentum change occurs. The net radiation pressure caused by the momentum change is aligned with the direction of movement of the incident laser beam. As a result, the microparticle in the evanescent field can be moved along the surface of the prism.

If an evanescent field generated along a channel waveguide is used, microparticles can be simultaneously moved along the waveguide [63]. Figure 23 illustrates the particle propulsion along a channel waveguide. The microparticle is attracted and propelled on the waveguide with the evanescent field. Figure 24 shows a photograph sequence taken at 5 s intervals, illustrating the motion of a 5.1-μm-diameter latex sphere along a waveguide. The net radiation pressure exerted on a microparticle in an evanescent field is calculated by using Maxwell's stress tensor based on electromagnetic theory and ray optics [63, 64]. In addition, by using the evanescent-wave driving method, not only dielectric but also metallic microparticles were moved simultaneously [63, 65].

Recently, counterpropagating evanescent waves have been utilized for guiding and trapping numerous microparticles [66]. This method allows an easy switching between trapping and guiding, thereby creating an optical conveyor belt for sampling/sorting of large-scale assemblies of micro-objects. Optical manipulation and large-scale trapping of microparticle arrays using cavity enhanced evanescent waves produced with a dielectric resonator was also demonstrated [67]. The forces associated with near-field optical micro-manipulation can be greatly increased through the use of cavity-enhanced evanescent waves.

Such propulsion of multiple microparticles using evanescent fields has unique features such as selective transportation near the surface and optical driving in an absorbing medium, because the evanescent field is localized at the surface. In the near future, if an optical waveguide is attached below a microchannel, living cells and microbes can be transported along a microchannel. This approach is useful for all optically controlled biochips without laser scanning optics or SLMs.

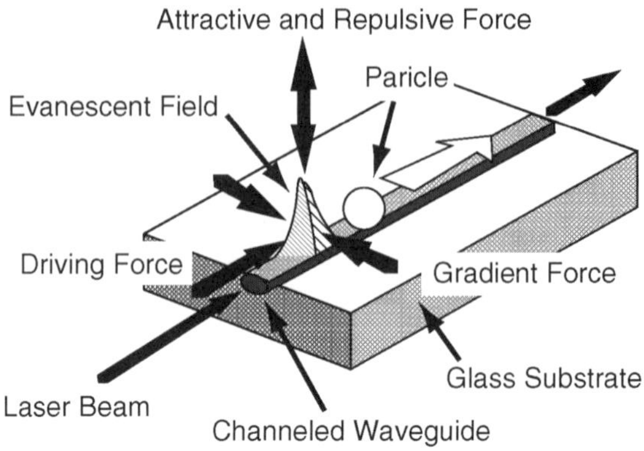


**Fig. 23.** Particle transportation along a waveguide using evanescent fields [63]

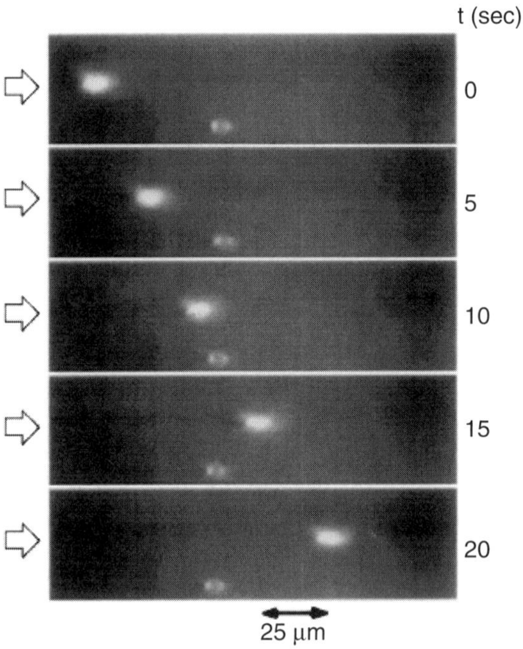


**Fig. 24.** Photograph sequence taken at 5 s intervals, illustrating the motion of a 5.1-μm-diameter latex sphere along a waveguide channel

# 4. Optically Driven Microfluidic Components

## 4.1. Particle Sorter Using an Optical Lattice

Optical trapping is a suitable method for manipulating microparticles inside a sealed space such as a microfluidic circuit or a microchamber. In normal optical trapping techniques, single or multiple focused laser beams are used to trap and transport multiple microparticles. By contrast, a 3D interferometric pattern of multiple laser beams, being called an optical lattice, enables sorting of numerous microparticles in a microchannel. Korda et al. demonstrated that microparticles interacting strongly with multiple optical traps can be pushed to one side by an appropriately tuned array [68]. In their experiments, an optical lattice is generated using holographic optical tweezers with an SLM. By turning the SLM, the optical lattice was also turned in a microchannel. As a result, the direction of the movement of microparticles was also slightly modified. This method can be applied for separation of mixture of microparticles. MacDonald et al. demonstrated an optical sorter for microscopic particles that exploits the interaction of particles with an extended, interlinked, dynamically reconfigurable 3D optical lattice [69]. Although the principle of their method is similar to Korda's approach, they utilized an extended, 3D optical lattice of interlinked sites, with linkages established along the desired channeling direction. The strength of the interaction with the lattice sites depends on the optical polarizability of the particles, giving tunable selection criteria. Figure 25 shows the concept of optical fractionation using a 3D optical lattice. In this case, the flow inside microchannel is laminar due to its low Reynolds number. Therefore all particles from chamber B would flow into chamber D. By introducing a 3D optical lattice, the selected particles are introduced into chamber C. MacDonald et al. demonstrated both sorting by size (of protein microcapsule drug delivery agents) and sorting by refractive index (of other colloidal particle streams). Figure 26 shows an example of the optical fractionation by size. Black crosses represent the frame-by-frame positions of two protein microcapsules (diameter: 2 μm) as they flow from right to left across the optical lattice. A significant angular deflection of the protein microcapsules is achieved, while a coflowing capsule (diameter: 4 μm) of the same sort flows nearly straight through (white dots). MacDonald et al. asserted that the sorting efficiency of this method approaches nearly 100%. This powerful, noninvasive technique is suited for sorting and fractionation within integrated microfluidic systems, and can be applied in colloidal, molecular, and biological research.

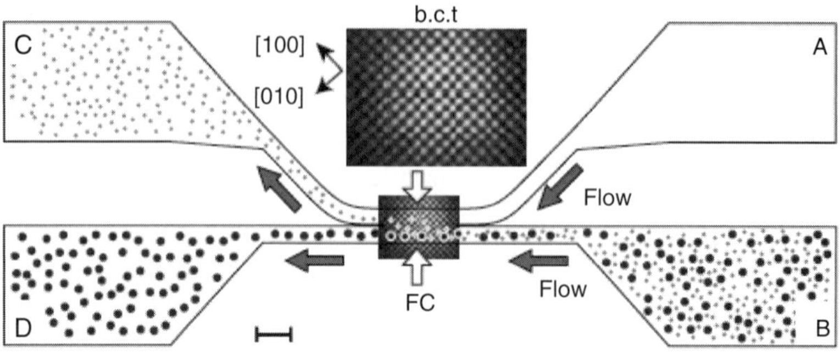


**Fig. 25.** The concept of optical fractionation using a 3D optical lattice [69]

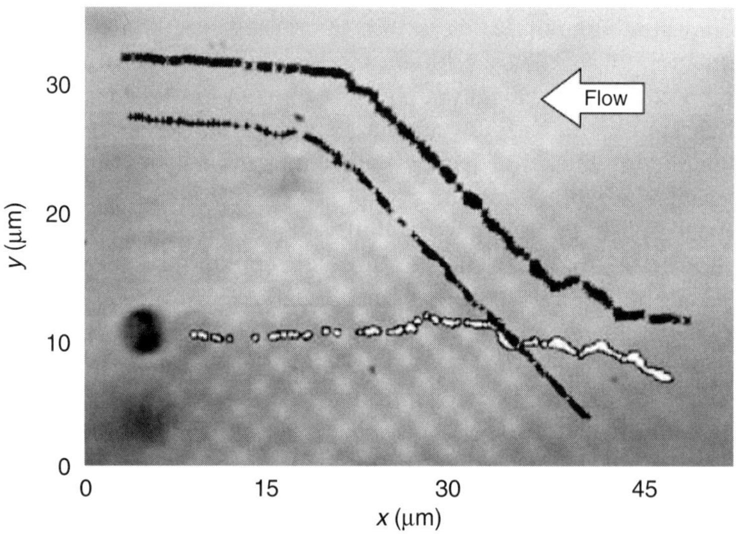


**Fig. 26.** Demonstration of optical fractionation by sample size [69]

### 4.2. Optically Driven Micropump and Microvalve with Colloidal Structures

Colloidal microspheres were used as tools for the transport and control of microfluidic flow. Terray et al. created micropumps and microvalves using optically trapped microspheres [70]. Figure 27 shows a gear pump with four microspheres that are simultaneously manipulated by time-divided laser scanning. A peristaltic pump was also demonstrated as shown in Fig. 28. Recently another type of micropump using birefringence microbeads has been proposed [71]. In this pump, liquid between two spinning microbeads is carried out owing to the viscous force in ultra-low Reynolds number flow. This micropump is suitable for transporting biological samples such as cells without any damage.

In addition, Terray et al. demonstrated that microparticles dispersed in an aqueous solution containing a photoinitiator and an aqueous monomer can be connected linearly just by focusing a laser beam [19]. The linearly connected microspheres were utilized as a microvalve inside a microchannel. Figure 29 shows the operation of the microvalve controlled by optical trapping. The valve allows to divide microparticles into desired and undesired samples. The in situ fabrication of microcomponents such as microvalves and microstirrers is a powerful method to construct further functional biochips.

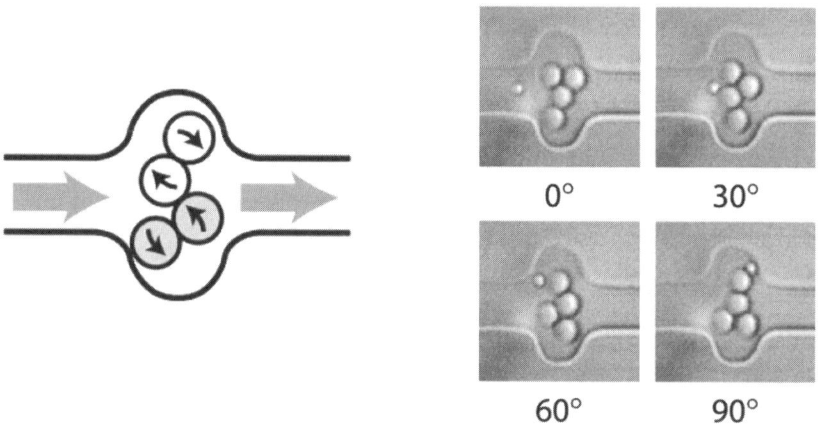


**Fig. 27.** Optically driven gear micropump using colloidal microspheres [70]

Such colloid-based microfluidic devices make it possible to pump not only electrolytes but also organic and inorganic solvents. In addition, the optical driving of microfluidic elements provides disposable functional biochips without the need for expensive actuators and sensors. For these reasons, they could be extensively used in chemical and biochemical reactions and cell analyses in the near future.

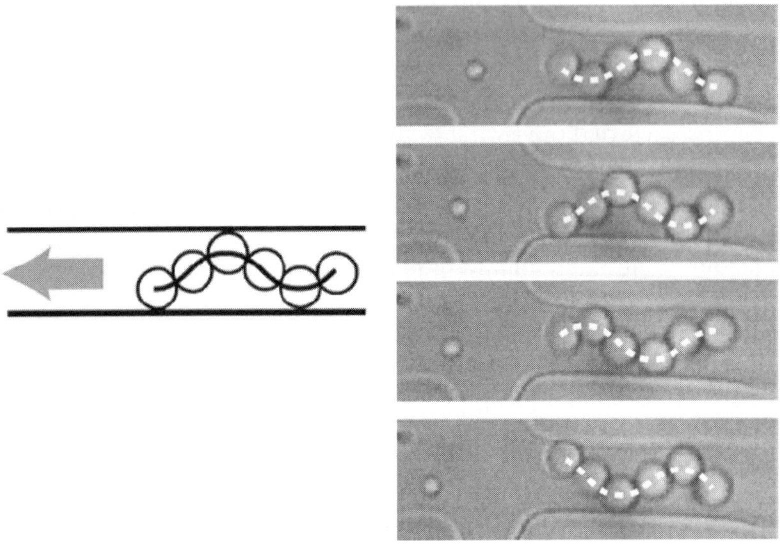

**Fig. 28.** Optically driven peristaltic micropump using colloidal microspheres [70]

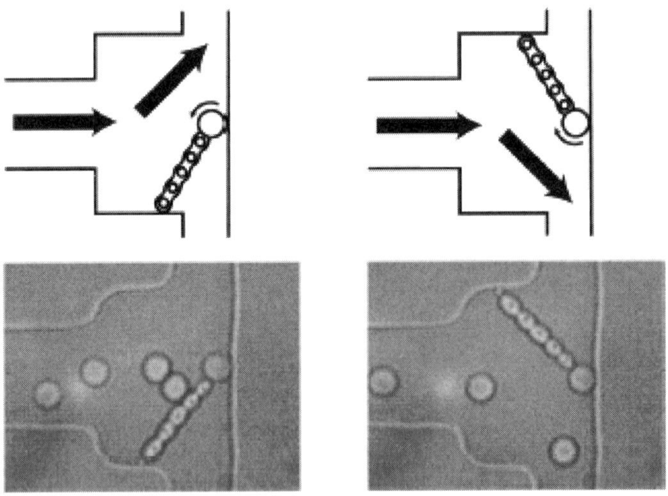

**Fig. 29.** Optically controlled microvalve made of microspheres [19]

### 4.3. Optically Driven Micropump Produced by Two-Photon Microstereolithography

Although colloidal particles are simply used for all-optical microfluidic control devices such as micropumps and microvalves [19, 70, 71], microfabricated devices are well suited for high-precision microfluidic control. This is because the microfabricated structures are designed and produced in the optimal shape to reduce leak in pumps and valves.

In recent years, an optically driven micropump, in which two lobed rotors are incorporated into a microchannel, was developed using two-photon microstereolithography [72, 73]. Two-photon microstereolithography has the ability to produce arbitrary 3D microstructures with nearly 100 nm resolution [37, 38, 74]. In addition, this technique makes it possible to produce 3D movable microstructures through the direct writing of a femtosecond laser beam [75]. Optically driven microturbines and micro-tweezers have been developed by using the direct drawing method [76, 77].

Unlike colloid-based microfluidic devices, the lobed micropump is confined to a microchannel by its own shaft, thus preventing the rotors from moving unless continuously irradiated with light. This allows the micropump to be easily integrated into a practical biochip. Figure 30 shows a schematic diagram of the optically driven lobed micropump. This pump is driven by means of radiation pressure generated by focusing a laser beam. When the laser beam is focused on the side of the rotor, the net radiation pressure applied to the rotor points toward the focus, allowing the rotor to be controlled by changing the trajectory of the scanning laser beam. Using a time-divided laser-scanning method allows to control the two rotors simultaneously. While the rotors are tightly engaged and rotated, fluid is trapped in the spaces between them and carried around in them. To reduce the leakage in the micropump, the two rotors need to be meshed tightly by radiation pressure while the laser beam is scanning.

Figure 31 shows sequential images of the micropump being driven. The tracer particle was brought into the microchamber by the pumped flow. When the rotation direction was reversed, the tracer particle was pushed to the outlet. The velocity of the particle is proportional to the rotational speed of the rotors. The flow rate was estimated to be less than 1 pL min$^{-1}$ Ultralow flow rates such as this have potential applications in micro/nano fluidic devices. The micropump is suitable for the transport of highly viscous fluids, since it is a kind of positive-displacement pump. The viscous drag force exerted on the rotor is proportional to its rotation speed in the Stokes' approximation for low-Reynolds number flow. Therefore, when the rotation speed of the rotors is low, highly viscous fluids can be moved. The refractive indices of usable liquids are limited by that of the rotors, which must be higher than that of the surrounding liquid to enable trapping. The use of photopolymers with high refractive indices will expand the range of usable liquids.

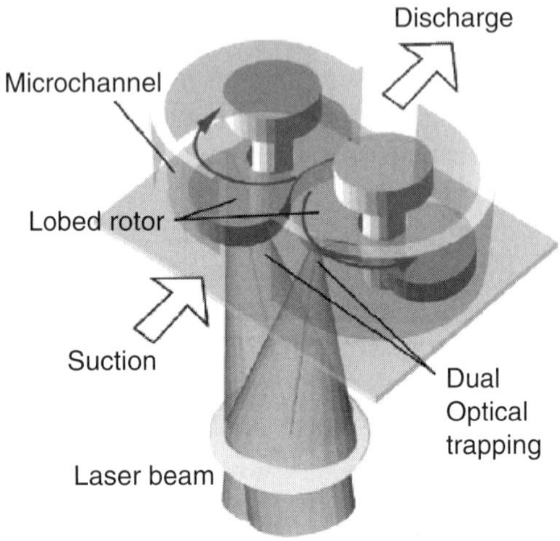


**Fig. 30.** Optically driven lobed micropump produced by two-photon microstereolithography [73]

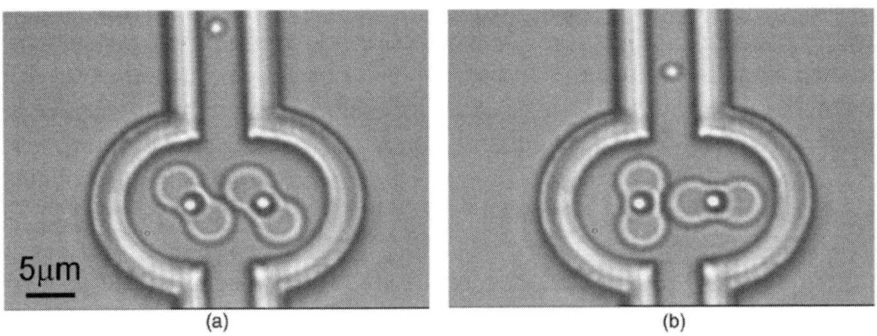


**Fig. 31.** Sequential images taken while driving the optically driven micropump [73]

### 4.4. Optically Controlled Micromanipulators Produced by Two-Photon Microstereolithography

Figure 32 shows the concept of all-optically controlled biochip proposed by Maruo [78]. In this chip, microtools such as microtweezers and micropumps are cooperatively controlled by scanning a single laser beam. Target biological samples are transported into a chamber by pressure-driven flow, and examined with the microtweezers. To analyze the biological sample, chemical reagents are supplied to it using micropumps and microvalves. Since the optically controlled biochip does not need expensive and sophisticated microactuators, it is suitable for disposable use.

The optically controlled microtools were fabricated by using an assembly-free process based on two-photon microstereolithography [76, 77]. Since two-photon microstereolithography can provide 3D microstructures by a direct laser scanning inside photopolymer, both microtools and a microchannel are easily fabricated together. The microtools can be also built in a microchannel made of glass or other plastics. The integration of optically controlled microtools into a mass-produced microchannels can provide low-cost functional biochips.

Figure 33 shows an SEM image of nanotweezers produced by two-photon microstereolithography [76]. The diameter of the probe is 250 nm. Figure 34 shows particle manipulation with optically controlled micromanipulators in a liquid [79]. The probes are contrived and designed to grasp a microobject tightly. The two arms are alternately trapped and driven by a single laser beam without on-off control of laser power. A glass bead of 5 μm diameter was successfully held with the two manipulator arms. In addition, three-hands micromanipulators were also fabricated and driven successfully, as shown in Fig. 35. Such cooperative control of multiple microtools enables various operations in biological research.

Figure 36 shows the operation of a prototype of an all-optically controlled biochip to perform a series of operations for biochemical and biological research [78]. In this biochip, both micromanipulators and a microseparator are integrated into a y-branch microchannel (width: 13 μm, height: 20 μm). Target biological samples are selected to the upper channel and examined by the micromanipulators.

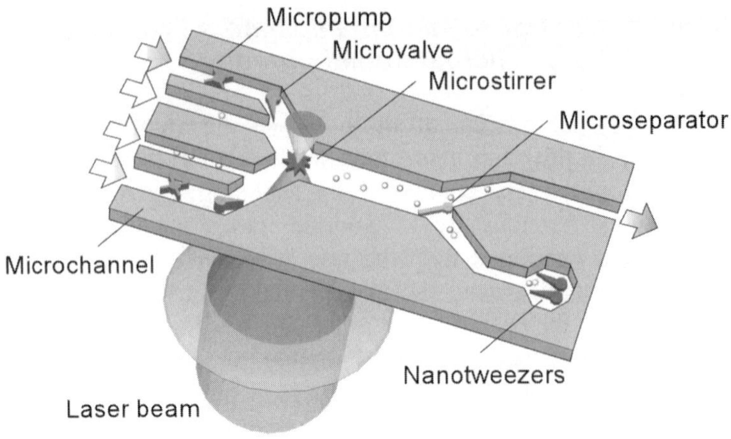


**Fig. 32.** Concept of an all-optically controlled biochip [78]

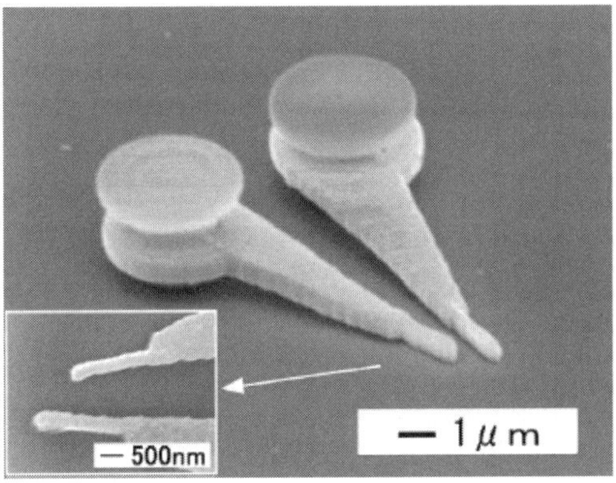


**Fig. 33.** Nanotweezers produced by two-photon microstereolithography [76]

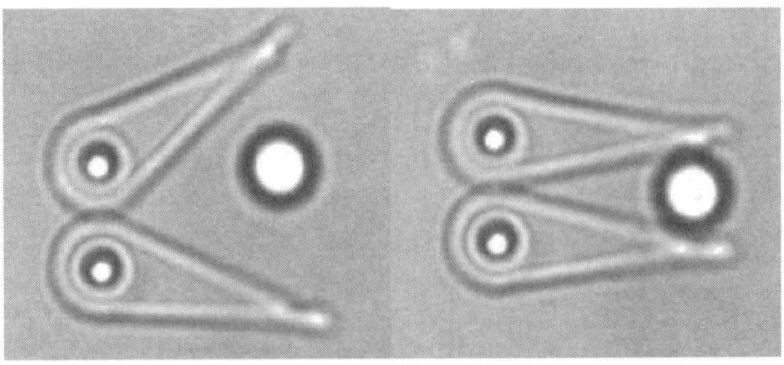

**Fig. 34.** Manipulation of a glass microbead (diameter: 5 μm) using optically controlled micromanipulators [79]

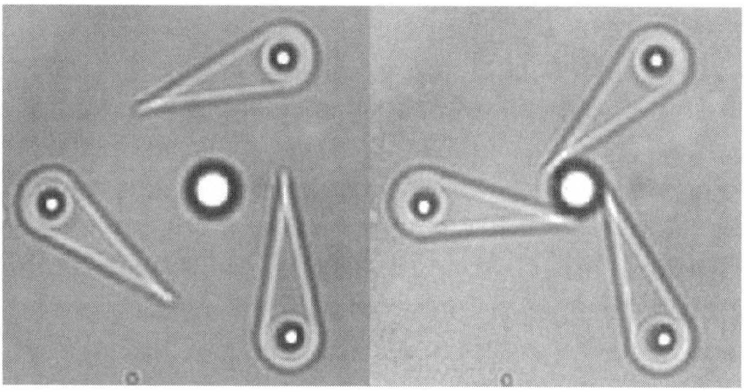

**Fig. 35.** Holding of a glass microbead (diameter: 5 μm) using three-hand micromanipulators [79]

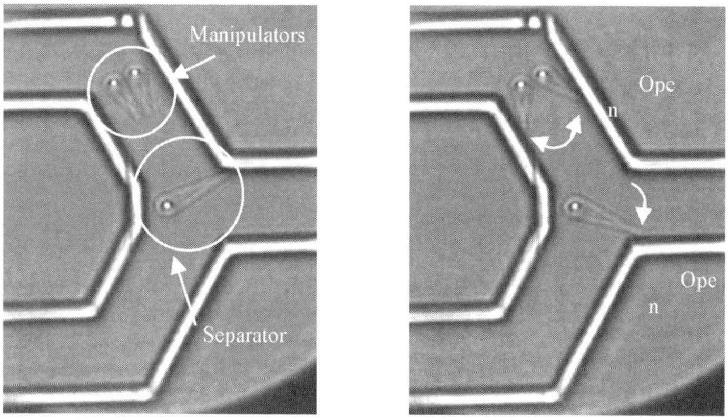


**Fig. 36.** A prototype of an all-optically controlled biochip [78]

## 5. Bio-manipulation Based on Optical Tweezers

### *5.1. Cell Stretcher Using Optical Radiation Pressure*

Optical trapping based on a single focused laser beam allows to trap biological samples such as cells and bacteria with less optical damage by selecting a laser beam of appropriate wavelength and power [80]. If the biological samples are sufficiently elastic, they can be deformed by radiation pressure. This phenomenon is utilized to analyze the elasticity of biological samples such as human blood cells. Kaneta et al. evaluated the elasticity, which is dependent on the age of a cell, with erythrocyte cells

[81]. The deformation of the cells was observed with a charge-coupled-device (CCD) camera. They use a loosely focused laser beam rather than that of previous work [80] to reduce the extent of degradation of the cells.

By contrast, when a dielectric microobject is placed between two opposed, nonfocused laser beams, the total force acting on the microobject is zero, because the two-beam trap geometry is symmetric and all the resulting surface forces cancel. Nevertheless, if the microobject is sufficiently elastic, the surface forces stretch the microobject along the beam axis [82, 83]. Guck et al. constructed a device, called an optical stretcher, that can be used to measure the viscoelastic properties of biological samples such as cells. Figure 37 shows a schematic of the stretching of a cell trapped in the optical stretcher. In this system, two optical fibers are used to generate counterpropagating laser beams. They succeeded in deforming human red blood cells and mouse fibroblasts. The surface stress profiles caused by irradiation of laser beams were also calculated by using ray optics analysis. In this technique, radiation damage is avoided since a double-beam trap does not require focusing for stable trapping.

### *5.2. Manipulation of Biomolecules with Optically Trapped Micro/nano Particles*

Optical trapping is a powerful tool for the manipulation of not only cells but also biomolecules such as DNA and motor proteins. Since the magnitude of optical trapping force is on the order of pN, optical manipulation is suitable for measuring mechanical properties of biomolecules. Recent progress in measurement of the physical properties of the DNA double helix is summarized in a review paper [7, 84]. In standard techniques of

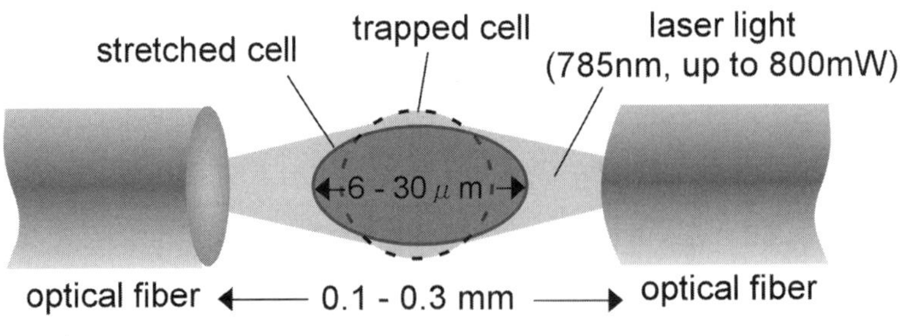


**Fig. 37.** Schematic of the stretching of a cell trapped in the optical stretcher [83]

single molecule analysis, a microbead is attached to one end of a single biomolecule, and a micropipette or an AFM probe is attached to the other end. The measurement of the displacement or rotation of the microbead allows to determine various physical properties such as the elasticity [85] or torsional strain [86] and to study reactions of biomolecules such as folding–unfolding transitions [87].

In case two microbeads are attached to both ends of a single biomolecule, dual optical trapping of the two microbeads allows to knot the molecule [88]. In recent years, a novel method for the manipulation of a single biomolecule at a desired position was recently proposed by Hirano et al. [89]. In their method, the use of a bead cluster formed by laser trapping makes it possible to manipulate a single DNA molecule at any point on the molecule without the need for prior chemical modification as in DNA–bead complex techniques. Figure 38 shows sequential images of the manipulation of a single DNA molecule, tandem lambda DNA, using laser clustering of 0.2-μm latex beads by laser trapping. By aggregating nanoparticles containing a single DNA at a focus, the desired point along the DNA strand can be trapped without any chemical modification unlike conventional manipulation of microbeads attached to DNA strands with chemical treatments.

Another approach for manipulating a long DNA molecule has been demonstrated by Washizu's group [90]. In their method, a microfabricated structure like a hook is used for reeling a long DNA molecule. The reeled molecule can be disentangled by rotating the microhook reversely. This technique is a powerful tool for DNA analysis without fragmentation.

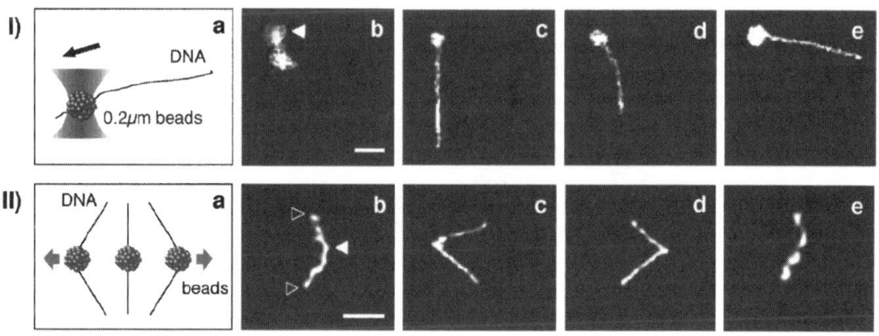


**Fig. 38.** Sequential images of the manipulation of a single DNA molecule, tandem lambda DNA, using laser clustering of 0.2 μm latex beads by optical trapping [89]

### 5.3. Optically Controlled Microtools for Biological Samples

In conventional micromanipulation using optically trapped microparticles, the particles are supplied into a chamber or a microchannel. By contrast, Ichikawa et al. proposed an in situ formation of gel microbeads made of a thermoreversible hydrogel [91]. They demonstrated indirect manipulation of a yeast cell using an in situ fabricated hydrogel microbead as shown in Figure 39. Since the grown gel microbead is trapped and manipulated to adhere to the target cell, the cell can be manipulated indirectly. Direct optical manipulation using a high power laser beam often causes optical damage to specimens for a number of reasons including transient local heating and photochemical reactions, whereas this method is better since it avoids the direct irradiation of the high power laser beam to the target cell. The thermosensitive hydrogel is also used for fixation of living cells in a microfluidic circuit [92]. In this case, a near-infrared laser beam is used for both optical trapping and locally thermosensitive hydrogelation. This method is feasible for rapid manipulation, immobilization, cleaning, isolation, and extraction of a single cell in a biochip.

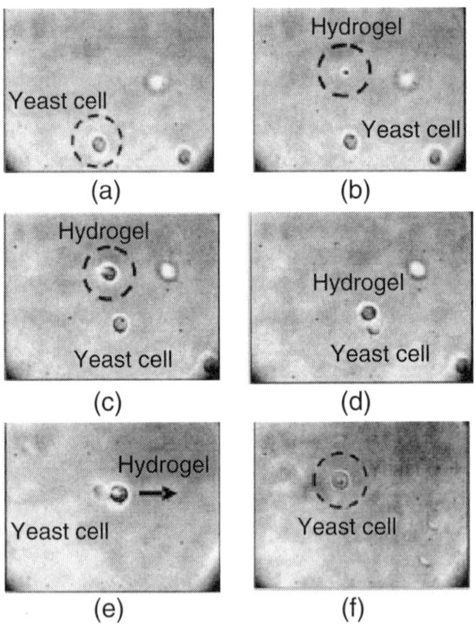


**Fig. 39.** Indirect laser manipulation of a yeast cell using a gel microbead [91]. (**a**) Target yeast cell (diameter: 4 μm) before irradiation. (**b**) Hydrogel generation (formation of core). (**c**) Hydrogel generation (27 s later). (**d**) Adhesion to the cell. (**e**) Transport by laser micromanipulation. (**f**) Releasing in the cage, resolution of hydrogel

## 6. Conclusions and Outlook

Optical micromanipulation is a promising method to manipulate various micro and nanoscale objects including microparticles, micromachined structures, and biological samples. The unique features of optical manipulation are as follows: (1) remote control in a sealed environment, (2) 3D manipulation with perfect freedom, (3) parallel manipulation of multiple microparticles, and (4) gentle manipulation with an ultrasmall force of the order of pico-Newton.

As a promising application utilizing the remote control in a sealed environment, all optically controlled biochips have attracted much attention in recent years for their ability of manipulating biological samples and fluids [69, 70–73, 78, 79, 92, 93]. Although conventional optical microscopes and lasers are used for optical trapping in most of the current systems, on-chip lasers or optical fiber systems are ideal for practical use in medical diagnosis and palm-top analysis systems. Some pioneer works have been reported in this context. For example, optical fibers have been utilized to manipulate microparticles inside a microchannels [94, 95]. In addition, semiconductor lasers are integrated into a microchannel for the sequential manipulation of microparticles [96]. In the near future, all optically controlled biochips that are fully integrated not only with light sources but also with detectors could be created by advanced micromachining techniques.

## References


1. Ashkin A, Dziedzic JM, Bjorkholm JE, Chu S (1986) Observation of a single-beam gradient force optical trap for dielectric particles. Opt. Lett. 11: 288–290
2. Ashkin A (2000) History of optical trapping and manipulation of small-neutral particle, atoms, and molecules. IEEE J. Sel. Top. Quantum Electron. 6: 841–856
3. Grier DG (2003) A revolution in optical manipulation. Nature 424: 810–816
4. Neuman KC, Block SM (2004) Optical trapping. Review of Scientific Instruments 75: 2787–2809
5. Molloy JE, Padgett MJ (2002) Lights, action: optical tweezers. Contemporary Physics 43: 241–258
6. Ashkin A (1992) Forces of a single-beam gradient laser trap on a dielectric sphere in the ray optics regime. J. Biophys. 61: 569–582
7. Bustamante C, Bryant Z, Smith SB (2003) Ten years of tension: single-molecule DNA mechanics. Nature 421: 423–427
8. Arai Y, Yasuda R, Akashi K, Harada Y, Miyata H, Kinosita K, Itoh H (1999) Tying a molecular knot with optical tweezers. Nature 399: 446–448

9. Tanaka H, Homma K, Iwane AH, Katayama E, Ikebe R, Saito J, Yanagida T, Ikebe M (2002) Themotor domain determines the large step ofmyosin-V. Nature 415: 192–195
10. Ghislain LP, Webb WW (1993) Scanning-force microscope based on an optical trap. Opt. Lett. 18: 1678–1680
11. Kawata S, Inouye Y, Sugiura T (1994) Near-field scanning optical microscope with a Laser trapped probe. Jpn. J. Appl. Phys., Part 2 33: L1725–L1727
12. Sugiura T, Okada T, Inouye Y, Nakamura O, Kawata S (1997) Gold-bead scanning near-field optical microscope with laser force position control. Opt. Lett. 22: 1663–1665
13. Fujiwara H, Sasaki K (1999) Upconversion lasing of a thulium-ion-doped fluorozirconate glass microsphere. J. Appl. Phys. 86: 2385–2388
14. Simmons RM, Finer JT, Chu S, Spudich JA (1996) Quantitative measurements of force and displacement using an optical trap. Biophys. J. 70: 1813–1822
15. Gittes F, Schmidt CF (1998) Interference model for back-focal-plane displacement detection in optical tweezers. Opt. Lett. 23: 7–9
16. Nugent-Glandorf L, Perkins TT (2004) Measuring 0.1-nm motion in 1 ms in an optical microscope with differential back-focal-plane detection. Opt. Lett. 29: 2611–2613
17. Ota T, Sugiura T, Kawata S (2002) Surface-force measurement with a laser-trapped microprobe in solution. Appl. Phys. Lett. 80: 3448–3450
18. Gauthier RC (1997) Theoretical investigation of the optical trapping force and torque on cylindrical micro-objects. J. Opt. Soc. Am. B 14: 3323–3333
19. Terray A, Oakey J, Marr DWM (2002) Fabrication of linear colloidal structures for microfluidic applications. Appl. Phys. Lett. 81: 155–157
20. Sasaki K, Koshioka M, Misawa H, Kitamura N, Masuhara H (1991) Optical trapping of a metal particle and a water droplet by a scanning laser beam. Appl. Phys. Lett. 60: 807–809
21. Gu M, Morrish D (2002) Three-dimensional trapping of Mie metallic particles by the use of obstructed laser beams. J. Appl. Phys. 91: 1606–1612
22. Sato S, Harada Y, Waseda Y (1994) Optical trapping of microscopic metal particles. Opt. Lett. 19: 1807–1809
23. Furukawa H, Yamaguchi I (1998) Optical trapping of metallic particles by a fixed Gaussian beam. Opt. Lett. 23: 216–218
24. Ke PC, Gu M (1999) Characterization of trapping force on metallic Mie particles. Appl. Opt. 38: 160–167
25. Svoboda K, Block SM (1994) Optical trapping of metallic Rayleigh particles. Opt. Lett. 19: 930–932
26. Yoshikawa H, Matsui T, Masuhara H (2004) Reversible assembly of gold nanoparticles confined in an optical microcage. Phys. Rev. E 70: 061406
27. Ito S, Yoshikawa H, Masuhara H (2002) Laser manipulation and fixation of single gold nanoparticles in solution at room temperature. Appl. Phys. Lett. 80: 482–484

28. Beth RA (1936) Mechanical detection and measurement of the angular momentum of light. Phys. Rev. 50: 115–125
29. Poynting JH (1909) The wave motion of a revolving shaft, and a suggestion as to the angular momentum in a beam of circularly polarised light. Proc. Roy. Soc. London Ser. A 82: 560–567
30. Friese MEJ, Nieminen TA, Heckenberg NR, Rubinsztein-Dunlop H (1998) Optical alignment and spinning of laser-trapped microscopic particles. Nature 394: 348–350
31. Higurashi E, Sawada R, Ito T (1998) Optically induced angular alignment of birefringent micro-objects by linear polarization. Appl. Phys. Lett. 73: 3034–3036
32. Born M, Wolf E (1980) Principle of Optics, 6th edn. Pergamon Press, Oxford, pp 705–708
33. Neale SL, Macdonald MP, Dholakia K, Krauss TF (2005) All-optical control of microfluidic components using form birefringence. Nat. Mat. 4: 530–533
34. Higurashi E, Ukita H, Tanaka H, Ohguchi O (1994) Optically induced rotation of anisotropic micro-objects fabricated by surface micromachining. Appl. Phys. Lett. 64: 2209–2210
35. Higurashi E, Ohguchi O, Tamamura T, Ukita H, Sawada R (1997) Optically induced rotation of dissymmetrically shaped fluorinated polyimide micro-objects in optical traps. J. Appl. Phys. 82: 2773–2779
36. Higurashi E, Sawada R, Ito T (1998) Optically induced rotation of a trapped micro-object about an axis perpendicular to the laser beam axis. Appl. Phys. Lett. 72: 2951–2953
37. Maruo S, Nakamura O, Kawata S (1997) Three-dimensional microfabrication with two-photon absorbed photopolymerization. Opt. Lett. 22: 132–134
38. Kawata S, Sun HB, Tanaka T, Takada K (2001) Finer features for functional microdevices. Nature 412: 697–698
39. Galajda P, Ormos P (2001) Complex micromachines produced and driven by light. Appl. Phys. Lett. 78: 249–251
40. Galajda P, Ormos P (2002) Rotors produced and driven in laser tweezers with reversed direction of rotation. Appl. Phys. Lett. 80: 4653–4655
41. Ukita H, Kanehira M (2002) A shuttlecock optical rotator – Its design, fabrication and evaluation for a microfluidic mixer. IEEE Journal of Selected Topics in Quantum Electronics 8: 111–117
42. Sasaki K, Koshioka M, Misawa H, Kitamura N, Masuhara H (1991) Pattern formation and flow control of fine particles by laser-scanning micromanipulation. Opt. Lett. 16: 1463–1465
43. Arai F, Yoshikawa K, Sakami T, Fukuda T (2004) Synchronized laser micromanipulation of multiple targets along each trajectory by single laser. Appl. Phys. Lett. 85: 4301–4303
44. Mio C, Gong T, Terray A, Marr DWM (2000) Design of a scanning laser optical trap for multiparticle manipulation. Rev. Sci. Instrum. 71: 2196–2200
45. Faucheux LP, Bourdieu LS, Kaplan PD, Libchaber AJ (1995) Optical thermal ratchet. Phys. Rev. Lett. 74: 1504–1507

46. Lee SH, Ladavac K, Polin M, Grier DG (2005) Observation of flux reversal in a symmetric optical thermal ratchet. Phys. Rev. Lett. 94: 110601
47. Ricárdez-Vargas I, Rodríguez-Montero P, Ramos-García R (2006) Modulated optical sieve for sorting of polydisperse microparticles. Appl. Phys. Lett. 88: 121116
48. Durnin J (1987) Exact-solutions for nondiffracting beams. I. The scalar theory. J. Opt. Soc. Am. A 4: 651–654
49. Arlt J, Garces-Chavez V, Sibbett W, Dholakia K (2001) Optical micromanipulation using a Bessel light beam. Opt. Commun. 197: 239–245
50. Herman RM, Wiggins TA (1991) Production and uses of diffractionless beams. J. Opt. Soc. Am. A 8: 932–942
51. Garcés-Chávez V, McGloin D, Melville H, Sibbett W, Dholakia K (2002) Simultaneous micromanipulation in multiple planes using a self-reconstructing light beam. Nature 419: 145–147
52. Dufresne ER, Spalding GC, Dearing MT, Sheets SA, Grier DG (2001) Computer-generated holographic optical tweezer arrays. Rev. Sci. Instrum. 72: 1810–1816
53. Eriksen RL, Daria VR, Glückstad J (2002) Fully dynamic multiple-beam optical tweezers. Opt. Express 10: 597–602
54. Leach J, Sinclair G, Jordan P, Courtial J, Padgett MJ, Cooper J, Laczik ZJ (2004) 3D manipulation of particles into crystal structures using holographic optical tweezers. Opt. Express 12: 220–226
55. Rodrigo PJ, Eriksen RL, Daria VR, Glückstad J (2002) Interactive light-driven and parallel manipulation of inhomogeneous particles. Opt. Express 10: 1550–1556
56. Rodrigo PJ, Daria VR, Glückstad J (2004) Real-time interactive optical micromanipulation of a mixture of high- and low-index particles. Opt. Express 12: 1417–1425
57. Agarwal R, Ladavac K, Roichman Y, Yu G, Lieber CM, Grier DG (2005) Manipulation and assembly of nanowires with holographic optical traps. Opt. Express 13: 8906–8912
58. Melville H, Milne GF, Spalding GC, Sibbett W, Dholakia K, McGloin D (2003) Optical trapping of three-dimensional structures using dynamic holograms. Opt. Express 11: 3562–3567
59. Bingelyte V, Leach J, Courtial J, Padgett MJ (2003) Optically controlled three-dimensional rotation of microscopic objects. Appl. Phys. Lett. 82: 829–831
60. Curtis JE, Grier DG (2003) Structure of optical vortices. Phys. Rev. Lett. 90: 133901
61. Guo CS, Liu X, He JL, Wang HT (2004) Optimal annulus structures of optical vortices. Opt. Express 12: 4625–4634
62. Kawata S, Sugiura T (1992) Movement of micrometer-sized particles in the evanescent field of a laser beam. Opt. Lett. 17: 772–774
63. Kawata S, Tani T (1996) Optically driven Mie particles in an evanescent field along a channeled waveguide. Opt. Lett. 21: 1768–1770

64. Lester M, Nieto-Vesperinas M (1999) Optical forces on microparticles in an evanescent laser field. Opt. Lett. 24: 936–938
65. Ng LN, Zervas MN, Wilkinson JS, Luff BJ (2000) Manipulation of colloidal gold nanoparticles in the evanescent field of a channel waveguide. Appl. Phys. Lett. 76: 1993–1995
66. Garcés-Cháveza V, Dholakia K, Spalding GC (2005) Extended-area optically induced organization of microparticles on a surface. Appl. Phys. Lett. 86: 031106
67. Reece PJ, Garcés-Chávez V, Dholakia K (2006) Near-field optical micro-manipulation with cavity enhanced evanescent waves. Appl. Phys. Lett. 88: 221116
68. Korda PT, Taylor MB, Grier DG (2002) Kinetically locked-in colloidal transport in an array of optical tweezers. Phys. Rev. Lett. 89: 128301
69. MacDonald MP, Spalding GC, Dholakia K (2003) Microfluidic sorting in an optical lattice. Nature 426: 421–424
70. Terray A, Oakey J, Marr DWM (2002) Microfluidic control using colloidal devices. Science 296: 1841–1844
71. Leach J, Mushfique H, Leonardo R, Padgett M, Cooper J (2006) An optically driven pump for microfluidics. Lab Chip 6: 735–739
72. Maruo S, Inoue H (2005) Optically driven micropump produced by two-photon microstereolithography. Proceedings of Micro Total Analysis Systems 2005: 590–592
73. Maruo S, Inoue H (2006) Optically driven micropump produced by three-dimensional two-photon microfabrication. Appl. Phys. Lett. 89: 144101
74. Maruo S, Kawata S (1998) Two-photon-absorbed near-infrared photo-polymerization for three-dimensional microfabrication. J. Microelectromech. Syst. 7: 411–415
75. Maruo S, Ikuta K (2000) Three-dimensional microfabrication by use of single-photon-absorbed polymerization. Appl. Phys. Lett. 76: 2656–2658 (2000)
76. Maruo S, Ikuta K, Korogi H (2003) Submicron manipulation tools driven by light in a liquid. Appl. Phys. Lett. 82: 133–135
77. Maruo S, Ikuta K, Korogi H (2003) Force-controllable, optically driven micromachines fabricated by single-step two-photon microstereolithography. J. Microelectromech. Syst. 12: 533–539
78. Maruo S, Hiratsuka Y (2005) All optically controlled micromanipulation systems. Proceedings of Micro Total Analysis Systems 2005: 1206–1208
79. Maruo S, Hiratsuka Y (2007) Micromanipulation tools driven by synchronized laser scanning. Opt. Express (submitted)
80. Ashkin A, Dziedzic JM, Yamane T (1987) Optical trapping and manipulation of single cells using infrared laser beams. Nature 330: 769–771
81. Kaneta T, Makihara J, Imasaka T (2001) An "optical channel": A technique for the evaluation of biological cell elasticity. Anal. Chem. 73: 5791–5795
82. Guck J, Ananthakrishnan R, Moon TJ, Cunningham CC, Käs J (2000) Optical Deformability of soft biological dielectrics. Phys. Rev. Lett. 84: 5451–5454

83. Guck J, Ananthakrishnan R, Mahmood H, Moon TJ, Cunningham CC, Käs J (2001) The optical stretcher: A novel laser tool to micromanipulate cells. Biophys. J. 81: 767–784
84. Ishii Y, Ishijima A, Yanagida T (2001) Single molecule nanomanipulation of biomolecules. Trends Biotech. 19: 211–216
85. Quake SR, Babcock H, Chu S (1997) The dynamics of partially extended single molecules of DNA. Nature 388: 151–154
86. Bryant Z, Stone MD, Gore J, Smith SB, Cozzarelli NR, Bustamante C (2003) Structural transitions and elasticity from torque measurements on DNA. Nature 424: 338–341
87. Kellermayer MSZ, Smith SB, Granzier HL, Bustamante C (1997) Folding-Unfolding Transitions in Single Titin Molecules Characterized with Laser Tweezers. Science 276: 1112–1116
88. Arai Y, Yasuda R, Akashi K, Harada Y, Miyata H, Kinosita K, Itoh H (1999) Tying a molecular knot with optical tweez ers. Nature 399: 446–448
89. Hirano K, Baba Y, Matsuzawa Y, Mizuno A (2002) Manipulation of a Coiled DNA Molecule Using Laser Clustering of Microparticles. Appl. Phys. Lett. 80: 515–517
90. Terao K, Kabata H, Oana H, Washizu M (2005) Complete extension of chromosomal DNA and its manipulation using optically-driven micro-fabricated hooks. Proceedings of Micro Total Analysis Systems 2005: 715–717
91. Ichikawa A, Arai F, Yoshikawa K, Uchida T, Fukuda T (2005) In situ formation of a gel microbead for indirect laser micromanipulation of micro-organisms. Appl. Phys. Lett. 87: 191108
92. Arai F, Ng C, Maruyama H, Ichikawa A, El-Shimy H, Fukuda T (2005) On chip single-cell separation and immobilization using optical tweezers and thermosensitive hydrogel. Lab Chip 5: 1399–1403
93. Glückstad J (2004) Microfluidics – Sorting particles with light. Nat. Mat. 3: 9–10
94. Domachuk P, Cronin-Golomb M, Eggleton BJ, Mutzenich S, Rosengarten G, Mitchell A (2005) Application of optical trapping to beam manipulation in optofluidics. Opt. Express 13: 7265–7275
95. Jensen-McMullin C, Lee HP, Lyons ER (2005) Demonstration of trapping, motion control, sensing and fluorescence detection of polystyrene beads in a multi-fiber optical trap. Opt. Express 13: 2634–2642
96. Cran-McGreehin SJ, Dholakia K, Krauss TF (2006) Monolithic integration of microfluidic channels and semiconductor lasers. Opt. Express 14: 7723–7729

# Chapter 8
# Dielectrophoretic Microfluidics

Eric Cummings[1] and Boris Khusid[2]

[1]LabSmith, Inc., 7665 Hawthorne Ave, Livermore, CA 94550, USA

[2]New Jersey Institute of Technology, University Heights, Newark, NJ 07102, USA

## 1. Introduction

*Dielectrophoresis* (DEP), the term coined by Pohl [1, 2], is the motion of an object under forces resulting from electric field gradients. The applications of DEP are based on differentiation of dielectric and conducting properties of objects to allow their separation and identification. DEP has been demonstrated to be useful for separating polydisperse particle suspensions into homogeneous subpopulations, manipulating and concentrating biologically relevant molecules, distinguishing dead and living cells or ill and healthy cells, assembling carbon nanotubes, etc. Numerous books and reviews address the fundamentals and various applications of DEP: Pohl [2], Pethig [3], Mizuno and Washizu [4], Jones [5], Fuhr et al. [6], Pethig [7], Pethig and Markx [8], Koch et al. [9], Gascoyne and Vykoukal [10], Hughes [11, 12], Morgan and Green [13], Burke [14], and Gonzalez and Remcho [15].

Early experiments on DEP were conducted on simple electrode configurations, typically pin–plate, pin–pin, and wire–wire. Microfabrication technologies borrowed from the semiconductor industry facilitate the fabrication of novel and more capable dielectrophoretic microfluidic systems. For example, micro- and nanoscaled electrodes can produce high fields of up to several kV $mm^{-1}$ from the application of only a few volts. Precisely contoured arrays of thousands or millions of field-shaping structures can be patterned and replicated economically. Moreover, undesired effects like electro-convection and Joule heating can be suppressed in tiny microfluidic devices.

Since particles undergoing DEP acquire a dipole moment, they interact with one another through dipole–dipole forces. It is well known [2, 16] that, because of dipolar interactions, polarized particles have a tendency to align rapidly along the field direction and to form chains that subsequently coalesce into thicker columns. This phenomenon governs the behavior of electrorheolgical fluids [17, 18], most commonly colloidal suspensions, which undergo fluid-viscoelastic solid transition exhibiting a yield stress at low shear rates. Recent experiments on DEP reveal that, because of the particle confinement at the microscale, combining dielectrophoretic and dipole forces suggests new ways to control and manipulate particles in microdevices. In particular, the reviews by Hughes [12] and Burke [14] clearly indicate that these techniques could extend the use of DEP to the nanoscale for applications ranging from nano-biotechnology to molecular electronics.

Compared to other available methods, DEP is becoming one of the major techniques for micro- and emerging nanoscale systems because the technique requires no moving parts, can be incorporated more favorably into microsystems than more conventional methods, and employs polarization forces that are insensitive to charge, which is difficult to control. In view of the promising potential of the DEP, we will summarize the main results of currently available theories for the quantification of dielectrophoretic phenomena at the microscale, and consider the basics of the design and operation of dielectrophoretic microdevices, particularly insulator-based or iDEP devices, a relatively new class that use patterned insulators to produce tightly controlled electric field distributions.

We review theoretical and practical aspects of DEP, rather than details of currently used microdevices. Through the knowledge of principles underlying the operation of DEP miniaturized systems, the reader will develop a better understanding of how to employ dielectrophoretic and dipole forces in a wide variety of challenging biological and clinical applications.

## 2. Quantification of Dielectrophoretic Micro-Fluidics

### 2.1. Electric Force Acting on an Individual Particle

The force exerted by a field on a particle consists of two terms [19]

$$\mathbf{F} = Q\mathbf{E} + (\mathbf{P} \cdot \nabla)\mathbf{E}, \tag{1}$$

where $Q$ is the particle charge and $\mathbf{P}$ is the particle dipole moment. The first term in (1) is the well-known *electrophoretic force,* which causes a charged particle subject to a direct current (DC) field to travel along the electric lines, whereas the other term is referred to as the *polarization force*. In contrast to the electrophoretic force, the latter acts on both neutral and charged particles. Equation (1) represents the first terms in the expansion of the net electric force acting on the induced charges in a polarized particle as a power series in the field gradient [19].

The polarization of a not perfectly insulating or perfectly conducting particle depends on the time variation of the applied electric field over some characteristic time interval in the recent past. In particular, for a suspended sphere when the speed with which charge carriers rearrange themselves in response to an applied field is independent of the field strength, [19, 20]

$$\mathbf{P} = 3v_{\mathrm{p}}\varepsilon_0\varepsilon_{\mathrm{f}} \int_{-\infty}^{t} \bar{\beta}(t-t')\mathbf{E}(t')\,\mathrm{d}t', \tag{2}$$

where $v_p$ is the particle volume, $\varepsilon_0$ is the vacuum permittivity, $\varepsilon_f$ is the fluid dielectric constant, and $\bar{\beta}(t)$ describes the dielectric relaxation. The real and imaginary parts of the Fourier transform of $\bar{\beta}$

$$\mathrm{Re}(\beta^*) + i\,\mathrm{Im}(\beta^*) = \int_{0}^{\infty} \bar{\beta}(t)e^{-i\omega t}\,\mathrm{d}t, \tag{3}$$

describe the interaction of a particle with an oscillating (AC) field of the frequency $\omega = 2\pi\nu$, where $\nu$ is given in Hz. A model of a perfect dielectric [19, 20] assumes that the polarization responds instantaneously to an applied field so that $\bar{\beta}(t) = C_1\delta(t)$ where $C_1$ is some constant and $\delta(t)$ is the Dirac delta function. In this case, (3) yields $\mathrm{Re}(\beta^*) = C_1$ and $\mathrm{Im}(\beta^*) = 0$. The opposite extreme of a perfectly conducting material corresponds to the polarization proportional to the rate of the time variation of an applied field so that $\bar{\beta}(t) = C_2\delta'(t)$ where $C_2$ is some constant and $\delta'(t)$ is the derivative of the delta-function. In this case, $\mathrm{Re}(\beta^*) = 0$ and $\mathrm{Im}(\beta^*) = \omega C_2$. When the electric properties of the particle and that of the suspending fluid (i.e., their dielectric constants, $\varepsilon_{\mathrm{p}}$ and $\varepsilon_{\mathrm{f}}$, and conductivities, $\sigma_{\mathrm{p}}$ and $\sigma_{\mathrm{f}}$), are frequency independent, the Maxwell-Wagner theory [21, 22] yields

$$\mathrm{Re}(\beta^{*}) = \frac{\varepsilon_{\mathrm{p}} - \varepsilon_{\mathrm{f}}}{\varepsilon_{\mathrm{p}} + 2\varepsilon_{\mathrm{f}}} + \frac{3(\varepsilon_{\mathrm{f}}\sigma_{\mathrm{p}} - \varepsilon_{\mathrm{p}}\sigma_{\mathrm{f}})(\sigma_{\mathrm{p}} + 2\sigma_{\mathrm{f}})}{\left[(\sigma_{\mathrm{p}} + 2\sigma_{\mathrm{f}})^{2} + \omega^{2}\varepsilon_{0}^{2}(\varepsilon_{\mathrm{p}} + 2\varepsilon_{\mathrm{f}})^{2}\right](\varepsilon_{\mathrm{p}} + 2\varepsilon_{\mathrm{f}})} \tag{4}$$

$$\mathrm{Im}(\beta^{*}) = -\frac{3\omega\varepsilon_{0}(\varepsilon_{\mathrm{f}}\sigma_{\mathrm{p}} - \varepsilon_{\mathrm{p}}\sigma_{\mathrm{f}})}{(\sigma_{\mathrm{p}} + 2\sigma_{\mathrm{f}})^{2} + \omega^{2}\varepsilon_{0}^{2}(\varepsilon_{\mathrm{p}} + 2\varepsilon_{\mathrm{f}})^{2}} \,. \tag{5}$$

Because the dipole moment **P** in (1) is proportional to the field strength, it follows from (2) that the polarization force is proportional to the square of the applied field. This implies that under the action of an electric field,

$$\mathbf{E} = \mathbf{E}_{01}(\mathbf{r})\cos\omega t + \mathbf{E}_{02}(\mathbf{r})\sin\omega t \,, \tag{6}$$

the polarization force yields a non-zero time-average value [23, 24].

For a single-phase applied electric field which contains only one component in (6), the polarization force **F**, often referred to as the dielectrophoretic force, is related to the instantaneous component of polarization [5] by means of

$$\langle \mathbf{F}_{\mathrm{diel}} \rangle = \frac{3}{2}\varepsilon_{0}\varepsilon_{\mathrm{f}}v_{\mathrm{p}}\,\mathrm{Re}\left[\beta^{*}(\omega)\right]\nabla E_{\mathrm{rms}}^{2}(\mathbf{r}) \,, \tag{7}$$

where $\langle\ \rangle$ denotes averaging over the period of the field oscillations and $E_{\mathrm{rms}}$ is the root mean square of the electric field strength. This force causes a particle to move to the regions of high field strength (*positive* DEP) or low field strength (*negative* DEP), depending on whether the particle is more ($\mathrm{Re}(\beta^{*}) > 0$) or less ($\mathrm{Re}(\beta^{*}) < 0$) polarizable than the suspending fluid. Equation (4) shows that particles respond dielectrophoretically according to their dielectric constant at sufficiently high-frequencies and their conductivity at sufficiently low frequencies or zero frequency (DC). While (7) was derived using charge polarization considerations, the equations reveal that the dielectrophoretic force depends on electrical currents, either conduction or displacement.

When an electrode array is energized with a series of sequentially phase-shifted oscillatory voltages a traveling electric wave is produced, which contains both components in (6) so that $\mathbf{E}_{01} \approx \mathbf{E}_{0}\cos\varphi(\mathbf{r})$ and $\mathbf{E}_{02} \approx \mathbf{E}_{0}\sin\varphi(\mathbf{r})$ where the spatial variation of the phase $\varphi(\mathbf{r})$ is significantly larger than that of the amplitude $E_0$. In this case, the polarization force $\mathbf{F}$, often referred to as the traveling-wave dielectrophoretic force, is related to the rate-dependent component of polarization [23, 24] by means of

$$\langle \mathbf{F}_{\mathrm{tw}} \rangle \approx -\frac{3}{2} \varepsilon_0 \varepsilon_{\mathrm{f}} v_{\mathrm{p}} \operatorname{Im}\left[ \beta^*(\omega) \right] E_0^2 \nabla \varphi(\mathbf{r}) \; . \tag{8}$$

Depending on the sign of $\operatorname{Im}(\beta^*)$ in (8), this force directs a particle to the regions where the phase is larger or smaller. When the value of $\operatorname{Im}(\beta^*)$ is given by (5), the use of a traveling-wave enables one to separate particles with different ratios of $\varepsilon_{\mathrm{p}} / \sigma_{\mathrm{p}}$. Thus, combinations of the dielectrophoretic and traveling wave forces and different applied-field frequencies offer multiple dimensions of selectivity.

There have been numerous theoretical and experimental efforts to study various effects on the polarization of particles and macromolecules subject to spatially uniform electric fields and find the relation between the particle polarization and its electro-kinetic properties [3, 20–22, 25–31]. Unfortunately, the use of the macroscopic expressions (2) and (3) for *a priory* quantitative simulation of the polarization of fine particles and macromolecules is limited because $\beta^*$ is a strong function of the molecular structure, molecule–solvent interactions, and solvent properties. No simple model exists containing a few adjustable parameters to characterize the dependence of $\operatorname{Re}(\beta^*)$ and $\operatorname{Im}(\beta^*)$ on the applied frequency, molecular geometry, electrical and physical properties of the solvent, etc.

Dielectric relaxation spectroscopy [20, 25], using spatially uniform low fields (~1 V mm$^{-1}$), has been used to measure the polarization of a colloid in bulk to obtain data on the frequency dependence of an individual particle's polarizability, $\beta^*$, [32–34]. These measurements enable quantitative predictions of particle DEP in strong fields (~kV mm$^{-1}$). Specifically, $\beta^*$ is calculated from experimental data on the concentration and frequency dependence of the complex dielectric permittivity of a suspension, $\varepsilon_{\mathrm{s}}^*$, and the suspending fluid, $\varepsilon_{\mathrm{f}}^*$, using the Clausius-Mosotti relation,

$$(\varepsilon_{\mathrm{s}}^* - \varepsilon_{\mathrm{f}}^*) \big/ (\varepsilon_{\mathrm{s}}^* + 2\varepsilon_{\mathrm{f}}^*) = \beta^*(\omega) \; c, \tag{9}$$

which is applicable for a relatively low particle volume fraction, $c$ [20–22].

### 2.2. Field Driven Phase Transitions

Dipole–dipole interparticle interactions decay with the inverse cube of separation distance and, depending on the relative orientation of the dipoles, vary from attraction to repulsion [19]. When the dielectric constants and conductivities of the particles and the suspending fluid are frequency independent, the time-average energy of the dipole–dipole interaction between two spheres subjected to an oscillatory field is [35]

$$W^{\mathrm{dd}} = 2\sqrt{2}\pi a^3 \varepsilon_0 \varepsilon_{\mathrm{f}} \Psi_\omega^{\mathrm{d}} \mathrm{E}_{rms}^2 (a/\mathrm{R})^3 (1-3\cos^2\theta), \qquad (10)$$

where $\mathrm{a}$ is the particle radius, $\mathbf{R}$ is the vector between the particle centers, $R$ is the distance between the particle centers, $\theta$ is the angle between the vectors $\mathbf{R}$ and $\mathbf{E}$, $E_{\mathrm{rms}}$ is the root mean square of the electric field strength, and the function $\Psi_\omega^{\mathrm{d}}$ describes the electric energy of interparticle interactions, which depend on frequency and the electric properties of the particles and suspending fluid. A rough estimate yields $\Psi_\omega^{\mathrm{d}} \sim 3\mathrm{Re}(\beta^*)^2$.

When the dipolar interparticle interactions, (10) exceed their thermal randomizing motion, an electrically induced aggregation of the particles occurs. A thermodynamic theory developed by Khusid and Acrivos [35, 36], predicts the threshold of the field strength above which polarizable particles start aggregating. Specifically, this theory yields the following expressions for the chemical potential, $\mu_{\mathrm{p}}$, and the osmotic pressure, $\Pi_{\mathrm{p}}$, of the particles averaged over the period of the field oscillations:

$$\mu_{\mathrm{p}} = (k_{\mathrm{B}}T/v_{\mathrm{p}})(\mathrm{d}f_0/\mathrm{d}c) - \varepsilon_0 (\partial\varepsilon_{\mathrm{s}}'/\partial c)_{\omega t_{\mathrm{c}}} \langle E^2/2\rangle \qquad (11)$$

$$\Pi_{\mathrm{p}} = k_{\mathrm{B}}TcZ/v_{\mathrm{p}} + \varepsilon_0 \left[\varepsilon_{\mathrm{s}}' - c(\partial\varepsilon_{\mathrm{s}}'/\partial c)_{\omega t_{\mathrm{c}}}\right] \langle E^2/2\rangle \qquad (12)$$

where $f_0 = c\,\ln\frac{c}{e} + c\int_0^c \frac{Z-1}{c}\mathrm{d}c$, $k_{\mathrm{B}}T$ is the thermal energy, $c$ is the particle volume fraction, $\langle\ \rangle$ denotes averaging over the oscillation period, $Z$ is the suspension compressibility factor, and $(\partial\varepsilon_{\mathrm{s}}'/\partial c)_{\omega t_{\mathrm{c}}}$ is the derivative of the real part of the suspension's dielectric permittivity, $\varepsilon_{\mathrm{s}}^*$, taken at a constant value of $\omega t_{\mathrm{s}}$ with $t_{\mathrm{s}}$ being the relaxation time of the dielectric phenomena [35]. The first terms in (11) and (12), refer, respectively, to the free energy and the osmotic pressure in the absence of the electric field. The other terms represent the interaction of particles with the electric field and particle–particle dipole interactions (10). The concentration dependence of the compressibility factor was described by the Carnahan-Starling equation for a suspension of spheres in a disordered state [22], $Z = (1+c+c^2-c^3)/(1-c)^3$ when $0 < c \le 0.5$ and by $Z \approx A/(c_{\mathrm{m}} - c)$, where $A \sim 2.34$ when $c \to c_{\mathrm{m}}$, where $c_{\mathrm{m}} \sim 0.68$ corresponds to a random close packing. The Clausius-Mosotti expression is used to compute $\varepsilon_{\mathrm{s}}^*$. When $\mathrm{Im}(\varepsilon_{\mathrm{f}}^*)$ and $\mathrm{Im}(\beta^*)$ are small (9) yields

$$(\partial\varepsilon_{\mathrm{s}}'/\partial c)_{\omega t_{\mathrm{c}}} \approx 3\varepsilon_{\mathrm{f}}\beta/(1-c\beta)^2,\ \beta = \mathrm{Re}(\beta^*),\ \varepsilon_{\mathrm{f}} = \mathrm{Re}(\varepsilon_{\mathrm{f}}^*). \quad (13)$$

Equations (11) and (12) predict the critical conditions beyond which a random arrangement of particles is unstable in a spatially uniform field in terms of particle volume fraction $c$ and a key parameter $\lambda \equiv \varepsilon_0\varepsilon_{\mathrm{f}} \langle E^2 \rangle v_{\mathrm{p}}/k_{\mathrm{B}}T$, a ratio of the electric and thermal energies. For example, $\lambda \sim 2.2\times10^4$ for a micrometer-sized particle in water subject to $E$ = 0.5 kV mm$^{-1}$. The particle concentration versus field strength phase diagram is similar to the diagram of concentration versus temperature for phase separation of two liquids, which are completely miscible only above a certain temperature [36]. The simplest suspension phase diagram is sketched in Fig. 1, in which the spinodal separates the metastable, **M**, from the unstable, **U**, regime in the two-phase region. This phase diagram corresponds to the case of small $\mathrm{Im}(\beta^*)$ for which $c_1 \to 0$ and $c_2 \to c_{\mathrm{m}}$ for $\lambda \to \infty$. Equation $\partial\mu_{\mathrm{p}}/\partial c = 0$ represents the spinodal curve in the phase diagram (Fig. 1). The critical point on the spinodal curve, $c_{\mathrm{cr}}$ and $\lambda_{\mathrm{cr}}$, coincides with the inflection point of $\mu_{\mathrm{p}}$, as a function of $c$. The single-phase region of the phase diagram includes the random spatial arrangement of particles, whereas the two-phase region (at $\lambda > \lambda_{\mathrm{cr}}$) corresponds to the appearance of a field-induced phase separation. On the coexistence curve separating the two-phase from the single-phase region (Fig. 1), the chemical potentials, and the osmotic pressures of particles in the coexisting phases at concentrations, $c_1$ and $c_2$, are equal, thus

$$\mu_{\mathrm{p}}(c_1) = \mu_{\mathrm{p}}(c_2) \quad \text{and} \quad \Pi_{\mathrm{p}}(c_1) = \Pi_{\mathrm{p}}(c_2). \quad (14)$$

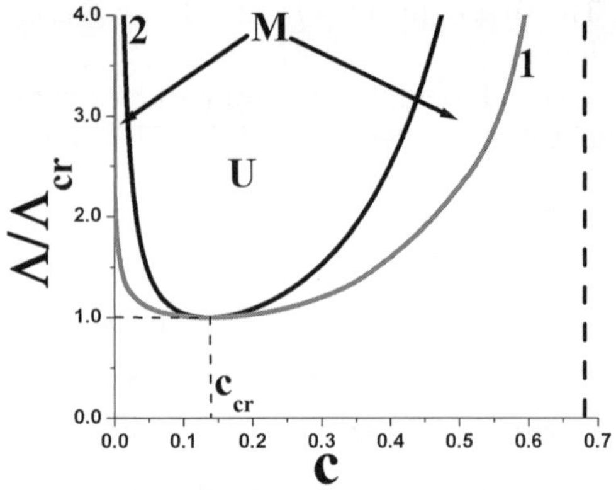


**Fig. 1.** Phase diagram of particle concentration, $c$, vs. relative field strength, $\Lambda = 3\beta^2\lambda$ with $\lambda \equiv \varepsilon_0\varepsilon_f \langle E^2 \rangle v_p / k_B T$, plotted for a suspension of particles whose polarizability $\beta = 0.1$. Curves 1 and 2 are the coexistence and spinodal curves, respectively. With increasing $\beta$ from −0.5 to 1, $\Lambda_{cr}(\beta)$ decreases from 25.2 to 11.7 whereas $c_{cr}(\beta)$ increases from 0.11 to 0.27 with $\Lambda_{cr}(0.1) = 20.4$, and $c_{cr}(0.1) = 0.14$ [35–37]

When a spatially nonuniform field applied to a suspension is so strong that the local values of the particle concentration and of the field strength in certain regions of the dielectrophoretic device fall into the two-phase region of the suspension phase diagram in Fig. 1, dielectrophoresis of the particles could be accompanied by the field-induced phase transition due to interparticle dipolar interactions [38]. The kinetics of the phase transition depends on the ratio of two time scales, $\tau_a$ and $\tau_d$. The former refers to the characteristic time for the field-induced structure formation in a suspension subjected to a spatially uniform field [35, 39, 40]; the latter measures the time for a particle to move over a characteristic length of the system, $d$, dielectrophoretically [32, 33]. These time scales are

$$\tau_a \approx \frac{\eta_f d^2}{\varepsilon_0\varepsilon_f \left[\mathrm{Re}(\beta^*) V_{rms}\right]^2}\xi \quad \text{and} \quad \tau_d = \frac{3\eta_f d^4}{a^2\varepsilon_0\varepsilon_f \left|\mathrm{Re}(\beta^*)\right| V_{rms}^2}, \tag{15}$$

hence

$$\tau_d/\tau_a \approx 3d^2 \left|\mathrm{Re}(\beta^*)\right| / a^2\xi(c), \tag{16}$$

where $\eta_f$ is the viscosity of the suspending fluid, $\xi \approx 0.4\left[(\pi / 6c)^{5/3} - 1\right]$ is a coefficient that depends on the volume fraction of the particles, $a$ is the particle radius, and $V_{rms}$ is the root mean square of the applied voltage. The regime $\tau_d << \tau_a$ corresponds to *heterogeneous aggregation* [33, 34]. Under these conditions typical of dilute suspensions, the particles are forced to accumulate in certain areas of the device and then to aggregate. The regime $\tau_d >> \tau_a$ corresponds to *homogeneous aggregation* [33, 34], in which macroscopic particle redistribution during phase transition is negligible.

## 2.3. Electro-Hydrodynamic Models

To enable the dielectrophoretic manipulation of particles and macromolecules, the applied field should be sufficiently strong to dominate the thermal Brownian motion. However, a high electric field also gives rise to undesirable processes in the suspending fluid, such as electrolysis, electro-osmosis, thermal convection, and electro-thermal flow caused by the variation in the fluid conductivity, permittivity, and density due to Joule heating [41–44]. Estimation of the orders of magnitude of the various forces experienced by a particle and a fluid in dielectrophoretic microfluidics [41, 42] indicates that dielectrophoresis would prevail in small-scale devices and for high field frequencies.

Below we briefly outline two electro-hydrodynamic models, which enable simulations of dielectrophoretic phenomena without using any fitting parameters. One of them is a single-particle model, which is appropriate for a dilute suspension, whereas the other describes the effects of the interparticle electric and hydrodynamic interactions on the field driven dielectrophoresis and particle aggregation.

### 2.3.1. Single-Particle Model

A single-particle model takes into account only the Stokes drag force, the dielectrophoretic force, (7), and the gravity force exerted on flowing particles and ignores the electrical and hydrodynamic interparticle interactions, the effects of the migration of the particles on the fluid flow and the field configuration, and the particle Brownian motion being typically insignificant for particles of micrometer and even submicrometer sizes. This model requires no fitting parameters provided that properties of the particles (polarizability, size distribution, etc.) are measured independently. With these assumptions, the force balance for a spherical particle flowing through a channel yields

$$6\pi\eta_{\mathrm{f}}a\left[\mathbf{u}-\mathbf{v}_{\mathrm{f}}(\mathbf{r})\right]=2\pi\varepsilon_0\varepsilon_{\mathrm{f}}a^3\,\mathrm{Re}(\beta^*)\nabla E_{\mathrm{rms}}^2(\mathbf{r})+\frac{4}{3}\pi(\rho_{\mathrm{p}}-\rho_{\mathrm{f}})a^3\mathbf{g},$$

$$\frac{\mathrm{d}\mathbf{r}}{\mathrm{d}t}=\mathbf{u},\quad \mathbf{r}\big|_{t=0}=\mathbf{r}_0, \tag{17}$$

where $a$ is the particle radius, $\eta_{\mathrm{f}}$ is the fluid viscosity, $\mathbf{u}$ is the local particle velocity, $\mathbf{v}_{\mathrm{f}}$ is the local fluid velocity along the channel, $\rho_{\mathrm{p}}$ and $\rho_{\mathrm{f}}$ are the particle and fluid densities, respectively, $\mathbf{g}$ is the gravitational acceleration, and $\mathbf{r}(t)$ is the displacement of a particle initially at $\mathbf{r}_0$. To find the local values of the fluid velocity, $\mathbf{v}_{\mathrm{f}}(\mathbf{r})$, and the field gradient, $\nabla E_{\mathrm{rms}}^2(\mathbf{r})$, Navier-Stokes' equations for the flow and Laplace's equation for the electric field should be solved for the fluid in the channel. The force balance equation (17) then yields the particle velocity **u**. The field-induced displacement of a particle, $\mathbf{r}(t)$, being initially at $\mathbf{r}_0$ is obtained by integrating the equation for the particle trajectory (17).

The single-particle model is applicable both in the single-phase region of the suspension phase diagram (Fig. 1) as well as in the two-phase region provided that the channel characteristics meet the requirement $\tau_{\mathrm{d}} << \tau_{\mathrm{a}}$ in (16). The main nondimensional parameters of (17) for particles flowing through a dielectrophoretic device are the following:

1. The ratio of the dielectrophoretic time (15) to the average fluid residence time, $\tau_{\mathrm{d}}Q/(AL)$, where $L$ is the channel length, $Q$ is the volumetric flow rate, and $A$ is the channel cross sectional area
2. The channel length to the electrode width ratio, $L/d$
3. The ratio of the gravitational and dielectrophoretic forces $G=2(\rho_{\mathrm{p}}-\rho_{\mathrm{f}})gd^3/3\varepsilon_0\varepsilon_{\mathrm{f}}\left|\mathrm{Re}(\beta^*)\right|V_{\mathrm{rms}}^2$

Predictions of (17) for the motion and segregation of polarized particles flowing through a dielectrophoretic device were found to be quantitatively consistent with the experimental data for dilute particle suspensions in both ground-based [32–34] and microgravity environments (aboard the NASA research aircraft KC-135, [45]. These predictions used values for $\mathrm{Re}(\beta^*)$ calculated from measurements of the dielectric permittivity of the suspension in low fields (9).

Predicting the dielectrophoretic migration and rotation of a nonspherical polarizable particle requires considering the balance of hydrodynamic and electric forces and torques exerted on the particle by an electric field and a flowing fluid. The comprehensive discussion of the hydrodynamic force and torque acting on a nonspherical particle is given by Happel and Brenner [46], whereas the electrical force and torque are discussed by Jones [5]. The quantification of the dielectrophoretic migration of nonspherical particles was addressed recently by Evoy et al. [47], Liu and Bau [48], Kim and Han [49], and Riegelman et al. [50].

One may expect that small microelectrodes, which can produce large field gradients at low voltages, are favorable for precise and controlled manipulation of particles. However, experimental data [51] clearly demonstrate that precision in positioning particles on energized electrodes can decrease with electrode size due to the formation of multiparticle aggregates. Field-induced interactions of particles collected on electrodes do not affect their spatial distribution only when the dipole–dipole force, $F_{dd}$, between particles is significantly smaller than the dielectrophoretic force (7) exerted on the particles. Using the expression for $F_{dd}$ (cf. 10), we find that the electrode size should be limited according to

$$F_{dd}/F_{dep} \sim 6\,\mathrm{Re}(\beta^*)(a/d)^3 << 1, \qquad (18)$$

where $a$ is the particle radius. This requirement imposes limitations on the miniaturization of electrode arrays in microdevices and is also a requirement for the validity of the single-particle model (17). A similar minimum size constraint exists for insulator-based devices.

### 2.3.2. Model for Collective Phenomena

Sufficiently strong interparticle electric interactions cause a suspension to separate into low- and high-concentrated phases by rearranging randomly distributed particles into various aggregation patterns. Several experimental studies have demonstrated that interparticle interactions dramatically affect suspension behavior in dielectrophoretic devices. Bennett et al. [52] and James et al. [53] observed formation of a "bolus" characterized by a sharp concentration front when flowing particles were confined by negative DEP to the entrance region of a dielectrophoretic gate (Fig. 2a). The resulting field-driven arrangement of the particles formed a distinct boundary between regions enriched with and depleted of particles. Kumar et al. [54] reported formation of a field-induced front between the suspension and suspending fluid in suspensions of neutrally buoyant particles undergoing negative DEP (Fig. 2b). This front was found to move slowly away from high field regions and to confine the particles to a thin column in the low field region (Fig. 2b). The phenomenon of front formation driven by interparticle interactions suggests a new method for strongly concentrating particles in focused regions of micro-devices.

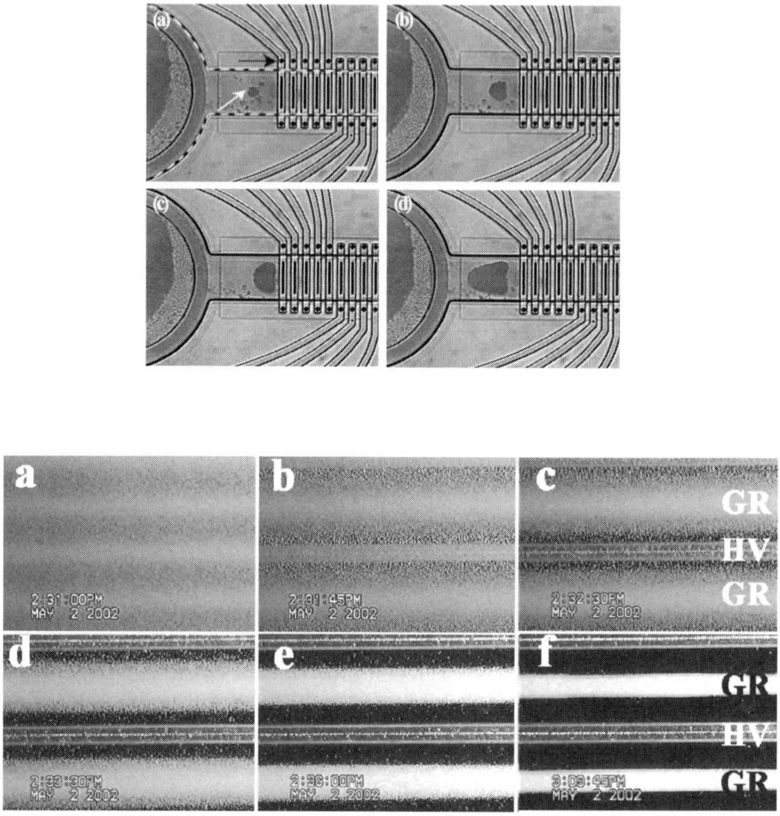


**Fig. 2.** Field-induced formation of concentration fronts: (**a**) 1-μm polystyrene beads in deionized water form a sharp-boundaried "bolus" (white arrow) when negative DEP at 30 MHz exclude them from a high-field "dielectrophoretic gate" [52] shown at times (a) 10, (b) 70, (c) 120, and (d) 180 s. The dashed line outlines the fluidic channel and electrodes. The black arrow indicates the energized electrode. The flow rate of the 0.1% v/v suspension is 0.24–9.6 pl $s^{-1}$. The scale bar is 20 μm. (**b**) A distinct concentration front forms when a spatially uniform 10% v/v suspension of neutrally buoyant 87-μm-diameter poly-α olefin particles (white) dispersed in corn oil undergoes negative DEP [54]; (a) before and (b) 45 s, (c) 90 s, (d) 150 s, (e) 300 s, and (f) 39 min after the application of 5 kVrms at 100 Hz. The electrode width is 1.6 mm. HV and GR refer to the high-voltage and grounded electrodes, respectively. A low-field strength region exists above the GR electrode [54].

These collective phenomena are modeled [52, 54] using coupled quasi-steady electric field equations plus momentum and continuity balance equations for a suspension averaged over the field oscillations, with the suspension approximated as a Newtonian fluid having a concentration dependent viscosity. The bulk electric force on a suspension and particle velocity relative to the suspending fluid were expressed in terms of the particles' chemical potential $\mu_{\mathrm{p}}$, given by (11).

When local values of the particle concentration and electric field lie within the single-phase region of the suspension phase diagram in Fig. 1, we have

$$\nabla \cdot \left[ \varepsilon_{\mathrm{s}}^{*}(\omega, c) \nabla \varphi_{\omega}^{*} \right] = 0 \text{ with } \mathbf{E}_{\omega}^{*} = -\nabla \varphi_{\omega}^{*}; \text{ and } \nabla \cdot \mathbf{v} = 0$$

$$\rho_{\mathrm{s}} \left( \frac{\partial \mathbf{v}}{\partial t} + \mathbf{v} \cdot \nabla \mathbf{v} \right) = -\nabla p + \nabla \cdot \eta_{\mathrm{s}}(c) \left( \nabla \mathbf{v} + \nabla \mathbf{v}^{\mathrm{T}} \right) - c \nabla \mu_{\mathrm{p}} + c \left( \rho_{\mathrm{p}} - \rho_{\mathrm{f}} \right) \mathbf{g}$$

$$\frac{\partial c}{\partial t} + \nabla \cdot \left( c \mathbf{v} + \mathbf{j}_{\mathrm{p}} \right) = 0, \tag{19}$$

where $\mathbf{j}_{\mathrm{p}} = \dfrac{c(1-c)^{2} v_{\mathrm{p}}}{6\pi a \eta_{\mathrm{s}}(c)} \left[ -\nabla \mu_{\mathrm{p}} + (\rho_{\mathrm{p}} - \rho_{\mathrm{f}}) \mathbf{g} \right]$, $\mathbf{E}_{\omega}^{*}$ and $\varphi_{\omega}^{*}$ are the Fourier components of the electric field and potential, respectively, $\varepsilon_0$ is the vacuum permittivity, $\varepsilon_{\mathrm{s}}^{*}$ is the dielectric permittivity of the suspension at the field frequency $\omega$, $c$ is the particle volume fraction, $\mathbf{v}$ is the bulk velocity of the suspension, $\mathbf{g}$ is the gravitational acceleration, $p$ is the pressure, $v_{\mathrm{p}}$ is the particle volume, $a$ is the particle radius. The suspension density is $\rho_{\mathrm{s}} = (1-c)\rho_{\mathrm{f}} + c\rho_{\mathrm{p}}$ where $\rho_{\mathrm{p}}$ and $\rho_{\mathrm{f}}$ are the particle and fluid densities. The suspension viscosity is $\eta_{\mathrm{s}} = \eta_{\mathrm{f}} \hat{\eta}(c)$, where $\eta_{\mathrm{f}}$ is the suspending fluid viscosity and $\hat{\eta} = (1 - c/c_{\mathrm{m}})^{-2}$, where $c_{\mathrm{m}} \sim 0.68$ is the concentration of spheres at random close packing. The particle chemical potential, $\mu_p$, is a function of the particle concentration, $c$, and oscillation-averaged field strength, $\langle E^2 \rangle = \left| \mathbf{E}_{\omega}^{*}(\mathbf{r}) \right|^2 / 2$. The factor $(1-c)^2$ in the expression for the particle flux, $\mathbf{j}_{\mathrm{p}}$, represents the hindrance function.

Next, we consider the suspension equations for local values of the particle concentration and electric field in the two-phase region of the phase diagram for relatively fast local aggregation, i.e., $\tau_{\mathrm{d}} >> \tau_{\mathrm{a}}$. Under these conditions, the value of the chemical potential is the same in both phases propagation using independently measured data for the fluid and particle properties are quantitatively consistent with experimental data [52, 54]. Analysis of the (19) reveals two limiting regimes for the dielectrophoretic formation of a sharp concentration front. One of them is the quasi-equilibrium regime in which DEP is accompanied by field-driven phase separation leading to an abrupt change in particle concentration [38]. The position and the magnitude of the concentration front are determined by the balance of the dielectrophoretic, dipole–dipole, and steric forces. The other is a hydrodynamic mechanism that operates through the balance of the dielectrophoretic and hydrodynamic forces [37]. The position and the magnitude of the concentration jump is governed by the rapid local growth of the suspension viscosity due to the particle accumulation near the front, which slows down the field-driven particle motion by decreasing their mobility, proportional to $(1-c)^2/\hat{\eta}(c)$. The concentration front forms in both the single-phase and the two-phase states of a suspension and travels with the speed of the front increasing with increasing the applied voltage. In contrast to the front formation due to the field driven phase transition, the hydrodynamic mechanism can operate only in an unsteady regime of DEP and will cease when the particle spatial distribution reaches the equilibrium.

## 3. Microfluidic Applications of Dielectrophoresis

Dielectrophoresis has been applied by rescarchers primarily for separating and concentrating particles, but dielectrophoretic techniques are now being explored for directing or mediating self assembly of particles into micro- and nanostructured materials. The ability to use dielectrophoresis to induce phase transitions in colloids potentially opens even more applications.

To date, most applications of dielectrophoresis have employed oscillating applied fields at ~100 kHz frequencies or higher applied directly to immersed electrodes. Numerous such device architectures and configurations have been developed. A typical dielectrophoretic device comprises a flow channel equipped with an array of closely spaced energized and grounded electrodes to generate a spatially nonuniform field [10, 11, 14, 15]. The maximum electric field is located at the edges of the electrodes and the minimum electric field in the interelectrode gaps. The electrode structures are typically between fractions of a micrometer and a few micrometers thick. The electrode arrays are typically fabricated by producing patterns of conducting material onto an insulating substrate. Other devices have been made that employ nonplanar or multiple-layers of electrodes

[55]. These "electrode-based" microfluidic devices have been well documented in numerous text books (e.g., [11–13]).

Electrodeless or "Insulator-based" dielectrophoretic (iDEP) microfluidic devices use spatially nonuniform insulators to produce electric field gradients to drive DEP (e.g., [56–60]). These devices have not been as thoroughly documented, yet show tremendous promise as stand-alone processors as well as interstage elements in complicated microfluidic systems. Furthermore, the dielectrophoretic phenomena exploited in iDEP devices often manifest as anomalous behavior in poorly designed devices for conveying particles electrokinetically. For these reasons, the remainder of this chapter focuses mainly on iDEP architectures.

### 3.1. Primary Flows

Practical applications of DEP generally rely on a competition between DEP as a "secondary flow" and a "primary" transport mechanism, e.g.,

1. Diffusion or thermal (Brownian) motion of the particles, which acts to resist concentration gradients. Diffusion scales inversely with particle size and produces random displacements typically on the scale of 0.1 μm in 1 s for bacteria.
2. Advection or pressure-driven flow that drags particles in the flow direction according to the particle shape, size, and orientation. Advective flow speeds can range from $<1\ \mu m\ s^{-1}$ to $>10\ mm\ s^{-1}$ in typical DEP applications.
3. Sedimentation or inertial and gravitational flows moves particles in proportion to the particle size and density difference between the particle and the immersion fluid. Typical sedimentation speeds of microscale particles are $\sim 0.1\ \mu m\ s^{-1}$ to $10\ \mu m\ s^{-1}$.
4. Electrokinesis is the superposition of electrophoresis: flow produced by an electric field acting on a charged particle and electro-osmosis and flow produced by an electric field acting on net-charged regions of fluid. Electrokinetic speeds scale linearly with the applied field and the difference of the net charge density on the surfaces of the particle and the boundary of the immersion fluid. Typical electrokinetic speeds used in DEP applications are $10\ \mu m\ s^{-1}$ to $>10\ mm\ s^{-1}$.

For example, the primary transport mechanism may be a pressure driven flow that carries particles down a channel. DEP then acts as a secondary flow that perturbs or restrains from their primary motion.

The dependences of DEP and the primary flow on particle properties means different particles flow differently. The principal aim of dielectrophoretic device design is to exploit this particle specificity to achieve one or more practical functions, e.g., selectively concentrating particles, spatially segregating particles, temporally segregating particles, etc.

The dielectrophoretic component of a particle's flow velocity, $\mathbf{u}_{DEP}$, can be empirically related to the local electric field, **E**, by $\mathbf{u}_{DEP} = \mu_{DEP}\, \nabla(\mathbf{E}\cdot\mathbf{E})$. The constitutive parameter $\mu_{DEP}$, called the dielectrophoretic mobility, depends on the particle geometry and electrical properties relative to the immersion fluid (c.f. (7)). Primary transport mechanisms are similarly related empirically to forcing fields via a constitutive mobility. For example, consider the electrokinetic particle flow velocity $\mathbf{u}_{EK} = \mu_{EK}\, \mathbf{E}$, where $\mu_{EK}$, is the electrokinetic mobility of the particle, which depends on the relative charge density of the particle and channel boundary surfaces and also on the particle and channel geometries when these dimensions are comparable to the Debye length.

A dielectrophoretic particle trap occurs where the dielectrophoretic velocity successfully opposes the primary flow velocity. That is, the component of the dielectrophoretic velocity along a primary flow streamline is equal or greater than the negative of the primary flow velocity. When the primary flow is electrokinetic, this means $\mathbf{u}_{DEP}\cdot(-\mathbf{u}_{EK}) \geq \mathbf{u}_{EK}\cdot\mathbf{u}_{EK}$, or

$$(\mu_{DEP}/\mu_{EK})\, (\nabla I/I) \cdot \mathbf{E} \leq -1, \tag{21}$$

where $I$ is the electric field intensity, $\mathbf{E}\cdot\mathbf{E}$. Equation (21) decomposes the dependencies on of particle, fluid, and boundary physical properties ($\mu_{DEP}/\mu_{EK}$), the relative nonuniformity of the electric field ($\nabla I/I$), and the electric field scaling (linear). These terms are typically design constraints, design-time parameters, and run-time parameters, respectively.

### *3.2. Non-uniform Electric Field Generators*

All DEP applications share one characteristic: the need for a spatially nonuniform electric field i.e., nonzero ($\nabla I/I$), in the fluid. There are several ways to produce such a nonuniform field. The first and most common technique is to apply voltages directly to immersed (complex) conductors. These conductors couple conduction and displacement current into the fluid, depending on the applied field frequency (and amplitude, near DC) and whether the conductors contact or are insulated from the fluid.

A second approach is to apply an electric field through a fluid in a channel having spatially nonuniform boundaries. The field is generally applied via electrodes immersed in the fluid; however as mentioned above, the electrodes do not directly produce the field nonuniformities used for DEP. Rather, features like obstacles, constrictions or inclusions that bound the fluid force the electrical current to follow a nonuniform path and produce a nonuniform electric field. These features must have a different (complex) conductivity from the fluid and are typically insulators. This approach is called insulator-based DEP (iDEP) or electrodeless DEP (eDEP) in the literature. This arrangement is favorable in part because the features can be made lithographically in the same step as the channels, insulators are less susceptible to fouling than electrodes, allow operation near DC without bubble formation in the device, and are compatible with electrokinetic primary flows.

A third approach is to apply an electric field through a fluid in a channel whose boundaries have nonuniform (complex) conductivity. Conduction or displacement current flowing into and out of the boundary produces nonuniform electric fields without the need for geometrical features.

### 3.3. Modes of Operation

Device operation is often characterized as either "batch" or "continuous." In purely batch operation, fluids and particles are placed in the device, the device is operated for a period of time, then the resulting redistribution of fluids and particles has some utility or function, e.g., displaying a collection of abnormal cells through a microscope to a pathologist, providing a concentrated plug of particles to a subsequent device stage, providing filtered fluid to a subsequent state, etc.

Batch operation generally employs dielectrophoretic "traps," in which DEP overcomes other flow mechanisms to immobilize particles. Such traps localize particles, allowing a high concentration factor. However, interparticle interactions near traps can limit selectivity. Moreover, trap capacity and selectivity have an inverse relation as shown in the insulator-based dielectrophoretic device in Fig. 3 [61]. In the device, bright and dimly fluorescent bacteria flow electrokinetically at essentially identical speeds from right to left. However, brightly fluorescent bacteria are immobilized against this flow by negative DEP from the high-field regions between glass posts. These particles are trapped in potential wells of finite size and capacity. Figure 3 shows traps at saturation in a steady state: on average, when a new bacterium is retained by the trap, it displaces another one. To trap more particles, the potential well can be deepened by increasing the applied field as in Fig. 3; however, now both types of bacteria are collected. It is possible to design tailored structures to improve trap selectivity and capacity; however, the difficulties of treating interparticle interactions, explicit hydrodynamic effects, and finite-particle-size effects limits our current ability to apply theory to this optimization.

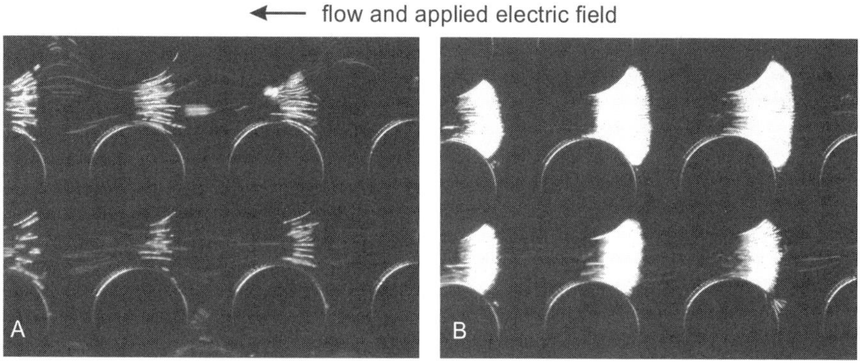


**Fig. 3.** A simple iDEP sorter in steady-state operation. Circles are glass posts on 200-μm centers spanning the ~20-μm-deep, covered channel. Bright particles are *Bacillus cereus* and dim particles are *Bacillus subtilis*. (**a**) At an applied field of ~50 V mm$^{-1}$, a small number of *b. Cereus* occupy each trap. (**b**) At an applied field of ~75 V mm$^{-1}$, both bacteria types are trapped. The trap capacity is higher, but selectivity is worse. Source: Y Fintschenko, Sandia

In purely continuous operation, fluid and particles flow through the device without accumulating over time. Separation or concentration occurs by redirecting the particle paths from the primary flow. For example, a purely continuous device could accept a stream of blood cells, and provide separate or spatially segregated output streams of the various cell types. By analogy with optics, such a device is called a continuous cell spectrometer. Alternatively, a dilute input stream of particles can be continuously processed into a concentrated output stream and a further-diluted output stream, a process called continuous concentration.

Continuous operation generally employs DEP to form and maintain particle "streams" that are propelled by the primary flow. Figure 4a shows an example of dielectrophoretic streams of 200-nm polystyrene particles [60]. The black objects are the end of an array of 50 rows of glass posts on 63-μm centers. The primary flow is electrokinetic from top to bottom. Positive DEP forms the streams in a competition between repulsion from the low-field region at the top and bottom stagnation points of the posts and attraction along the field concentrations at the sides of the posts. For this post shape, attractive forces win, and the columns of posts produce concentrated streams. This phenomenon is reproduced qualitatively in the simulation of Fig. 4b, that treats electrokinesis (EK), DEP, and Brownian motion, but neglects hydrodynamic forces and steric effects. Theory can be more readily applied for these flows because the particles are not highly concentrated and are not diverted significantly from their electrokinetic streamlines.

and is specified as a function of $\lambda$ by the coexistence curve (14) of the phase diagram. The particle concentrations in both phases, $c_1$ and $c_2$, are also specified by (14) as functions of $\lambda$. The volume fraction of the high concentration phase, $\psi$, is given by the lever rule, $c_2\psi + c_1(1-\psi) = c$, as a function of $c$ and $\psi$. For simplicity, the model ignores the explicit dependence on $\psi$ of the complex dielectric permittivity, viscosity, and hindrance function for this zone. These parameters are obtained the same way as in the single-phase region. Under these conditions, the two-phase model consists of (19), in which the chemical potential is now specified as a function of the field strength, $\mu_{\mathrm{p}} = \mu_{\mathrm{p}}(\lambda)$, through solution of (14). Notice that the spatial gradient of the particle chemical potential, $\nabla\mu_{\mathrm{p}}$, considered as a function of the local particle concentration and the relative field strength, experiences an abrupt change when the suspension undergoes the local field-driven phase transition, i.e., when the point $c(\mathbf{r},t)$, $\lambda(\mathbf{r},t)$ representing the local suspension composition in the phase diagram (Fig. 1), crosses the coexistence curve.

For the limiting case of a dilute suspension for which $c \to 0$, the first two equations of (19) reduce to the field and flow equations for the suspending fluid while the third yields

$$\frac{\partial c}{\partial t} + \nabla \cdot \left[ c\left(\mathbf{v} + \mathbf{u}_{\mathrm{e}}\right) \right] = D_{\mathrm{B}} \Delta c, \tag{20}$$

where $\mathbf{u}_{\mathrm{e}} = \dfrac{v_{\mathrm{p}}}{6\pi a \eta_{\mathrm{f}}} \left[ \dfrac{3}{2} \varepsilon_0 \varepsilon_{\mathrm{f}} \operatorname{Re}(\beta) \nabla E_{\mathrm{rms}}^2 + (\rho_{\mathrm{p}} - \rho_{\mathrm{f}}) \mathbf{g} \right]$ is the velocity of a particle relative to the fluid under the action of the dielectrophoretic and gravity forces, and $D_{\mathrm{B}} = k_{\mathrm{B}} T / 6\pi a \eta_{\mathrm{f}}$ is the particle Brownian diffusion coefficient. When Brownian motion is insignificant, (20) leads to the single-particle model, (17). Another limiting case is the quasi-equilibrium state of neutrally buoyant particles subject to a spatially nonuniform AC field [38], for which (19) yield $\mathbf{v} = \nabla p = 0$, and $\nabla \mu_{\mathrm{p}}(\mathbf{r}) = 0$, where $\mu_{\mathrm{p}}$ is given by (11). This spatial distribution of the particles is reached when the chemical potential of a particle becomes independent of position. This equation represents $c$ as a function of $\langle E^2 \rangle$ along the curve of constant $\mu_{\mathrm{p}}$ of the suspension phase diagram, Fig. 1.

Equations (19) require no fitting parameters provided that particle properties are measured independently. Simulations of front formation and

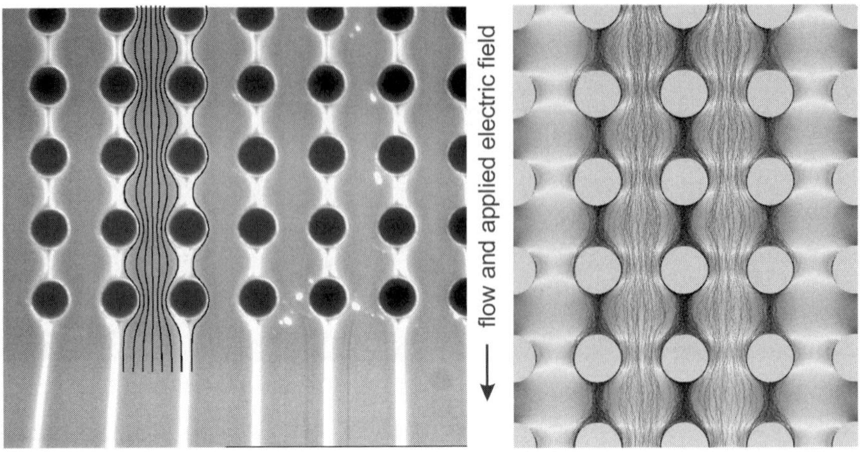


**Fig. 4.** Positive DEP forms streams of 200-nm particles in an array of circular posts on 63-μm centers. (**a**) Experiment and calculated EK streamlines. (**b**) Numerical simulation of EK, DEP, and Brownian motion. Source: Sandia Microfluidics Group

Dielectrophoretic devices are not limited to these extremes of operation, however. Most dielectrophoretic devices have a continuously flowing sample stream during processing and therefore are not purely batch devices even though the processed particles accumulate in a batch fashion. Most continuous devices end up accumulating particles either internally or at their exit ports because of the challenges of matching fluxes with subsequent devices. Moreover, hybrid continuous and batch devices, allow significant performance advantages, as is discussed later.

An often neglected element of DEP device design and operation is the ability to switch applied field magnitudes, directions, and frequency during the run time of a device to establish function and improve performance. Such field sequencing allows a generic device design to be used for a variety of different purposes by simply changing the field-sequence "program" [62]. These programs can be adapted in real-time in response to detector and sensor data. For example, the applied fields can be automatically adjusted to avoid saturation and interparticle effects when particle concentration is high and to allow higher flow rates when particle concentration is low.

Field sequencing can also limit complexity and risk at design time. Rather than designing a complicated system having a fixed number of sequential processing stages for selectivity against a high background of particles, one could design a device so that samples can be repeatedly cycled through the stage as needed for refinement. The best mode of device operation can be found and refined through field testing rather than at design time where specifications are generally sketchy and only a handful of comparatively low-fidelity simulations are practical.

A designer, researcher or even an automated sensor can refine or modify a field sequence as easily as compiling or running a program and can iterate rapidly (in minutes or seconds) and with minimal additional hardware cost. In contrast, the delay between a microfluidic hardware design change and its implementation is usually measured in weeks or months. Replication techniques also favor the use and reapplication of generic designs, since replication costs are negligible compared to nonrecurring costs of making molds and stamps. Even for nonreplication-based microsystems, however, nonrecurring costs of sequentially testing new design iterations can be prohibitive.

### 3.4. Depletion and Enhancement

Devices can be operated to employ dielectrophoretic repulsion or attraction to the objects that produce the nonuniform electric field. Dielectrophoretic repulsion causes "depletion" of particles or a lowered concentration of particles near the objects. Dielectrophoretic attraction causes "enhancement" of the concentration of particles near the objects. When competed against a primary flow, depletion and enhancement produce qualitatively different effects. For example, the depletion effect from negative DEP of the particles in Fig. 3 is enough to stop their downstream motion. The particles collect upstream of the posts. The same array would operate in enhancement for particles that exhibit positive DEP, which would accumulate downstream of the posts. Figure 4 shows a net enhancement of particles undergoing positive DEP. As discussed earlier, this net effect was a result of a competition in which attraction overcame repulsion. Figure 5, on the other hand shows essentially the same system as in Fig. 4a, but with square posts instead of round posts. The relatively large low field region along columns of posts repels particles enough to overcome the attraction particles experience briefly at the field concentration near the corners of the squares, resulting in a dramatic depletion effect. This effect persists when there is a slight angle between the applied field and the columns as in Fig. 5. Like in Fig. 3, particles accumulate upstream of the objects, but instead of being trapped, they are continuously diverted from the primary flow.

Figure 6 shows numerical simulations of point particles traveling down an array of posts under a slightly tilted electric field, including the effects of DEP, EK, and weak diffusion. Numbers at the bottom are proportional to the dielectrophoretic mobility: negative numbers correspond to negative DEP; positive numbers correspond to positive DEP. Dielectrophoretically neutral particles (0.0) follow the EK flow. Both enhancement and depletion cause particles to follow columns rather than the electric field.

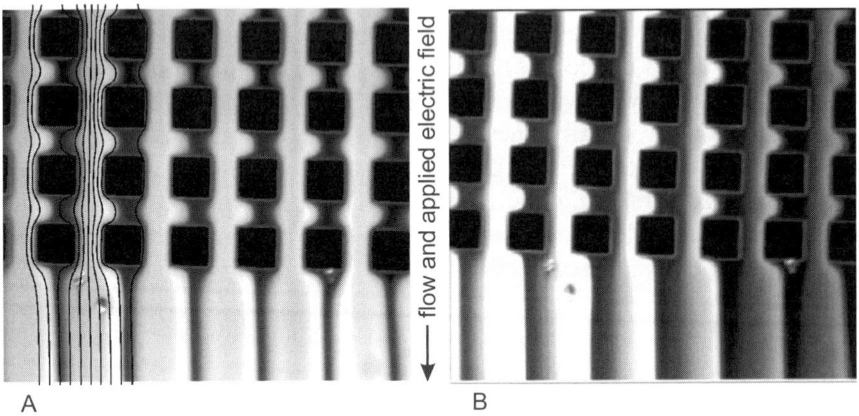


**Fig. 5.** (**a**) Positive DEP forms depleted streams of 200-nm particles in arrays of square posts on 63-μm centers. (**b**) The depletion effect persists at modest array tilts, so particles flow along array columns rather than the electric field. Source: Sandia Microfluidics Group

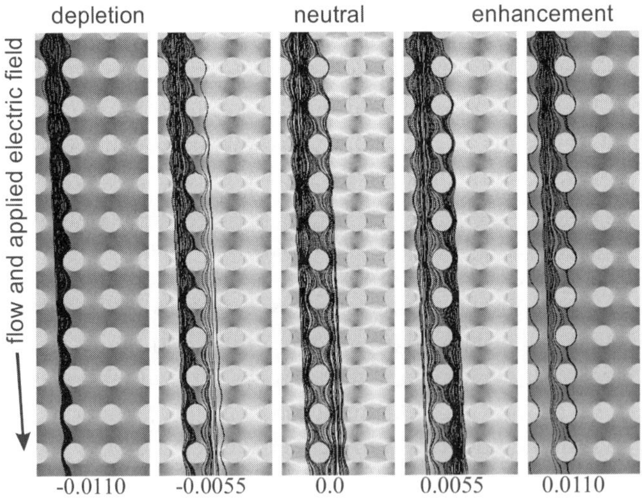


**Fig. 6.** Numerical simulations of particle trajectories under the influence of DEP, EK, and diffusion. The numbers are proportional to the dielectrophoretic mobility. Negative DEP produces depletion and positive DEP produces enhancement for this post shape

### 3.5. Architectural Considerations

With the range of primary-flow options, operating modes, and myriad of electrode, channel, and obstacle architectures, a designer has a lot of decisions to make. The impact of these decisions can be far-reaching and the cost of bad choices in performance and redesign considerable. The merit of any particular design or architecture depends sensitively on factors specific to the application. These factors include device and support equipment cost and performance factors such as throughput, concentration factor, selectivity, collection efficiency, and the ability to maintain these factors when transferring particles into and out of the device, if necessary. Such factors are relatively tractable to characterize at the design stage or after a modest amount of testing. Unfortunately, a designer must deal with far more difficult-to-characterize and analytically intractable factors like robustness and immunity from fouling, behavior over widely varying particle compositions and concentrations, etc.

#### 3.5.1. Fouling

To date, the issue of fouling of DEP and most other microfluidic devices is unresolved, but receiving considerable attention. Fouling problems will likely not be resolved by a single magic bullet or specific device geometry, but rather mitigated by a combination of improvements in materials and immersion fluid recipes, operating algorithms, and detailed device design.

One engineering approach to mitigate risk from fouling is to make devices routinely disposable, which increases the importance of fabrication cost factors and complicates automation. Another is to design the devices to facilitate manual or automated cleaning. Routine manual cleaning may be impractical in remote or unattended systems, while automated cleaning complicates material choices, may not be compatible with immersed electrodes, adds system complexity, and is typically only modestly effective at restoring performance. Moreover, the design may need to accord with requirements of regulation authorities like the American Food and Drug Administration (FDA), the Center for Disease Control (CDC), e.g., that all items that come into contact with body fluids or tissues be disposable or autoclavable. The tradeoff between single-use and reuse is only one of many in DEP device architecture and design: the message is that the specific electric-field distribution ($\nabla I/I$) is only one of many elements in DEP device design.

### 3.5.2. Throughput

Of the tractable factors affecting DEP devices, the required throughput has perhaps the largest impact. For examples, the speed of particles in traveling-wave DEP is generally of order 1–10 μm $s^{-1}$. A common and effective approach to cell-typing and sorting is to find a frequency at which the Clausius-Mossotti factor for the particle changes sign (e.g., [10]). This "cross-over frequency" is a characteristic of the particle and immersion fluid system and provides for good selectivity. However, near the cross-over frequency, dielectrophoresis is nearly nulled, thus particle speeds are generally well under 10 μm $s^{-1}$. At this speed, it takes about 30 min to process one microliter of sample in a channel having a width of 1 mm and a depth of 50 μm.

Raw blood typically contains about 5 million cells per microliter. If the system can process raw blood, the throughput is around 2,500 cells $s^{-1}$. By comparison, a bench-scale cell cytometer can process cells about an order of magnitude faster. To reach the sorting speed of a fast cytometer, the geometry of such a system could be modestly scaled, e.g., using a 10-mm- rather than 1-mm-wide channel. If however the migration speed is only 1 μm $s^{-1}$, geometrical scaling is much less convenient: a 100-mm-wide flow channel is approaching typical wafer scales, the use of which is likely to increase cost significantly. Moreover, maintaining uniform flow, injections, and elutions with such a wide channel poses serious design and interconnection challenges.

Now consider a typical sample-processing application for sensors for biological pathogens in the air. Air sampling is typically performed using a hydrodynamic aerosol collector whose output is a milliliter-scale aqueous suspension of particles, surfactants, and other additives every 30 min or so. The particle concentration is generally far lower than raw blood and, for sensitivity, it is desirable to process the entire sample. Even at a flow rate of 10 μm $s^{-1}$, the system of the previous example would take about a month to process the sample. Scaling the width of the channel to 10 mm reduces the processing time to two to three days, still generally unsuitable for sensing, especially since a biological sample is not static on this time frame. Scaling the cross-section to 100 mm brings the analysis time down to about 6 h, which is still over an order of magnitude too long to process samples at the rate provided by the collector.

One can increase throughput by scaling up the channel depth, but this must be approached judiciously. For device function to be preserved, the electrodes and electrode spacing should be adjusted as needed to produce the same electric field distribution as in the shallower channel. If the electrodes are fashioned to span the depth of the channel, no scaling of the field or electrodes may be necessary, since the electric field distribution does not vary throughout the depth of the channel. If, however, the electrodes are patterned on the channel surface, this scaling is proportional to the channel depth and the applied voltages must be scaled as the square of the depth to obtain the same electric field gradient and dielectrophoretic speeds. Higher driving voltages and deep channels compound the problem of Joule (Ohmic) heating of the immersion fluid in the channel. Deep channels also tend to be sensitive to unwanted hydrodynamic flows, particularly compared to flows of only 1–10 μm $s^{-1}$, produced by a variety of hard-to-control effects, e.g., liquid level imbalances in reservoirs, surfactant concentration differences in reservoirs, air currents, vibrations, motion, etc. Another factor to consider is the cost of deeper channels, which increases for etched channels since deep etches are time-consuming. On the other hand, channels having a depth of the order of a millimeter could be produced by conventional machining, injection molding, and stamping, possibly reducing cost.

A better tradeoff may be to employ a planar iDEP architecture for processing these larger sample volumes. The characteristic velocities of iDEP systems are in the 1 mm $s^{-1}$ to 10 mm $s^{-1}$ range. A 10 mm wide by 50-μm deep channel flowing at 10 mm $s^{-1}$ can process a 1 ml sample in about three and a half minutes. The downside of the iDEP approach is that it is often limited to probing the particles at or near DC because of difficulties applying high voltage at high-frequencies.

Consider another sample processing problem: detecting and processing microbes in drinking water. Compared to blood or aerosol concentrator effluent, the particle load is quite small. In fact, one would typically be concerned to find a single live microbe in as much as one liter of water. The current approach to water testing is to culture everything caught on a mechanical filter such as a porous film. The culture step typically takes two days, providing the bad news of contamination after there could be widespread exposure. Is it possible to improve on this detection time using DEP?

A planar iDEP channel that is 100-mm wide by 1-mm deep channel running at 10 mm $s^{-1}$ passes a liter of liquid in about 17 min. Thus it is possible, at least in principle, to pass enough liquid through such a device, but only at relatively extreme scales in each dimension and speed. It would be a safer bet to adopt an architecture whose flow cross-sectional area could be scaled up more favorably. One such architecture [63] is a wide flow channel packed with insulators. The nonuniform geometry of the

Figure 7b shows schematically a continuous flow system. If the particles do not cross, but flow along the active region out of the lower port, the concentration factor of this system is the ratio of flow volumes through the inlet and concentrate outlet. If however, the active region terminates in a dielectrophoretic trap near the port, the concentration factor can be much higher. The concentration factor of such a hybrid continuous-batch system is the ratio of the processed volume to the volume of the terminal trap. Having the trap in the immediate vicinity of a small-geometry port significantly reduces the loss in concentration factor upon elution. Such a hybrid system has demonstrated saturation-limited concentration of 6,000 × in 16 s, as will be discussed later.

While most microfluidic systems are planar, three-dimensional structures like the continuous device in Fig. 7c could be used to achieve ultra-high concentration factors. Operated as a hybrid continuous-batch concentrator, concentration factors far in excess of $10^6$ are possible even for relatively fragile biological particles.

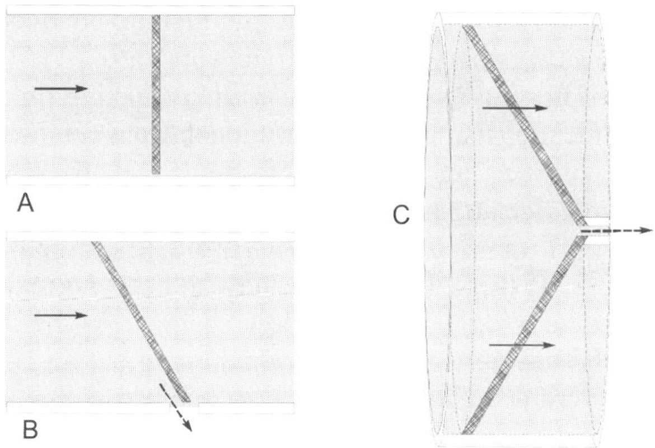


**Fig. 7.** Concentrator architectures. The hatched region dielectrophoretically inhibits particle passage. (**a**) One-dimensional batch concentrator. (**b**) One-dimensional continuous concentrator or two-dimensional continuous-batch hybrid. (**c**) Two-dimensional continuous concentrator or three-dimensional continuous-batch hybrid

Concentrators can be classified by the dimensionality of the reduction in uncertainty in particle position and time. For example, the device in Fig. 7a reduces this uncertainty in one dimension. The device in Fig. 7b operated continuously also is a one-dimensional concentrator, but operated in a hybrid continuous-batch mode is a two-dimensional concentrator. The device in Fig. 7c operated continuously is a two-dimensional concentrator and in a hybrid continuous-batch mode is a three-dimensional concentrator. If staged well, sequential concentrators can add their dimensionalities. As a guideline, one can respectively expect and hope to obtain of order 100 × and 1,000 × per dimension in concentrators made of impermeable, fouling resistant material, less with nonideal materials.

### 3.5.4. Heating

Other important considerations, especially at high throughput, are power requirements and effects of Joule heating of the sample. Most biological samples cannot be heated by more than a few tens of degrees without affecting their properties or behavior. Even if the sample is insensitive to temperature extremes, the sample must not be heated to vaporization. In shallow, planar geometries, heat transfer from the fluid to the channel boundaries can suppress the temperature rise from Joule heating. Glass and silica substrates are typically good thermal conductors, but unfilled plastic substrates are generally not. However, heat can be effectively transferred from channels made in a sufficiently thin low-thermal-conductivity plastic film laminated to a high-thermal-conductivity substrate.

The problem of heating is most severe with conductive fluids and high-throughput architectures that do not benefit much from conduction through the substrate. Consider an element of fluid that passes through a dielectrophoretic device at 1 mm $s^{-1}$ in which the root mean square applied electric field is 20 V $mm^{-1}$, typical for iDEP of ~2 μm cells. The worst case heating occurs if there is negligible heat transfer normal to the flow direction, e.g., in the middle of a wide flow channel. As the fluid element passes through the device it receives thermal energy at an approximately constant rate according to its electrical conductivity. Table 1 shows the distances the fluid element can propagate in such a field until its temperature rises by 10 K and until it boils. For low-conductivity ground water, there is no appreciable heating even if the high field is applied over many centimeters. High-conductivity ground water (equivalent to ~0.1 mM NaCl) can start to have an effect on the sample temperature when the interaction length is only a few millimeters. More highly conductive fluids like blood or seawater pose serious challenges to the design of both high- and low-throughput dielectrophoretic devices.

The heating problem is particularly severe for systems in which the fluid conducts the electricity from "remote" electrodes. Electrodes that have to be spaced by only 73 μm can hardly be considered remote for 10 μm red blood cells. One approach to increase electrode spacing without excessive heating is to constrict the cross-sectional area substantially in the region of the device that actively drives DEP. This localizes the high-field region and minimizes dissipation, also improving the electrical efficiency of the device.

**Table 1.** Adiabatic Joule heating effects of common solutions in a DEP device at a velocity of 1 mm $s^{-1}$ and electric field of 20 V $mm^{-1}$

| Immersion fluid | Conductivity (S $m^{-1}$) | Interaction length at 10 K rise (mm) | Interaction length at boiling (mm) |
|---|---|---|---|
| Ground water (low) | $0.5 \times 10^{-3}$ | 210 | $1.3 \times 10^{4}$ |
| Ground water (high) | $30 \times 10^{-3}$ | 3.5 | 210 |
| 1 mM NaCl | 0.37 | 0.28 | 17 |
| Blood plasma | 1.4 | 0.073 | 4.5 |
| Sea water | 5.0 | 0.021 | 1.3 |

The interaction length is the distance over which the field is applied.

### 3.6. Examples of Architectures

A wide body of literature exists on electrode-based devices and architectures; however, iDEP devices are a comparatively new and unexplored category. These examples will focus on unconventional and iDEP devices.

#### 3.6.1. Post-Based Devices

A simple class of iDEP devices employ posts that span the depth of a channel or similar channel constrictions to produce electric field concentrations [57, 59, 60]. An advantage of this approach is that it does not necessarily require any fabrication steps apart from the processing that makes the channels themselves. It is also amenable to a variety of replication approaches and has the feature that the posts can double as structural supports that prevent sagging of the channel lid and improve the strength of the lid bond. Post-based iDEP devices for concentrating and sorting microspheres [60], DNA [59], viruses [65], and bacteria [61, 66] generally operate via a "trap and release" or batch-type operation as shown in Fig. 3.

However, it is also possible to construct a post-based device for continuous concentration utilizing enhancement or depletion as in Figs. 4-6. Figure 8 [67] shows one such design that uses an array of posts tilted with respect to the flow direction. Particles that experience depletion or enhancement are diverted from the primary flow to follow the tilted columns of posts toward the bottom side of the channel until the particles spill into the small concentrate port at the bottom right of the design. The concentration factor of this device is 160 ×, specified at design time as the ratio of the inlet and outlet widths.

Because posts produce regions of relatively high and low electric field, particles always experience attraction and repulsion as they transit the posts. As shown earlier, the shape of the posts determines in part whether such competition results in a net attraction or repulsion. Unfortunately this competition severely limits the angular range that particles can be diverted from the primary flow. As a result, such devices must be long relative to their flow cross section. Since dielectrophoresis is sensitive to the absolute electric field, the required voltage and power dissipation of such a device scales with the device length, limiting the conductivity of the solutions that can be processed. Throughput is also limited. By judicious design of the array arrangement and post shape, this problem can be mitigated, but only by a limited extent.

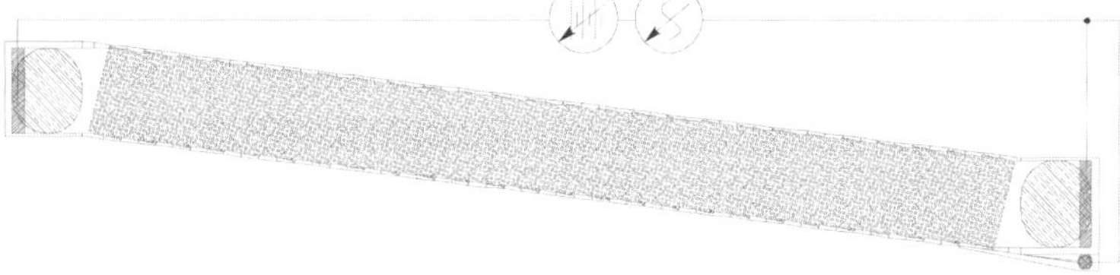

**Fig. 8.** A post-based iDEP device for continuous particle concentration. Sample from the left port is electrokinetically driven through the post-packed column. Because of DEP, particles follow columns of posts toward the bottom channel side where they exit to a concentrate port at the bottom left

### 3.6.2. Facet-Based Devices

With the luxury of two or more channel depths in a microsystem, a powerful methodology exists for designing devices from faceted channels having uniform electric fields within each facet. This "faceted" design methodology [68] is based on the fact that electric field gradients disappear everywhere except immediately at an interface between channels of different depths provided the interfaces are straight and the channel walls are angled to satisfy the condition $\tan \theta_1 / d_1 = \tan \theta_2 / d_2$, where $d$ is the channel depth, $\theta$ is the angle with respect to normal to the interface, and subscripts 1 and 2 represent the channel regions linked by the interface. This faceted design methodology allows the rational design of low-dispersion turns, splitters, expanders, and manifolds for electrokinetic flow, but has numerous applications in DEP devices [69].

Consider the facet diagrammed in Fig. 9. Because the electric field is uniform in the $y$ direction, we can express the electric field distribution as $\mathbf{E} = E_1 \cos \theta_1 R(x,z)\, \hat{\mathbf{e}}_x + E_1 \sin \theta_1\, \hat{\mathbf{e}}_y + E_1 \cos \theta_1 S(x,z)\, \hat{\mathbf{e}}_z$, where $E_1$ is the magnitude of the electric field in region *1*, $R(x,z)$ and $S(x,z)$ are dimensionless functions that describe the field variation at the interface between regions 1 and 2, and $\hat{\mathbf{e}}_x$, $\hat{\mathbf{e}}_y$, and $\hat{\mathbf{e}}_z$ are unit vectors in the $x$, $y$, and $z$ directions respectively. In order for particles to be inhibited from passing the interface, the $x$-directed component of the electrokinetic velocity, $\mu_{\mathrm{EK}} E_1 \cos\theta_1$, must be opposed by the dielectrophoretic velocity, yielding the inhibition condition:

$$(\mu_{\mathrm{DEP}}/\mu_{\mathrm{EK}})\, g\, E_1 \cos\theta_1 \leq -1, \quad \text{where} \tag{22}$$

$$g = 2(\partial R/\partial x + S/R\ \partial S/\partial x). \tag{23}$$

The details of the interface gradient function $g$ depend on the interface profile and depth ratio. If $\mu_{\mathrm{DEP}}$ is negative (negative DEP), Particles can only be inhibited where $g$ is positive, i.e., transitions from deep to shallow. Conversely, if $\mu_{\mathrm{DEP}}$ is positive (positive DEP) can only be inhibited where $g$ is negative, i.e., transitions from shallow to deep.

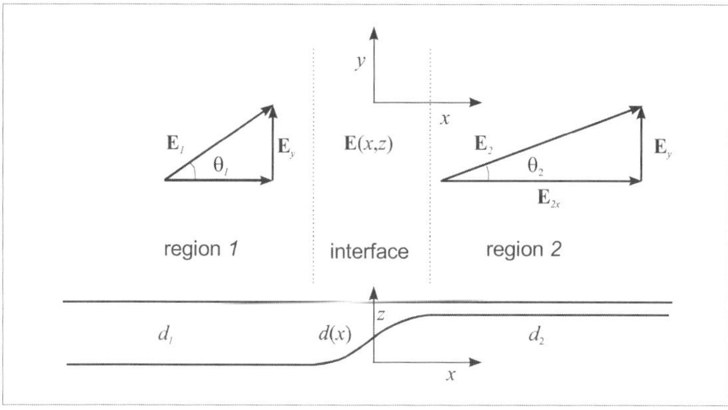


**Fig. 9.** Fields in a faceted microchannel. In region 1 of uniform depth $d_1$, the electric field is uniformly $E_1$. In region 2 of depth $d_2$, the electric field is uniformly $E_2$. The electric field gradient at the interface is useful for driving dielectrophoresis

Figure 10 shows a numerical simulation of a faceted design. In Fig. 10a, the gray scale depicts the electric field magnitude. Because the field is uniform in each facet, the color is piece-wise uniform. Figure 10b shows a simulation of a suspension of point particles having a distribution of dielectrophoretic mobilities indicated by grayscale. Negative DEP causes light-colored particles to flow along the leading edge of the facet to a trap. Positive DEP causes dark-colored particles to flow along the trailing edge to a trap. Dielectrophoretically neutral gray particles are unaffected. Such a device is a hybrid continuous and batch concentrators. Particles in each trap can be eluted into a very small volume by momentarily lowering the field, producing high concentration factors.

Unlike post-based continuous dielectrophoretic devices, particles experience a uniform attraction or repulsion as they flow tangentially along the interface. Particles can therefore be diverted by much larger angles with respect to the primary flow, providing for reduced device length, lower operating fields, and lower power dissipation.

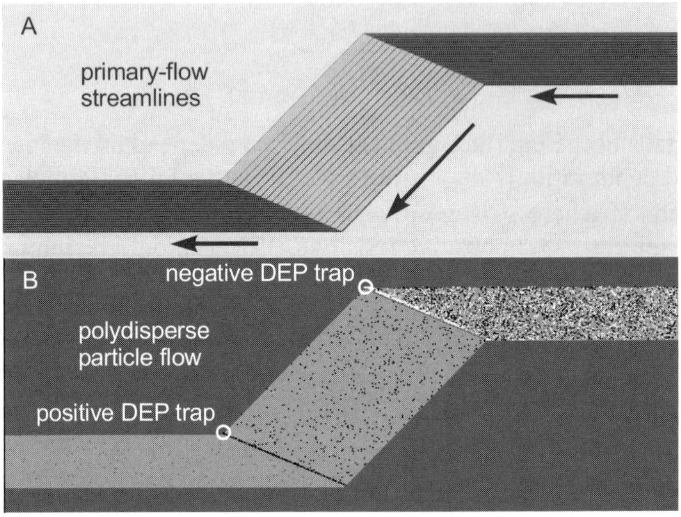


**Fig. 10.** Numerical simulation of a simple faceted DEP device. (**a**) Electrokinetic flow streamlines and electric field lines. (**b**) Simulated motion of a polydisperse suspension. White, gray, and black particles have negative, neutral, and positive DEP mobilities, respectively

A piece-wise uniform arrangement of electric fields is ideal for high-selectivity dielectrophoretic devices because particles throughout the width of the channel experience identical electric fields, so there is no dependence on the dielectrophoretic force experienced by a particle with position along a channel. A variation in dielectrophoretic force with particle depth remains however and can be severe, as shown in Fig. 11a. The sharp tip at the interface produces sharply peaked field gradients that reduces resolution because a particle initially along the bottom streamlines experiences a far stronger dielectrophoretic effect than one initially along the top streamline. The interface in Figure 11b produces a field gradient that has far less variation with depth. Unfortunately, the profile of interfaces usually depends exclusively on the fabrication technique or etching procedure, so a designer has limited options for tailoring a design for peak performance. Prototyping is also complicated because of the importance of the fabrication process on performance.

spaces between the insulators produces a distribution of nonuniform electric fields. The primary engineering controls of such a system are the length of the packing, the packing size distribution, and the packing electrical properties. Another approach [64] uses an insulating cloth, mesh, random or track-etched membrane to concentrate the electric field. The fluid is pushed through the porous insulator, which acts like any other element for producing $\nabla I/I$: while the electric field is turned on, particles that are less conductive than the immersion fluid are repelled from high-electric-field regions and if the repulsion is sufficient to overcome the fluid drag forces, the particles collect upstream of pores. Conversely, particles that are more conductive than the immersion fluid are attracted to high-electric-field regions. If this attraction is sufficiently high, these particles can overcome the drag force on them to collect downstream of the pores.

### 3.5.3. Concentration Factor

The concentration factor of a device can be considered the ratio of uncertainties in initial and final particle location in space and time. Consider the schematic device in Fig. 7a. Suppose a dilute solution of particles is flowed for a time past a dielectrophoretically active region shown by the hatching. If the active region traps the particles, the concentration factor ramps up with time until the traps are saturated. If only the particles' presence in the traps is needed, then the relevant concentration factor is the volume of liquid flowed through the system divided by the volume of the traps. By this metric, bacteria concentration factors of 3,000 [65] and more have been demonstrated over the course of several minutes. If the input sample is sufficiently dilute and particles can withstand fields within the traps indefinitely without damage, there is no upper bound on the concentration factor. However, complex biological particles like cells and bacteria do not survive high fields indefinitely. At the fields required for trapping, ~20–50 V mm$^{-1}$, most bacteria exhibit changes in their membrane conductivity after a period of only a few minutes. Thus the maximum concentration factor is limited by the volume that can be passed through the system before damaging particles. This limit is not generally improved by flowing the fluid faster, since trapping requires correspondingly high fields, which generally increases the damage rate and reduces the allowable run time more than proportionally.

Most often, the concentrated particles must be eluted to a port, channel, or other element. In this case, the relevant concentration factor is the ratio of the processed volume to the eluted volume. Because of hydrodynamic dispersion, the elution volume is generally many times the total volume of the traps.

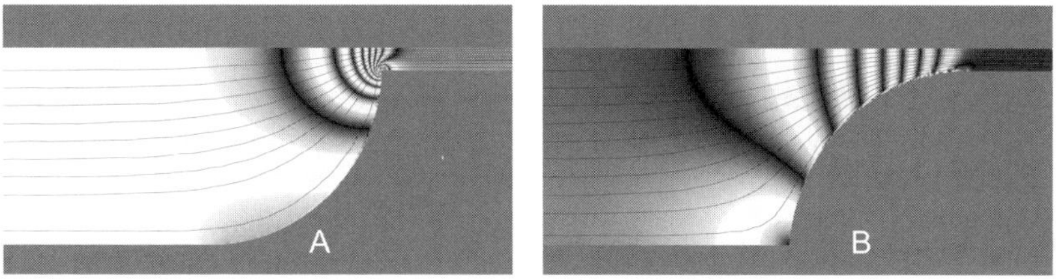


**Fig. 11.** Electric field lines and contours of electric field magnitude (fringes) near channel depth changes. (**a**) Isotropically etched steps produce strong, localized field peaks, reducing resolution and possibly damaging particles. (**b**) The inverse of this shape, obtained e.g., by molding or stamping from an isotropically etched master, concentrates the field more uniformly with depth

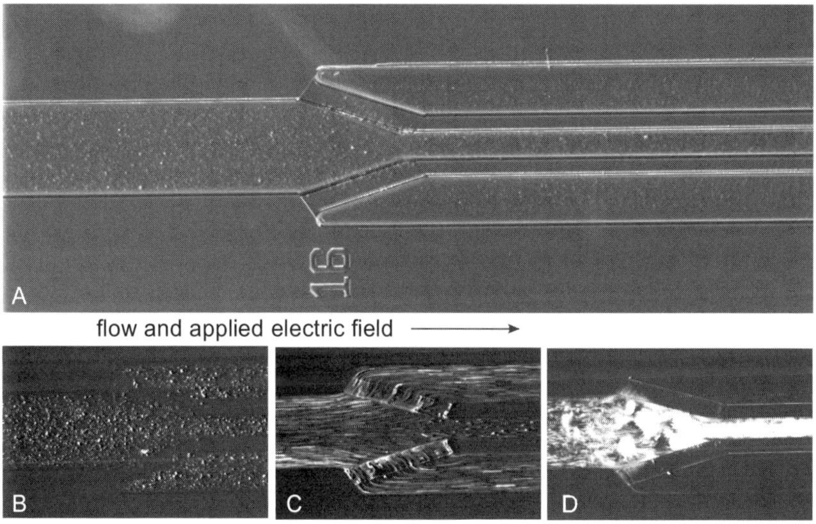


**Fig. 12.** Demonstration of a faceted dielectrophoretic device on 1-$\mu$m polystyrene particles. (**a**) Two-level microsystem image (**b**) electric field off (**c**) electric field low and (**d**) electric field high. Source: L.M. Barrett and A. J. Skulan, Sandia National Laboratories

### 3.6.3. Corduroy Devices

The corduroy design methodology employs slender ridges or valleys on a channel surface as shown in Fig. 13 to produce nonuniform electric fields. The name comes from the appearance of an array of such features. This methodology arose as a rational approach to the design of dielectrophoretic spectrometers, which are difficult to design under the rigorous faceted methodology. A geometrically similar approach using patterned electrodes has been developed by Dürr et al. [55].

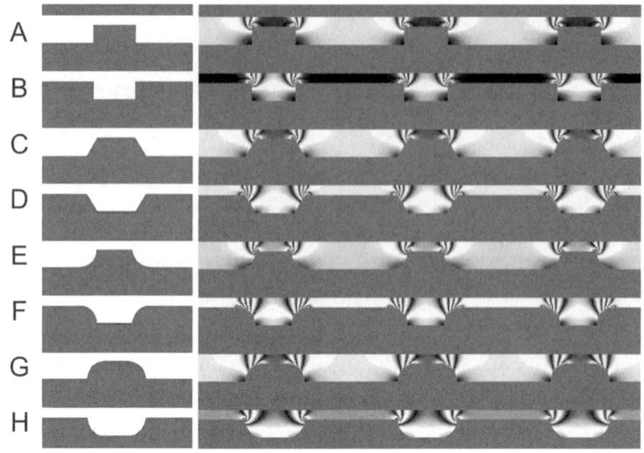


**Fig. 13.** Electric field concentrations from insulating ridges and valleys. Fringes in the channel are contours of constant electric field. These shapes can be obtained from standard etches or stamping with an etched master. A and B: Bosch etch ridge and valley; C and D: anisotropic etch ridge and valley; E: isotropically etched ridge; F: inverted isotropically etched valley; G: inverted isotropically etched ridge; H: isotropically etched valley

Like the faceted devices, the inhibition threshold of such a slender feature is approximately proportional to the cosine of the incidence angle of the electric field on the feature. Unlike the faceted approach, the electric field is not precisely uniform in each region, but this nonuniformity is small provided the ridges are narrow and the spatially averaged depth of the channel does not vary significantly across width of the channel. Cross-channel uniformity can be maintained by using only a sparse collection of slender ridges as in Fig. 14a or by maintaining a uniform density of ridges as in Fig. 14b and c.

As with faceted devices, there is an inhibition threshold above which particles are not able to cross the ridges. However, even below this threshold, particles are deflected slightly at each ridge, as shown in Fig. 14b. The angle of the deflection increases with the magnitude of the dielectrophoretic mobility until, at and above the inhibition threshold, particles simply follow the ridges, as shown in Fig. 14c. The angular dispersion of the corduroy array operated below the inhibition threshold is useful for continuous particle sorting.

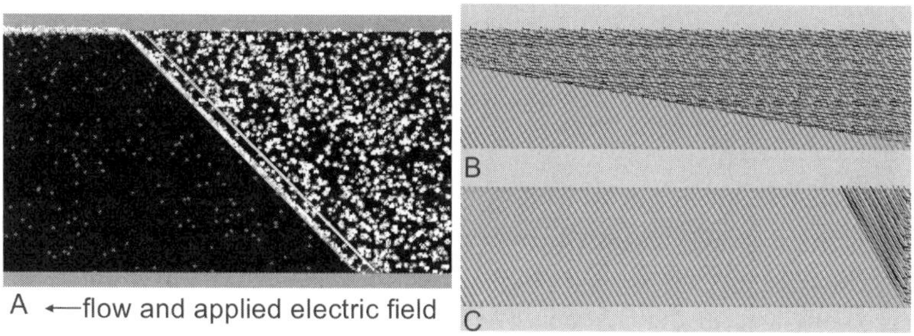


**Fig. 14.** Simulations of corduroy devices. (**a**) A simple concentrator operating on a polydisperse suspension. Color depicts DEP mobility: gray: zero; white highest positive. (**b**) Multiple-ridge concentrator operating on monodisperse particles below and (**c**) above the inhibition threshold

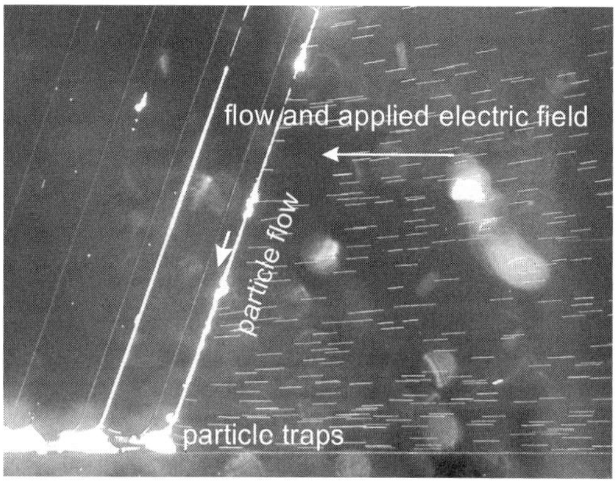


**Fig. 15.** Operation of a corduroy hybrid continuous-batch concentrator. The concentrator achieved 6,000 × concentration of 2-μm polystyrene particles in 16 s. The ridges in the 20-μm-deep by 3-mm-wide glass channel are 13-μm tall and 100-μm wide. The electrokinetic particle flow is from right to left at ~1 mm $s^{-1}$, produced by a field of 80 V $mm^{-1}$. Source: Sandia Microfluidics Group

Because of averaging, the net transport from many separate interactions provides better resolution than that of a single near-threshold interaction. Figure 15 shows such the operation of a hybrid continuous-batch corduroy concentrator. Unlike the rigorous faceted approach, a strength of the corduroy approach is the ability to employ curved ridges, which also support spectrometer functions as shown in Fig. 16.

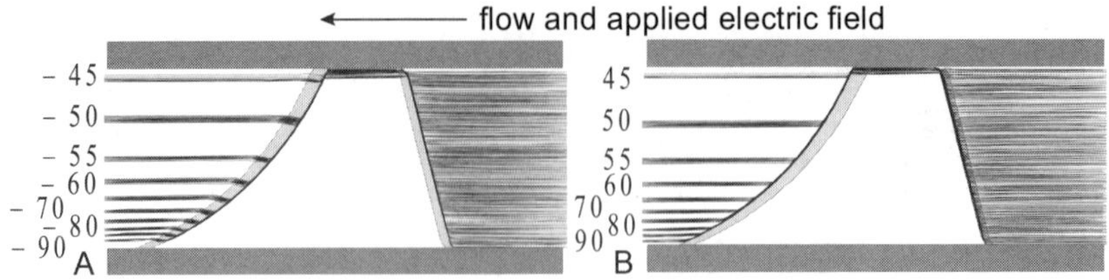


**Fig. 16.** Simulations of a simple corduroy spectrometer. The first ridge collects particles to one side of the channel. The curved ridge disperses particles because of the dependence of the inhibition threshold on ridge angle. Particles having negative (**a**) and positive (**b**) mobilities as indicated (arbitrary units) are tracked

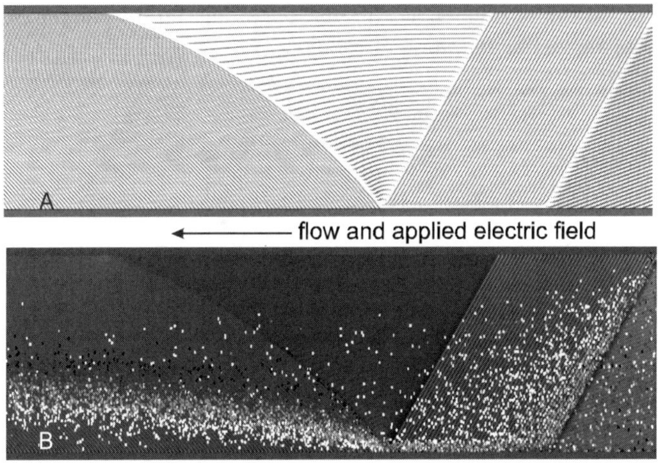


**Fig. 17.** A corduroy particle spectrometer operated below the inhibition threshold. (**a**) Design: ridges appear gray in the white channel. (**b**) Simulation results. Poor grid resolution of ridges in the upper left causes the black (high-DEP-mobility) particles to spill prematurely

The corduroy design methodology can take advantage of the ability to microfabricate and replicate elaborate arrays of features at the same cost as simple devices. Figure 17 shows an example of a particle spectrometer that operates below the inhibition threshold. Direct simulation of such a device (Fig. 17b) can pose greater challenges than fabrication.

## 4. Conclusion

Dielectrophoresis has been studied since Pohl's early pioneering work, [1, 70] but remains fresh and promising. New applications and architectures for DEP continue to produce vigorous interest and development.

Theoretical work into behavior like field-induced phase transitions at high particle concentration is producing insights that have direct bearing on practical applications like sorting dense collections of blood cells and seeking rare particles from a dense and unknown background.

The application of DEP to real-world sample processing and analysis problems also requires thoughtful device architecting and detail design. Many design pitfalls exist: it is humbling to realize how little design slack one has before a lack of rigor affects performance. Common mistakes are

1. Failing to produce a sufficiently "uniformly nonuniform" electric field throughout the channel
2. Failing to design for low hydrodynamic dispersion
3. Failing to minimize power dissipation and heating
4. Ignoring field concentrations outside the device, e.g., holes, corners
5. Ignoring as-etched geometries

These mistakes often arise from modular thinking derived from our macroscopic intuition. Fortunately, rigorous and semi-rigorous design methodologies and design-oriented simulation tools are being developed along with a generation of engineers who are learning to intuit the world from the perspective of an electrocuted bacterium.

## References

1. Pohl HA (1951) The motion and precipitation of suspenoids in divergent electric fields. J Appl Phys 22: 869–871
2. Pohl HA (1978) Dielectrophoresis: The behavior of neutral matter in nonuniform electric fields. Cambridge University Press, Cambridge
3. Pethig R (1979) Dielectric and electronic properties of biological materials. John Wiley & Sons, New York
4. Mizuno A, Washizu M (1995) Biomedical engineering. In: Chang J-S, Kelly AJ, Crowley JM (eds) Handbook of electrostatic processes. Marcel Dekker, New York, pp 653–686
5. Jones TB (1995) Electromechanics of particles. Cambridge University Press, Cambridge
6. Fuhr G, Zimmermann U, Shirley SG (1996) Cell motion in time-varying fields: Principles and potential. In: Zimmerman U, Neil G (eds) Electromanipulations of cells. CRC Press, Boca Raton, pp 259–327
7. Pethig R (1996) Dielectrophoresis: Using inhomogeneous AC electrical fields to separate and manipulate cells. Crit Rev Biotechnol 16: 331–348
8. Pethig R, Markx GH (1997) Applications of dielectrophoresis in biotechnology. Trends Biotechnol 15: 426–432

9. Koch M, Evans A, Brunnschweiler A (2000) Microfluidic technology and applications. Research Studies Press, Baldock
10. Gascoyne PRC, Vykoukal J (2002) Particle separation by dielectrophoresis (review). Electrophoresis 23: 1973–1983
11. Hughes MP (2002) Strategies for dielectrophoretic separation in laboratory-on-a-chip systems (review). Electrophoresis 23: 2569–2582
12. Hughes MP (2003) Nanoelectromechanics in engineering and biology. CRC Press, Boca Raton
13. Morgan H, Green N (2003) AC Electrokinetics: colloids and nanoparticles, Research Studies Press, Baldock
14. Burke PJ (2004) Nano-dielectrophoresis: Electronic nanotweezers. In: Nalwa HS (ed) Encyclopedia of nanoscience and nanotechnology, vd. 6, American Scientific Publishers, Stevenson Ranch, pp 623–641
15. Gonzalez CF, Remcho VT (2005) Harnessing dielectric forces for separations of cells, fine particles and macromolecules (review). J Chromatogr A 1079: 59–68
16. Priestley J (1769) The history and present state of electricity with original experiments, 2nd edn. J Dodsley, London
17. Havelka KO, Filisko FE (eds) (1995) Progress in electrorheology. Science and technology of electrorheological materials. Plenum Press, New York
18. Sproston JL, Stanway R (1995) Electrorheology. In: Chang J-S, Kelly AJ, Crowley JM (eds) Handbook of electrostatic processes. Marcel Dekker, New York, 399–412
19. Landau LD, Lifshitz LP, Pitaevski LP (1984) Electrodynamics of continuous media. Pergamon, Oxford
20. Bottcher CJF (1973) Theory of polarization, vd. I and II. Elsevier, Amsterdam
21. Dukhin SS, Shilov VN (1974) Dielectric phenomena and the double layer in disperse systems and polyelectrolytes. Wiley, New York
22. Russel WB, Saville DA, Schowalter WR (1989) Colloidal dispersions. Cambridge University Press, Cambridge
23. Wang X-B, Huang Y, Becker FF, Gascoyne PRC (1994) A unified theory of dielectrophoresis and travelling wave dielectrophoresis. J Appl Phys D 27: 1571–1574
24. Wang X-B, Huang Y, Becker FF, Gascoyne PRC (1995) Non-uniform spatial distributions of both the magnitude and phase of AC electric fields determine dielectrophoretic forces. Biochim Biophys Acta 1243: 185–194
25. Hill NE, Vaughan WE, Price AH, Davies M (1969) Dielectric properties and molecular behavior. Van Nostrand, London
26. Lyklema J (1995) Fundamentals of interface and colloid science: vol. II, Solid–liquid interfaces. Academic Press, London
27. Green NG, Morgan H (1999) Dielectrophoresis of submicrometer latex spheres. 1. Experimental results. J Phys Chem B 103: 41–50
28. Saville DA, Bellini T, Degiorgio V, Mantegazza F (2000) An extended Maxwell–Wagner theory for the electric birefringence of charged colloids. J Chem Phys 113: 6974–6983

29. Delgado AV, Delgado DV (2001) Interfacial electrokinetics and electrophoresis. Marcel Dekker, New York
30. Ermolina I, Morgan H (2005) The electrokinetic properties of latex particles: Comparison of electrophoresis and dielectrophoresis. J Colloid Interface Sci 285: 419–428
31. Zhou H, Preston MA, Tilton RD, White LR (2005) Calculation of the electric polarizability of a charged spherical dielectric particle by the theory of colloidal electrokinetics. J Colloid Interface Science 285: 845–856
32. Dussaud A, Khusid B, Acrivos A (2000) Particle segregation in suspensions subject to high-gradient ac electric fields. J Appl Phys 88: 5463–5473
33. Qiu Z, Markarian N, Khusid B, Acrivos A (2002) Positive dielectrophoresis and heterogeneous aggregation in high-gradient ac electric fields. J Appl Phys 92: 2829–2843
34. Markarian N, Yeksel M, Khusid B, Farmer K, Acrivos A (2003) Particle motions and segregation in dielectrophoretic micro-fluidics. J Appl Phys 94: 4160–4169
35. Khusid B, Acrivos A (1995) Effects of conductivity in electric-field-induced aggregation in electrorheological fluids. Phys Rev E 52: 1669–1693
36. Khusid B, Acrivos A (1999) Phase diagrams of electric-field-induced aggregation in conducting colloidal suspensions. Phys Rev E 60: 3015–3035
37. Kumar A, Acrivos A, Khusid B, Jacqmin D (2007) Electric field driven formation of particle concentration fronts in suspensions. Fluid Dynam Res 39: 169–192
38. Khusid B, Acrivos A (1996) Effects of interparticle electric interactions on dielectrophoresis in colloidal suspensions. Phys Rev E 54: 5428–5435
39. Shapiro M, Shalom AL, Lin IJ (1985) Dielectric flocculation of nondiffusive particles in a nonconducting fluid and a uniform electric field. J Appl Phys 58: 1028–1031
40. Hass KC (1993) Computer simulations of nonequilibrium structure formation in electrorheological fluids. Phys Rev E 47: 3362–3373
41. Castellanos A, Ramos A, Gonzalez A, Green NG, Morgan H (2003) Electrohydrodynamics and dielectrophoresis in microsystems: Scaling laws. J Phys D: Appl Phys 36: 2584–2597
42. Zheng, L, Li S, Burke PJ, Brody JP (2003), Towards single molecule manipulation with dielectrophoresis using nanoelectrodes. In: 2003 Third IEEE Conference on Nanotechnology vol-2, IEEE-NANO 2003, 437–440
43. Meinhart C, Wang D, Turner K (2003) Measurement of ac electrokinetic flows. Biomedical Microdevices 5: 139–145
44. Wang D, Sigurdson M, Meinhart CD (2005) Experimental analysis of particle and fluid motion in ac electrokinetics. Exp Fluids 38: 1–10
45. Markarian N, Yeksel M, Khusid B, Kumar A, Tin P (2004) Effects of clinorotation and positive dielectrophoresis on suspensions of heavy particles. Phys Fluids 16: 1826-1829
46. Happel J, Brenner H (1965) Low Reynolds number hydrodynamics. Prentice-Hall, Englewood Cliffs

47. Evoy S, DiLello N, Deshpande V, Narayanan A, Liu H, Riegelman M, Martin BR, Hailer B, Bradley J-C, Weiss W, Mayer TS, Gogotsi Y, Bau HH, Mallouk TE, Raman S (2004) Dielectrophoretic assembly and integration of nanowire devices with functional CMOS operating circuitry. Microelectronic Engineering 75: 31–42
48. Liu H, Bau HH (2004) The dielectrophoresis of cylindrical and spherical particles submerged in shells and in semi-infinite media. Phys Fluids 16: 1217–1228
49. Kim J-E, Han C-S (2005) Use of dielectrophoresis in the fabrication of an atomic force microscope tip with a carbon nanotube: A numerical analysis. Nanotechnology 16: 2245–2250
50. Riegelman M, Liu H, Bau HH (2006) Controlled nanoassembly and construction of nanofluidic devices. J Fluids Eng 128: 6–13
51. Markarian N, Yeksel M, Khusid B, Farmer K, Acrivos A (2003) Limitations on the scale of an electrode array for trapping particles in microfluidics by positive dielectrophoresis. Appl Phys Lett 82: 4839–4841
52. Bennett DJ, Khusid B, James CD, Galambos PC, Okandan M, Jacqmin D, Acrivos A (2003) Combined field-induced dielectrophoresis and phase separation for manipulating particles in microfluidics. Appl Phys Lett 83: 4866–4868
53. James CD, Okandan M, Galambos P, Mani SS, Bennett D, Khusid B, Acrivos A (2006) Surface micromachined dielectrophoretic gates for the front-end device of a biodetection system. J Fluids Eng 128: 14–19
54. Kumar A, Qiu Z, Acrivos A, Khusid B, Jacqmin D (2004) Combined negative dielectrophoresis and phase separation in nondilute suspensions subject to a high-gradient ac electric field. Phys Rev E 69: 021402-1-10
55. Dürr M, Kentsch J, Müller T, Schnelle T, Stelzle M (2003) Microdevices for manipulation and accumulation of micro- and nanoparticles by dielectrophoresis. Electrophoresis 24: 722–731
56. Masuda S, Washizu M, Nariba T (1989) Novel method of cell fusion in field constriction are in fluid integrated circuit. IEE Trans. Ind Appl 25: 732–737
57. Washizu M (1990) Electrostatic manipulation of biological objects, J. Electrostatics 25: 109–123
58. Lee SW, Yang SD, Kim YK (1994) Micromachined cell handling devices. Eng Med Biol Soc, Proc 16th Annual International Conf. IEEE 2: 1019–1020
59. Chou C, Tegenfeldt J, Bakajin O, Chan S, Cox E, Darnton N, Duke T, Austin R (2002) Electrodeless dielectrophoresis of single- and double-stranded DNA. Biophys J 83: 2170–2179
60. Cummings EB, Singh AK (2003) Dielectrophoresis in microchips containing arrays of insulating posts: Theoretical and experimental results. Anal Chem 77: 4323–4328
61. Lapizco-Encinas BH, Simmons BA, Cummings EB, Fintschenko Y (2004) Insulator-based dielectrophoresis for the selective concentration and separation of live bacteria in water. Electrophoresis 25: 1695–1704

62. Simmons BA, McGraw GJ, Davalos RV, Fiechtner GJ, Fintschenko Y, Cummings EB (2006) The development of polymeric devices as dielectrophoretic separators and concentrators. MRS Bulletin: 31
63. Suehiro J, Noutomi D, Shutou M, Hara M (2003) Selective detection of specific bacteria using dielectrophoretic impedance measurement method combined with an antigen-antibody reaction. J Electrostatics 58: 229–246
64. McGraw GJ, Simmons BA (2005) personal conversation
65. Lapizco-Encinas BH, Davalos RV, Simmons BA, Cummings EB, Fintschenko Y (2005) An insulator-based (electrodeless) dielectrophoretic concentrator for microbes in water. J Microbio Methods 62: 317–326
66. Lapizco-Encinas BH, Simmons BA, Cummings EB, Fintschenko Y (2004) Dielectrophoretic concentration and separation of live and dead bacteria in an array of insulators. Anal Chem 76: 1571–1579
67. Cummings EB (2003) Streaming dielectrophoresis for continuous-flow microfluidic devices. IEEE EMB-M: Nov–Dec, 75–84
68. Fiechtner GJ, Cummings EB (2003) Faceted design of channels for low-dispersion electrokinetic flows in microfluidic systems. Anal Chem 75: 4747–4755
69. Barrett LM, Skulan AJ, Singh AK, Cummings EB, Fiechtner GJ (2005) Dielectrophoretic manipulation of particles and cells using insulating ridges in faceted prism microchannels. Anal Chem 77: 6798–6804
70. Pohl H, Hawk L (1966) Separation of living and dead cells by dielectrophoresis. Science 152: 647–649

# Chapter 9
# Ultrasonic Particle Manipulation

Martyn Hill[1] and Nicholas R. Harris[2]

[1]School of Engineering Sciences
The University of Southampton, Southampton, SO17 1BJ, UK.
[2]Electronics and Computer Science
The University of Southampton, Southampton, SO17 1BJ, UK.

## 1. Introduction

Particles and other acoustic discontinuities (such as second phase fluids) within an ultrasonic field experience small forces [1, 2]. These forces are generally larger in an ultrasonic standing wave (USW) than in a progressive wave. Consider the system represented in Fig. 1, in which particles within a fluid are bounded by an ultrasonic transducer and a reflective boundary. If the transducer is driven so as to excite a resonance frequency of the cavity, a standing wave will be set up with associated pressure maxima and minima. Particles will generally be transported towards pres-

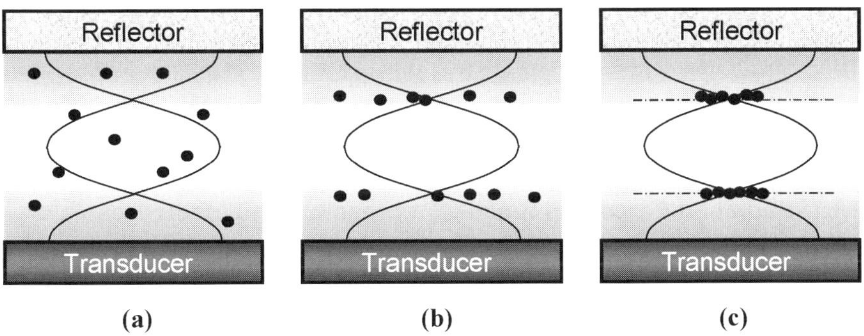


**Fig. 1.** Movement of particles within an ultrasonic standing wave (USW). (**a**) A standing wave is excited by driving a transducer at a thickness resonance of a fluid layer, (**b**) particles within the field move to pressure nodal planes due to the axial primary radiation force, and (**c**) a short time later particles aggregate within those planes due to lateral primary radiation forces and secondary radiation forces

sure nodes within the field, by axial radiation forces. Once particles have moved to these nodal planes, weaker lateral radiation forces, along with secondary radiation forces (due to particle–particle interaction) will cause particles to agglomerate at certain points within the nodal planes.

Particle handling is a standard operation that microfluidic analysis systems are required to perform [3] and the length scale of action of MHz frequency USWs makes the technology an attractive option, particularly at stages in which fine handling of individual particles is not required. In particular, USW fields are attractive in biotechnology applications [4] and there is substantial evidence of their ability to manipulate and hold cells without reducing cell viability (see section *cell viability*).

## 2. Theory

### *2.1. Radiation Forces*

Acoustic radiation force, or acoustic radiation pressure, is a phenomenon that has been known since the late *19th* century, being described by Kundt and Lehman [5], and studied in some detail by Rayleigh [6]. An acoustic discontinuity within a sound field will generally experience a force due to the scattering of the wave by the discontinuity. Such forces occur in both progressive waves and standing waves and are due to a nonlinear interaction, which has generated a substantial amount of debate and contradictory results in the literature [see for example 7, 8]. Indeed, in his 1977 review of the field, Beyer [9] describes radiation pressure as a '…phenomenon that the observer thinks he understands – for short intervals, and only every now and then'. The acoustic radiation forces generated on small particles (relative to the wavelength) within a USW are significantly higher than forces produced within progressive waves, and it is systems based around USWs that are of the primary interest for particle manipulation in microfluidic systems.

#### 2.1.1. Radiation Forces on Small Compressible and Incompressible Spheres

The first comprehensive calculation of acoustic radiation forces on a small particle within a standing wave was presented by King [10]. These calculations were performed for entirely rigid spheres in an inviscid fluid. King's theory predicts that particles will move toward either the node or antinode of the field depending, in this simplified theory, on what King refers to as 'the relative density factor' a function of the ratio of the fluid and particle densities. When the ratio of the particle density to the fluid density is less than 0.4 (and the particle is incompressible) the acoustic force will act towards the pressure antinode. For density ratios above 0.4, which will be the case for real near-rigid particles, the acoustic radiation force will act towards the pressure node of the standing wave. King also balances gravitational forces, Stokes drag forces and radiation pressure to predict the paths of small particles falling through acoustic fields. King's predictions were subsequently shown to be in good agreement with experimental results for solid spheres suspended in a gas [11], but the predictions show significant errors for spheres that can not be assumed to be rigid in comparison to the surrounding fluid.

The error is particularly significant in the case of bubbles within a resonant field, and this prompted Yosioka and Kawasima [12] to extend King's theory to allow for compressible spheres. They derived an expression for the time-averaged acoustic radiation force on a spherical particle of radius *a,* at position $x$ within a one-dimensional standing wave of acoustic energy density $\varepsilon$ as

$$F(x) = 4\pi k \varepsilon a^3 \, \Phi(\beta, \rho) \sin(2kx) \; . \tag{1}$$

The term $\Phi(\beta, \rho)$, often referred to as the *acoustic contrast factor*, is given by

$$\Phi(\beta, \rho) = \frac{\rho_\mathrm{p} + \frac{2}{3}(\rho_\mathrm{p} - \rho_\mathrm{f})}{2\rho_\mathrm{p} + \rho_\mathrm{f}} - \frac{\beta_\mathrm{p}}{3\beta_\mathrm{f}} \tag{2}$$

where $\beta$ and $\rho$ are the compressibility and the mass density of the fluid and the particle, indicated by subscripts $f$ and $p$ respectively. The wave number, $k$ is equal to $2\pi/\lambda$ where $\lambda$ is the wavelength and the compressibility is related to the speed of sound, $c$, by $\beta = 1/\rho c^2$.

As can be seen from Fig. 2, the direction of the force on rigid spheres (i.e. a compressibility ratio of zero) corresponds with King's theory, but as the particle compressibility increases, the force will act towards the pressure antinode over a much wider range of density ratios. For most applications of interest, including cells in aqueous solution, the acoustic force will act towards the pressure node; however, in the case of bubbles that are smaller than resonant size [13, 14] and certain two-phase fluid mixtures [15, 16], the bubble, or second phase fluid, will experience an acoustic force acting towards the pressure antinode [17].

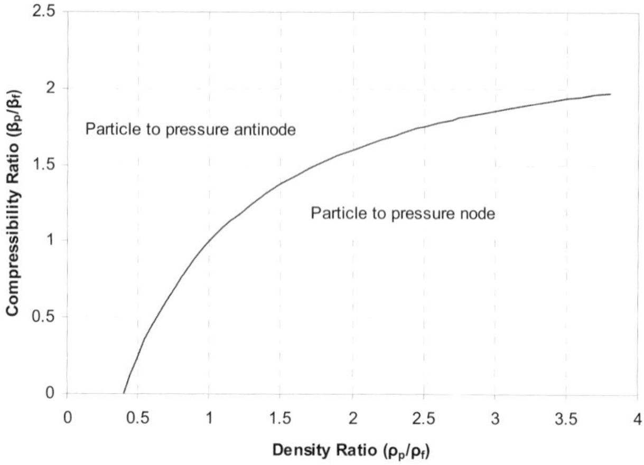


**Fig. 2.** Plot of $\Phi(\beta, \rho) = 0$ as a function of density and compressibility ratios

An alternative, but equivalent, formulation for acoustic radiation force, based on the energy stored within a one-dimensional plane standing wave, was derived by Gor`kov [18]. Gor`kov demonstrated that the force on a particle is dependent upon the spatial gradients of the time-averaged kinetic ($E_{\mathrm{kin}}(x)$) and potential ($E_{\mathrm{pot}}(x)$) energies, factored by functions of the fluid and particle densities and compressibilities:

$$F(x) = \frac{\partial}{\partial x}\left(\frac{4\pi a^3}{3}\left(\frac{3(\rho_{\mathrm{p}} - \rho_{\mathrm{f}})}{(2\rho_{\mathrm{p}} + \rho_{\mathrm{f}})}E_{\mathrm{kin}}(x) - \left(1 - \frac{\beta_{\mathrm{p}}}{\beta_{\mathrm{f}}}\right)E_{\mathrm{pot}}(x)\right)\right). \tag{3}$$

Substituting for the energy terms associated with a plane standing wave with a rigid boundary at $x$=0 [19]:

$$\begin{aligned} E_{\mathrm{kin}}(x) &= \varepsilon \sin^2(kx) \\ E_{\mathrm{pot}}(x) &= \varepsilon \cos^2(kx) \end{aligned} \tag{4}$$

leads directly to (1). The more general three-dimensional formulation of (3) quoted by Gor`kov is applicable to a wide variety of acoustic resonant field geometries.

### 2.1.1.1. Particle Positions Under Gravity and Diffusive Forces

The above discussion considers the direction in which the acoustic force will act on a particle in suspension. However, the equilibrium position taken up within the field will be determined by the resolution of this force and any other forces acting on the particle. For instance, in the case of a planar resonant field in which the positive $x$ axis points vertically upwards, (1) predicts that the particle will be in equilibrium when

$$3k\varepsilon\Phi(\beta,\rho)\sin(2kx)=(\rho_{\mathrm{p}}-\rho_{\mathrm{f}})g. \tag{5}$$

It is interesting to note that the equilibrium position, for this case in which the particle is small compared to the acoustic wavelength, is independent of the size of the particle, but a function the acoustic contrast factor along with the strength and frequency of the ultrasonic field and the acceleration due to gravity $g$. This offers a means of characterising the acoustic field, the pressure amplitude of which is often difficult to determine in microsystems. This is achieved by balancing acoustic and gravitational forces on an acoustically levitated particle, as particles that are not neutrally buoyant will have a minimum acoustic energy density required to hold the particle against buoyancy forces [20]. Note also that equilibrium positions associated with the pressure node and antinode exist, but for any particle/fluid combination only those associated with one of these (generally the pressure nodes) represent points of stable equilibrium. In many circumstances in which the application of ultrasonic manipulation appears attractive, including the manipulation of biological cells, the particles are close to neutrally buoyant and the equilibrium position will be very close to the pressure node. However for heavy particles, or for bubbles, the buoyancy offset may be significant.

The magnitude of the acoustic forces under consideration, along with the small size of the particles, means that diffusive forces on the particles due to Brownian motion may be significant. This is expressed by Higashitani et al. [21] using a probability distribution to describe particle concentration gradients associated with nodal planes.

### 2.1.1.2. The Effect of Viscosity

The theory outlined above assumes an ideal, non-viscous fluid. This theory has been extended to the case of viscous fluids by several investigators including Westervelt [22], who considered the relationship between viscous drag and forces due to scattering from a rigid sphere in both plane progressive and standing waves. Doinikov [23] derives a general expression for acoustic radiation forces in fluids that are viscous and thermally conducting and considers the special cases of plane progressive, plane standing and spherically diverging fields. Doinikov then shows that for both small rigid spheres [24] and small compressible liquid drops [25] in plane standing waves corrections to (1) for viscosity and thermal effects are minor. Further analysis by Danilov and Mironov [26] also concentrates on the case of a sphere that is small compared with the wavelength and demonstrates the importance of streaming effects in cases where the sphere is small compared to viscous boundary layer depths. Deviations of measured radiation forces from those predicted by equation (1) are noted by Yasuda and Kamakura [27] for particles smaller than 5 μm and were tentatively attributed to viscous boundary layer effects.

### 2.1.2. Some Practical Considerations

#### *2.1.2.1. The Influence of Particles on the Field*

In general, resonant chambers for particle manipulation (microfluidic or otherwise) are designed to provide particular acoustic characteristics when they are filled with the working fluid of interest. However, the chambers are designed to work when filled with a fluid–particle mix, which will have acoustic characteristics that differ from the particle-free fluid. Further, most applications will need to work over a range of different particle concentrations. This will then modify the acoustic field itself [2]. Even single particles may influence the resonance frequency [28] and as a uniform suspension of particles has those particles concentrated at the pressure nodes, the frequency will shift further [29], possibly necessitating automatic frequency control to lock on to the required resonance [30, 31]. In cases where the particles are very different acoustically from the fluid medium, agglomerations can cause localized breakdown of the acoustic field and a subsequent loss of the ordered collection of particles at the pressure nodes.

#### *2.1.2.2. Temperature Changes*

As an acoustic resonator is operated its temperature will, in general, rise due to dissipation in the transducer and acoustic energy absorption within the fluid and adjacent layers. The magnitude of the temperature rise will be dependent on the input power, the size of the resonator, the transducer material, and the material characteristics of the resonator layers, and hence there are significant differences in the thermal behaviour of resonators reported in the literature, with, for example, Bazou et al. [32] and Hawkes and Coakley [33] reporting less than 0.5 K and 1 K temperature increase respectively in microscale chambers, while Doblhoffdier et al. [34] required an integrated cooling circuit to avoid problematic temperature rises in their rather larger chamber designed for the retention of mammalian cells.

Temperature rises are likely to result in a change in the resonance frequency, possibly requiring frequency tracking as mentioned above. Particularly problematic in microfluidic devices is the tendency to generate bubbles at elevated temperatures. In addition, temperature induced convection can disrupt the required ordering of particles and in extreme cases temperature rises may lead to mechanical failure of devices, particularly at layer interfaces.

### 2.1.3. Lateral Forces and Secondary Radiation Forces

#### *2.1.3.1. Lateral Forces*

The above discussion considers plane standing waves in which variations in acoustic characteristics exist only in the field's axial direction. Real standing wave fields will also have variations in the lateral directions leading particles to move not just to axial pressure nodes (typically), but also to particular points within those energy nodal planes, referred to by Schram as acoustic 'hot spots' [35]. These inhomogeneities in the field may be caused by a number of phenomena:

1. Reduction in energy density towards the edges of localised axial fields [36, 37]
2. Near field effects [38] due to geometric interference patterns
3. Two- or three-dimensional acoustic modes within the enclosure [39]
4. The coupling of structural (for example bending) modes into the fluid [40]
5. Source inhomogeneity [1]

In some cases these lateral variations represent a desirable, inherent element of the device operation [37, 38], while in others, the variations are not desirable and may be detrimental to the device performance [39]. In real devices, lateral forces may well exist due to a combination of the above causes. Typically, particles that have been moved by the axial force into a pressure nodal plane will then move to areas of maximum acoustic kinetic energy density within that plane. The magnitude [41] and distribution [20] of the lateral forces within resonators has been measured and shown to compare well with calculations using Gor`kov's [18] formulation for radiation forces (3).

#### *2.1.3.2. Secondary Radiation Forces*

In addition to the radiation forces (both axial and lateral) due to the primary acoustic field, secondary acoustically generated forces will be produced between particles themselves. These particle–particle interactions, known as Bjerknes forces, were studied in bubbles in an acoustic field by Crum [42]. Weiser et al. [43] later investigated the phenomenon in the context of red blood cells in an acoustic field. Theses forces aid the formation of aggregates within a standing wave, but are negligible until the particles are in close proximity.

### *2.2. Acoustic Streaming*

A phenomenon that may modify the effects of acoustic radiation forces is that of acoustic streaming [44]. Acoustic streaming is a steady-state fluid flow generated by the oscillatory acoustic field. Probably the best known form of streaming is bulk, or Eckart streaming, in which a 'jet' of fluid is propelled away from an acoustic source due to attenuation of a propagating wave within the body of the fluid. In addition to Eckart streaming, streaming-induced vortices may be generated at a sub-wavelength scale. These 'microstreaming' phenomena include Rayleigh streaming, an ordered pattern of vortices associated with the side boundaries of a standing wave, and the much smaller scale Schlichting streaming, which occurs within the viscous boundary layer itself [45].

Acoustic streaming may provide distinct benefits in some microfluidic applications, but may also be problematic in others. On the positive side, streaming can be used to enhance beneficial mass transfer [46]. Examples that use radiation forces to hold particles and rely on streaming to move fluid or submicron particles to them can be seen under *trapping* below. Another potentially beneficial effect of streaming in microfluidic systems is the ability to initiate mixing in highly laminar systems [see for example 47, 48].

In other systems, streaming may limit the ability of radiation forces to order particles within a chamber [49]. In particular, Stokes drag forces become relatively more important than acoustic radiation forces as particle size decreases. This limits the lower particle size that can be manipulated by radiation forces. Kuznetsova et al. [50, 51] found that particles above a micron diameter tend to move with the radiation force while particles of a micron and below are increasingly dispersed by the streaming vortices, although it is likely that the size limit is dependent to some extent on the chamber design and excitation frequency. Kuznetsova also notes the appearance in these systems of 'wall-independent bulk suspension vortices, with circulation planes parallel to the transducer radiating surface', which do not conform to bulk, Rayleigh or Schlichting streaming, but are likely to be generated by lateral variations in the acoustic field strength.

It should be noted that modelling of streaming tends to be computationally challenging due to the length-scales and time-scales of importance [52]. Relatively low velocity flows are generated by very high frequency oscillations while Rayleigh vortices are of a significantly larger length scale than the viscous boundary layer thicknesses that also need to be modelled.

### 2.3. Modelling of Standing Waves for Ultrasonic Force Fields

The modelling of planar ultrasonic devices can be broken into two main aspects: *Field modelling* to predict the acoustic forces, and *Particle modelling* to predict the effect of these forces on the particle/fluid mix.

### 2.4. Field Modelling

Typically, planar resonators consist of a driving transducer driving several layers of material, usually terminating in an air-backed boundary, with one layer being fluid-filled. Referring to Fig. 3, which shows a typical structure, it can be seen that the fluid layer is bounded by a reflector layer and a carrier layer. The carrier layer (so-called because it carries the transducer) acts to isolate the transducer from the fluid and is often a structural component of the device. Figure 3 also details an adhesive layer, which may be required to bond the transducer to the carrier layer. Sometimes the carrier layer is referred to as the matching layer, as it can provide some impedance matching functionality to transmit the sound energy into the structure. The fluid layer is the layer in which manipulation of particles occurs.

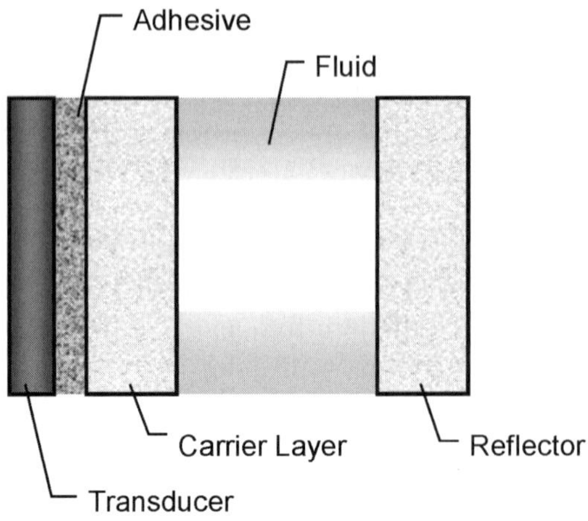


**Fig. 3.** A one-dimensional, multi-layer structure

A common approach to planar resonators is to treat the system as a one-dimensional problem, although recent work has been using finite element methods to analyse the problem in two dimensions [see for example 39, 53]. Nowotny and Benes [54, 55] developed a general transfer matrix method for analysing one-dimensional multilayer resonators. This has subsequently been used by a number of authors to model several particle manipulators, including Gröschl [1, 30] and Hawkes et al. [56]. In both of these examples the Nowotny model has shown a good match to experimental results and has been helpful in improving resonator design.

An alternative approach, using electro-acoustic analogues was developed by Hill et al. [57]. This model is more restricted than that of Nowotny in its transducer representation, but is otherwise similar in concept. This model was used in conjunction with an experimental programme to investigate the behaviour of several resonators and, in particular, demonstrated the influence that an adhesive bonding layer can have on resonator performance [58]. The model was also used to explore coincident resonances, in which two layers of the resonator have thickness modes at similar frequencies. Subsequently the model was used successfully in the design a number of microfabricated resonator systems [59–61].

A later attempt [62] to characterise the underlying properties of layered resonators allowed layer thicknesses to be selected for predetermined placement of pressure nodes, even for placement of nodes at the reflector boundary. The approach simplifies to solving

$$k_{\mathrm{f}}\, t_{\mathrm{f}} = \tan^{-1}\left(\frac{r_{\mathrm{f}}\left(r_{\mathrm{r}}^{\,2} + r_{0}^{\,2}\tan^{2}\left(k_{\mathrm{r}} t_{\mathrm{r}}\right)\right)}{r_{\mathrm{r}}\tan\left(k_{\mathrm{r}} t_{\mathrm{r}}\right)\left(r_{\mathrm{r}}^{\,2} - r_{0}^{\,2}\right)}\right) \tag{6}$$

where $t_{\mathrm{r}}$ and $t_{\mathrm{f}}$ are the thicknesses of the reflector and fluid layers respectively, $k$ is the wave number and $r_{\mathrm{f}}$ and $r_0$ are the acoustic impedances of the fluid and reflector layers. A given ratio of the thicknesses of the two layers (i.e a fixed design) is represented by a straight line passing through 0,0 on the graph in Fig. 4, which shows the first four solutions of (6). The crossing point of the resonator design line and the solutions to (6) provides the frequency of resonance of the combined structure, and from this the nodal position can be predicted.

The motion of the particles is also affected by the flow of the carrier fluid [33]. Full particle path prediction in a flowing medium requires both the electro-acoustic model data, flow data and the summation of other forces acting on the particle to be assessed [63]. Such modelling allows the path of particles to be predicted within a manipulation chamber and Fig. 5 shows both predicted particle paths, and a section of particle concentration with the depth of the chamber at a point in the channel, allowing visualisation of the formation of the concentrated stream in the device, thus allowing prediction of the performance of particle manipulators.

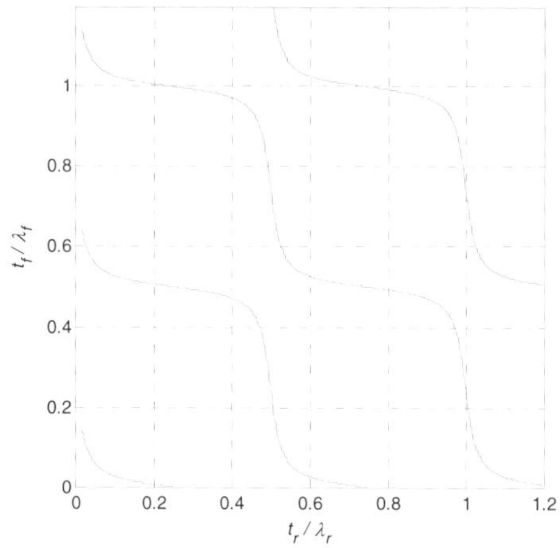


**Fig. 4.** The first four solutions of (6), based on [62]

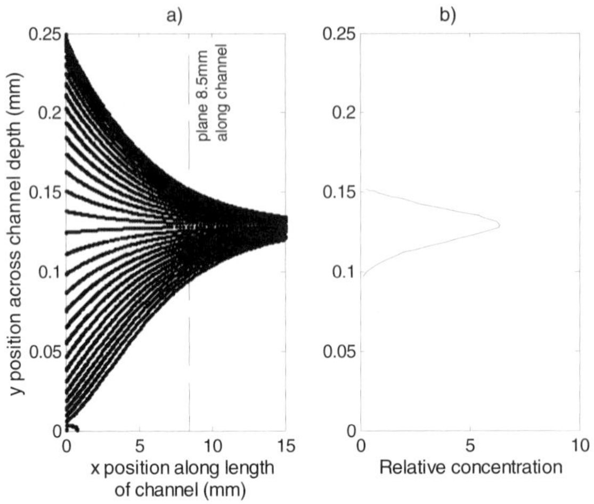


**Fig. 5.** Convergence of particles to a nodal plane, and particle concentration at a distance of 8.5 mm down the channel. Reprinted from [63] with permission from Elsevier

The modelling discussed to this point assumes a one-dimensional structure, which is clearly a major approximation, as shown by the section on lateral forces above, and these are currently under investigation to enhance the understanding of the complex effects within practical devices. In particular finite element modelling for two-dimensional structures has been successful at predicting the existence of lateral modes, and has allowed parameters that affect this to be adjusted [39].

## 3. Transduction Techniques

### *3.1. Direct Excitation of Bulk Acoustic Waves*

For most applications requiring the ultrasonic manipulation of particles within fluids, the generation of the ultrasound is achieved with a piezoelectric material. Piezoelectric materials are those that exhibit a strain with applied voltage, and conversely, an applied stress cause a charge to be released. Although there are other ways of generating ultrasound in fluids, it is the piezoelectric method that is by far the most common and so will be discussed here.

### 3.1.1. Bulk PZT

The most common and effective material for generating ultrasound is Lead Zirconate Titanate (PZT). Although a crystalline material, it has ferroelectric properties, which add to its usefulness. Ferroelectric materials have dipoles that can be switched allowing versatility in the shapes that can be made. The most common form is as plates or cylinders made from PZT powder that has been sintered. Once the material has been sintered, it requires polarisation to align the domains in the material, and this is achieved by applying heat and a high electric field strength.

There are many different grades of PZT, but these can be broken down into 'hard' and 'soft' grades [64]. The hard grades tend to be better optimised for high power operation, particularly for sonar applications, and the soft grades are better suited to act as receivers, being optimised for their ability to respond sensitively to incident acoustic energy (hydrophones). It must be said that there is no simple rule governing the choice of grade in many applications and at low powers, or for devices that combine a transmit and receive function, a soft grade may be better suited.

Important parameters that can affect the choice of material include the following:

1. *Operating temperature:* PZT has a temperature beyond which the piezoelectric activity does not occur. This is called the Curie temperature. Once sintered PZT has exceeded the Curie temperature, the piezoelectric effect is lost permanently, or until the material is repolarised.
2. *Coupling coefficient:* This provides an indication of the ability to convert electrical energy into mechanical energy, and vice versa.
3. *Piezoelectric coefficient:* ($d$ for voltage coefficient and $g$ for charge coefficient): This gives a direct indication of how much charge is generated for a given stress, or conversely how much extension is available for an applied voltage.
4. *Dielectric constant:* PZT has a very high dielectric constant and has been used to make capacitors. The dielectric constant has a direct effect on the value of capacitance for a given transducer and thus has an impact on time constants and gains in associated electronic circuitry.
5. *Loss factor:* This is a measure of the dissipation within the device.

A PZT plate will exhibit resonant characteristics. These can be modelled by treating the transducer as a series RLC circuit in parallel with a static capacitance [65]. More complex models can be used, for example three port models with distributed parameters, can model transient effects more accurately [66], but the standard RLC model is usually sufficient for narrow band continuous wave operation. The resonance in the thickness mode occurs for an isolated transducer when the thickness is equal to an integer number of half-wavelengths. This is modified by the addition of carrier layers, or by bonding the transducer to a structure.

Ultrasonic particle manipulation usually exploits a stationary, or pseudo stationary acoustic field. The simple case is that of the standing wave in which a single transducer generates an acoustic signal, which is then reflected back by a parallel reflector to set up a standing wave. If, for example, a standing wave is set up in a parallel cavity with a single node in the centre of the cavity, then a nodal plane will be established and particles flowing in the device will be concentrated into a sheet. If another transducer is then added at right angles, then a crossed standing wave system is established that has the effect of concentrating particles into a column in the direction of flow. This strategy can be extended to allow a third transducer which could in principle allow particles to be concentrated at a point.

If the reflector surface is replaced with another transducer, then a pseudo standing wave is established. The advantage of this is that by changing the relative phase of the opposing transducers, the position of the nodes can be changed [17, 67].

Although it can be effective, adhesive bonding of PZT plates to microfluidic devices is not ideal, especially if production is being contemplated, and so other methods of generation are being investigated.

### 3.1.2. Thick-Film PZT

An alternative approach to creating a piezoelectric actuator is to deposit by thick film [68]. The thick-film process involves depositing pastes onto a carrier layer and then firing the pastes to create a hardened film. Usually the pastes take the form of conductors, resistors or dielectrics, but various special purpose pastes have been developed. One of these is a piezoelectric paste based on powdered PZT. This allows the actuator to be printed directly onto the fluidic device without the need to use a glue layer, which can introduce variability. The thick film process imposes some limitations on the choice of materials that can be used as a carrier layer, as traditional thick film processing is a high temperature process (firing occurs at 800°C or above), but materials that have been successfully used include alumina ceramic, silicon, certain stainless steels and inconel.

Figure 6 shows a thick-film PZT actuator printed on silicon. In this case the actuator is a two layer device, which gives improved performance over a single layer of equivalent thickness [69].

(a) (b)

**Fig. 6.** A thick-film actuator shown (**a**) on a silicon/Pyrex resonator and (**b**) in cross section

An interesting extension to this is to produce free standing thick-film transducers, where the substrate is removed after firing [70], and a novel way of creating arrays of free standing transducers has been achieved by Lilliehorn et al. [71]. These transducers allow the creation of arrays as each can have dimensions of less than 500 × 500 μm and thickness resonances of between 10 and 15 MHz. The multilayer nature of these devices also results in lower drive voltages being necessary. Such transducers have been used for ultrasonic particle trapping systems [38].

There have been attempts to produce low temperature polymer-based PZT inks [72] but their piezoelectric activity is not high and so these are not as suited to act as generators. Other methods include spray depositing of PZT [73], and painting of PZT [74], but these are not currently able to deliver much acoustic power in the actuation of bulk waves.

### 3.2. Magnetostrictive Excitation

Magnetostrictive materials are similar to piezoelectric materials in that they can change their dimensions under the control of an electric signal. Magnetostrictive materials are affected by magnetic fields, which can be supplied by a current-carrying coil. In this respect, magnetostrictive materials can be viewed as current driven, whereas piezoelectric materials are voltage driven. Magnetostrictive materials have an advantage in that the driving coil can be wrapped around the material, and therefore does not need to be in direct contact with it. Such transducers have found application in NDT (non destructive testing) on pipes. However, magnetostrictive materials have traditionally had small extensions, compared with PZT, until the development of a class of material called massive magnetostrictives. An example of this is Terfenol. Extensions of the order of 1,000 ppm [75] are possible compared with 100 ppm with PZT, and so interest is increasing in these materials. Transducers have only been demonstrated in the 10 s of kHz at present [76].

### 3.3. Excitation via Leaky Surface Waves and Plate Waves

Another method of generating ultrasonic fields is to use structural waves (such as surface waves or flexural waves) within a solid boundary that couple into the fluid. Although the field generated is evanescent, this does not preclude operation when the fluid chamber is small, and this type of excitation is well suited to microfabricated devices [77]. Haake and Dual [78] excite bending waves in a glass plate that couples directly into a thin fluid layer. Under these conditions an acoustic wave is emitted into the fluid and is then reflected by the bottom of the channel, setting up a complex interaction between the modal behaviour in the fluid and the plate. By modeling dispersion relationships between the plate and fluid modes they establish conditions, in terms of the thickness of the plate and the operating frequency that are required to establish the required acoustic pattern in the fluid. In a later paper the authors discuss different methods of controlling the sound field [79] allowing particles to be levitated and then moved to specific lateral positions. The group have also demonstrated the movement of biological cells into lines and points without signs of significant cell damage [40, 80] and have developed a microfabricated device, which allows the manipulation of particles using this technique prior to harvesting using a microgripper [53].

### 3.4. Sol–Gel

The Sol–Gel method of thin film deposition can also be used for creating ultrasonic transducers. The sol–gel method allows a liquid carrying the active materials (a dispersion of colloidal particles in a liquid) to be deposited onto a substrate, and then further processing allows the liquid to gel [81]. Additional processing can then be used such as densification by heating if necessary. The layers tend to be quite thin giving rise to high frequency operation, although this can be adjusted by the mass loading of the substrate. For example in [82] a 2 μm thickness of PZT was deposited onto a 270 μm clamped plate structure, resulting in a resonance at 1 MHz.

### 3.5. Alternative Materials

Other materials that have been used for generating ultrasound include Lithium Niobate, Gallium Orthophosphate and Quartz, although other materials exist. Lithium Niobate and Gallium Orthophosphate tend to be used for high temperature applications (Niobate up to about 600°C [83] and Gallium Orthophosphate up to 700°C [84]), although Lithium Niobate has a larger coupling coefficient. Recent work has examined alternative single crystal materials [85]. Marin-Franch et al. [86] have reported that single crystal magnesium niobate-lead titanate has significant potential for use as a transducer as its piezoelectric coefficients are significantly larger than PZT, and they offer simulations of potential transducers, concluding that the single crystal material may offer improvement in bandwidth for high frequency arrays in particular.

## 4. Applications of Ultrasonic Particle Manipulation

Wide ranges of frequencies and powers have been reported in the literature for the manipulation of particles. For manipulation in liquid (rather than gaseous) media, frequencies are typically between 0.5 and 15 MHz. Lower frequencies increase the likelihood of generating acoustic cavitation at the pressures required for manipulation. Cavitation (the generation of bubbles during the low pressure phase of an acoustic cycle, which may subsequently collapse and generate shock waves) is generally undesirable in these systems.

For microfluidic applications, typical cavity dimensions will generally be compatible with frequencies above 1 MHz ($\lambda = 1.5$ mm in water). Energy densities also vary, and are often not quoted as they are difficult to measure. However, to give an order of magnitude impression, a standing wave energy density of about 30 $Jm^{-3}$ at 3 MHz would correspond to a pressure amplitude of about 0.5 MPa and generate a 100 pN radiation force on a 10 μm diameter polystyrene sphere [see, for examples of comparator data: 87, 88].

Particle sizes are typically above 1 μm, although manipulation of submicron particles is reported [27]. Many materials have been used in constructing chambers for ultrasonic resonators (the important properties being dimensional stability and a high acoustic $Q$ factor) including steel [89], brass [90], glass [91], quartz [27] and silicon [59].

### 4.1. Cell Viability

Many of the uses reported for USW manipulation of particles are based upon biological cells [4]. Hence the question of cell viability becomes an important issue. Doblhoffdier et al. [34] suspended mammalian cells in a large resonator at different power levels and reported no significant loss in viability, a result reinforced by a later more detailed study in a similar chamber [92]. Wang et al. demonstrated that mammalian cells held in a mesh by USW forces retained viability [93] and Gherardini et al. established that the viability of yeast cells was maintained for days following treatment [94]. More recently, Hultström et al. [91] trapped cells in a microfluidic chip and showed that their ability to proliferate thereafter was maintained.

### 4.2. Filtration and Concentration

A frequently reported use of ultrasonic radiation forces aims to filter particles, either to remove unwanted particles from a fluid medium or to concentrate particles prior to further analysis. Filtration has been reported in batch mode (a form of ultrasonic centrifugation) and in continuous (flow-through) mode. This section describes filtration techniques that rely primarily on ultrasonic radiation forces and does not deal with ultrasonically enhanced membrane filtration, although microfluidic versions of this have been developed [95].

Filtration and concentration have very obvious applications in microfluidic systems. Sample pre-treatment may be required for 'real world' samples contaminated by particulates. Barrier filtration, with the associated fouling and replacement of the barrier, limits the ability of microfluidic analysis systems to operate without intervention. USW filtration offers the opportunity to remove unwanted particles without barriers, but in a continuous, flow-through system. In addition, the potential to concentrate samples offers, for near neutrally buoyant samples in particular, a microfluidic equivalent of a centrifuge.

#### 4.2.1. Enhanced Sedimentation

The most commonly reported ultrasonic filtration technique up to the late 1990s relied on allowing particles to agglomerate under USW forces (axial, lateral and secondary) as shown schematically in Fig. 7. As agglomerates grow, the relative importance of gravitational forces over viscous drag forces grows until the clumps of particles sediment under gravity.

This technique lends itself to batch processing on a macro-scale. Schram [35] describes a number of applications of this technique, including an

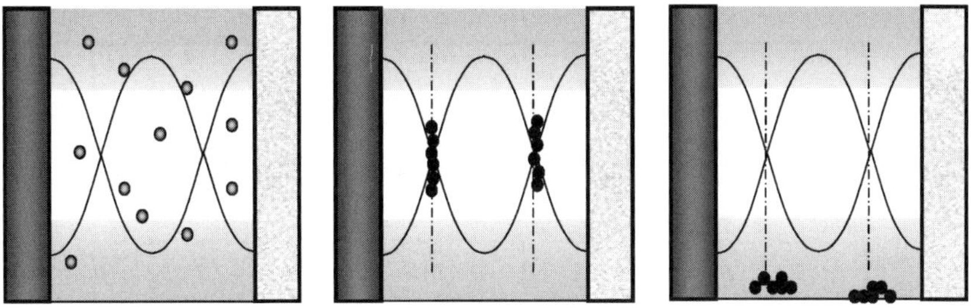

**Fig. 7.** Enhanced sedimentation. Particles within an ultrasonic field (*left*) are agglomerated by the field (*centre*) and then tend to sediment under gravity (*right*)

interesting use of orthogonal crossed standing waves to enhance agglomeration. A similar technique was used by Cousins et al. for preparation of blood plasma [96, 97] and for concentration of bacteria [98].

With careful fluidic design, this approach can be developed to work in a flow-through mode. With reference to Fig. 7, this could be achieved by an inlet at the bottom of the resonator and to one side, flow vertically upwards through the field, a clarified outlet at the top and a particulate output directly below the resonator and beneath the inlet. Such a system has been developed commercially, again on a macro-scale, for the retention of cells within a bioreactor [31].

A similar technique described by Hawkes and Coakley [99] shows high separation efficiencies (up to 99%) in a smaller device specifically designed to have a small dead volume and resulting in a concentration of yeast cells, which compared well with centrifugation. However, sedimentation of smaller bacterial cells was less successful, primarily as the sedimentation relies on particle agglomeration under USW forces overcoming the Stokes drag forces due to the fluid flow, which tend to dominate for smaller particles [100]. While this device is still larger than microfluidic scale, its simplicity and relative robustness to changes in flow characteristics make the principle attractive for pre-filtration for microfluidics.

### 4.2.2. Flow-Through Filtration

An alternative approach to filtration uses radiation forces to gather particles within part of a channel's cross section and then to draw those particles off from a concentrated outlet and the remaining fluid from a clarified outlet. This can be achieved on a large scale by holding particles within multiple nodal planes and allowing the flow to diverge through the resonator [57, 101]. However working on a microfluidic scale lends itself to working with channels that are less than a wavelength and therefore contain a single nodal plane [90, 102].

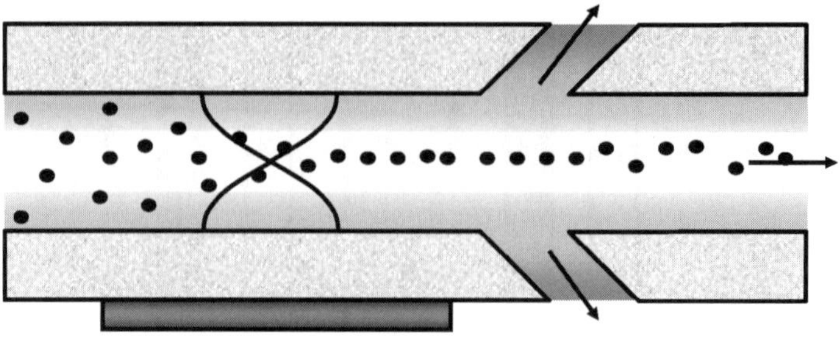


**Fig. 8.** Half wavelength laminar flow filter

Such a system is shown diagrammatically in Fig. 8. Particles within a fluid are carried into a channel within which a half-wavelength standing wave is excited. As particles move through the standing wave they are forced to the nodal plane at the centre of the chamber. This allows concentrated particles to be drawn off from the centre of the flow and clarified fluid to be drawn from either the upper or lower outlet, or from both as shown here. This technique was demonstrated by Hawkes and Coakley [33] using a channel of 250 μm deep, machined from stainless steel and driven at about 3 MHz. The filter was tested using latex particles of between 1.5 and 25 μm and with yeast cells. Clarifications of up to 1,000-fold were achieved with yeast.

The same principle has been used in a microfabricated flow through filter [59]. This used channels etched into a Pyrex and silicon structure. Variations on the same basic design used both bulk PZT [103] and thick film PZT excitation [104]. This microfabricated device achieved a 50-fold clarification of yeast cells, but only a 5-fold clarification of 1 μm latex beads. Development of this filter also involved an examination of microchannel geometries and their effect on the system's operation [105].

This half-wave filtration principle was extended by Petersson et al. [106, 107] in a device designed to filter lipid particles from whole blood. This application makes use of the fact that red blood cells in blood plasma will move to pressure nodes, while lipids will move to the pressure antinodes. This allows a form of fractionation in which lipids are driven to the upper and lower faces of a channel such as that of Fig. 8, while the red cells are concentrated in the centre. Over 70% of the red cells were collected in one third of the original fluid volume and more than 80% of the lipid particles were removed. The channels, of 350 μm width, were anisotropically etched into silicon, and, in contrast to the layered resonators shown in this chapter, the transducer is orthogonal to the direction of the standing wave enabling several channels to be excited from a single transducer. The authors envisage a scaled-up system of between 100 and 200 channels to achieve surgically useful throughput.

Using an alternative geometry, Goddard and Kaduchak [108] demonstrated a concentration of particles at a pressure nodal line using a standing wave field generated within a glass tube (2 mm internal diameter).

#### 4.2.3. Ultrasound Within a Porous Mesh

An interesting alternative to the filtration techniques described above is to use the radiation forces within a USW field to modify the filtration characteristics of a porous mesh [93, 109, 110]. In this approach particles within a fluid flow through a high porosity polyester mesh. As the pore size is up to two orders of magnitude larger than the particles, they pass through the mesh without being trapped. However when the mesh is excited ultrasonically, acoustic radiation forces tend to hold particles on the mesh elements, creating an effective filter. Cleaning the mesh may then be achieved by turning off the ultrasonic signal and releasing the particles. Devices described so far have been large, but it is a principle that may have potential for scaling down for use as a microfluidic pre-filter stage.

### *4.3. Particle Trapping*

An application relying on lateral radiation forces based upon a focussed USW is described by Hertz [111]. This device was investigated in an attempt to overcome some of the limitations of using an optical trap with optically absorbing materials. Focussing ultrasound, using two opposing concave transducers in this example, creates a steep lateral energy gradient in the standing wave. Particles within the field agglomerate at a pressure node, but are also trapped laterally at the focus of the field, and can be held against a flow of the surrounding fluid. Acoustic traps tend to lack the precision to deal with individual cells or particles, but have great potential for aggregating and trapping a population of cells within a microfluidic chamber [112]. This approach outlined by Hertz has subsequently been developed to hold samples for laser fluorescence analysis within a capillary [113] or in a 96-well plate [61]. These applications all use a focussed beam to trap particles, but the same group use a different approach when integrating a USW trap into a dielectrophoretic (DEP) chip [88]. In this case a bulk PZT transducer is operated in thickness mode and its energy is transmitted into the DEP microfluidic chip through a PMMA wedge, which refracts the beam into the required orientation and couples it into a microchannel formed by a Pyrex/silicon–borosilicate sandwich. This sets up a USW field with nodes running parallel to the direction of flow through the chip. The USW field effects large-scale alignment of particles prior to fine manipulation within the more localised DEP field. Different strategies for controlling these fields independently allow complex manipulation of 10 μm particles.

An alternative approach uses the lateral variations in a transducer near-field to trap particles in a microfluidic chip [38]. The 10 MHz transducers used were multilayered with lateral dimensions of about 800 μm square and a layer thickness of 36 μm. The size of the transducers allows the creation of individually controllable USW traps at different points within a microfluidic channel [114].

Yasuda [115] observed trapping, aligning and mixing in a microchamber driven at different voltages, and reports minimal cell damage even at mixing amplitudes when the chamber was driven at 3.5 MHz.

### 4.3.1. Trapping to Enhance Particle–Particle Interaction

The scale of action of USW trapping, along with secondary radiation forces tending to draw particles together allows it to be used to enhance the interaction between particles within the trap. This has been used to increase the speed of an agglutination test for meningitis through the interaction of antigen coated latex beads [116–119].

It has also been used in sub wavelength depth chambers for cell–cell interaction studies [36, 120]. Cells can be brought together in monolayers that are levitated within the field so their interaction is not modified by adjacent boundaries. Spengler and Coakley [121] discuss the suitability of different traps (USW, DEP or Optical) depending on the number cells and the size of cell aggregates required.

### 4.3.2. Trapping to Enhance Particle–Fluid Interaction

In addition to trapping particle agglomerates, USWs will also induce streaming around those agglomerates, as discussed under 'Acoustic Streaming' earlier. Both static suspension within a flowing fluid and streaming effects can increase mass transfer from the surrounding fluid into the aggregate. This was used by Morgan et al. [37] to accelerate the testing of the toxic effects of a flowing toxicant on a multi-cellular spheroid.

Such an approach has also been used to increase the take-up of retroviruses, which are too small to be held by radiation forces, into target cells for genetic transfection [122]. At higher powers, transfection can be enhanced by generating cavitation-induced rupture of the cell membranes (sonoporation) [89, 123]. In this case, it might be expected that the target cells will be concentrated at the pressure node, while the small bubbles in the contrast agent used to initiate cavitation would be forced away from the cells to the pressure antinodes. However at high acoustic pressure amplitudes, these bubbles oscillate about the nodal plane [124] allowing them to generate sonoporation and increase genetic transfection.

## 4.4. Sensor Enhancement

A fluid filled chamber with flat, parallel, rigid boundaries will resonate when driven at frequencies corresponding to an integer number of half wavelengths across the chamber. In these cases the boundaries will coincide with pressure maxima and velocity minima. However real boundaries are not fully rigid, and if the fluid is bounded by a wall of finite extent with its own resonant characteristic it will see a boundary acoustic impedance that has a reactive component. In this case, it is possible that the boundary will not be at a point of maximum pressure and in fact the boundary could coincide with a pressure node [62]. In such a case, which will occur when the boundary is at a half wavelength resonance, particles near the boundary will be forced towards it rather than to a nodal plane away from the surface. This has the potential to transport particles to a sensor surface and significantly enhance the sensor system's sensitivity. This was demonstrated by Hawkes et al. [125] using an arrangement similar in principle to that shown in Fig. 9. Particles within a fluid flow into the chamber shown from the right. The chamber had a fluid layer of about 200 μm, a steel carrier layer and a glass reflector. A quarter wavelength resonance is excited

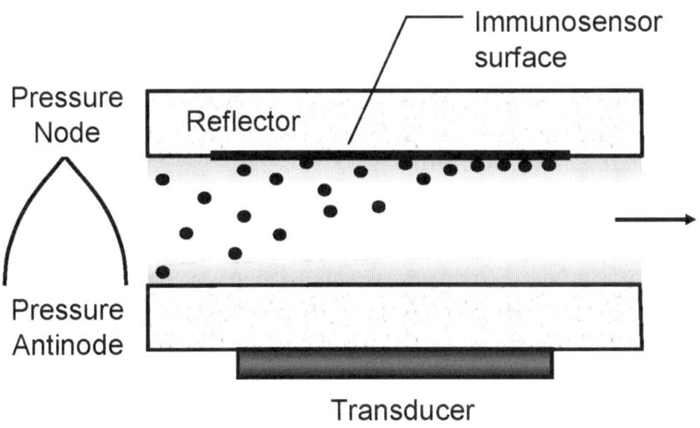


**Fig. 9.** Sensor enhancement by forcing particles onto a wall

in the chamber with a pressure node at the boundary with the reflector. This boundary is coated with an antibody chosen to capture bacterial particles of interest. The work showed a significant increase in the capture of spores within the USW field and also showed that capture increased as the transducer driving voltage increased. Later work [126] extended this principle by fabricating an optical metal-clad leaky waveguide on the reflector that was in turn coated with an antibody.

Modelling of quarter wave resonators [62] suggested that a layered resonator having a quarter wave resonance would generally also have resonances approximating to a half wavelength. This was demonstrated experimentally by Harris et al. [60] who fabricated a resonator from Pyrex and silicon that could move particles to the reflector at one frequency and to the centre of the fluid layer when driven at a different frequency. This has the potential not just to move particles to the surface, but also to remove unbound particles from that surface by switching frequency.

Another study of immunosensor capture enhanced by a quarter wavelength USW is described by Martin et al. [87]. This study showed that relatively small changes in the thickness of the reflector made a significant difference to the position of the pressure node within the fluid layer. In this study, the ultrasound enhanced particle detection by a factor of 70.

### *4.5. Particle Washing – Exchange of Containing Medium*

This application makes use of the laminar flow characteristics inherent in microfluidic devices. Two fluids can be brought into contact with each other and can flow alongside each other with no appreciable mixing, as indicated in Fig. 10. Particles in one fluid (fluid B) can then be moved by radiation forces into the second fluid (fluid A), assuming that acoustic nodes are positioned appropriately. The fluids can then be separated at the outlets and the net result is that particles have been moved from one fluid to another on a continuous basis.

Hawkes et al. [127] demonstrated the approach using yeast particles, and a washing rate of about 80% could be achieved. In the device described, higher drive powers did not result in higher efficiency of washing due to mixing effects of the two fluids. This mixing was ascribed to entrapment of fluid with the particles along with acoustic streaming and it was noted that it may be useful for mixing at a microfluidic scale.

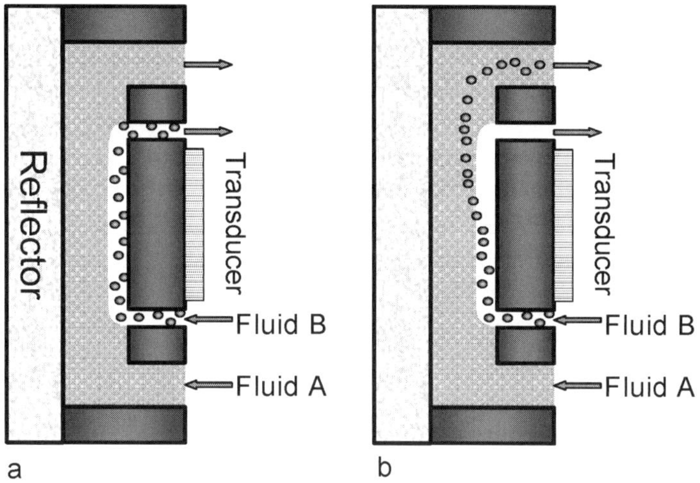


**Fig. 10.** The principle of cell washing. In (**a**) no sound is applied and only diffusion mixing applies. In (**b**) radiation forces are used to drive particles across the fluid boundary (based on [127])

A similar approach was taken by Petersson et al. [128], who demonstrated the technique for the purpose of blood washing. Bovine blood was used and 95% of the red cells were 'washed' into clean plasma, with 98% of the contaminants in the original plasma being removed. For the chamber used in this experiment, the flow rate was of the order of 0.27 ml $min^{-1}$ or less giving a Reynolds number of about 20. The channels used were etched in silicon and had two particulate-mix fluid inlets on each side of the target fluid into which particles were concentrated before being extracted from a single central channel.

### *4.6. Particle Fractionation*

Particle fractionation allows a mix of different particles to be sorted according to a physical property such as size or density. Ultrasound can be used to fractionate on a continuous basis, by exploiting the feature that different size particles move at different speeds in a given acoustic field [129]. One process that has been used with some success is field flow fractionation, where the residence time of a population of particles is proportional to their size [130].

Ultrasonic devices have been used to fractionate on a continuous basis. The technique makes use of the fact that the magnitude of the radiation force experienced by a particle (1) is proportional to its radius cubed and is also dependent on its material properties.

Therefore, for a given particle material, different sized particles will experience different forces. Larger particles experience larger forces. However, once a particle starts to move, it will also be affected by drag forces, which are also dependent on the size of the particle. Smaller particles will have lower terminal velocities than larger ones, so a differential velocity is created, leading to separation. Aboobaker et al. make some predictions and measurements of particle velocity against size [131]. Hence for a mixture of different sized particles, it is possible to differentiate particles by their velocity in a standing wave system as indicated in Fig. 11.

For example in [132] a continuous fractionator is described where an ultrasonic field is used to drive a concentration of particles near one wall across the width of the chamber, similar to Fig. 11. Note it is important that the particles all start off from the same point, so some mechanism for introducing the particles at the bottom of the chamber is necessary. At the outlet of the chamber, the particles are evenly spread across the width if the residence time is correct. This technique requires a balance between residence time in the chamber and lateral particle velocity. If the residence time is too long, then all the particles will arrive at the same wall, albeit that the smaller ones will have taken longer. A variation of this technique, which is similar to the particle washing concept above, is described in [133]. Larger particles are forced into a second fluid by a USW and are

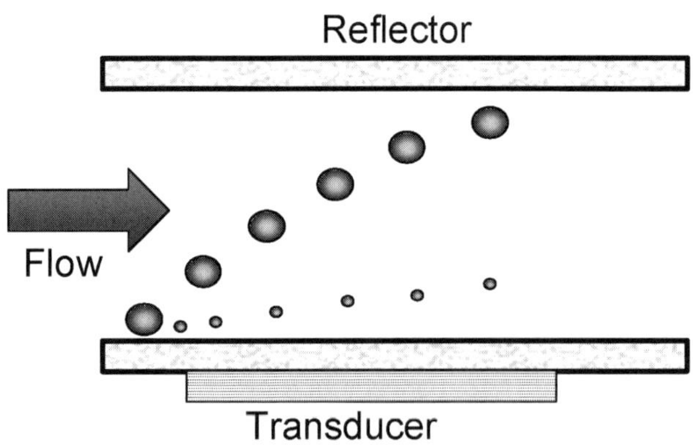


**Fig. 11.** The principle of fractionation. Large particles experience larger forces than than small particles

hence separated from smaller particles. In this implementation two transducers are used and the phasing between the two altered to allow the node to be positioned. This work demonstrated that large red blood cells could be separated from smaller cells in a plasma.

Other methods of fractionation have been tried where the acoustic force is balanced against a force other than gravity. In particular, electrostatic forces were used in addition to acoustic forces by Yasuda et al. [134], in which 10 μm and 20 μm polystyrene particles were separated from each other by balancing electrostatic forces against radiation forces. Wiklund et al. successfully used capillary electrophoresis combined with ultrasound to achieve separation for improving the concentration limit of detection in small samples over capillary electrophoresis [135].

## 5. The Future of Ultrasonic Particle Manipulation

While ultrasonic radiation forces have been known for many years, their practical use has been limited, amongst other things, by the low magnitude of the forces produced and the poor match of their lengths of action to most macro-scale applications. On a microfluidic scale, however, the scale of action of ultrasonic forces from waves in the low MHz region is relatively large. This gives tremendous potential for trapping and manipulating larger particles and groups of particles, perhaps prior to finer manipulation from other techniques. In addition, the relatively low temperature rises associated with carefully designed resonators and the growing evidence of their ability to hold cells without a reduction in cell viability suggest that USW manipulation will become a significant complement to other microfluidic particle handling techniques.

## Acknowledgement

The authors gratefully acknowledge the assistance given by Rosemary Townsend in preparation of this chapter.

## References

1. Gröschl, M (1998) Ultrasonic separation of suspended particles, Part I: Fundamentals. Acustica 84 (3), 432–447
2. Wang, TG, and Lee, CP (1998) Radiation pressure and acoustic levitation. In *Nonlinear Acoustics* (Hamilton, M.F., and Blackstock, D.T., eds), pp 177–205, Academic press, New York

3. Vilkner, T, Janasek, D, and Manz, A (2004) Micro total analysis systems. Recent developments. Analytical Chemistry 76 (12), 3373–3385
4. Coakley, WT (1997) Ultrasonic separations in analytical biotechnology. Trends in Biotechnology 15 (12), 506–511
5. Kundt, A, and Lehman, O (1874) Longitudinal vibrations and acoustic figures in cylindrical columns of liquids. Annal Physik 153, 1–11
6. Rayleigh, JW (1902) On the pressure of vibrations. Philosophical Magazine 3, 338–346
7. Westervelt, PJ (1957) Acoustic radiation pressure. Journal of the Acoustical Society of America 29 (1), 26–29
8. Chu, BT, and Apfel, RE (1982) Acoustic radiation pressure produced by a beam of sound. Journal of the Acoustical Society of America 72 (6), 1673–1687
9. Beyer, RT (1978) Radiation pressure – History of a mislabeled tensor. Journal of the Acoustical Society of America 63 (4), 1025–1030
10. King, LV (1934) On the acoustic radiation pressure on spheres. Proceedings of Royal Society of London A147, 212–240
11. Leung, E, Jacobi, N, and Wang, T (1981) Acoustic radiation force on a rigid sphere in a resonance chamber. Journal of the Acoustical Society of America 70 (6), 1762–1767
12. Yosioka, K, and Kawasima, Y (1955) Acoustic radiation pressure on a compressible sphere. Acustica 5, 167–173
13. Leighton, TG (1994) The acoustic bubble. Academic Press, New York
14. Eller, A (1968) Force on a bubble in a standing acoustic wave. Journal of the Acoustical Society of America 43 (1), 170–171
15. Crum, LA (1971) Acoustic force on a liquid droplet in an acoustic stationary wave. Journal of the Acoustical Society of America 50 (1), 157–163
16. Marston, PL, and Thiessen, DB (2004) Manipulation of fluid objects with acoustic radiation pressure. In *Transport Phenomena in Microgravity*, pp 414–434, Annals of the New York Academy of Sciences, New York
17. Abe, Y, Kawaji, M, and Watanabe, T (2002) Study on the bubble motion control by ultrasonic wave. Experimental Thermal and Fluid Science 26 (6–7), 817–826
18. Gor'kov, LP (1962) On the forces acting on a small particle in an acoustical field in an ideal fluid. Sovlet Physics Doklady 6, 773–775
19. Kinsler, LE, Frey, AE, Coppens, AB, and Saunders, JV (1982) Fundamentals of acoustics. Wiley, New York
20. Whitworth, G, and Coakley, WT (1992) Particle column formation in a stationary ultrasonic-field. Journal of the Acoustical Society of America 91 (1), 79–85
21. Higashitani, K, Fukushima, M, and Matsuno, Y (1981) Migration of suspended particles in plane stationary ultrasonic field. Chemical Engineering Science 36 1877–1882

22. Westervelt, PJ (1951) The theory of steady forces caused by sound waves. Journal of the Acoustical Society of America 23 (4), 312–315
23. Doinikov, AA (1997) Acoustic radiation force on a spherical particle in a viscous heat-conducting fluid. 1. General formula. Journal of the Acoustical Society of America 101 (2), 713–721
24. Doinikov, AA (1997) Acoustic radiation force on a spherical particle in a viscous heat-conducting fluid. 2. Force on a rigid sphere. Journal of the Acoustical Society of America 101 (2), 722–730
25. Doinikov, AA (1997) Acoustic radiation force on a spherical particle in a viscous heat-conducting fluid. 3. Force on a liquid drop. Journal of the Acoustical Society of America 101 (2), 731–740
26. Danilov, SD, and Mironov, MA (2000) Mean force on a small sphere in a sound field in a viscous fluid. Journal of the Acoustical Society of America 107 (1), 143–153
27. Yasuda, K, and Kamakura, T (1997) Acoustic radiation force on micrometer-size particles. Applied Physics Letters 71 (13), 1771–1773
28. Leung, E, Lee, CP, Jacobi, N, and Wang, TG (1982) Resonance frequency-shift of an acoustic chamber containing a rigid sphere. Journal of the Acoustical Society of America 72 (2), 615–620
29. Kwiatkowski, CS, and Marston, PL (1998) Resonator frequency shift due to ultrasonically induced microparticle migration in an aqueous suspension: Observations and model for the maximum frequency shift. Journal of the Acoustical Society of America 103 (6), 3290–3300
30. Gröschl, M (1998) Ultrasonic separation of suspended particles, Part II: Design and operation of separation devices. Acustica 84 (4), 632–642
31. Gaida, T, DoblhoffDier, O, Strutzenberger, K, Katinger, H, Burger, W, Groschl, M, Handl, B, and Benes, E (1996) Selective retention of viable cells in ultrasonic resonance field devices. Biotechnology Progress 12 (1), 73–76
32. Bazou, D, Kuznetsova, LA, and Coakley, WT (2005) Physical environment of 2-D animal cell aggregates formed in a short pathlength ultrasound standing wave trap. Ultrasound in Medicine and Biology 31 (3), 423–430
33. Hawkes, JJ, and Coakley, WT (2001) Force field particle filter, combining ultrasound standing waves and laminar flow. Sensors and Actuators B:Chemical 75 (3), 213–222
34. Doblhoffdier, O, Gaida, T, Katinger, H, Burger, W, Groschl, M, and Benes, E (1994) A novel ultrasonic resonance field device for the retention of animal-cells. Biotechnology Progress 10 (4), 428–432
35. Schram, CJ (1991) Manipulation of particles in an acoustic field. In *Advances in Sonochemistry* (Mason, T.J., ed), Elsevier, Amsterdam
36. Bazou, D, Foster, GA, Ralphs, JR, and Coakley, WT (2005) Molecular adhesion development in a neural cell monolayer forming in an ultrasound trap. Molecular Membrane Biology 22 (3), 229–240
37. Morgan, J, Spengler, JF, Kuznetsova, L, Coakley, WT, Xu, J, and Purcell, WM (2004) Manipulation of in vitro toxicant sensors in an ultrasonic standing wave. Toxicology In Vitro 18 (1), 115–120

38. Lilliehorn, T, Simu, U, Nilsson, M, Almqvist, M, Stepinski, T, Laurell, T, Nilsson, J, and Johansson, S (2005) Trapping of microparticles in the near field of an ultrasonic transducer. Ultrasonics 43 (5), 293–303
39. Townsend, RJ, Hill, M, Harris, NR, and White, NM (2006) Investigation of two-dimensional acoustic resonant modes in a particle separator Ultrasonics 44, e467–e471
40. Haake, A, Neild, A, Radziwill, G, and Dual, J (2005) Positioning, displacement, and localization of cells using ultrasonic forces. Biotechnology and Bioengineering 92 (1), 8–14
41. Woodside, SM, Bowen, BD, and Piret, JM (1997) Measurement of ultrasonic forces for particle–liquid separations. Aiche Journal 43 (7), 1727–1736
42. Crum, LA (1975) Bjerknes forces on bubbles in a stationary sound field. Journal of the Acoustical Society of America 57 (6), 1363–1370
43. Weiser, MAH, Apfel, RE, and Neppiras, EA (1984) Interparticle forces on red-cells in a standing wave field. Acustica 56 (2), 114–119
44. Riley, N (2001) Steady streaming. Annual Review of Fluid Mechanics 33, 43–65
45. Zarembo, LK (1971) Acoustic streaming. In *High Intensity Ultrasonic Fields* (Rozenberg, L.D., ed), Plenum Press, New York
46. Pollet, BG, Lorimer, JP, Hihn, JY, Phull, SS, Mason, TJ, and Walton, DJ (2002) The effect of ultrasound upon the oxidation of thiosulphate on stainless steel and platinum electrodes. Ultrasonics Sonochemistry 9 (5), 267–274
47. Bengtsson, M, and Laurell, T (2004) Ultrasonic agitation in microchannels. Analytical and Bioanalytical Chemistry 378 (7), 1716–1721
48. Sritharan, K, Strobl, CJ, Schneider, MF, Wixforth, A, and Guttenberg, Z (2006) Acoustic mixing at low Reynold's numbers. Applied Physics Letters 88 (5), 054102
49. Spengler, JF, Coakley, WT, and Christensen, KT (2003) Microstreaming effects on particle concentration in an ultrasonic standing wave. Aiche Journal 49 (11), 2773–2782
50. Kuznetsova, LA, and Coakley, WT (2004) Microparticle concentration in short path length ultrasonic resonators: Roles of radiation pressure and acoustic streaming. Journal of the Acoustical Society of America 116 (4), 1956–1966
51. Kuznetsova, LA, Martin, SP, and Coakley, WT (2005) Sub-micron particle behaviour and capture at an immuno-sensor surface in an ultrasonic standing wave. Biosensors & Bioelectronics 21, 940–948
52. Haydock, D, and Yeomans, JM (2001) Lattice Boltzmann simulations of acoustic streaming. Journal of Physics A: Mathematical and General 34 (25), 5201–5213
53. Neild, A, Oberti, S, Beyeler, F, Dual, J, and Nelson, BJ (2006) A microparticle positioning technique combining an ultrasonic manipulator and a microgripper. Journal of Micromechanics and Microengineering 16 (8), 1562–1570

54. Nowotny, H, and Benes, E (1987) General one-dimensional treatment of the layered piezoelectric resonator with 2 electrodes. Journal of the Acoustical Society of America 82 (2), 513–521
55. Nowotny, H, Benes, E, and Schmid, M (1991) Layered piezoelectric resonators with an arbitrary number of electrodes (general one-dimensional treatment). Journal of the Acoustical Society of America 90 (3), 1238–1245
56. Hawkes, JJ, Gröschl, M, Nowotny, H, Armstrong, S, Tasker, P, Coakley, WT, and Benes, E (2002) Single half wavelength ultrasonic particle filter: Predictions of the transfer matrix multi-layer resonator model and experimental filtration results. Journal of the Acoustical Society of America 111, 1259–1266
57. Hill, M, and Wood, RJK (2000) Modelling in the design of a flow-through ultrasonic separator. Ultrasonics 38 (1–8), 662–665
58. Hill, M, Shen, Y, and Hawkes, JJ (2002) Modelling of layered resonators for ultrasonic separation. Ultrasonics 40, 385–392
59. Harris, NR, Hill, M, Beeby, SP, Shen, Y, White, NM, Hawkes, JJ, and Coakley, WT (2003) A silicon microfluidic ultrasonic separator. Sensors and Actuators B 95 (1–3), 425–434
60. Harris, NR, Hill, M, Shen, Y, Townsend, RJ, Beeby, SP, and White, NM (2004) A dual frequency, ultrasonic, microengineered particle manipulator. Ultrasonics 42, 139–144
61. Wiklund, M, Toivonen, J, Tirri, M, Hanninen, P, and Hertz, HM (2004) Ultrasonic enrichment of microspheres for ultrasensitive biomedical analysis in confocal laser-scanning fluorescence detection. Journal of Applied Physics 96 (2), 1242–1248
62. Hill, M (2003) The selection of layer thicknesses to control acoustic radiation force profiles in layered resonators. JASA 114 (5), 2654–2661
63. Townsend, RJ, Hill, M, Harris, NR, and White, NM (2004) Modelling of particle paths passing through an ultrasonic standing wave. Ultrasonics 42, 319–324
64. Rosen, CZ, Hiremath, BV, and Newnham, RE (1992) Piezoelectricity (Key Papers in Physics). AIP, New York
65. Zelenka, J (1986) *Piezoelectric Resonators and their Applications.* Elsevier, Amsterdam
66. Sherrit, S, Leary, S., Dolgin, B., Bar-Cohen, Y (1999) Comparison of the Mason and KLM equivalent circuits for piezoelectric resonators in the thickness mode. In *1999 IEEE Ultrasonics Symposium*, pp 921–926, IEEE
67. Whitworth, G, Grundy, MA, and Coakley, WT (1991) Transport and harvesting of suspended particles using modulated ultrasound. Ultrasonics 29 (6), 439–444
68. Harris, NR, Koch, M, Beeby, SP, White, NM, and Evans, AGR (1998) Thick-Film printing of PZT onto silicon for micromechanical applications. In Proceedings of *Micromechanics Europe*, 78–81
69. Harris, NR, Hill, M, White, NM, and Beeby, SP (2004) Acoustic power output measurements for thick-film PZT transducers. Electronics Letters 40 (10), 636–637

70. Stecher, G (1987) Free Supporting Structures in Thick-film technology: A substrate integrated pressure sensor. In *6th European Micrelectronics Conference*, 421–427
71. Lilliehorn, T, and Johansson, S (2004) Fabrication of multilayer 2D ultrasonic transducer microarrays by green machining. Journal of Micromechanics and Microengineering 14 (5), 702–709
72. White, NM, Papakostas, T (2000) Screen printable polymer piezoelectrics. Sensor Review 20 (2), 135–138
73. Jian, L, Jiaru, C, Wenhao, H, and Zhimin, P (2002) Preparation of thick Pb(Zr,Ti)O3 (PZT) film by electrostatic spray deposition (ESD) for application in micro-system technology. Japanese Journal of Applied Physics 41, 4317–4320
74. White, JR, De Poumeyrol, B, Hale, JM, and Stephenson, R (2004) Piezoelectric paint: Ceramic–polymer composites for vibration sensors. Journal of Materials Science 39 (9), 3105–3114
75. Claeyssen, F, Colombani, D, Tessereau, A, and Ducros, B (1991) Giant dynamic magnetostrain in rare-earth iron magnetostrictive materials. IEEE Transactions on Magnetics 27 (6), 5343–5345
76. Or, SW, Yung, CS, and Lo, CY (2006) A 64-kHz sandwich transducer fabricated using pseudo 1-3 magnetostrictive composite. IEEE Transactions on Magnetics 42 (1), 47–50
77. Black, JP, White, RM, and Grate, JW (2002) Microsphere capture and perfusion in microchannels using flexural plate wave structures. In: *Proceddings of the IEEE Ultrasonic Symposium*, Vol.1, 475–479.
78. Haake, A, and Dual, J (2004) Positioning of small particles by an ultrasound field excited by surface waves. Ultrasonics 42 (1–9), 75–80
79. Haake, A, and Dual, J (2005) Contactless micromanipulation of small particles by an ultrasound field excited by a vibrating body. Journal of the Acoustical Society of America 117 (5), 2752–2760
80. Haake, A, Neild, A, Kim, DH, Ihm, JE, Sun, Y, Dual, J, and Ju, BK (2005) Manipulation of cells using an ultrasonic pressure field. Ultrasound in Medicine and Biology 31 (6), 857–864
81. Hench, LL, and West, JK (1990) The Sol–gel process. Chemical Reviews (90), 33–72
82. Muralt, P, Ledermann, N, Baborowski, J, Barzegar, A, Gentil, S, Belgacem, B, Petitgrand, S, Bosseboeuf, A, and Setter, N (2005) Piezoelectric micromachined ultrasonic transducers based on PZT thin films. IEEE Transactions on Ultrasonics Ferroelectrics and Frequency Control 52 (12), 2276–2288
83. McNab, A, Kirk, KJ, and Cochran, A (1998) Ultrasonic transducers for high temperature applications. IEE Proceedings: Science Measurement and Technology 145 (5), 229–236
84. Kazys, R, Voleisis, A, Sliteris, L, Mazeika, L, Van Nieuwenhove, R, Kupschus, P, and Abderrahim, HA (2005) High temperature ultrasonic transducers for imaging and measurements in a liquid Pb/Bi eutectic alloy. IEEE Transactions on Ultrasonics Ferroelectrics and Frequency Control 52 (4), 525–537

85. Park, SEE, and Hackenberger, W (2002) High performance single crystal piezoelectrics: Applications and issues. Current Opinion in Solid State & Materials Science 6 (1), 11–18
86. Marin-Franch, P, Cochran, S, and Kirk, K (2004) Progress towards ultrasound applications of new single crystal materials. Journal of Materials Science: Materials in Electronics 15 (11), 715–720
87. Martin, SP, Townsend, RJ, Kuznetsova, LA, Borthwick, KAJ, Hill, M, McDonnell, MB, and Coakley, WT (2005) Spore and micro-particle capture on an immunosensor surface in an ultrasound standing wave system. Biosensors and Bioelectronics 21 (5), 758–767
88. Wiklund, M, Günther, C, Lemor, R, Jäger, MGF, and Hertz, HM (2006) Ultrasonic standing wave manipulation technology integrated into a dielectrophoretic chip. Lab on a Chip 6, 1537–1544
89. Khanna, S, Amso, NN, Paynter, SJ, and Coakley, WT (2003) Contrast agent bubble and erythrocyte behavior in a 1.5-MHz standing ultrasound wave. Ultrasound in Medicine and Biology 29 (10), 1463–1470
90. Hawkes, JJ, Barrow, D, and Coakley, WT (1998) Microparticle manipulation in millimetre scale ultrasonic standing wave chambers. Ultrasonics 36 (9), 925–931
91. Hultström, J, Manneberg, O, Dopf, K, Hertz, HM, Brismar, H, and Wiklund, M (2006) Proliferation and viability of adherent cells manipulated by standing-wave ultrasound in a microfluidic chip. Ultrasound Meidicine and Biology 33, 175–181
92. Bohm, H, Anthony, P, Davey, MR, Briarty, LG, Power, JB, Lowe, KC, Benes, E, and Groschl, M (2000) Viability of plant cell suspensions exposed to homogeneous ultrasonic fields of different energy density and wave type. Ultrasonics 38 (1–8), 629–632
93. Wang, ZW, Grabenstetter, P, Feke, DL, and Belovich, JM (2004) Retention and viability characteristics of mammalian cells in an acoustically driven polymer mesh. Biotechnology Progress 20 (1), 384–387
94. Gherardini, L, Cousins, CM, Hawkes, JJ, Spengler, J, Radel, S, Lawler, H, Devcic-Kuhar, B, and Groschl, M (2005) A new immobilisation method to arrange particles in a gel matrix by ultrasound standing waves. Ultrasound in Medicine and Biology 31 (2), 261–272
95. Caton, PF, and White, RM (2001) MEMS microfilter with acoustic cleaning. In: Technical Bigest MEMs2007: 14th IEEE International Conference on Micro Electro Mechanical Systems, Piscataway, NJ, IEEE press, pp 479–482
96. Cousins, CM, Holownia, P, Hawkes, JJ, Limaye, MS, Price, CP, Keay, PJ, and Coakley, WT (2000) Plasma preparation from whole blood using ultrasound. Ultrasound in Medicine and Biology 26 (5), 881–888
97. Cousins, CM, Holownia, P, Hawkes, JS, Price, CP, Keay, P, and Coakley, WT (2000) Clarification of plasma from whole human blood using ultrasound. Ultrasonics 38 (1–8), 654–656
98. Cousins, CM, Melin, JR, Venables, WA, and Coakley, WT (2000) Investigation of enhancement of two processes, sedimentation and conjugation, when

bacteria are concentrated in ultrasonic standing waves. Bioseparation 9 (6), 343–349

99. Hawkes, JJ, and Coakley, WT (1996) A continuous flow ultrasonic cell-filtering method. Enzyme and Microbial Technology 19 (1), 57–62
100. Hawkes, JJ, Limaye, MS, and Coakley, WT (1997) Filtration of bacteria and yeast by ultrasound-enhanced sedimentation. Journal of Applied Microbiology 82 (1), 39–47
101. Frank, A, Bolek, W, Groschl, M, Burger, W, and Benes, E (1993) Separation of suspended particles by use of the inclined resonator concept. In *Ultrasonics International 93 – Conference Proceedings*, 519–522
102. Hawkes, JJ, Barrow, D, Cefai, J, and Coakley, WT (1998) A laminar flow expansion chamber facilitating downstream manipulation of particles concentrated using an ultrasonic standing wave. Ultrasonics 36 (8), 901–903
103. Harris, NR, Hill, M, Townsend, RJ, White, NM, and Beeby, SP (2005) Performance of a micro-engineered ultrasonic particle manipulator. Sensors and Actuators B: Chemical 111, 481–486
104. Harris, NR, Hill, M, Torah, RN, Townsend, RJ, Beeby, SP, White, NM, and Ding, J (2006) A multilayer thick-film PZT actuator for MEMs applications. Sensors and Actuators A: Physical 132, 311–316
105. Townsend, RJ, Hill, M, Harris, NR, White, NM, Beeby, SP, and Wood, RJK (2005) Fluid modelling of microfluidic separator channels. Sensors and Actuators B: Chemical 111, 455–462
106. Petersson, F, Nilsson, A, Holm, C, Jonsson, H, and Laurell, T (2004) Separation of lipids from blood utilizing ultrasonic standing waves in microfluidic channels. Analyst 129 (10), 938–943
107. Petersson, F, Nilsson, A, Holm, C, Jonsson, H, and Laurell, T (2005) Continuous separation of lipid particles from erythrocytes by means of laminar flow and acoustic standing wave forces. Lab on a Chip 5 (1), 20–22
108. Goddard, G, and Kaduchak, G (2005) Ultrasonic particle concentration in a line-driven cylindrical tube. Journal of the Acoustical Society of America 117 (6), 3440–3447
109. Grossner, MT, Belovich, JM, and Feke, DL (2005) Transport analysis and model for the performance of an ultrasonically enhanced filtration process. Chemical Engineering Science 60 (12), 3233–3238
110. Gupta, S, and Feke, DL (1998) Filtration of particulate suspensions in acoustically driven porous media. AIChE Journal 44 (5), 1005–1014
111. Hertz, HM (1995) Standing-wave acoustic trap for nonintrusive positioning of microparticles. Journal of Applied Physics 78 (8), 4845–4849
112. Johann, RM (2006) Cell trapping in microfluidic chips. Analytical And Bioanalytical Chemistry 385 (3), 408–412
113. Wiklund, M, Nilsson, S, and Hertz, HM (2001) Ultrasonic trapping in capillaries for trace-amount biomedical analysis. Journal of Applied Physics 90 (1), 421–426
114. Lilliehorn, T, Nilsson, M, Simu, U, Johansson, S, Almqvist, M, Nilsson, J, and Laurell, T (2005) Dynamic arraying of microbeads for bioassays in microfluidic channels. Sensors And Actuators B: Chemical 106 (2), 851–858

115. Yasuda, K (2000) Non-destructive, non-contact handling method for biomaterials in micro-chamber by ultrasound. Sensors and Actuators B: Chemical 64 (1–3), 128–135
116. Barnes, RA, Jenkins, P, and Coakley, WT (1998) Preliminary clinical evaluation of meningococcal disease and bacterial meningitis by ultrasonic enhancement. Archives of Disease in Childhood 78 (1), 58–60
117. Gray, SJ, Sobanski, MA, Kaczmarski, EB, Guiver, M, Marsh, WJ, Borrow, R, Barnes, RA, and Coakley, WT (1999) Ultrasound-enhanced latex immunoagglutination and PCR as complementary methods for non-culture-based confirmation of meningococcal disease. Journal of Clinical Microbiology 37 (6), 1797–1801
118. Jenkins, P, Barnes, RA, and Coakley, WT (1997) Detection of meningitis antigens in buffer and body fluids by ultrasound-enhanced particle agglutination. Journal of Immunological Methods 205 (2), 191–200
119. Sobanski, MA, Vince, R, Biagini, GA, Cousins, C, Guiver, M, Gray, SJ, Kaczmarski, EB, and Coakley, WT (2002) Ultrasound enhanced detection of individual meningococcal serogroups by latex immunoassay. Journal of Clinical Pathology 55 (1), 37–40
120. Bazou, D, Coakley, WT, Meek, KM, Yang, M, and Pham, DT (2004) Characterisation of the morphology of 2-D particle aggregates in different electrolyte concentrations in an ultrasound trap. Colloid and Surface A: Physicochemical Engineering Aspects 243 (1–3), 97–104
121. Spengler, JF, and Coakley, WT (2003) Ultrasonic trap to monitor morphology and stability of developing microparticle aggregates. Langmuir 19 (9), 3635–3642
122. Lee, YH, and Peng, CA (2005) Enhanced retroviral gene delivery in ultrasonic standing wave fields. Gene Therapy 12 (7), 625–633
123. Khanna, S, Hudson, B, Pepper, CJ, Amso, NN, and Coakley, WT (2006) Fluorescein isothiocynate-dextran uptake by chinese hamster ovary cells in a 1.5 MHz ultrasonic standing wave in the presence of contrast agent. Ultrasound In Medicine and Biology 32 (2), 289–295
124. Kuznetsova, LA, Khanna, S, Amso, NN, Coakley, WT, and Doinikov, AA (2005) Cavitation bubble-driven cell and particle behavior in an ultrasound standing wave. Journal of the Acoustical Society of America 117 (1), 104–112
125. Hawkes, JJ, Long, MJ, Coakley, WT, and McDonnell, MB (2004) Ultrasonic deposition of cells on a surface. Biosensors & Bioelectronics 19 (9), 1021–1028
126. Zourob, M, Hawkes, JJ, Coakley, WT, Brown, BJT, Fielden, PR, McDonnell, MB, and Goddard, NJ (2005) Optical leaky waveguide sensor for detection of bacteria with ultrasound attractor force. Analytical Chemistry 77 (19), 6163–6168
127. Hawkes, JJ, Barber, RW, Emerson, DR, and Coakley, WT (2004) Continuous cell washing and mixing driven by an ultrasound standing wave within a microfluidic channel. Lab on a Chip 4 (5), 446–452

128. Petersson, F, Nilsson, A, Jonsson, H, and Laurell, T (2005) Carrier medium exchange through ultrasonic particle switching in microfluidic channels. Analytical Chemistry 77 (5), 1216–1221
129. Masudo, T, and Okada, T (2006) Particle separation with ultrasound radiation force. Current Analytical Chemistry 2 (2), 213–227
130. Giddings, JC (1993) Field-flow fractionation: Analysis of macromolecular, colloidal, and particulate materials. Science 260, 1456–1465
131. Aboobaker, N, Meegoda, JN, and Blackmore, D (2003) Fractionation and segregation of suspended particles using acoustic and flow fields. ASCE Journal of Environmental Engineering 129 (5), 427–434
132. Kumar, M, Feke, DL, and Belovich, JM (2005) Fractionation of cell mixtures using acoustic and laminar flow fields. Biotechnology and Bioengineering 89 (2), 129–137
133. Kapishnikov, S, Kantsler, V, and Steinberg, V (2006) Continuous particle size separation and size sorting using ultrasound in a microchannel. Journal of Statistical Mechanics: Theory and Experiment
134. Yasuda, K, Umemura, S, and Takeda, K (1996) Particle separation using acoustic radiation force and electrostatic force. Journal of the Acoustical Society of America 99 (4), 1965–1970
135. Wiklund, M, Spegel, P, Nilsson, S, and Hertz, HM (2003) Ultrasonic-trap-enhanced selectivity in capillary electrophoresis. Ultrasonics 41 (4), 329–333

# Chapter 10
# Electrophoresis in Microfluidic Systems

Victor M. Ugaz and Jennifer L. Christensen

Artie McFerrin Department of Chemical Engineering, Texas A&M University, College Station, TX 77843, USA.

## 1. Introduction

Many chemical and biochemical analysis methods involve performing a sequence of processes that can be broadly classified in terms of sample preparation, reactions, and product analysis. Since the reaction products often contain mixtures of multiple chemical species, subsequent analytical steps must be capable of separating and identifying the individual components. Electrophoresis, which relies on inducing detectable differences in migration behavior between charged species under the influence of an applied electric field, has proven to be a highly versatile analytical technique owing to a favorable combination of characteristics including relatively simple hardware design and compatibility with a wide range of analytes including biological macromolecules (e.g., DNA, proteins). More recently, there has been considerable interest in adapting electrophoresis technology to miniaturized microfluidic formats with the aim of producing portable low-cost versions of conventional benchtop-scale instrumentation. Ultimately, it is envisioned that these efforts will enable electrophoresis to become an integral component of self-contained "lab-on-a-chip" devices capable of putting the power to perform a variety of sophisticated chemical, biological, and biomedical assays directly in the hands of those who need the information most.

On the macroscale, the development of capillary electrophoresis (CE) marked a breakthrough that overcame many limitations of early slab gel

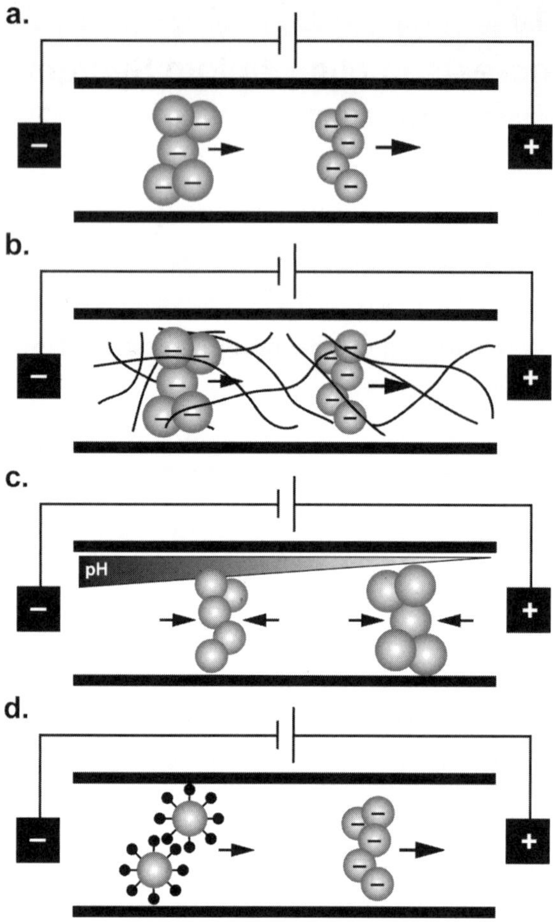


**Fig. 1.** Electrophoresis modes employed in miniaturized systems include (**a**) free solution electrophoresis, (**b**) gel electrophoresis, (**c**) isoelectric focusing, and (**d**) micellar electrokinetic chromatography (negatively charged analytes are depicted)

instruments. The use of ultra-narrow capillaries (typically 50–100 μm inner diameter) provided more efficient heat dissipation and allowed higher electric fields to be applied while minimizing temperature nonuniformities arising from Joule heating effects. This helped make it possible to achieve faster analyte mobilities and shorter run times without sacrificing separation performance. In addition to these benefits, the capillary format inherently requires reduced sample and reagent quantities and offers a greater capacity for automation. Since capillary cross-sectional dimensions are on the same scale as typical microfluidic channels, it is not surprising that many miniaturization efforts have focused on direct scale-down of proven

CE technologies. A number of different microscale implementations of this fundamental phenomenon have been explored in order to suit the requirements of specific analytes and applications, including adaptations of slab gel and capillary-based techniques as well as chromatographic methods [16, 30, 59, 61, 91, 106, 112, 146, 153, 175, 200, 206, 301, 342]. Some of the most common variations that have been reported include the following (Fig. 1).

### *1.1. Free Solution Electrophoresis*

This technique involves separation of molecular analytes on the basis of their migration speed in response to an applied electric field (typically expressed in terms of electrophoretic mobility $\mu = v/E$ where $v$ is the migration speed and $E$ is the electric field strength). When analytes are suspended in an ionic buffer environment at a specific pH, each species migrates with a different mobility allowing them to be resolved as distinct zones and separated on the basis of size and charge (Fig. 1a). Depending on the substrate material used to construct the separation channel, surface charge effects may also induce a bulk electroosmotic flow component superimposed on the analytes' electrophoretic migration. This can be advantageous or detrimental to the separation depending on the relative magnitudes of electrophoretic and electroosmotic effects. Consequently, the ability to precisely control surface charge and buffer pH is critical to achieving optimal separation performance.

### *1.2. Gel Electrophoresis*

This approach differs from free solution methods by the introduction of a sieving matrix material (often a polymer gel) into the separation channel, and is frequently employed for separation of DNA and proteins. The use of a sieving matrix is essential in the case of DNA, where the free draining molecular configuration adopted in free solution results in mobilities that are independent of fragment size. The gel matrix reintroduces a size dependence to the electrophoretic migration as analytes travel through the porous gel network, with smaller fragments experiencing less resistance and eluting faster (Fig. 1b). Gel electrophoresis also offers advantages associated with decreased diffusional broadening of the separated zones and a reduction in electroosmotic flow due to adsorption of the gel matrix along the microchannel walls that neutralizes surface charges. The need to load or cast a viscous polymer gel into a small-diameter microchannel, however, can pose challenges.

### 1.3. Isoelectric Focusing (IEF)

This method is often applied in protein separations and involves a process whereby analytes are separated on the basis of their isoelectric points. Here, migration occurs in free solution through a pH gradient established along the length of the separation channel. Electrophoretic migration continues until each analyte reaches a location where the local pH renders it neutrally charged (Fig. 1c). The quality and reproducibility of the pH gradient are key factors that determine achievable separation performance associated with IEF techniques.

### 1.4. Micellar Electrokinetic Chromatography (MEKC)

This is a hybrid technique involving a combination of electrophoretic and chromatographic principles that can be applied to the separation of both neutral and charged species. Here, negatively charged micelles are formed by adding an appropriate concentration of surfactant to the buffer solution. Since the micelles are formed by association of the hydrophobic and hydrophilic groups comprising individual surfactant molecules, the hydrophobic or hydrophilic character of the analyte dictates the extent of interactions with the micelles. Consequently, an analyte's mobility is determined by an interplay between charge, size, and hydrophobicity. Hydrophobic analytes will preferentially interact with the hydrophobic micellar interior resulting in net migration speeds close to that of the micelles, while the migration of hydrophilic species will not be strongly affected. Separation of neutral species can be achieved by superimposing a bulk electroosmotic flow.

It should be emphasized that the techniques described above, while encompassing the majority of methods adapted to microfabricated systems, represent a subset of the much broader field of electrophoresis technology. Miniaturized electrochromatography techniques represent another important class of analytical techniques, and are reviewed separately in an accompanying chapter. Here we focus on developments reported in refereed journals, with the understanding that additional studies may be documented in sources such as conference proceedings and patent literature.

## 2. Electrophoresis in Microfabricated Systems

The design of most microfabricated electrophoresis systems is relatively simple, consisting of the following fundamental elements: (1) a sample injection zone, (2) an electrophoresis separation channel, and (3) a system for detection of the migrating analytes (Fig. 2). Despite this underlying simplicity, advances in micro- and nanofabrication technology have enabled greater sophistication to be incorporated through integration with other on-chip processes to achieve enhanced throughput and efficiency.

In terms of economics, the benefits of miniaturization include reduced reagent consumption and an increased capacity for automation so that the costs associated with performing chemical and biochemical reactions can be dramatically lowered. In terms of hardware, the use of batch photolithography-based microfabrication and micromachining technology allows hundreds or thousands of devices to be produced simultaneously at virtually the same cost per wafer. Consequently, as has been repeatedly demonstrated in the microelectronics industry, the cost benefits of microfabrication become most compelling when the device size becomes as small as possible. For electrophoresis applications, this means that the ability to construct separation columns of sufficient length to deliver the required level of resolution and sensitivity while occupying the most compact allowable on-chip area is of critical importance.

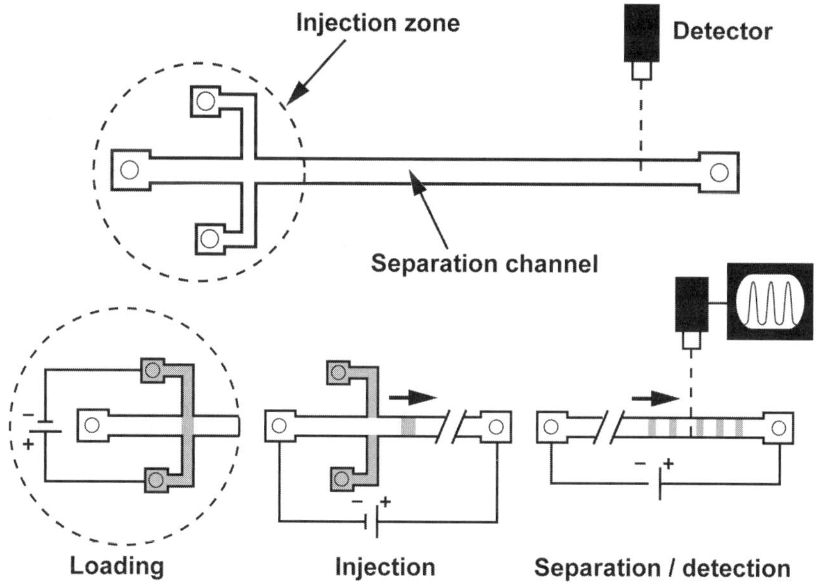


**Fig. 2.** A typical microscale electrophoresis run begins by electrokinetically introducing a sample into the device, after which the voltage is switched so that a narrow band is injected into the separation channel. Different species migrate with different mobilities and separate into distinct zones that are detected downstream

### 2.1. Injection and Separation

To effectively separate comigrating species, the difference in their electrophoretic mobilities must allow them to move apart and separate into distinct zones at a faster rate than the zones broaden due to the cumulative effects of diffusion and dispersion during the elution time. Often, a finish-line mode of detection is employed whereby an optical or electronic signal is generated as the migrating species travel past a fixed downstream detection point. The ability to distinguish neighboring zones or peaks can then be quantified in terms of a parameter called the *separation resolution R* that expresses the ratio of the distance between peaks ($t_2 - t_1$) to the sum of their half-widths at the base [95] (Fig. 3). If the zones follow a Gaussian profile, the half width of each peak can be taken as twice its standard deviation $\sigma$, yielding the following expression

$$R = \frac{t_2 - t_1}{2(\sigma_1 + \sigma_2)}. \tag{1}$$

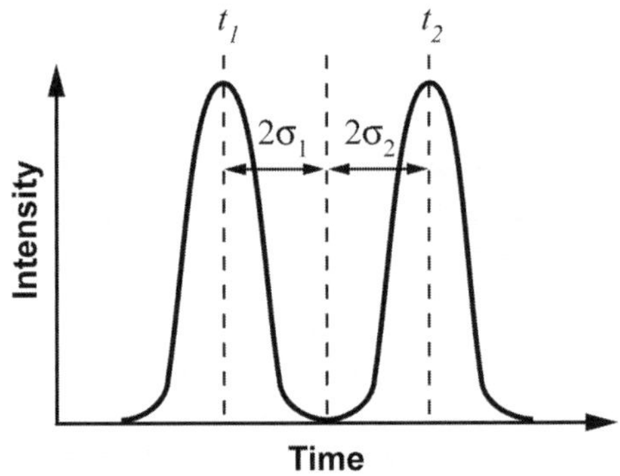


**Fig. 3.** In a typical electrophoresis experiment, the intensity of fluorescently labeled migrating analytes is recorded as they travel past a fixed detection point in the separation channel. The ability to distinguish two neighboring zones can be expressed in terms of the separation between peaks relative to their widths

Peak widths may also be expressed in terms of their full width at half maximum (FWHM), which can be related to the standard deviation according to FWHM = $\left[2\sqrt{2\ln 2}\right]\sigma$ such that (1) becomes

$$R = \left[\sqrt{2\ln 2}\right]\frac{t_2 - t_1}{(\mathrm{FWHM}_1 + \mathrm{FWHM}_2)}. \tag{2}$$

By convention, a value of $R \geq 0.5$ is often taken as a criterion to indicate that two neighboring peaks are clearly distinguishable.

The observed widths of the migrating zones reflect contributions from a number of factors including diffusional broadening, the initial width of the injected sample plug, and the finite detection volume. Considering only the diffusional contribution, the quantities in (1) can be expressed in terms of mobility (i.e., the migration speed) and a diffusion coefficient (i.e., the rate of band broadening during electrophoretic migration) yielding the following equivalent expression for separation resolution [18, 188]

$$R = \frac{1}{4}\left(\frac{\Delta\mu}{\mu}\right)\frac{L}{\sqrt{2D(L/\mu E)}}. \tag{3}$$

Here, the term $\Delta\mu/\mu$ is the *selectivity* (i.e., the relative mobilities between neighboring zones), $\mu$ is the average mobility of the neighboring zones, and $L$ is the separation length (i.e., the distance between injection and detection points). $D$ is a coefficient characterizing the rate of longitudinal zone broadening and is generally larger than the field-free diffusion coefficient. An important point to note from (3) is that resolution scales with separation length as $R \propto \sqrt{L}$, establishing a limit on the extent to which the length of a separation column can be reduced. Separation lengths can vary widely (millimeters to tens of centimeters) depending on the specific application, sample type, and resolution requirements. In cases where $L$ cannot be reduced without adversely affecting performance, longer separation columns can be folded into spiral or serpentine geometries, often with specially designed turns to minimize dispersion (Fig. 4). These designs permit a more compact arrangement on the chip surface (Table 1).

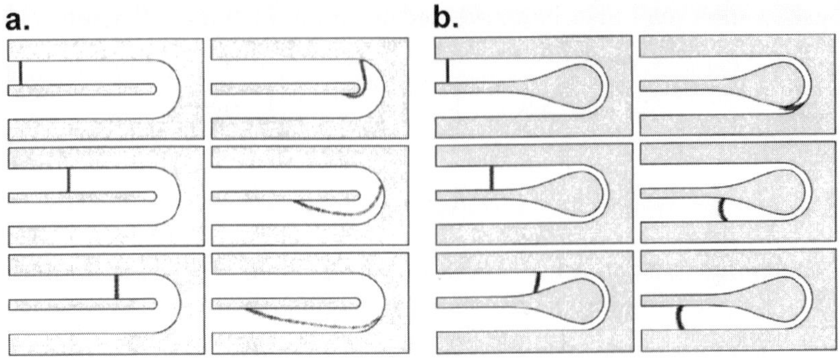


**Fig. 4.** Numerically simulated analyte transport through a U-shaped microchannel segment with **(a)** a uniform cross-section and **(b)** a low-dispersion turn geometry to minimize distortion of the migrating zone [100] (reproduced with permission, copyright 2001 American Chemical Society)

**Table 1.** Electrophoresis microchips incorporating separation channels folded into compact geometries

| Compact geometry method | References |
|---|---|
| Serpentine | [21, 44, 126, 131, 135, 136, 143, 187, 215, 238, 269, 273, 287] |
| Serpentine (low dispersion geometry) | [15, 58, 99, 100, 166, 207, 226–228, 236, 268, 297, 337] |
| Serpentine (surface modification) | [139, 169] |
| Spiral | [45, 98] |

The process by which samples are injected into the separation channel is also important. Injection of a nonconcentrated and unfocused sample zone not only requires a long separation distance in order to distinguish each component, but the corresponding signal from each species may fall below the detectable range as the zones spread by diffusion. Injection of a concentrated and focused sample zone allows each component to be detected in a considerably shorter separation distance. A variety of schemes are possible, the majority of which involve a perpendicularly crossed channel geometry where analytes are electrokinetically transported across the separation channel after which the voltage is switched such that only the sample volume at the intersection is injected [71, 107, 108, 131, 135, 140, 151, 154, 159, 167, 179, 194, 239, 241, 272, 305, 339] (Fig. 2). A sampling of other techniques includes bulk flow [123], capillary [271], diffusion-based [270], pressure-driven [170], hydrodynamic [8], dielectrophoretic trapping [2, 3, 40], nanocapillary array interconnects [28], combined electrophoretic and electroosmotic processes at microchannel–nanochannel intersections [49, 56, 318], and on-chip microelectrode arrays [19, 65, 177, 178, 262].

### 2.2. Sieving Gels

A variety of sieving matrix formulations have been employed in electrophoresis microdevices, most notably in applications involving DNA and protein separations where analyte mobilities in free solution are either not size dependent or only weakly so [1, 330] (Table 2). Polyacrylamide gels (both

**Table 2.** Sieving matrix materials used in electrophoresis microchips

| Gel matrix | References |
|---|---|
| Agarose | [119, 120, 155, 300] |
| Bacterial cellulose fibrils | [283] |
| GeneScan polymer | [264, 286, 303] |
| Hydroxyethyl cellulose (HEC) | [77, 118, 125, 142, 144, 163–165, 202, 203, 217, 226, 249, 263, 265, 271, 285, 288–291, 321, 322, 324–326] |
| Hydroxypropyl cellulose (HPC) | [70, 74, 250, 251] |
| HPC/HEC combination | [313] |
| Hydroxypropylmethyl cellulose (HPMC) | [316, 328, 329] |
| HPMC/latex nanoparticles | [284] |
| Methylcellulose | [52, 235, 274, 294–296, 341] |
| Pluronic | [278, 300] |
| Polyacrylamide (noncrosslinked; LPA) | [15, 27, 34, 72, 73, 96, 101, 151, 157, 166, 183, 184, 187, 204, 205, 207, 211, 220, 221, 227, 228, 239, 240, 248, 253–256, 271, 273, 287, 309, 310, 323, 332, 337] |
| LPA-*co*-alkylacrylamide | [38] |
| LPA-*co*-dihexylacrylamide | [39] |
| LPA nanogel | [58, 261] |
| Polyacrylamide (crosslinked, photopolymerized) | [19, 32, 115, 116, 177, 178, 222, 229, 262, 298–300] |
| Polyacrylamide (crosslinked, chemically polymerized) | [24, 306] |
| Polydimethylacrylamide (PDMA) | [68, 75, 152, 190, 192, 252, 292, 319, 320] |
| Polyethylene glycol (PEG)/polylactic acid (PLA) nanospheres | [281] |
| Polyethylene oxide (PEO) | [293, 331] |
| POP-6 | [6, 7, 85–87] |
| Polyvinylpyrrolidone (PVP) | [12, 154, 217, 235, 244, 297, 344] |
| SDS 14-200 | [260] |
| SNAP solution | [44] |

crosslinked and noncrosslinked) are widely used owing to favorable properties including optical transparency, electro-neutrality, and overall separation performance. Crosslinked gels generally yield a denser pore network however, the necessity to polymerize a gel inside the separation channel means the matrix is generally not replaceable, limiting device reusability. Photopolymerization techniques have been used to precisely position gels inside microchannel networks by using a photomask to selectively illuminate the device with UV light after which the unpolymerized reagents can be easily removed. Noncrosslinked polyacrylamide gels (i.e., linear polyacrylamide; LPA) that form entangled networks at concentrations exceeding a critical threshold are also useful for electrophoretic separations and are replaceable. Since formation of this entanglement network is accompanied by an increase in viscosity, the pressure required to inject the gel matrix into the microseparation channel is one factor that limits the available range of polymer concentrations. Thermoreversible gels such as agarose (a polysaccharide) and Pluronic (an amphiphillic block copolymer) have also been investigated as a means to aid loading and allow for replacement of the sieving matrix. These formulations undergo a thermally induced transition from gel-like to liquid-like behavior over a specific temperature range. Other sieving gels that have been studied include hydroxyethyl cellulose, hydroxypropyl cellulose, methylcellulose, polydimethylacrylamide, and polyvinylpyrrolidone.

### *2.3. Detection*

Another critical component of electrophoresis microdevices is the ability to detect migration of the separated zones (Table 3). Typically, a laser-induced fluorescence (LIF) system is employed whereby fluorescently labeled species are illuminated by a laser (e.g., argon ion) that excites fluorophores conjugated with the migrating analytes. The resulting fluorescence signal is filtered to block background illumination from the excitation source and recorded using a photodetector (e.g., photomultiplier tube). In some cases, optical microscope systems are employed whereby the electrophoresis chip is mounted on a stage and the fluorescently labeled migrating zones are observed and recorded under magnification using a digital (e.g., CCD) camera. Under sufficiently high magnification, it is even possible to observe the dynamics of individual DNA molecules as they migrate through the separation channel (Fig. 5).

**Table 3.** Detection techniques employed in electrophoresis microchips

| Detection technique | References |
|---|---|
| Absorbance | [122, 176, 197] |
| Chemiluminescence | [124] |
| Electrochemical | [8, 11, 77, 93, 94, 150, 155, 168, 193, 198, 199, 279, 326] |
| Infrared | [231] |
| Laser-induced fluorescence (LIF) | [4, 6, 7, 9, 12, 15, 21, 22, 27, 34–39, 42–48, 51, 52, 58, 68, 70–75, 78–88, 92, 96–98, 101, 102, 107–109, 113, 116, 118, 125, 126, 130–138, 142, 143, 144, 147, 149, 151, 152, 154, 157, 160–167, 181, 183–187, 190, 192, 194–196, 201, 203–205, 207, 211, 215, 217, 218, 223, 224, 226–228, 230, 235–237, 239, 240, 244–246, 248–258, 260, 261 263–265, 268, 269, 271, 273, 274, 285–293, 303, 306, 309–313, 315, 316, 319, 320, 322–325, 332, 333, 335–337, 340, 344, 345] |
| Microscope/digital camera | [5, 10, 19, 25, 26, 29, 32, 41, 64, 104, 105, 114, 115, 119–121, 141, 156, 177–179, 202, 208, 213, 214, 220, 221, 229, 238, 242, 243, 262, 266, 267, 276, 278, 294–296, 298–300, 317, 327, 334, 338] |
| Mass spectrometry | [55, 145, 173, 174, 182, 219] |
| On-chip photodiode | [24, 222, 321] |
| Raman spectroscopy | [314] |
| Single molecule imaging | [39, 66, 69, 103, 127, 209, 297, 307] |
| Two beam fluorescence cross correlation spectroscopy | [20] |

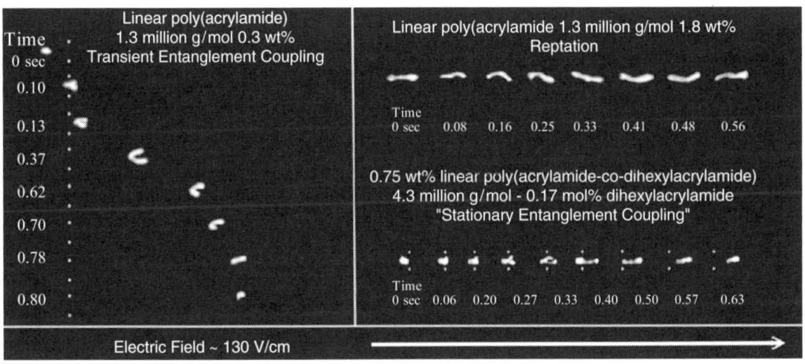


**Fig. 5.** Images of single λ-DNA molecules migrating through three different gel matrices [39]. The gel concentration and composition both influence the dynamics of interactions with the matrix during electrophoresis (reproduced with permission, copyright 2006 American Chemical Society)

Optical detection techniques generally require an illumination source and photodetection components that in many cases contribute significantly to the overall size of the system. These issues have stimulated interest in the development of miniaturized detection technology, as has been demonstrated through construction of silicon-based microelectrophoresis chips incorporating on-chip integration of photodiode circuitry. Nonoptical detection techniques have also been investigated as a means of providing more compact device designs. Electrochemical detection, for example, involves monitoring electrochemical potential variations as analytes migrate past a working electrode positioned within the separation channel [304].

### 2.4. Device Construction

The majority of electrophoresis microchips are constructed using silicon or glass substrates (Table 4). Typical microchannel cross-sectional dimensions range from tens to hundreds of microns in width to 5–100 μm in depth. The substrate containing the microchannel network is then bonded to a corresponding flat piece of glass or silicon in order to form a sealed enclosure. In the case of electrophoresis, at least one of the substrates is typically glass or quartz in order to provide optical transparency for analyte detection. Access holes are drilled in order to inject surface treatments, sieving gels, and samples into the microchannel. Subsequent packaging may involve adding connections to external liquid handling hardware and electronic voltage or temperature control components. The use of silicon substrates can be advantageous by offering the potential to incorporate on-chip electrodes, heaters, temperature sensors, and photodetection circuitry.

Plastic substrates have also been explored for use in electrophoresis. One of the most widely used plastic substrates in microfluidics is polydimethylsiloxane (PDMS), a silicone rubber possessing excellent optical transparency and the ability to form strong bonds with both glass and PDMS surfaces. The use of plastic substrates can considerably simplify fabrication by allowing molding or *soft lithography* processes to be employed whereby microchannel structures are cast against a rigid master mold. PDMS is an elastomeric material, a property that can also be exploited to construct mechanical fluidic control components such as on-chip valves and pumps. Unfortunately, many plastics are gas permeable making it difficult to polymerize gels inside the microchannel because any oxygen present within the substrate inhibits the reaction.

**Table 4.** Substrate materials used to construct electrophoresis microchips

| Substrate material | References |
|---|---|
| Acrylic | [114, 202] |
| Calcium fluoride | [231] |
| Glass, fused silica | [4, 6, 7, 9, 11, 12, 17, 21, 22, 27, 34–39, 42–48, 55, 58, 68, 71–73, 75–82, 84–88, 90, 92, 96–98, 101, 102, 107, 108, 115, 116, 118, 121, 125, 126, 128–133, 136–138, 142, 142–144, 150–152, 154, 157, 160–169, 173, 174, 176, 181–186, 190, 191–196, 201, 203–205, 207, 211, 215–218, 223, 226–228, 230, 236–240, 242–244, 248–251, 253–258, 263–268, 271, 276, 279, 280, 285, 286, 288–292, 303, 309, 310, 314–316, 319, 320, 322–326, 332, 333, 336, 337, 340, 343, 344] |
| Glass/polyester film | [122] |
| Polydimethylsiloxane (PDMS) | [64, 94, 127, 198, 199, 208, 209, 213, 214, 224, 327] |
| PDMS/glass | [8, 15, 69, 70, 74, 113, 119, 124, 147, 155, 156, 162, 269, 334] |
| Polyethylene terephthalate glycol (PETG) | [111] |
| Polymethylmethacrylate (PMMA) | [50–54, 89, 139, 149, 179, 187, 219–221, 235, 246, 252, 260, 261, 273, 274, 287, 293, 295, 311–313, 328, 329, 331, 335, 341] |
| PMMA/glass | [20] |
| PMMA/PDMS | [120] |
| Polyvinyl chloride embedded capillary | [306] |
| Quartz | [10, 13, 42, 83, 93, 134, 135, 141, 197, 294, 296, 297, 345] |
| Silicon | [26] |
| Silicon/glass | [5, 19, 23–25, 29, 32, 66, 103–105, 109, 177, 178, 222, 225, 229, 262, 298–300, 307, 338] |
| Silicon/parylene | [321] |
| Silicon/silicon oxide | [41] |
| Thermoplastic elastomer | [278] |
| Vivak | [317] |
| Zeonor | [145] |

While glass substrates are desirable for use in electrophoresis applications due to their optical transparency, they also possess a characteristic negative surface charge at pH values above ~3 due to deprotonation of surface silanol (Si-OH) groups to form silanoate ($Si–O^-$). The resulting negative surface charge attracts positively charged counterions dispersed in the buffer solution such that a thin layer near the channel walls is formed where their concentration is locally in excess of the bulk value. When an electric field is applied, this positively charged counterion layer migrates toward the negative electrode in a sheath-like manner, transporting the bulk liquid with it (Fig. 6). This electroosmotic flow (EOF) phenomenon may be advantageous in terms of providing a means to pump liquids and reagents in microchannels, but is generally undesirable for DNA separations because the EOF direction is opposite to that of the electrophoretic migration of negatively charged DNA.

The magnitude and reproducibility of EOF can also be challenging to precisely control because it is highly sensitive to surface chemistry and buffer pH. Electroosmotic flow phenomena have been extensively studied in conventional CE systems where glass and fused silica capillaries are routinely used. These efforts have resulted in the development of robust coating procedures capable of effectively neutralizing surface charges [14, 57, 60] (Table 5). The most widely used process, originally developed by Hjertén [117], involves covalently polymerizing a thin layer of polyacrylamide or other passivating polymer along the inner surface of the capillary or microchannel. Electroosmotic effects are not as problematic in gel electrophoresis because the gel matrix often acts to neutralize or screen the negative surface charge. Characterization and control of EOF is less well studied in other substrate materials (e.g., plastics), but is an active area of ongoing research.

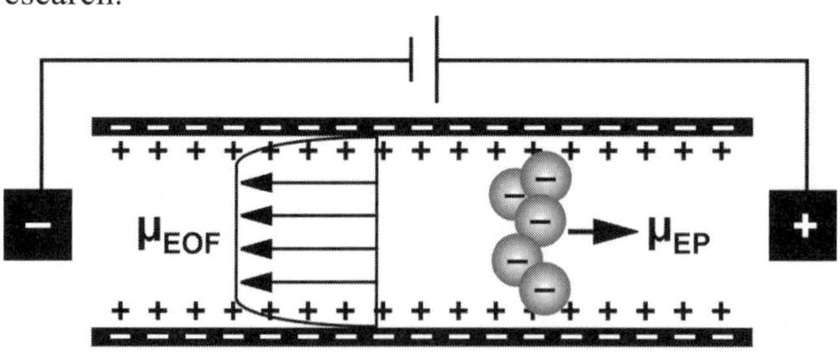


**Fig. 6.** The intrinsically negative surface charge along the walls of a glass microchannel induces formation of a positively charged counterion layer that is transported toward the cathode when an electric field is applied. This layer acts as a sheath around the bulk liquid, generating an electroosmotic flow with characteristic speed $\mu_{EOF}$. These EOF effects are superimposed on an analyte's electrophoretic migration ($\mu_{EP}$) and may act to either reinforce or oppose it

**Table 5.** Surface passivation techniques used in electrophoresis microchips

| Surface passivation technique | References |
|---|---|
| [Acryloylamino)propyl] trimethylammonium chloride (BCQ) | [55, 173, 174] |
| Bio-Rad run buffer | [152] |
| HCl | [119] |
| HCl, NaOH | [92, 113, 264] |
| Hydroxyethyl cellulose (HEC) | [249] |
| Hjertén process | [15, 27, 34, 58, 68, 75, 77, 96, 101, 125, 134, 136, 142, 144, 151, 157, 163–166, 183, 184, 197, 203–205, 207, 211, 217, 218, 227, 228, 248, 253–256, 263, 265, 285, 292, 294, 296, 319, 320, 322, 336, 337] |
| $HNO_3$ | [70] |
| $HNO_3$/NaOH | [43] |
| Hydroxypropylmethyl cellulose (HPMC) | [261] |
| LPA (in PDMS microchannel) | [327] |
| LPA (photopolymerized) | [115, 116] |
| Methanol | [224] |
| Methyl hydroxyethyl cellulose (MHEC) | [260] |
| NaOH | [21, 22, 35, 42, 48, 102, 138, 143, 147, 154, 162, 167, 181, 182, 185, 213, 214, 230, 242, 276, 317, 332, 333, 345] |
| $O_2$ plasma | [120] |
| Polydimethylacrylamide (PDMA) | [44, 64, 84] |
| PDMA/alkyl glycidyl ether | [208, 209] |
| Polydimethylsiloxane (PDMS) | [9] |
| PDMS/Pluronic | [201] |
| Polyethylene glycol (PEG) | [220] |
| Polyhydroxyethylacrylamide | [38, 39] |
| Polyvinylpyrrolidone (PVP) | [12, 118, 141, 244, 288–290, 344] |
| Silane | [323] |

## 3. Applications of Microchip Electrophoresis

Microfabricated electrophoresis systems have been used to analyze samples ranging from small molecules, to DNA, to proteins (Table 6). Analysis of both single- and double-stranded DNA is of particular interest because of the central role it plays in a variety of biomedical and molecular biology applications. Separation of single-stranded DNA fragments labeled with covalently attached fluorophores are typically performed under denaturing conditions generated by addition of chemical denaturants and/or running at an elevated temperature (~50°C) to inhibit secondary structure formation that would alter migration behavior. Fragment sizes typically range from oligonucleotides tens of bases in length up to around one kilobase. Double-stranded DNA separations are not performed under denaturing conditions and fragments can be labeled with intercalating dyes that do not require covalent attachment. Fragment sizes that can be separated extend from around 100 bases to the kilobase size range. DNA molecules in the 40–50 kilobase vicinity and longer are large enough to be directly observable using an optical microscope, enabling migration of individual molecules to be recorded. Protein separations differ somewhat from DNA because they possess a variable charge to size ratio. Two dimensional configurations are often employed to deal with this added level of complexity whereby species are first separated in one dimension

**Table 6.** Analytes separated using electrophoresis microchips

| Analyte separated | References |
|---|---|
| DNA (double-stranded) | [5, 7, 10, 12, 19, 24–26, 29, 32, 34, 39, 41, 44, 52, 64, 66, 68–70, 74, 75, 77, 78, 85–88, 101, 103–105, 118–121, 125, 127–129, 136, 141, 142, 144, 148, 151, 152, 154, 155, 163–166, 177–179, 190, 202, 203, 208, 209, 212, 216, 217, 222, 226, 229, 233, 235, 240, 244, 250–252, 263–265, 271, 274, 278, 281–288, 290, 291, 293, 295, 297, 300, 307, 311, 313, 316, 319–322, 324–326, 328, 329, 331, 338, 341, 343, 344] |
| DNA (single-stranded) | [6, 15, 58, 71, 72, 76, 78, 96, 157, 183, 184, 187, 192, 204, 205, 207, 211, 227, 228, 239, 248, 253–256, 261, 271, 273, 289, 292, 294, 296, 298–300, 302, 303, 323, 332, 337] |
| Proteins | [9, 17, 27, 35, 38, 42, 84, 90, 97, 98, 114–116, 122, 124, 138, 156, 185, 186, 197, 220, 221, 234, 236, 249, 260, 327, 335] |
| Small molecules | [4, 8, 20–22, 36, 37, 43–48, 50, 51, 53–55, 71, 79–83, 92–94, 102, 107–109, 113, 126, 130–135, 137, 143, 145, 147, 149, 150, 155, 160–162, 167, 168, 173, 174, 176, 181, 182, 193–196, 198, 199, 201, 213–215, 218, 219, 223, 224, 230, 231, 237, 238, 242, 243, 246, 257, 258, 266–269, 276, 279, 306, 309, 310, 312, 314, 315, 317, 326, 333, 334, 336, 340, 345] |

according to charge (e.g., using isoelectric focusing in a pH gradient), after which each constant-charge zone is separated according to size. Finally, applications involving separation of neutral or weakly charged analytes can be accomplished using MEKC techniques.

### 3.1. Advanced Electrophoresis Methods

Recent advancements have resulted in the development of separation technologies offering capabilities beyond those found in adaptations of conventional electrophoresis techniques [180] (Table 7). One notable example is the use of micro- and nanofabricated pillar arrays as separation matrix structures in place of polymer gels (Fig. 7a). These post arrays can be easily mass produced using micromachining technology and possess inherently uniform monodisperse two-dimensional pore morphologies. By manipulating the size, shape, and spacing of the pillars, separation properties can be tuned to make them compatible with a wide range of samples and analyte sizes. In addition, since the sieving structures are prefabricated inside the microchannel, the gel loading process is eliminated. The detailed physics associated with collisions and interactions that occur between the migrating analytes and the array of obstacles are responsible for introducing size-dependent mobilities. The size of the nanostrucutred pillars is limited by the resolution of photolithographic patterning, typically in the vicinity of a few hundred nanometers. Consequently, these designs are most effective for separation of large analytes (e.g., kilobase sized DNA). Another class of micromachined sieving structures that have been employed for electrophoresis involves patterning of periodically spaced nanometer size gaps between the floor and ceiling of the separation channel (Fig. 7b). Unlike pillars, the gaps are vertically spaced and can be fashioned relatively straightforwardly by controlled etching. The physics of DNA electrophoresis through these nanogap arrays can be described in terms of an entropic trapping mechanism whereby excursions into the nanogaps produce migration characterized by rapid hops between gaps, with larger fragments experiencing a higher probability of excursions into the nanogaps and thereby eluting faster than smaller ones.

A novel and entirely different class of fabricated sieving structure consists of magnetic microspheres that become assembled into a packed matrix upon application of an external magnetic field (Fig. 7c). Here, the pore network morphology depends on an interplay between the size and concentration of the particles and the separation channel geometry. This arrangement allows sieving matrices with tunable reproducible pore sizes to be formed in-situ, and the aggregated particle network can be instantaneously dispersed once the magnetic field is removed. Another novel separation phenomenon of interest to miniaturized systems is surface electrophoresis of DNA. Here, size dependent mobility is induced via interactions between the substrate (typically silicon) and DNA chains adsorbed on its surface. In this way, surface electrophoresis can be used to perform size selective fractionation in the absence of conventional sieving gels.

**Table 7.** Variations of electrophoretic separation adapted to the microchip format

| Separation mechanism | References |
|---|---|
| Atomic force microscope (AFM) | [302] |
| Electric field flow fractionation (EFFF) | [33] |
| Entropic recoil | [26] |
| Entropic trapping | [31, 62, 103–105, 247, 277] |
| Isoelectric focusing (IEF) | [114, 122, 124, 197, 249] |
| Isotachophoresis (ITP) | [314] |
| Magnetic bead array | [63, 64, 208–210] |
| Micellar electrokinetic chromatography (MEKC) | [45, 160, 215, 236, 238, 243, 260, 279, 310, 315] |
| Microfabricated pillar array | [5, 10, 25, 29, 41, 65–67, 121, 127, 141, 225, 232, 307, 308, 334, 338] |
| Microstructured periodic cavities | [69] |
| Optical fractionation | [191] |
| Single molecule fluorescence bursts | [88] |
| Surface electrophoresis | [171, 172, 189, 233, 259] |

Finally, an intriguing adaptation of electrophoresis for separation of short oligonucleotide fragments has been demonstrated using an electric field imposed parallel to an atomic force microscope (AFM) tip [302] (Fig. 7d). The DNA fragments to be separated are first electrokinetically collected at the base of the AFM cantilever, then released by a periodically reversing electric field applied between the cantilever and substrate. The DNA fragments display a size-dependent mobility as they migrate along the length of the AFM tip, and the inherently small dimensions allow very high electric fields to be applied with minimal Joule heating. By tuning the frequency at which the electric field is reversed, desired fragment sizes can be transferred from the tip to the surface of a substrate while slower moving fragments are driven back to the reservoir before reaching the tip.

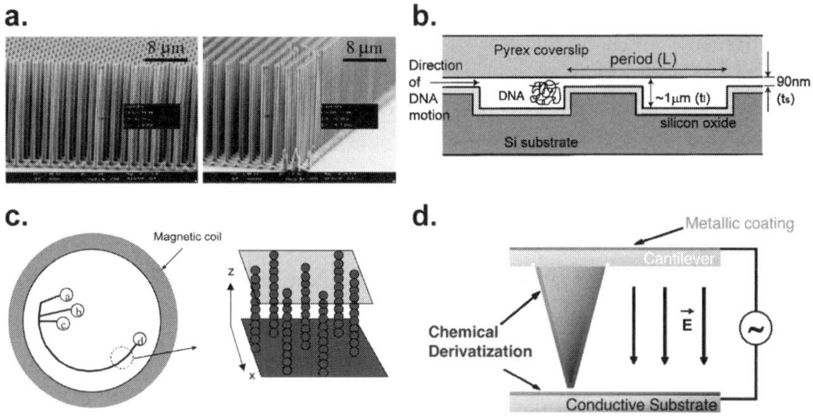


**Fig. 7.** (**a**) Scanning electron micrograph of microfabricated high-aspect ratio pillar arrays in two different arrangements [29]. The structures are etched into a silicon substrate using a deep reactive ion etching (DRIE) process capable of producing pillar diameters and spacings of order 500 nm (reproduced with permission, copyright 2006 IOP Publishing LTD). (**b**) Illustration of a typical geometry used to perform separations based on entropic trapping effects [103] (reproduced with permission, copyright 1999 The American Physical Society). (**c**) Schematic of an arrangement incorporating a separation channel inside a magnetic coil (left) [64]. A columnar network morphology is formed in a suspension of 1 μm dia superparamagnetic particles when a magnetic field is applied (right) that can function as a sieving matrix for DNA separations (reproduced with permission, copyright 2002 American Association for the Advancement of Science). (**d**) Concept of separating DNA using atomic force microscopy [302]. When an electric field is applied between an AFM cantilever and a substrate, DNA fragments initially at the base of the AFM tip migrate toward the substrate with size-dependent mobilities and can be selectively deposited by tuning the electric field reversal frequency (reproduced with permission, copyright 2006 American Institute of Physics)

### 3.2. Integrated Systems

Considerable efforts have also been directed toward combining electrophoresis-based analysis with other operations to produce integrated lab-on-a-chip systems (Table 8). Many genomic analysis applications, for example, involve using the polymerase chain reaction (PCR) to chemically amplify minute DNA samples to detectable levels prior to analysis. The PCR process involves repeatedly cycling a reagent mixture (template DNA, primers, dNTPs, a thermostable polymerase enzyme, and other buffering additives) through temperatures corresponding to *denaturation* of the double-stranded template (~95°C), *annealing* of single-stranded oligonucleotide primers at locations adjacent to the target region (~50–65°C), and enzyme directed synthesis (*extension*) of the complementary target strand (~72°C). The inherently high efficiency of this process, yielding an exponential increase in the number of copies with each cycle, has helped make PCR one of the key enabling technologies in molecular biology. Thus, it is not surprising that considerable efforts have been directed toward integrating PCR with electrophoretic separation of the amplified DNA fragments. Miniaturization of PCR is challenging, however, due to the potential for evaporation when minute volumes of aqueous reagents are repeatedly heated to temperatures in the vicinity of 95°C, and due to the potential for non-specific adsorption of reagents at the reactor walls in high surface to volume microchannels that can inhibit the reaction [158, 245, 275].

**Table 8.** Integrated microfluidic devices incorporating electrophoresis

| Integrated functions | References |
|---|---|
| 2-D separation | [98, 236, 238, 260] |
| Band capture | [73, 177, 178] |
| Cell lysis | [201, 319] |
| Immunoassay | [37] |
| Isothermal amplification | [24, 110, 285] |
| Labeling | [35, 42, 83, 97, 102, 132, 133, 186, 264] |
| Polymerase chain reaction (PCR) | [68, 70, 119, 152, 163–166, 228, 229, 319, 324, 344] |
| Purification | [15, 228] |
| Restriction digest | [97, 136, 229] |
| Sanger sequencing | [15] |

Examples of other operations integrated with microchip electrophoresis include cell lysis, sample purification, multi-dimensional separations, labeling, and hybridization. Progress has also been made toward incorporating multiple sample processing and analysis steps in a single microfluidic device, as demonstrated in a hybrid glass–silicon design developed in the Burns group capable of amplifying a 106 bp DNA target from a bacterial genomic template via an isothermal strand displacement amplification (SDA) process followed by gel electrophoresis with integrated photodetection of the fluorescently labeled reaction products [24]. Subsequent work

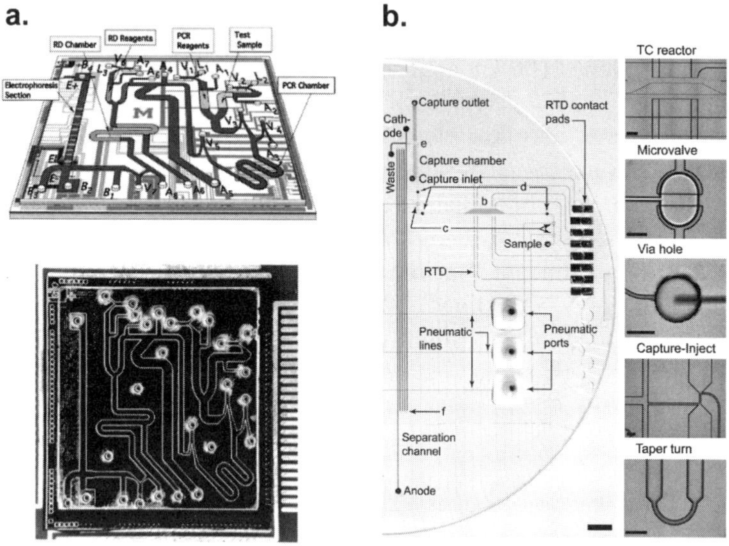


**Fig. 8.** (**a**) Schematic (upper) and photograph (lower) of a 1.5 × 1.6 cm microfluidic device capable of performing a sequential PCR, restriction enzyme digestion, and gel electrophoresis [229] (reproduced with permission, copyright 2006, The Royal Society of Chemistry). (**b**) Layout of a microfluidic bioprocessor for performing Sanger cycle sequencing of DNA (scale bar = 5 mm) [15] (reproduced with permission, copyright 2006, The National Academy of Sciences of the USA)

has resulted in development of an ultra-compact 1.5 × 1.6 cm hybrid glass/silicon microfluidic chip capable of performing a PCR-RFLP assay for influenza A virus detection involving two sequential reactions followed by an electrophoretic separation [229] (Fig. 8a). Most recently, the Mathies group has reported a multilayer glass/PDMS design capable of performing integrated Sanger cycle sequencing of DNA (a reaction requiring similar thermal cycling parameters as PCR) followed by product purification and electrophoretic separation in a 30-cm long folded channel [15]. Read lengths of 556 bases were achieved with 99% accuracy (Fig. 8b).

Commercial DNA analysis systems based on microchip capillary electrophoresis technology have also appeared on the market (Table 9). For example, the Agilent 2100 Bioanalyzer, the most widely used commercial chip-based DNA analysis device, is based on a Caliper LabChip design incorporating arrays of glass electrophoresis microchannels that can be loaded with a low viscosity gel-dye mixture. This system uses interchangeable electrophoresis chips that interface with a benchtop power supply and optical detection system. The Hitachi SV1100 Microchip CE system is based on a similar concept, but the electrophoresis microchips are constructed using PMMA. The BioMEMS-768 DNA sequencer from Network Biosystems/Shimadsu takes a slightly different approach by using microfabrication technology to construct arrays of hundreds of 40 cm long separation channels in 50 × 25 cm glass substrates to provide higher throughput and integration with conventional automated liquid handling systems.

**Table 9.** Commercial microchip electrophoresis systems

| Commercial system | References |
|---|---|
| Agilent 2100 | [17, 76, 90, 128, 129, 212, 216, 343] |
| Hitachi SV1100 | [50, 51, 53, 54, 76, 110, 129, 148, 234, 281–284, 328, 329, 331, 341] |

## 4. Summary and Outlook

Considerable strides have been made since microfabricated electrophoresis devices were first developed in the early 1990s [194]. These advancements have been made possible through a combination of improvements in micromachining technology, sieving matrix materials, and instrumentation. Despite these advancements, the current generation of miniaturized systems generally have yet to demonstrate benefits in cost or performance that are compelling enough to make them seriously competitive with conventional benchtop-scale CE technology. Some of the challenges that remain to be addressed include reducing device size to the point where the enormous cost savings associated with photolithographic microfabrication can be fully realized, enhancing integration, improving the capability to interface with the external macroscale environment, and developing miniaturized detection technology. Systematic fundamental studies are also important in order to provide a more complete understanding of the physics of electrophoresis in micro- and nanoscale environments. Future developments in these areas over the next few years are likely to lay the foundation for a new generation of rapid, sensitive, and inexpensive instrumentation with separation performance exceeding that in many of the conventional benchtop-scale analytical systems available today.

## References


1. Albarghouthi MN, Barron AE (2000) Polymeric matrices for DNA sequencing by capillary electrophoresis. Electrophoresis 21: 4096–4111
2. Asbury CL, van den Engh G (1998) Trapping of DNA in nonumiform oscillating electric fields. Biophysical Journal 74: 1024–1030
3. Asbury CL, Diercks AH, van den Engh G (2002) Trapping of DNA by dielectrophoresis. Electrophoresis 23: 2658–2666
4. Attiya S, Jemere AB, Tang T, Fitzpatrick G, Seiler K, Chiem N, Harrison DJ (2001) Design of an interface to allow microfluidic electrophoresis chips to drink from the fire hose of the external environment. Electrophoresis 22: 318–327
5. Baba M, Sano T, Iguchi N, Iida K, Sakamoto T, Kawaura H (2003) DNA size separation using artificially nanostructured matrix. Applied Physics Letters 83: 1468–1470
6. Backhouse C, Caamano M, Oaks F, Nordman E, Carrillo A, Johnson B, Bay S (2000) DNA sequencing in a monolithic microchannel device. Electrophoresis 21: 150–156
7. Backhouse CJ, Gajdal A, Pilarski LM, Crabtree HJ (2003) Improved resolution with microchip-based enhanced field inversion electrophoresis. Electrophoresis 24: 1777–1786
8. Backofen U, Matysik FM, Lunte CE (2002) A chip-based electrophoresis system with electrochemical detection and hydrodynamic injection. Analytical Chemistry 74: 4054–4059
9. Badal MY, Wong M, Chiem N, Salimi-Moosavi H, Harrison DJ (2002) Protein separation and surfactant control of electroosmotic flow in poly (dimethylsiloxane)-coated capillaries and microchips. Journal of Chromatography A 947: 277–286
10. Bakajin O, Duke TAJ, Tegenfeldt J, Chou CF, Chan SS, Austin RH, Cox EC (2001) Separation of 100-kilobase DNA molecules in 10 seconds. Analytical Chemistry 73: 6053–6056
11. Baldwin RP, Roussel TJ, Crain MM, Bathlagunda V, Jackson DJ, Gullapalli J, Conklin JA, Pai R, Naber JF, Walsh KM, Keynton RS (2002) Fully integrated on-chip electrochemical detection for capillary electrophoresis in a microfabricated device. Analytical Chemistry 74: 3690–3697
12. Barta C, Ronai Z, Nemoda Z, Szekely A, Kovacs E, Sasvari-Szekely M, Guttman A (2001) Analysis of dopamine D4 receptor gene polymorphism using microchip electrophoresis. Journal of Chromatography A 924: 285–290

13. Becker H, Lowack K, Manz A (1998) Planar quartz chips with submicron channels for two-dimensional capillary electrophoresis applications. Journal of Micromechanics and Microengineering 8: 24–28
14. Belder D, Ludwig M (2003) Surface modification in microchip electrophoresis. Electrophoresis 24: 3595–3606
15. Blazej RG, Kumaresan P, Mathies RA (2006) Microfabricated bioprocessor for integrated nanoliter-scale sanger DNA sequencing. Proceedings of the National Academy of Sciences of the United States of America 103: 7240–7245
16. Bousse L, Cohen C, Nikiforov T, Chow A, Kopf-Sill AR, Dubrow R, Parce JW (2000) Electrokinetically controlled microfluidic analysis systems. Annual Review of Biophysics and Biomolecular Structure 29: 155–181
17. Bousse L, Mouradian S, Minalla A, Yee H, Williams K, Dubrow R (2001) Protein sizing on a microchip. Analytical Chemistry 73: 1207–1212
18. Brahmasandra SN, Burke DT, Mastrangelo CH, Burns MA (2001) Mobility, diffusion and dispersion of single-stranded DNA in sequencing gels. Electrophoresis 22: 1046–1062
19. Brahmasandra SN, Ugaz VM, Burke DT, Mastrangelo CH, Burns MA (2001) Electrophoresis in microfabricated devices using photopolymerized polyacrylamide gels and electrode-defined sample injection. Electrophoresis 22: 300–311
20. Brister PC, Weston KD (2006) Evaluation of two-beam fluorescence cross correlation spectroscopy for electrophoretic analysis of protein digests. Analyst 131: 303–310
21. Bromberg A, Mathies RA (2003) Homogeneous immunoassay for detection of TNT and its analogues on a microfabricated capillary electrophoresis chip. Analytical Chemistry 75: 1188–1195
22. Burke BJ, Regnier FE (2001) Electrophoretically mediated microanalysis of β-galactosidase on microchips. Electrophoresis 22: 3744–3751
23. Burns MA, Mastrangelo CH, Sammarco TS, Man FP, Webster JR, Johnson BN, Foerster B, Jones D, Fields Y, Kaiser AR, Burke DT (1996) Microfabricated structures for integrated DNA analysis. Proceedings of the National Academy of Sciences of the United States of America 93: 5556–5561
24. Burns MA, Johnson BN, Brahmasandra SN, Handique K, Webster JR, Krishnan M, Sammarco TS, Man PM, Jones D, Heldsinger D, Mastrangelo CH, Burke DT (1998) An integrated nanoliter DNA analysis device. Science 282: 484–487
25. Cabodi M, Chen YF, Turner SWP, Craighead HG, Austin RH (2002) Continuous separation of biomolecules by the laterally asymmetric diffusion array with out-of-plane sample injection. Electrophoresis 23: 3496–3503
26. Cabodi M, Turner SWP, Craighead HG (2002) Entropic recoil separation of long DNA molecules. Analytical Chemistry 74: 5169–5174
27. Callewaert N, Contreras R, Mitnik-Gankin L, Carey L, Matsudaira P, Ehrlich D (2004) Total serum protein *N*-glycome profiling on a capillary electrophoresis-microfluidics platform. Electrophoresis 25: 3128–3131

28. Cannon DM, Kuo TC, Bohn PW, Sweedler JV (2003) Nanocapillary array interconnects for gated analyte injections and electrophoretic separations in multilayer microfluidic architectures. Analytical Chemistry 75: 2224–2230
29. Chan YC, Lee YK, Zohar Y (2006) High-throughput design and fabrication of an integrated microsystem with high aspect-ratio sub-micron pillar arrays for free-solution micro capillary electrophoresis. Journal of Micromechanics and Microengineering 16: 699–707
30. Chen G, Lin YH, Wang J (2006) Monitoring environmental pollutants by microchip capillary electrophoresis with electrochemical detection. Talanta 68: 497–503
31. Chen Z, Escobedo FA (2003) Simulation of chain-length partitioning in a microfabricated channel via entropic trapping. Molecular Simulation 29: 417–425
32. Chen Z, Burns MA (2005) Effect of buffer flow on DNA separation in a microfabricated electrophoresis system. Electrophoresis 26: 4718–4728
33. Chen Z, Chauhan A (2005) DNA separation by EFFF in a microchannel. Journal of Colloid and Interface Science 285: 834–844
34. Cheng J, Waters LC, Fortina P, Hvichia G, Jacobson SC, Ramsey JM, Kricka LJ, Wilding P (1998) Degenerate oligonucleotide primed polymerase chain reaction and capillary electrophoretic analysis of human DNA on microchip-based devices. Analytical Biochemistry 257: 101–106
35. Cheng SB, Skinner CD, Taylor J, Attiya S, Lee WE, Picelli G, Harrison DJ (2001) Development of a multichannel microfluidic analysis system employing affinity capillary electrophoresis for immunoassay. Analytical Chemistry 73: 1472–1479
36. Chiem N, Harrison DJ (1997) Microchip-based capillary electrophoresis for immunoassays: Analysis of monoclonal antibodies and theophylline. Analytical Chemistry 69: 373–378
37. Chiem NH, Harrison DJ (1998) Microchip systems for immunoassay: An integrated immunoreactor with electrophoretic separation for serum theophylline determination. Clinical Chemistry 44: 591–598
38. Chiesl TN, Shi W, Barron AE (2005) Poly(acrylamide-*co*-alkylacrylamides) for electrophoretic DNA purification in microchannels. Analytical Chemistry 77: 772–779
39. Chiesl TN, Putz KW, Babu M, Mathias P, Shaikh KA, Goluch ED, Liu C, Barron AE (2006) Self-associating block copolymer networks for microchip electrophoresis provide enhanced DNA separation via "inchworm" chain dynamics. Analytical Chemistry 78: 4409–4415
40. Chou C-F, Tegenfeldt JO, Bakajin O, Chan SS, Cox EC, Darnton N, Duke T, Austin RH (2002) Electrodeless dielectrophoresis of single- and double-stranded DNA. Biophysical Journal 83: 2170–2179
41. Chou CF, Bakajin O, Turner SWP, Duke TAJ, Chan SS, Cox EC, Craighead HG, Austin RH (1999) Sorting by diffusion: An asymmetric obstacle course for continuous molecular separation. Proceedings of the National Academy of Sciences of the United States of America 96: 13762–13765

42. Colyer CL, Mangru SD, Harrison DJ (1997) Microchip-based capillary electrophoresis of human serum proteins. Journal of Chromatography A 781: 271–276
43. Crabtree HJ, Cheong ECS, Tilroe DA, Backhouse CJ (2001) Microchip injection and separation anomalies due to pressure effects. Analytical Chemistry 73: 4079–4086
44. Culbertson CT, Jacobson SC, Ramsey JM (1998) Dispersion sources for compact geometries on microchips. Analytical Chemistry 70: 3781–3789
45. Culbertson CT, Jacobson SC, Ramsey JM (2000) Microchip devices for high-efficiency separations. Analytical Chemistry 72: 5814–5819
46. Culbertson CT, Ramsey RS, Ramsey JM (2000) Electroosmotically induced hydraulic pumping on microchips: Differential ion transport. Analytical Chemistry 72: 2285–2291
47. Culbertson CT, Jacobson SC, Ramsey JM (2002) Diffusion coefficient measurements in microfluidic devices. Talanta 56: 365–373
48. Culbertson CT, Tugnawat Y, Meyer AR, Roman GT, Ramsey JM, Gonda SR (2005) Microchip separations in reduced-gravity and hypergravity environments. Analytical Chemistry 77: 7933–7940
49. Dai JH, Ito T, Sun L, Crooks RM (2003) Electrokinetic trapping and concentration enrichment of DNA in a microfluidic channel. Journal of the American Chemical Society 125: 13026–13027
50. Dang F, Zhang L, Hagiwara H, Mishina Y, Baba Y (2003) Ultrafast analysis of oligosaccharides on microchip with light-emitting diode confocal fluorescence detection. Electrophoresis 24: 714–721
51. Dang FQ, Zhang LH, Jabasini M, Kaji N, Baba Y (2003) Characterization of electrophoretic behavior of sugar isomers by microchip electrophoresis coupled with videomicroscopy. Analytical Chemistry 75: 2433–2439
52. Dang FQ, Tabata O, Kurokawa M, Ewis AA, Zhang LH, Yamaoka Y, Shinohara S, Shinohara Y, Ishikawa M, Baba Y (2005) High-performance genetic analysis on microfabricated capillary array electrophoresis plastic chips fabricated by injection molding. Analytical Chemistry 77: 2140–2146
53. Dang FQ, Kakehi K, Cheng JJ, Tabata O, Kurokawa M, Nakajima K, Ishikawa M, Baba Y (2006) Hybrid dynamic coating with *n*-dodecyl β-D-maltoside and methyl cellulose for high-performance carbohydrate analysis on poly(methyl methacrylate) chips. Analytical Chemistry 78: 1452–1458
54. Dang FQ, Kakehi K, Nakajima K, Shinohara Y, Ishikawa M, Kaji N, Tokeshi M, Baba Y (2006) Rapid analysis of oligosaccharides derived from glycoproteins by microchip electrophoresis. Journal of Chromatography A 1109: 138–143
55. Deng YZ, Henion J, Li JJ, Thibault P, Wang C, Harrison DJ (2001) Chip-based capillary electrophoresis/mass spectrometry determination of carnitines in human urine. Analytical Chemistry 73: 639–646
56. Dhopeshwarkar R, Li SA, Crooks RM (2005) Electrokinetic concentration enrichment within a microfluidic device using a hydrogel microplug. Lab on a Chip 5: 1148–1154

57. Doherty EAS, Meagher RJ, Albarghouthi MN, Barron AE (2003) Microchannel wall coatings for protein separations by capillary and chip electrophoresis. Electrophoresis 24: 34–54
58. Doherty EAS, Kan CW, Paegel BM, Yeung SHI, Cao ST, Mathies RA, Barron AE (2004) Sparsely cross-linked "nanogel" matrixes as fluid, mechanically stabilized polymer networks for high-throughput microchannel DNA sequencing. Analytical Chemistry 76: 5249–5256
59. Dolnik V, Liu SR, Jovanovich S (2000) Capillary electrophoresis on microchip. Electrophoresis 21: 41–54
60. Dolnik V (2004) Wall coating for capillary electrophoresis on microchips. Electrophoresis 25: 3589–3601
61. Dolnik V, Liu SR (2005) Applications of capillary electrophoresis on microchip. Journal of Separation Science 28: 1994–2009
62. Dorfman KD, Brenner H (2002) Modeling DNA electrophoresis in microfluidic entropic trapping devices. Biomedical Microdevices 4: 237–244
63. Dorfman KD, Viovy JL (2004) Semiphenomenological model for the dispersion of DNA during electrophoresis in a microfluidic array of posts. Physical Review E 69: 011901
64. Doyle PS, Bibette J, Bancaud A, Viovy JL (2002) Self-assembled magnetic matrices for DNA separation chips. Science 295: 2237–2237
65. Duke T, Monnelly G, Austin RH, Cox EC (1997) Sequencing in nanofabricated arrays: A feasibility study. Electrophoresis 18: 17–22
66. Duke TAJ, Austin RH, Cox EC, Chan SS (1996) Pulsed-field electrophoresis in microlithographic arrays. Electrophoresis 17: 1075–1079
67. Duke TAJ, Austin RH (1998) Microfabricated sieve for the continuous sorting of macromolecules. Physical Review Letters 80: 1552–1555
68. Dunn WC, Jacobson SC, Waters LC, Kroutchinina N, Khandurina J, Foote RS, Justice MJ, Stubbs LJ, Ramsey JM (2000) PCR amplification and analysis of simple sequence length polymorphisms in mouse DNA using a single microchip device. Analytical Biochemistry 277: 157–160
69. Duong TT, Kim G, Ros R, Streek M, Schmid F, Brugger J, Anselmetti D, Ros A (2003) Size-dependent free solution DNA electrophoresis in structured microfluidic systems. Microelectronic Engineering 67-8: 905–912
70. Easley CJ, Karlinsey JM, Landers JP (2006) On-chip pressure injection for integration of infrared-mediated DNA amplification with electrophoretic separation. Lab on a Chip 6: 601–610
71. Effenhauser CS, Manz A, Widmer HM (1993) Glass chips for high-speed capillary electrophoresis separations with submicrometer plate heights. Analytical Chemistry 65: 2637–2642
72. Effenhauser CS, Paulus A, Manz A, Widmer HM (1994) High-speed separation of antisense oligonucleotides on a micromachined capillary electrophoresis device. Analytical Chemistry 66: 2949–2953
73. Effenhauser CS, Manz A, Widmer HM (1995) Manipulation of sample fractions on a capillary electrophoresis chip. Analytical Chemistry 67: 2284–2287

74. Effenhauser CS, Bruin GJM, Paulus A, Ehrat M (1997) Integrated capillary electrophoresis on flexible silicone microdevices: Analysis of DNA restriction fragments and detection of single DNA molecules on microchips. Analytical Chemistry 69: 3451–3457
75. Emrich CA, Tian HJ, Medintz IL, Mathies RA (2002) Microfabricated 384-lane capillary array electrophoresis bioanalyzer for ultrahigh-throughput genetic analysis. Analytical Chemistry 74: 5076–5083
76. Endo Y, Zhang L, Katashima R, Itakura M, Doherty EAS, Barron AE, Baba Y (2005) Effect of polymer matrix and glycerol on rapid single-strand conformation polymorphism analysis by capillary and microchip electrophoresis for detection of mutations in *K-ras* gene. Electrophoresis 26: 3380–3386
77. Ertl P, Emrich CA, Singhal P, Mathies RA (2004) Capillary electrophoresis chips with a sheath-flow supported electrochemical detection system. Analytical Chemistry 76: 3749–3755
78. Fabrizio EF, Nadim A, Sterling JD (2003) Resolution of multiple ssDNA structures in free solution electrophoresis. Analytical Chemistry 75: 5012–5021
79. Fan ZH, Harrison DJ (1994) Micromachining of capillary electrophoresis injectors and separators on glass chips and evaluation of flow at capillary intersections. Analytical Chemistry 66: 177–184
80. Fang O, Xu GM, Fang ZL (2002) A high-throughput continuous sample introduction interface for microfluidic chip-based capillary electrophoresis systems. Analytical Chemistry 74: 1223–1231
81. Fister JC, Jacobson SC, Davis LM, Ramsey JM (1998) Counting single chromophore molecules for ultrasensitive analysis and separations on microchip devices. Analytical Chemistry 70: 431–437
82. Fister JC, Jacobson SC, Ramsey JM (1999) Ultrasensitive cross correlation electrophoresis on microchip devices. Analytical Chemistry 71: 4460–4464
83. Fluri K, Fitzpatrick G, Chiem N, Harrison DJ (1996) Integrated capillary electrophoresis devices with an efficient postcolumn reactor in planar quartz and glass chips. Analytical Chemistry 68: 4285–4290
84. Foote RS, Khandurina J, Jacobson SC, Ramsey JM (2005) Preconcentration of proteins on microfluidic devices using porous silica membranes. Analytical Chemistry 77: 57–63
85. Footz T, Wunsam S, Kulak S, Crabtree HJ, Glerum DM, Backhouse CJ (2001) Sample purification on a microfluidic device. Electrophoresis 22: 3868–3875
86. Footz T, Somerville MJ, Tomaszewski R, Sprysak KA, Backhouse CJ (2003) Heteroduplex-based genotyping with microchip electrophoresis and DHPLC. Genetic Testing 7: 283–293
87. Footz T, Somerville MJ, Tomaszewski R, Elyas B, Backhouse CJ (2004) Integration of combined heteroduplex/restriction fragment length polymorphism analysis on an electrophoresis microchip for the detection of hereditary haemochromatosis. Analyst 129: 25–31

88. Foquet M, Korlach J, Zipfel W, Webb WW, Craighead HG (2002) DNA fragment sizing by single molecule detection in submicrometer-sized closed fluidic channels. Analytical Chemistry 74: 1415–1422
89. Ford SM, Kar B, McWhorter S, Davies J, Soper SA, Klopf M, Calderon G, Saile V (1998) Microcapillary electrophoresis devices fabricated using polymeric substrates and x-ray lithography. Journal of Microcolumn Separations 10: 413–422
90. Forrer K, Hammer S, Helk B (2004) Chip-based gel electrophoresis method for the quantification of half-antibody species in IgG4 and their by- and degradation products. Analytical Biochemistry 334: 81–88
91. Gao QF, Shi YN, Liu SR (2001) Multiple-channel microchips for high-throughput DNA analysis by capillary electrophoresis. Fresenius Journal of Analytical Chemistry 371: 137–145
92. Gao Y, Shen Z, Wang H, Dai ZP, Lin BC (2005) Chiral separations on multichannel microfluidic chips. Electrophoresis 26: 4774–4779
93. Gavin PF, Ewing AG (1996) Continuous separations with microfabricated electrophoresis-electrochemical array detection. Journal of the American Chemical Society 118: 8932–8936
94. Gawron AJ, Martin RS, Lunte SM (2001) Fabrication and evaluation of a carbon-based dual-electrode detector for poly(dimethylsiloxane) electrophoresis chips. Electrophoresis 22: 242–248
95. Giddings JC (1991) Unified separation science. Wiley, New York
96. Goedecke N, McKenna B, El-Difrawy S, Carey L, Matsudaira P, Ehrlich D (2004) A high-performance multilane microdevice system designed for the DNA forensics laboratory. Electrophoresis 25: 1678–1686
97. Gottschlich N, Culbertson CT, McKnight TE, Jacobson SC, Ramsey JM (2000) Integrated microchip-device for the digestion, separation and post-column labeling of proteins and peptides. Journal of Chromatography B 745: 243–249
98. Gottschlich N, Jacobson SC, Culbertson CT, Ramsey JM (2001) Two-dimensional electrochromatography/capillary electrophoresis on a microchip. Analytical Chemistry 73: 266–2674
99. Griffiths SK, Nilson RH (2000) Band spreading in two-dimensional microchannel turns for electrokinetic species transport. Analytical Chemistry 72: 5473–5482
100. Griffiths SK, Nilson RH (2001) Low dispersion turns and junctions for microchannel systems. Analytical Chemistry 73: 272–278
101. Haab BB, Mathies RA (1999) Single-molecule detection of DNA separations in microfabricated capillary electrophoresis chips employing focused molecular streams. Analytical Chemistry 71: 5137–5145
102. Hadd AG, Jacobson SC, Ramsey JM (1999) Microfluidic assays of acetylcholinesterase inhibitors. Analytical Chemistry 71: 5206–5212
103. Han J, Turner SW, Craighead HG (1999) Entropic trapping and escape of long DNA molecules at submicron size constriction. Physical Review Letters 83: 1688–1691

104. Han J, Craighead HG (2000) Separation of long DNA molecules in a microfabricated entropic trap array. Science 288: 1026–1029
105. Han JY, Craighead HG (2002) Characterization and optimization of an entropic trap for DNA separation. Analytical Chemistry 74: 394–401
106. Handal MI, Ugaz VM (2006) DNA mutation detection and analysis using miniaturized microfluidic systems. Expert Review of Molecular Diagnostics 6: 29–38
107. Harrison DJ, Manz A, Fan ZH, Ludi H, Widmer HM (1992) Capillary electrophoresis and sample injection systems integrated on a planar glass chip. Analytical Chemistry 64: 1926–1932
108. Harrison DJ, Fluri K, Seiler K, Fan ZH, Effenhauser CS, Manz A (1993) Micromachining a miniaturized capillary electrophoresis-based chemical-analysis system on a chip. Science 261: 895–897
109. Harrison DJ, Glavina PG, Manz A (1993) Towards miniaturized electrophoresis and chemical-analysis systems on silicon – An alternative to chemical sensors. Sensors and Actuators B: Chemical 10: 107–116
110. Hataoka Y, Zhang LH, Mori Y, Tomita N, Notomi T, Baba Y (2004) Analysis of specific gene by integration of isothermal amplification and electrophoresis on poly(methyl methacrylate) microchips. Analytical Chemistry 76: 3689–3693
111. Henry AC, Waddell EA, Shreiner R, Locascio LE (2002) Control of electroosmotic flow in laser-ablated and chemically modified hot imprinted poly(ethylene terephthalate glycol) microchannels. Electrophoresis 23: 791–798
112. Henry CS (ed) (2006) Microchip capillary electrophoresis. Methods and protocols. Humana Press, Totowa, NJ
113. Heo YS, Chung S, Cho K, Chung C, Han DC, Chang JK (2003) Effects of peak anomalies with the hydrophilic or hydrophobic properties of reservoirs during serial injection on a capillary electrophoresis microchip. Journal of Chromatography A 1013: 111–122
114. Herr AE, Molho JI, Drouvalakis KA, Mikkelsen JC, Utz PJ, Santiago JG, Kenny TW (2003) On-chip coupling of isoelectric focusing and free solution electrophoresis for multidimensional separations. Analytical Chemistry 75: 1180–1187
115. Herr AE, Singh AK (2004) Photopolymerized cross-linked polyacrylamide gels for on-chip protein sizing. Analytical Chemistry 76: 4727–4733
116. Herr AE, Throckmorton DJ, Davenport AA, Singh AK (2005) On-chip native gel electrophoresis-based immunoassays for tetanus antibody and toxin. Analytical Chemistry 77: 585–590
117. Hjerten S (1985) High-performance electrophoresis. Elimination of electroendosmosis and solute adsorption. Journal of Chromatography 347: 191–198
118. Hofgartner WT, Huhmer AFR, Landers JP, Kant JA (1999) Rapid diagnosis of herpes simplex encephalitis using microchip electrophoresis of PCR products. Clinical Chemistry 45: 2120–2128

119. Hong JW, Fujii T, Seki M, Yamamoto T, Endo I (2001) Integration of gene amplification and capillary gel electrophoresis on a polydimethylsiloxane–glass hybrid microchip. Electrophoresis 22: 328–333
120. Hong JW, Hosokawa K, Fujii T, Seki M, Endo I (2001) Microfabricated polymer chip for capillary gel electrophoresis. Biotechnology Progress 17: 958–962
121. Huang LR, Tegenfeldt JO, Kraeft JJ, Sturm JC, Austin RH, Cox EC (2002) A DNA prism for high-speed continuous fractionation of large DNA molecules. Nature Biotechnology 20: 1048–1051
122. Huang TM, Pawliszyn J (2002) Microfabrication of a tapered channel for isoelectric focusing with thermally generated pH gradient. Electrophoresis 23: 3504–3510
123. Huang XJ, Pu QS, Fang ZL (2001) Capillary electrophoresis system with flow injection sample introduction and chemiluminescence detection on a chip platform. Analyst 126: 281–284
124. Huang XY, Ren JC (2005) On-line chemiluminescence detection for isoelectric focusing of heme proteins on microchips. Electrophoresis 26: 3595–3601
125. Huang Z, Munro N, Huhmer AFR, Landers JP (1999) Acousto-optical deflection-based laser beam scanning for fluorescence detection on multichannel electrophoretic microchips. Analytical Chemistry 71: 5309–5314
126. Hutt LD, Glavin DP, Bada JL, Mathies RA (1999) Microfabricated capillary electrophoresis amino acid chirality analyzer for extraterrestrial exploration. Analytical Chemistry 71: 4000–4006
127. Inatomi K, Izuo S, Lee SS, Ohji H, Shiono S (2003) Electrophoresis of DNA in micro-pillars fabricated in polydimethylsiloxane. Microelectronic Engineering 70: 13–18
128. Jabasini M, Zhang LH, Dang FQ, Xu F, Almofli MR, Ewis AA, Lee JW, Nakahori Y, Baba Y (2002) Analysis of DNA polymorphisms on the human Y-chromosome by microchip electrophoresis. Electrophoresis 23: 1537–1542
129. Jabasini M, Ewis AA, Xu F, Mohamadi MR, Ping GC, Shinka T, Nakahori Y, Baba Y (2005) Multiplex PCR with multichannel microchip electrophoresis: An ultrafast analysis for genetic diseases. Journal of Chromatographic Science 43: 221–225
130. Jacobson SC, Hergenroder R, Koutny LB, Ramsey JM (1994) High-speed separations on a microchip. Analytical Chemistry 66: 1114–1118
131. Jacobson SC, Hergenroder R, Koutny LB, Warmack RJ, Ramsey JM (1994) Effects of injection schemes and column geometry on the performance of microchip electrophoresis devices. Analytical Chemistry 66: 1107–1113
132. Jacobson SC, Hergenroder R, Moore AW, Ramsey JM (1994) Precolumn reactions with electrophoretic analysis integrated on a microchip. Analytical Chemistry 66: 4127–4132
133. Jacobson SC, Koutny LB, Hergenroder R, Moore AW, Ramsey JM (1994) Microchip capillary electrophoresis with an integrated postcolumn reactor. Analytical Chemistry 66: 3472–3476

134. Jacobson SC, Moore AW, Ramsey JM (1995) Fused quartz substrates for microchip electrophoresis. Analytical Chemistry 67: 2059–2063
135. Jacobson SC, Ramsey JM (1995) Microchip electrophoresis with sample stacking. Electrophoresis 16: 481–486
136. Jacobson SC, Ramsey JM (1996) Integrated microdevice for DNA restriction fragment analysis. Analytical Chemistry 68: 720–723
137. Jacobson SC, Culbertson CT, Daler JE, Ramsey JM (1998) Microchip structures for submillisecond electrophoresis. Analytical Chemistry 70: 3476–3480
138. Jin LJ, Giordano BC, Landers JP (2001) Dynamic labeling during capillary or microchip electrophoresis for laser-induced fluorescence detection of protein–sds complexes without pre- or postcolumn labeling. Analytical Chemistry 73: 4994–4999
139. Johnson TJ, Ross D, Gaitan M, Locascio LE (2001) Laser modification of preformed polymer microchannels: Application to reduce band broadening around turns subject to electrokinetic flow. Analytical Chemistry 73: 3656–3661
140. Jung B, Bharadwaj R, Santiago JG (2003) Thousandfold signal increase using field-amplified sample stacking for on-chip electrophoresis. Electrophoresis 24: 3476–3483
141. Kaji N, Tezuka Y, Takamura Y, Ueda M, Nishimoto T, Nakanishi H, Horiike Y, Baba Y (2004) Separation of long DNA molecules by quartz nanopillar chips under a direct current electric field. Analytical Chemistry 76: 15–22
142. Kamei T, Paegel BM, Scherer JR, Skelley AM, Street RA, Mathies RA (2003) Integrated hydrogenated amorphous Si photodiode detector for microfluidic bioanalytical devices. Analytical Chemistry 75: 5300–5305
143. Kamei T, Paegel BM, Scherer JR, Skelley AM, Street RA, Mathies RA (2004) Fusion of a-Si:H sensor technology with microfluidic bioanalytical devices. Journal of Non-Crystalline Solids 338–40: 715–719
144. Kamei T, Toriello NM, Lagally ET, Blazej RG, Scherer JR, Street RA, Mathies RA (2005) Microfluidic genetic analysis with an integrated a-Si:H detector. Biomedical Microdevices 7: 147–152
145. Kameoka J, Craighead HG, Zhang HW, Henion J (2001) A polymeric microfluidic chip for CE/MS determination of small molecules. Analytical Chemistry 73: 1935–1941
146. Kan CW, Fredlake CP, Doherty EAS, Barron AE (2004) DNA sequencing and genotyping in miniaturized electrophoresis systems. Electrophoresis 25: 3564–3588
147. Karlinsey JM, Monahan J, Marchiarullo DJ, Ferrance JP, Landers JP (2005) Pressure injection on a valved microdevice for electrophoretic analysis of submicroliter samples. Analytical Chemistry 77: 3637–3643
148. Kataoka M, Inoue S, Kajimoto K, Sinohara Y, Baba Y (2004) Usefulness of microchip electrophoresis for reliable analyses of nonstandard DNA samples and subsequent on-chip enzymatic digestion. European Journal of Biochemistry 271: 2241–2247

149. Kelly RT, Woolley AT (2003) Thermal bonding of polymeric capillary electrophoresis microdevices in water. Analytical Chemistry 75: 1941–1945
150. Keynton RS, Roussel TJ, Crain MM, Jackson DJ, Franco DB, Naber JF, Walsh KM, Baldwin RP (2004) Design and development of microfabricated capillary electrophoresis devices with electrochemical detection. Analytica Chimica Acta 507: 95–105
151. Khandurina J, Jacobson SC, Waters LC, Foote RS, Ramsey JM (1999) Microfabricated porous membrane structure for sample concentration and electrophoretic analysis. Analytical Chemistry 71: 1815–1819
152. Khandurina J, McKnight TE, Jacobson SC, Waters LC, Foote RS, Ramsey JM (2000) Integrated system for rapid PCR-based DNA analysis in microfluidic devices. Analytical Chemistry 72: 2995–3000
153. Khandurina J, Guttman A (2003) Microscale separation and analysis. Current Opinion in Chemical Biology 7: 595–602
154. Kim DK, Kang SH (2005) On-channel base stacking in microchip capillary gel electrophoresis for high-sensitivity DNA fragment analysis. Journal of Chromatography A 1064: 121–127
155. Kim JH, Kang CJ, Jeon D, Kim YS (2005) A disposable capillary electrophoresis microchip with an indium tin oxide decoupler/amperometric detector. Microelectronic Engineering 78–79: 563–570
156. Kim SM, Burns MA, Hasselbrink EF (2006) Electrokinetic protein preconcentration using a simple glass/poly(dimethylsiloxane) microfluidic chip. Analytical Chemistry 78: 4779–4785
157. Koutny L, Schmalzing D, Salas-Solano O, El-Difrawy S, Adourian A, Buonocore S, Abbey K, McEwan P, Matsudaira P, Ehrlich D (2000) Eight hundred base sequencing in a microfabricated electrophoretic device. Analytical Chemistry 72: 3388–3391
158. Kricka LJ, Wilding P (2003) Microchip PCR. Analytical and Bioanalytical Chemistry 377: 820–825
159. Kurnik RT, Boone TD, Nguyen U, Ricco AJ, Williams SJ (2003) Use of floating electrodes in transient isotachophoresis to increase the sensitivity of detection. Lab on a Chip 3: 86–92
160. Kutter JP, Jacobson SC, Ramsey JM (1997) Integrated microchip device with electrokinetically controlled solvent mixing for isocratic and gradient elution in micellar electrokinetic chromatography. Analytical Chemistry 69: 5165–5171
161. Kutter JP, Ramsey RS, Jacobson SC, Ramsey JM (1998) Determination of metal cations in microchip electrophoresis using on-chip complexation and sample stacking. Journal of Microcolumn Separations 10: 313–319
162. Lacher NA, de Rooij NF, Verpoorte E, Lunte SM (2003) Comparison of the performance characteristics of poly(dimethylsiloxane) and Pyrex microchip electrophoresis devices for peptide separations. Journal of Chromatography A 1004: 225–235
163. Lagally ET, Simpson PC, Mathies RA (2000) Monolithic integrated microfluidic DNA amplification and capillary electrophoresis analysis system. Sensors and Actuators B: Chemical 63: 138–146

164. Lagally ET, Emrich CA, Mathies RA (2001) Fully integrated PCR-capillary electrophoresis microsystem for DNA analysis. Lab on a Chip 1: 102–107
165. Lagally ET, Medintz I, Mathies RA (2001) Single-molecule DNA amplification and analysis in an integrated microfluidic device. Analytical Chemistry 73: 565–570
166. Lagally ET, Scherer JR, Blazej RG, Toriello NM, Diep BA, Ramchandani M, Sensabaugh GF, Riley LW, Mathies RA (2004) Integrated portable genetic analysis microsystem for pathogen/infectious disease detection. Analytical Chemistry 76: 3162–3170
167. Lapos JA, Ewing AG (2000) Injection of fluorescently labeled analytes into microfabricated chips using optically gated electrophoresis. Analytical Chemistry 72: 4598–4602
168. Lapos JA, Manica DP, Ewing AG (2002) Dual fluorescence and electrochemical detection on an electrophoresis microchip. Analytical Chemistry 74: 3348–3353
169. Lee GB, Fu LM, Lin CH, Lee CY, Yang RJ (2004) Dispersion control in microfluidic chips by localized zeta potential variation using the field effect. Electrophoresis 25: 1879–1887
170. Lee NY, Yamada M, Seki M (2004) Pressure-driven sample injection with quantitative liquid dispensing for on-chip electrophoresis. Analytical Sciences 20: 483–487
171. Li BQ, Fang XH, Luo HB, Petersen E, Seo YS, Samuilov V, Rafailovich M, Sokolov J, Gersappe D, Chu B (2006) Influence of electric field intensity, ionic strength, and migration distance on the mobility and diffusion in DNA surface electrophoresis. Electrophoresis 27: 1312–1321
172. Li BQ, Fang XH, Luo HB, Seo YS, Petersen E, Ji Y, Rafailovich M, Sokolov J, Gersappe D, Chu B (2006) Separation of DNA with different configurations on flat and nanopatterned surfaces. Analytical Chemistry 78: 4743–4751
173. Li JJ, Kelly JF, Chemushevich I, Harrison DJ, Thibault P (2000) Separation and identification of peptides from gel-isolated membrane proteins using a microfabricated device for combined capillary electrophoresis/nanoelectrospray mass spectrometry. Analytical Chemistry 72: 599–609
174. Li JJ, Tremblay TL, Wang C, Attiya S, Harrison DJ, Thibault P (2001) Integrated system for high-throughput protein identification using a microfabricated device coupled to capillary electrophoresis/nanoelectrospray mass spectrometry. Proteomics 1: 975–986
175. Li SFY, Kricka LJ (2006) Clinical analysis by microchip capillary electrophoresis. Clinical Chemistry 52: 37–45
176. Liang ZH, Chiem N, Ocvirk G, Tang T, Fluri K, Harrison DJ (1996) Microfabrication of a planar absorbance and fluorescence cell for integrated capillary electrophoresis devices. Analytical Chemistry 68: 1040–1046
177. Lin RS, Burke DT, Burns MA (2003) Selective extraction of size-fractioned DNA samples in microfabricated electrophoresis devices. Journal of Chromatography A 1010: 255–268

178. Lin RS, Burke DT, Burns MA (2005) Addressable electric fields for size-fractioned sample extraction in microfluidic devices. Analytical Chemistry 77: 4338–4347
179. Lin YC, Ho HC, Tseng CK, Hou SQ (2001) A poly-methylmethacrylate electrophoresis microchip with sample preconcentrator. Journal of Micromechanics and Microengineering 11: 189–194
180. Lin YW, Huang MF, Chang HT (2005) Nanomaterials and chip-based nanostructures for capillary electrophoretic separations of DNA. Electrophoresis 26: 320–330
181. Liu BF, Hisamoto H, Terabe S (2003) Subsecond separation of cellular flavin coenzymes by microchip capillary electrophoresis with laser-induced fluorescence detection. Journal of Chromatography A 1021: 201–207
182. Liu J, Tseng K, Garcia B, Lebrilla CB, Mukerjee E, Collins S, Smith R (2001) Electrophoresis separation in open microchannels. A method for coupling electrophoresis with MALDI-MS. Analytical Chemistry 73: 2147–2151
183. Liu SR, Shi YN, Ja WW, Mathies RA (1999) Optimization of high-speed DNA sequencing on microfabricated capillary electrophoresis channels. Analytical Chemistry 71: 566–573
184. Liu SR, Ren HJ, Gao QF, Roach DJ, Loder RT, Armstrong TM, Mao QL, Blaga I, Barker DL, Jovanovich SB (2000) Automated parallel DNA sequencing on multiple channel microchips. Proceedings of the National Academy of Sciences of the United States of America 97: 5369–5374
185. Liu YJ, Foote RS, Culbertson CT, Jacobson SC, Ramsey RS, Ramsey JM (2000) Electrophoretic separation of proteins on microchips. Journal of Microcolumn Separations 12: 407–411
186. Liu YJ, Foote RS, Jacobson SC, Ramsey RS, Ramsey JM (2000) Electrophoretic separation of proteins on a microchip with noncovalent, postcolumn labeling. Analytical Chemistry 72: 4608–4613
187. Llopis SD, Stryjewski W, Soper SA (2004) Near-infrared time-resolved fluorescence lifetime determinations in poly(methylmethacrylate) microchip electrophoresis devices. Electrophoresis 25: 3810–3819
188. Luckey J, Norris T, Smith L (1993) Analysis of resolution in DNA sequencing by capillary gel electrophoresis. Journal of Physical Chemistry 97: 3067–3075
189. Luo HB, Gersappe D (2002) Modeling the dynamics of DNA electrophoresis on a flat surface. Electrophoresis 23: 2690–2696
190. Ma RB, Crabtree HJ, Backhouse CJ (2005) A rejuvenation method for poly(*N,N*-dimethylacrylamide)-coated glass microfluidic chips. Electrophoresis 26: 2692–2700
191. MacDonald MP, Spalding GC, Dholakia K (2003) Microfluidic sorting in an optical lattice. Nature 426: 421–424
192. Manage DP, Zheng Y, Somerville MJ, Backhouse CJ (2005) On-chip HA/SSCP for the detection of hereditary haemochromatosis. Microfluidics and Nanofluidics 1: 364–372

193. Manica DP, Mitsumori Y, Ewing AG (2003) Characterization of electrode fouling and surface regeneration for a platinum electrode on an electrophoresis microchip. Analytical Chemistry 75: 4572–4577
194. Manz A, Harrison DJ, Verpoorte EMJ, Fettinger JC, Paulus A, Ludi H, Widmer HM (1992) Planar chips technology for miniaturization and integration of separation techniques into monitoring systems – capillary electrophoresis on a chip. Journal of Chromatography 593: 253–258
195. Manz A, Verpoorte E, Effenhauser CS, Burggraf N, Raymond DE, Harrison DJ, Widmer HM (1993) Miniaturization of separation techniques using planar chip technology. HRC-Journal of High Resolution Chromatography 16: 433–436
196. Manz A, Effenhauser CS, Burggraf N, Harrison DJ, Seiler K, Fluri K (1994) Electroosmotic pumping and electrophoretic separations for miniaturized chemical-analysis systems. Journal of Micromechanics and Microengineering 4: 257–265
197. Mao QL, Pawliszyn J (1999) Demonstration of isoelectric focusing on an etched quartz chip with UV absorption imaging detection. Analyst 124: 637–641
198. Martin RS, Gawron AJ, Fogarty BA, Regan FB, Dempsey E, Lunte SM (2001) Carbon paste-based electrochemical detectors for microchip capillary electrophoresis/electrochemistry. Analyst 126: 277–280
199. Martin RS, Ratzlaff KL, Huynh BH, Lunte SM (2002) In-channel electrochemical detection for microchip capillary electrophoresis using an electrically isolated potentiostat. Analytical Chemistry 74: 1136–1143
200. Mastrangelo CH, Burns MA, Burke DT (1998) Microfabricated devices for genetic diagnostics. Proceedings of the IEEE 86: 1769–1787
201. McClain MA, Culbertson CT, Jacobson SC, Allbritton NL, Sims CE, Ramsey JM (2003) Microfluidic devices for the high-throughput chemical analysis of cells. Analytical Chemistry 75: 5646–5655
202. McCormick RM, Nelson RJ, AlonsoAmigo MG, Benvegnu J, Hooper HH (1997) Microchannel electrophoretic separations of DNA in injection-molded plastic substrates. Analytical Chemistry 69: 2626–2630
203. Medintz I, Wong WW, Sensabaugh G, Mathies RA (2000) High speed single nucleotide polymorphism typing of a hereditary haemochromatosis mutation with capillary array electrophoresis microplates. Electrophoresis 21: 2352–2358
204. Medintz I, Wong WW, Berti L, Shiow L, Tom J, Scherer J, Sensabaugh G, Mathies RA (2001) High-performance multiplex SNP analysis of three hemochromatosis-related mutations with capillary array electrophoresis microplates. Genome Research 11: 413–421
205. Medintz IL, Berti L, Emrich CA, Tom J, Scherer JR, Mathies RA (2001) Genotyping energy-transfer-cassette-labeled short-tandem-repeat amplicons with capillary array electrophoresis microchannel plates. Clinical Chemistry 47: 1614–1621
206. Medintz IL, Paegel BM, Blazej RG, Emrich CA, Berti L, Scherer JR, Mathies RA (2001) High-performance genetic analysis using microfabri-

cated capillary array electrophoresis microplates. Electrophoresis 22: 3845–3856
207. Medintz IL, Paegel BM, Mathies RA (2001) Microfabricated capillary array electrophoresis DNA analysis systems. Journal of Chromatography A 924: 265–270
208. Minc N, Futterer C, Dorfman K, Bancaud A, Gosse C, Goubault C, Viovy JL (2004) Quantitative microfluidic separation of DNA in self-assembled magnetic matrixes. Analytical Chemistry 76: 3770–3776
209. Minc N, Bokov P, Zeldovich KB, Futterer C, Viovy JL, Dorfman KD (2005) Motion of single long DNA molecules through arrays of magnetic columns. Electrophoresis 26: 362–375
210. Minc N, Viovy JL, Dorfman KD (2005) Non-markovian transport of DNA in microfluidic post arrays. Physical Review Letters 94: 198105
211. Mitnik L, Carey L, Burger R, Desmarais S, Koutny L, Wernet O, Matsudaira P, Ehrlich D (2002) High-speed analysis of multiplexed short tandem repeats with an electrophoretic microdevice. Electrophoresis 23: 719–726
212. Mohamadi MR, Kataoka M, Mahmoudian L, Jabasini M, Shinohara Y, Baba Y (2005) Analysis of sources of error in quantification of purified DNA fragments and unpurified PCR products by DNA microchip electrophoresis. Chromatographia 61: 339–344
213. Monahan J, Fosser KA, Gewirth AA, Nuzzo RG (2002) A split microchannel design and analytical model to compensate for electroosmotic instabilities in micro-separations. Lab on a Chip 2: 81–87
214. Monahan J, Gewirth AA, Nuzzo RG (2002) Indirect fluorescence detection of simple sugars via high-pH electrophoresis in poly(dimethylsiloxane) microfluidic chips. Electrophoresis 23: 2347–2354
215. Moore AW, Jacobson SC, Ramsey JM (1995) Microchip separations of neutral species via micellar electrokinetic capillary chromatography. Analytical Chemistry 67: 4184–4189
216. Mueller O, Hahnenberger K, Dittmann M, Yee H, Dubrow R, Nagle R, Ilsley D (2000) A microfluidic system for high-speed reproducible DNA sizing and quantitation. Electrophoresis 21: 128–134
217. Munro NJ, Snow K, Kant EA, Landers JP (1999) Molecular diagnostics on microfabricated electrophoretic devices: From slab gel- to capillary- to microchip-based assays for T- and B-cell lymphoproliferative disorders. Clinical Chemistry 45: 1906–1917
218. Munro NJ, Huang ZL, Finegold DN, Landers JP (2000) Indirect fluorescence detection of amino acids on electrophoretic microchips. Analytical Chemistry 72: 2765–2773
219. Musyimi HK, Guy J, Narcisse DA, Soper SA, Murray KK (2005) Direct coupling of polymer-based microchip electrophoresis to online MALDI-MS using a rotating ball inlet. Electrophoresis 26: 4703–4710
220. Nagata H, Tabuchi M, Hirano K, Baba Y (2005) High-speed separation of proteins by microchip electrophoresis using a polyethylene glycol-coated plastic chip with a sodium dodecyl sulfate-linear polyacrylamide solution. Electrophoresis 26: 2687–2691

221. Nagata H, Tabuchi M, Hirano K, Baba Y (2005) Microchip electrophoretic protein separation using electroosmotic flow induced by dynamic sodium dodecyl sulfate-coating of uncoated plastic chips. Electrophoresis 26: 2247–2253
222. Namasivayam V, Lin RS, Johnson B, Brahmasandra S, Razzacki Z, Burke DT, Burns MA (2004) Advances in on-chip photodetection for applications in miniaturized genetic analysis systems. Journal of Micromechanics and Microengineering 14: 81–90
223. Ocvirk G, Tang T, Harrison DJ (1998) Optimization of confocal epifluorescence microscopy for microchip-based miniaturized total analysis systems. Analyst 123: 1429–1434
224. Ocvirk G, Munroe M, Tang T, Oleschuk R, Westra K, Harrison DJ (2000) Electrokinetic control of fluid flow in native poly(dimethylsiloxane) capillary electrophoresis devices. Electrophoresis 21: 107–115
225. Ohji H, Izuo S, French PJ, Tsutsumi K (2002) Pillar structures with a submicron space fabricated by macroporous-based micromachining. Sensors and Actuators A: Physical 97–8: 744–748
226. Paegel BM, Hutt LD, Simpson PC, Mathies RA (2000) Turn geometry for minimizing band broadening in microfabricated capillary electrophoresis channels. Analytical Chemistry 72: 3030–3037
227. Paegel BM, Emrich CA, Weyemayer GJ, Scherer JR, Mathies RA (2002) High throughput DNA sequencing with a microfabricated 96-lane capillary array electrophoresis bioprocessor. Proceedings of the National Academy of Sciences of the United States of America 99: 574–579
228. Paegel BM, Yeung SHI, Mathies RA (2002) Microchip bioprocessor for integrated nanovolume sample purification and DNA sequencing. Analytical Chemistry 74: 5092–5098
229. Pal R, Yang M, Lin R, Johnson BN, Srivastava N, Razzacki SZ, Chomistek KJ, Heldsinger DC, Haque RM, Ugaz VM, Thwar PK, Chen Z, Alfano K, Yim MB, Krishnan M, Fuller AO, Larson RG, Burke DT, Burns MA (2005) An integrated microfluidic device for influenza and other genetic analyses. Lab on a Chip 5: 1024–1032
230. Palmer J, Burgi DS, Munro NJ, Landers JP (2001) Electrokinetic injection for stacking neutral analytes in capillary and microchip electrophoresis. Analytical Chemistry 73: 725–731
231. Pan T, Kelly RT, Asplund MC, Woolley AT (2004) Fabrication of calcium fluoride capillary electrophoresis microdevices for on-chip infrared detection. Journal of Chromatography A 1027: 231–235
232. Patel PD, Shaqfeh ESG (2003) A computational study of DNA separations in sparse disordered and periodic arrays of posts. Journal of Chemical Physics 118: 2941–2951
233. Pernodet N, Samuilov V, Shin K, Sokolov J, Rafailovich MH, Gersappe D, Chu B (2000) DNA electrophoresis on a flat surface. Physical Review Letters 85: 5651–5654

234. Ping GC, Zhu BM, Jabasini M, Xu F, Oka H, Sugihara H, Baba Y (2005) Analysis of lipoproteins by microchip electrophoresis with high speed and high reproducibility. Analytical Chemistry 77: 7282–7287
235. Qi SZ, Liu XZ, Ford S, Barrows J, Thomas G, Kelly K, McCandless A, Lian K, Goettert J, Soper SA (2002) Microfluidic devices fabricated in poly (methyl methacrylate) using hot-embossing with integrated sampling capillary and fiber optics for fluorescence detection. Lab on a Chip 2: 88–95
236. Ramsey JD, Jacobson SC, Culbertson CT, Ramsey JM (2003) High-efficiency, two-dimensional separations of protein digests on microfluidic devices. Analytical Chemistry 75: 3758–3764
237. Richter T, Shultz-Lockyear LL, Oleschuk RD, Bilitewski U, Harrison DJ (2002) Bi-enzymatic and capillary electrophoretic analysis of non-fluorescent compounds in microfluidic devices – Determination of xanthine. Sensors and Actuators B: Chemical 81: 369–376
238. Rocklin RD, Ramsey RS, Ramsey JM (2000) A microfabricated fluidic device for performing two-dimensional liquid-phase separations. Analytical Chemistry 72: 5244–5249
239. Roddy ES, Lapos JA, Ewing AG (2003) Rapid serial analysis of multiple oligonucleotide samples on a microchip using optically gated injection. Journal of Chromatography A 1004: 217–224
240. Roddy ES, Price M, Ewing AG (2003) Continuous monitoring of a restriction enzyme digest of DNA on a microchip with automated capillary sample introduction. Analytical Chemistry 75: 3704–3711
241. Roddy ES, Xu HW, Ewing AG (2004) Sample introduction techniques for microfabricated separation devices. Electrophoresis 25: 229–242
242. Rodriguez I, Lee HK, Li SFY (1999) Microchannel electrophoretic separation of biogenic amines by micellar electrokinetic chromatography. Electrophoresis 20: 118–126
243. Rodriguez I, Jin LJ, Li SFY (2000) High-speed chiral separations on microchip electrophoresis devices. Electrophoresis 21: 211–219
244. Ronai Z, Barta C, Sasvari-Szekely M, Guttman A (2001) DNA analysis on electrophoretic microchips: Effect of operational variables. Electrophoresis 22: 294–299
245. Roper MG, Easley CJ, Landers JP (2005) Advances in polymerase chain reaction on microfluidic chips. Analytical Chemistry 77: 3887–3894
246. Ross D, Ivory CF, Locascio LE, Van Cott KE (2004) Peak compression and resolution for electrophoretic separations in diverging microchannels. Electrophoresis 25: 3694–3704
247. Sakaue T (2006) DNA electrophoresis in designed channels. European Physical Journal E 19: 477–487
248. Salas-Solano O, Schmalzing D, Koutny L, Buonocore S, Adourian A, Matsudaira P, Ehrlich D (2000) Optimization of high-performance DNA sequencing on short microfabricated electrophoretic devices. Analytical Chemistry 72: 3129–3137

249. Sanders JC, Huang ZL, Landers JP (2001) Acousto-optical deflection-based whole channel scanning for microchip isoelectric focusing with laser-induced fluorescence detection. Lab on a Chip 1: 167–172
250. Sanders JC, Breadmore MC, Mitchell PS, Landers JP (2002) A simple PDMS-based electro-fluidic interface for microchip electrophoretic separations. Analyst 127: 1558–1563
251. Sanders JC, Breadmore MC, Kwok YC, Horsman KM, Landers JP (2003) Hydroxypropyl cellulose as an adsorptive coating sieving matrix for DNA separations: Artificial neural network optimization for microchip analysis. Analytical Chemistry 75: 986–994
252. Sassi AP, Paulus A, Cruzado ID, Bjornson T, Hooper HH (2000) Rapid, parallel separations of D1S80 alleles in a plastic microchannel chip. Journal of Chromatography A 894: 203–217
253. Schmalzing D, Koutny L, Adourian A, Belgrader P, Matsudaira P, Ehrlich D (1997) DNA typing in thirty seconds with a microfabricated device. Proceedings of the National Academy of Sciences of the United States of America 94: 10273–10278
254. Schmalzing D, Adourian A, Koutny L, Ziaugra L, Matsudaira P, Ehrlich D (1998) DNA sequencing on microfabricated electrophoretic devices. Analytical Chemistry 70: 2303–2310
255. Schmalzing D, Koutny L, Chisholm D, Adourian A, Matsudaira P, Ehrlich D (1999) Two-color multiplexed analysis of eight short tandem repeat loci with an electrophoretic microdevice. Analytical Biochemistry 270: 148–152
256. Schmalzing D, Tsao N, Koutny L, Chisholm D, Srivastava A, Adourian A, Linton L, McEwan P, Matsudaira P, Ehrlich D (1999) Toward real-world sequencing by microdevice electrophoresis. Genome Research 9: 853–858
257. Seiler K, Harrison DJ, Manz A (1993) Planar glass chips for capillary electrophoresis – Repetitive sample injection, quantitation, and separation efficiency. Analytical Chemistry 65: 1481–1488
258. Seiler K, Fan ZHH, Fluri K, Harrison DJ (1994) Electroosmotic pumping and valveless control of fluid-flow within a manifold of capillaries on a glass chip. Analytical Chemistry 66: 3485–3491
259. Seo YS, Luo H, Samuilov VA, Rafailovich MH, Sokolov J, Gersappe D, Chu B (2004) DNA electrophoresis on nanopatterned surfaces. Nano Letters 4: 659–664
260. Shadpour H, Soper SA (2006) Two-dimensional electrophoretic separation of proteins using poly(methyl methacrylate) microchips. Analytical Chemistry 78: 3519–3527
261. Shah JJ, Geist J, Locascio LE, Gaitan M, Rao MV, Vreeland WN (2006) Capillarity induced solvent-actuated bonding of polymeric microfluidic devices. Analytical Chemistry 78: 3348–3353
262. Shaikh FA, Ugaz VM (2006) Collection, focusing, and metering of DNA in microchannels using addressable electrode arrays for portable low-power bioanalysis. Proceedings of the National Academy of Sciences of the United States of America 103: 4825–4830

263. Shi YN, Simpson PC, Scherer JR, Wexler D, Skibola C, Smith MT, Mathies RA (1999) Radial capillary array electrophoresis microplate and scanner for high-performance nucleic acid analysis. Analytical Chemistry 71: 5354–5361
264. Sieben VJ, Backhouse CJ (2005) Rapid on-chip postcolumn labeling and high-resolution separations of DNA. Electrophoresis 26: 4729–4742
265. Simpson PC, Roach D, Woolley AT, Thorsen T, Johnston R, Sensabaugh GF, Mathies RA (1998) High-throughput genetic analysis using microfabricated 96-sample capillary array electrophoresis microplates. Proceedings of the National Academy of Sciences of the United States of America 95: 2256–2261
266. Sirichai S, de Mello AJ (1999) A capillary electrophoresis microchip for the analysis of photographic developer solutions using indirect fluorescence detection. Analyst 125: 133–137
267. Sirichai S, de Mello AJ (2001) A capillary electrophoresis chip for the analysis of print and film photographic developing agents in commercial processing solutions using indirect fluorescence detection. Electrophoresis 22: 348–354
268. Skelley AM, Mathies RA (2003) Chiral separation of fluorescamine-labeled amino acids using microfabricated capillary electrophoresis devices for extraterrestrial exploration. Journal of Chromatography A 1021: 191–199
269. Skelley AM, Scherer JR, Aubrey AD, Grover WH, Ivester RHC, Ehrenfreund P, Grunthaner FJ, Bada JL, Mathies RA (2005) Development and evaluation of a microdevice for amino acid biomarker detection and analysis on Mars. Proceedings of the National Academy of Sciences of the United States of America 102: 1041–1046
270. Slentz BE, Penner NA, Regnier F (2002) Sampling bias at channel junctions in gated flow injection on chips. Analytical Chemistry 74: 4835–4840
271. Smith EM, Xu HW, Ewing AG (2001) DNA separations in microfabricated devices with automated capillary sample introduction. Electrophoresis 22: 363–370
272. Song S, Singh AK, Kirby BJ (2004) Electrophoretic concentration of proteins at laser-patterned nanoporous membranes in microchips. Analytical Chemistry 76: 4589–4592
273. Soper SA, Ford SM, Xu YC, Qi SZ, McWhorter S, Lassiter S, Patterson D, Bruch RC (1999) Nanoliter-scale sample preparation methods directly coupled to polymethylmethacrylate-based microchips and gel-filled capillaries for the analysis of oligonucleotides. Journal of Chromatography A 853: 107–120
274. Soper SA, Henry AC, Vaidya B, Galloway M, Wabuyele M, McCarley RL (2002) Surface modification of polymer-based microfluidic devices. Analytica Chimica Acta 470: 87–99
275. Spitzack KD, Ugaz VM (2005) Polymerase chain reaction in miniaturized systems: Big progress in little devices. In: Minteer SD (ed) Microfluidic techniques: Reviews and protocols. Humana Press, Totowa, NJ, Chapter 10

276. Starkey DE, Han A, Bao JJ, Ahn CH, Wehmeyer KR, Prenger MC, Halsall HB, Heineman WR (2001) Fluorogenic assay for β-glucuronidase using microchip-based capillary electrophoresis. Journal of Chromatography B 762: 33–41
277. Streek M, Schmid F, Duong TT, Ros A (2004) Mechanisms of DNA separation in entropic trap arrays: A brownian dynamics simulation. Journal of Biotechnology 112: 79–89
278. Sudarsan AP, Ugaz VM (2004) Printed circuit technology for fabrication of plastic-based microfluidic devices. Analytical Chemistry 76: 3229–3235
279. Suljak SW, Thompson LA, Ewing AG (2004) Improving resolution for channel-format chip-based electrophoresis with electrochemical array detection. Journal of Separation Science 27: 13–20
280. Swinney K, Bornhop DJ (2002) Quantification and evaluation of joule heating in on-chip capillary electrophoresis. Electrophoresis 23: 613–620
281. Tabuchi M, Ueda M, Kaji N, Yamasaki Y, Nagasaki Y, Yoshikawa K, Kataoka K, Baba Y (2004) Nanospheres for DNA separation chips. Nature Biotechnology 22: 337–340
282. Tabuchi M, Baba Y (2005) A triple-injection method for microchip electrophoresis. Electrophoresis 26: 376–382
283. Tabuchi M, Baba Y (2005) Design for DNA separation medium using bacterial cellulose fibrils. Analytical Chemistry 77: 7090–7093
284. Tabuchi M, Katsuyama Y, Nogami K, Nagata H, Wakuda K, Fujimoto M, Nagasaki Y, Yoshikawa K, Kataoka K, Baba Y (2005) A design of nanosized pegylated-latex mixed polymer solution for microchip electrophoresis. Lab on a Chip 5: 199–204
285. Tang T, Badal MY, Ocvirk G, Lee WE, Bader DE, Bekkaoui F, Harrison DJ (2002) Integrated microfluidic electrophoresis system for analysis of genetic materials using signal amplification methods. Analytical Chemistry 74: 725–733
286. Taylor P, Manage DP, Helmle KE, Zheng Y, Glerum DM, Backhouse CJ (2005) Analysis of mitochondrial DNA in microfluidic systems. Journal of Chromatography B: Analytical Technologies in the Biomedical and Life Sciences 822: 78–84
287. Thomas G, Sinville R, Sutton S, Farquar H, Hammer RP, Soper SA, Cheng YW, Barany F (2004) Capillary and microelectrophoretic separations of ligase detection reaction products produced from low-abundant point mutations in genomic DNA. Electrophoresis 25: 1668–1677
288. Tian HJ, Brody LC, Landers JP (2000) Rapid detection of deletion, insertion, and substitution mutations via heteroduplex analysis using capillary- and microchip-based electrophoresis. Genome Research 10: 1403–1413
289. Tian HJ, Jaquins-Gerstl A, Munro N, Trucco M, Brody LC, Landers JP (2000) Single-strand conformation polymorphism analysis by capillary and microchip electrophoresis: A fast, simple method for detection of common mutations in BRCA1 and BRCA2. Genomics 63: 25–34
290. Tian HJ, Brody LC, Fan SJ, Huang ZL, Landers JP (2001) Capillary and microchip electrophoresis for rapid detection of known mutations by combin-

ing allele-specific DNA amplification with heteroduplex analysis. Clinical Chemistry 47: 173–185

291. Tian HJ, Landers JP (2002) Hydroxyethylcellulose as an effective polymer network for DNA analysis in uncoated glass microchips: Optimization and application to mutation detection via heteroduplex analysis. Analytical Biochemistry 309: 212–223
292. Tian HJ, Emrich CA, Scherer JR, Mathies RA, Andersen PS, Larsen LA, Christiansen M (2005) High-throughput single-strand conformation polymorphism analysis on a microfabricated capillary array electrophoresis device. Electrophoresis 26: 1834–1842
293. Tseng WL, Lin YW, Chen KC, Chang HT (2002) DNA analysis on microfabricated electrophoretic devices with bubble cells. Electrophoresis 23: 2477–2484
294. Ueda M, Kiba Y, Abe H, Arai A, Nakanishi H, Baba Y (2000) Fast separation of oligonucleotide and triplet repeat DNA on a microfabricated capillary electrophoresis device and capillary electrophoresis. Electrophoresis 21: 176–180
295. Ueda M, Nakanishi H, Tabata O, Baba Y (2000) Imaging of a band for DNA fragment migrating in microchannel on integrated microchip. Materials Science & Engineering C: Biomimetic and Supramolecular Systems 12: 33–36
296. Ueda M, Endo Y, Abe H, Kuyama H, Nakanishi H, Arai A, Baba Y (2001) Field-inversion electrophoresis on a microchip device. Electrophoresis 22: 217–221
297. Ueda M, Hayama T, Takamura Y, Horiike Y, Dotera T, Baba Y (2004) Electrophoresis of long deoxyribonucleic acid in curved channels: The effect of channel width on migration dynamics. Journal of Applied Physics 96: 2937–2944
298. Ugaz VM, Brahmasandra SN, Burke DT, Burns MA (2002) Cross-linked polyacrylamide gel electrophoresis of single-stranded DNA for microfabricated genomic analysis systems. Electrophoresis 23: 1450–1459
299. Ugaz VM, Burke DT, Burns MA (2002) Microdevice-based measurements of diffusion and dispersion in cross-linked and linear polyacrylamide DNA sequencing gels. Electrophoresis 23: 2777–2787
300. Ugaz VM, Lin RS, Srivastava N, Burke DT, Burns MA (2003) A versatile microfabricated platform for electrophoresis of double- and single-stranded DNA. Electrophoresis 24: 151–157
301. Ugaz VM, Elms RD, Lo RC, Shaikh FA, Burns MA (2004) Microfabricated electrophoresis systems for DNA sequencing and genotyping applications: Current technology and future directions. Philosophical Transactions of the Royal Society of London Series A: Mathematical Physical and Engineering Sciences 362: 1105–1129
302. Unal K, Frommer J, Wickramasinghe HK (2006) Ultrafast molecule sorting and delivery by atomic force microscopy. Applied Physics Letters 88: 183105

303. Vahedi G, Kaler C, Backhouse CJ (2004) An integrated method for mutation detection using on-chip sample preparation, single-stranded conformation polymorphism, and heterduplex analysis. Electrophoresis 25: 2346–2356
304. Vandaveer WR, Pasas-Farmer SA, Fischer DJ, Frankenfeld CN, Lunte SM (2004) Recent developments in electrochemical detection for microchip capillary electrophoresis. Electrophoresis 25: 3528–3549
305. Vazquez M, McKinley G, Mitnik L, Desmarais S, Matsudaira P, Ehrlich D (2002) Electrophoretic injection within microdevices. Analytical Chemistry 74: 1952–1961
306. Vegvari A, Hjerten S (2002) A hybrid microdevice for electrophoresis and electrochromatography using UV detection. Electrophoresis 23: 3479–3486
307. Volkmuth WD, Austin RH (1992) DNA electrophoresis in microlithographic arrays. Nature 358: 600–602
308. Volkmuth WD, Duke T, Wu MC, Austin RH, Szabo A (1994) DNA electrodiffusion in a 2D array of posts. Physical Review Letters 72: 2117–2120
309. vonHeeren F, Verpoorte E, Manz A, Thormann W (1996) Characterization of electrophoretic sample injection and separation in a gel-filled cyclic planar microstructure. Journal of Microcolumn Separations 8: 373–381
310. vonHeeren F, Verpoorte E, Manz A, Thormann W (1996) Micellar electrokinetic chromatography separations and analyses of biological samples on a cyclic planar microstructure. Analytical Chemistry 68: 2044–2053
311. Wabuyele MB, Ford SM, Stryjewski W, Barrow J, Soper SA (2001) Single molecule detection of double-stranded DNA in poly(methylmethacrylate) and polycarbonate microfluidic devices. Electrophoresis 22: 3939–3948
312. Wainright A, Williams SJ, Ciambrone G, Xue QF, Wei J, Harris D (2002) Sample pre-concentration by isotachophoresis in microfluidic devices. Journal of Chromatography A 979: 69–80
313. Wainright A, Nguyen UT, Bjornson T, Boone TD (2003) Preconcentration and separation of double-stranded DNA fragments by electrophoresis in plastic microfluidic devices. Electrophoresis 24: 3784–3792
314. Walker PA, Morris MD, Burns MA, Johnson BN (1998) Isotachophoretic separations on a microchip. Normal Raman spectroscopy detection. Analytical Chemistry 70: 3766–3769
315. Wallenborg SR, Bailey CG (2000) Separation and detection of explosives on a microchip using micellar electrokinetic chromatography and indirect laser-induced fluorescence. Analytical Chemistry 72: 1872–1878
316. Wang H, Qin JH, Dai ZP, Wang L, Bai JL, Lin BC (2003) Highly efficient separation of dsDNA fragments on glass chips by using an ultralow viscosity sieving matrix. Journal of Separation Science 26: 869–874
317. Wang SC, Morris MD (2000) Plastic microchip electrophoresis with analyte velocity modulation. Application to fluorescence background rejection. Analytical Chemistry 72: 1448–1452
318. Wang YC, Stevens AL, Han JY (2005) Million-fold preconcentration of proteins and peptides by nanofluidic filter. Analytical Chemistry 77: 4293–4299

319. Waters LC, Jacobson SC, Kroutchinina N, Khandurina J, Foote RS, Ramsey JM (1998) Microchip device for cell lysis, multiplex PCR amplification, and electrophoretic sizing. Analytical Chemistry 70: 158–162
320. Waters LC, Jacobson SC, Kroutchinina N, Khandurina J, Foote RS, Ramsey JM (1998) Multiple sample PCR amplification and electrophoretic analysis on a microchip. Analytical Chemistry 70: 5172–5176
321. Webster JR, Burns MA, Burke DT, Mastrangelo CH (2001) Monolithic capillary electrophoresis device with integrated fluorescence detector. Analytical Chemistry 73: 1622–1626
322. Woolley AT, Mathies RA (1994) Ultra-high-speed DNA fragment separations using microfabricated capillary array electrophoresis chips. Proceedings of the National Academy of Sciences of the United States of America 91: 11348–11352
323. Woolley AT, Mathies RA (1995) Ultra-high-speed DNA-sequencing using capillary electrophoresis chips. Analytical Chemistry 67: 3676–3680
324. Woolley AT, Hadley D, Landre P, deMello AJ, Mathies RA, Northrup MA (1996) Functional integration of PCR amplification and capillary electrophoresis in a microfabricated DNA analysis device. Analytical Chemistry 68: 4081–4086
325. Woolley AT, Sensabaugh GF, Mathies RA (1997) High-speed DNA genotyping using microfabricated capillary array electrophoresis chips. Analytical Chemistry 69: 2181–2186
326. Woolley AT, Lao KQ, Glazer AN, Mathies RA (1998) Capillary electrophoresis chips with integrated electrochemical detection. Analytical Chemistry 70: 684–688
327. Xiao DQ, Van Le T, Wirth MJ (2004) Surface modification of the channels of poly(dimethylsiloxane) microfluidic chips with polyacrylamide for fast electrophoretic separations of proteins. Analytical Chemistry 76: 2055–2061
328. Xu F, Jabasini M, Baba Y (2002) DNA separation by microchip electrophoresis using low-viscosity hydroxypropylmethylcellulose-50 solutions enhanced by polyhydroxy compounds. Electrophoresis 23: 3608–3614
329. Xu F, Jabasini M, Liu SQ, Baba Y (2003) Reduced viscosity polymer matrices for microchip electrophoresis of double-stranded DNA. Analyst 128: 589–592
330. Xu F, Baba Y (2004) Polymer solutions and entropic-based systems for double-stranded DNA capillary electrophoresis and microchip electrophoresis. Electrophoresis 25: 2332–2345
331. Xu F, Jabasini M, Baba Y (2005) Screening of mixed poly(ethylene oxide) solutions for microchip separation of double-stranded DNA using an orthogonal design approach. Electrophoresis 26: 3013–3020
332. Xu HW, Roddy ES, Roddy TP, Lapos JA, Ewing AG (2004) Parallel separations of oligonucleotides with optically gated sample introduction on multichannel microchips. Journal of Separation Science 27: 7–12
333. Xu XM, Roddy TP, Lapos JA, Ewing AG (2002) Parallel analysis with optically gated sample introduction on a multichannel microchip. Analytical Chemistry 74: 5517–5522

334. Xu Y, Zhang CX, Janasek D, Manz A (2003) Sub-second isoelectric focusing in free flow using a microfluidic device. Lab on a Chip 3: 224–227
335. Xue QF, Wainright A, Gangakhedkar S, Gibbons I (2001) Multiplexed enzyme assays in capillary electrophoretic single-use microfluidic devices. Electrophoresis 22: 4000–4007
336. Yang H, Chien RL (2001) Sample stacking in laboratory-on-a-chip devices. Journal of Chromatography A 924: 155–163
337. Yeung SHI, Greenspoon SA, McGuckian A, Crouse CA, Emrich CA, Ban J, Mathies RA (2006) Rapid and high-throughput forensic short tandem repeat typing using a 96-lane microfabricated capillary array electrophoresis microdevice. Journal of Forensic Sciences 51: 740–747
338. Yi S, Seo KS, Cho YH (2005) A DNA trapping and extraction microchip using periodically crossed electrophoresis in a micropillar array. Sensors and Actuators A: Physical 120: 429–436
339. Zhang CX, Manz A (2001) Narrow sample channel injectors for capillary electrophoresis on microchips. Analytical Chemistry 73: 2656–2662
340. Zhang L, Yin XF, Fang ZL (2006) Negative pressure pinched sample injection for microchip-based electrophoresis. Lab on a Chip 6: 258–264
341. Zhang LH, Dang FQ, Baba Y (2002) Stepwise gradient of linear polymer matrices in microchip electrophoresis for high-resolution separation of DNA. Electrophoresis 23: 2341–2346
342. Zhang LH, Dang FQ, Baba Y (2003) Microchip electrophoresis-based separation of DNA. Journal of Pharmaceutical and Biomedical Analysis 30: 1645–1654
343. Zhelev Z, Bakalova R, Ewis A, Ohba H, Ishikawa M, Baba Y (2005) Nonradioactive telomerase activity assay by microchip electrophoresis: Privileges to the classical gel electrophoresis assay. Electrophoresis 26: 3021–3024
344. Zhou ZM, Liu DY, Zhong RT, Dai ZP, Wu DP, Wang H, Du YG, Xia ZN, Zhang LP, Mei XD, Lin BC (2004) Determination of SARS-coronavirus by a microfluidic chip system. Electrophoresis 25: 3032–3039
345. Zugel SA, Burke BJ, Regnier FE, Lytle FE (2000) Electrophoretically mediated microanalysis of leucine aminopeptidase using two-photon excited fluorescence detection on a microchip. Analytical Chemistry 72: 5731–5735

# Chapter 11
# Chromatography in Microstructures

Ómar Gústafsson and Jörg P. Kutter

MIC – Department of Micro and Nanotechnology, Technical University of Denmark, Lyngby, Denmark

## 1. Introduction

### 1.1. Background

The possibility to separate a (complex) mixture of chemicals in a sample to gain understanding of the composition of the sample and the amounts of the individual components within the sample, or both, is among the most important and often used analytical techniques. Whenever specific enough sensors are not available or when many components of the sample are of interest, or when these components are contained within a demanding matrix (e.g., proteins in blood), separation techniques are the only available choice to prepare the samples before the next step in the analytical process. Typically, an aliquot of the sample is injected into the separation system, where – if successful – the individual components are separated in time and/or space and, subsequently, monitored, detected, identified, and/or quantified. It is important, in this context, to remember that steps preceding and succeeding the actual separation always also contribute to the overall performance of the systems, which is why they always have to be included in an overall evaluation. Electrophoresis, chromatography, and related variants of these two are among the predominant separation technologies used today.

Calvin Giddings, undoubtedly one of the dominating figures in developing and describing modern separation sciences, succinctly summarized separations as "the art to maximize differential (separative) transport, while minimizing dispersive transport" [1]. It is in this little sentence that all of the challenges of realizing high performance separations are contained. And it is also this statement where some of the advantages of performing separations in microstructures can be derived from. Because, while phenomena pertaining to the separative transport are in essence identical in microstructures compared to more conventional formats, the strengths of microsystems lie in the possibility to – literally – miniaturize the effects of dispersion, and hence, arrive at high performance separations. Let us get back to this important aspect further down.

First, when using the term miniaturization in this chapter, it is not meant in the sense of purely shrinking capillaries or tubes further and further. Instead, a more important aspect is that as many parts as possible of the separation systems are fabricated and integrated on a typically flat substrate, e.g., a piece of silicon, glass, or polymer. The final consequence of this idea is to have all required functions of a separation, or indeed of the entire analytical process, combined on that one piece of real estate, a so-called monolithic integration. This avoids external manipulation of the sample between different steps, unnecessary transfer lines, and any sort of connectors, which all can add to diminishing the performance of the analytical system. At the same time, shrinking dimensions can add other benefits, such as low sample, reagent, and energy consumption, low waste production, high surface to volume ratios, faster analysis times, higher throughputs, and more reliable analysis through parallelization and easily built-in redundancy. Of course, these and related advantages have already been discussed elsewhere in this book and the final products are labeled micrototal analysis systems or lab-on-a-chip systems.

This chapter will concern itself with microsystems invoking chromatography as the separating technique [2, 3]. We use the term chromatography to describe separations that require at least two different phases to be present. Often, one of them is stationary, while the other is mobile and pushed, moved or percolated through or past the stationary phase. Different components within the sample are then moved along with the mobile phase and will be delayed to different extents during their passage through the stationary phase. This is achieved by exploiting various physicochemical phenomena, such as (reversible) adsorption to the stationary phase or differential distribution between the two phases. A range of interactions have been used to achieve chromatographic separations, and thus make this technique very universal, i.e., available for charged and noncharged molecules, for highly polar and very nonpolar molecules, and for small molecules and all the way to complex biomolecules and even polymers. A more in-depth discussion of the many variants of chromatography and the mechanisms behind are beyond the scope of this chapter. The reader is referred to the many monographs on this topic (e.g., [1, 4, 5]). Also, purely electrophoretic separation techniques, where charged molecules are separated in an electric field according to their charge to mass ratio, are discussed in a different chapter in this book. Hybrid techniques, such as electrokinetic chromatography (EKC), which have attributes of both electrophoretic and chromatographic techniques, are also not discussed further here – the reader is referred to a recent monograph on this topic containing a chapter detailing EKC on chips [6].

## 1.2. Short Overview of Some Variants of Chromatography

### 1.2.1. Gas Chromatography (GC)

Here, the mobile phase is normally an inert gas (such as nitrogen or helium) that is pushed through a column or capillary containing the stationary phase by overpressure from a gas storage container. In modern GC, the stationary phase is typically a thin highly viscous film on the capillary walls. Interactions exploited in this technique are manifold, but are largely restricted to interactions with the stationary phase. The mobile phase is almost exclusively a transporting medium only. In GC, temperature is frequently used as a means to tune separations.

### 1.2.2. Pressure-Driven Liquid Chromatography (LC)

In this variant, the mobile phase is a liquid and the stationary phase is either a solid or a material with properties of a liquid immobilized on a solid carrier. Here, interactions take place with both the stationary and the mobile phase, i.e., the differential distribution of a sample component over the phases is exploited. Historically, the first LC separations were performed such that the stationary phase was hydrophilic (e.g., silica), while the mobile phase was hydrophobic (e.g., hexane). This was referred to as *normal-phase LC*, while the opposite case (stationary phase is hydrophobic and the mobile phase is hydrophilic) was termed *reversed-phase LC*. However, nowadays, the latter is much more common than the former, mainly due to its simpler and more reliable operation and the possibility to immediately use aqueous samples, as often obtained in biochemistry and medicine. The mobile phase is driven by overpressure, which in most cases is generated by a pump. The stationary phase can be realized in many ways, as a packing of porous or nonporous beads, as a surface coating, as a porous in situ polymerized monolithic bed, or as an array of micromachined pillars, to name but the most important variants. In most cases, the above-mentioned solutions only constitute carriers, where the actual interacting media need to be added by proper surface modification afterwards.

### 1.2.3. Electrochromatography (EC)

The main difference to LC is that here the mobile phase is moved through or past the stationary phase by means of electric fields and by exploiting electroosmosis as a phenomenon to induce motion of bulk electrolyte solutions under an applied voltage. This poses different requirements to equipment, mobile phase composition, and, in some cases, detection as compared to pressure-driven techniques. An often cited advantage of electrodriven techniques is the fact that the electroosmotic flow has a flat flow velocity profile over the cross section of the capillary or channel, whereas pressure-driven flows are characterized by a parabolic flow velocity profile, which has immediate consequences in terms of dispersion of chromatographic bands (see further down). On the other hand, pumping by pressure is fairly independent of the composition of the mobile phase, whereas the electroosmotic flow is directly and strongly dependent on pH, ionic strength, and percentage of added nonpolar modifiers, and thus reacts strongly to changes in the composition of the mobile phase, whether by design or by accident. As far as the selection of the stationary phase is concerned, both LC and EC can exploit a variety of interaction mechanisms, such as, e.g., hydrophobic interactions, ion exchange interactions, size exclusion mechanisms, and more specialized interactions, such as charge-transfer or affinity interactions. This technique is for historical reasons often better known as capillary electrochromatography (CEC).

### 1.2.4. Miscellaneous

This category includes techniques such as shear-driven chromatography and hydrodynamic chromatography. In shear-driven chromatography, neither pressure nor electric fields are used to drive the mobile phase. Instead, two adjacent walls enclosing the mobile phase are moved relative to each other, thereby inducing shear in the mobile phase, which again leads to relative movement between the mobile and the stationary phase and thus allows chromatography. This technique obviously comes with its own challenges with regard to instrumentation and the realization of proper injection and detection protocols. Hydrodynamic chromatography is, despite its name, not a true chromatographic technique as it does not employ two different phases. It makes use of a parabolic flow velocity profile and very small channel dimensions, so that differently sized molecules will sample different percentages of the parabolic profile and hence spend different amounts of time traversing the system. Larger molecules are faster in this technique than smaller molecules, since the latter can get closer to the wall regions with their center of mass and therefore also experience the slower velocity regions near the walls. On average, they will therefore move more slowly than larger molecules. This technique works best if the inner diameter of the capillary or the depth of the channel are only a bit larger than the largest radii of the molecules to be separated.

### *1.3. Some Theoretical Considerations*

This section cannot be a replacement for an in-depth treatment of chromatographic theory, but we need to prepare at least a foundation for some discussions concerning performance, to define figures of merit for comparison and to understand scaling effects when going from conventional formats to microchips. Further, we will restrict ourselves to discussing elution chromatography, where a defined sample aliquot is injected, as opposed to, e.g., frontal chromatography, where sample is continuously infused.

The injection plug is ideally infinitesimally narrow, but has of course a certain width in the longitudinal direction and is often approximated by a top-hat function. As the separation proceeds, different types of analytes within the sample plug will be retained within the system to different degrees resulting in discrete zones or bands containing, ideally, only one type of analyte (separative or differential transport). At the same time, however, these bands widen in the longitudinal direction due to a number of processes collectively called band broadening mechanisms (dispersive transport). At the same time, the concentration profile of the original top-hat shapes evolve into Gaussian shapes. The standard deviation, $\sigma$, i.e., half the width of the Gaussian shape at the point of inflection, can now be used as a means to describe the effects of band broadening, and hence the efficiency of a separation process. Two figures of merit – in essence borrowed from distillation theory – are important, the plate height, $H$, and the plate number, $N$. $H$ is defined as the variance, $\sigma^2$, per unit separation length, $L$, while $N$ then computes as $L/H$. The overall variance of a system is typically the sum of a number of variances related to various band broadening mechanisms, such as, e.g., the injection plug length, the detection width, longitudinal diffusion, Joule heating effects, influences arising from the parabolic flow profile in pressure-driven systems, and so forth. Both, $N$ and $H$, are easily accessible from experimental data, such as the measured arrival times of the centers of mass of the bands at a detector and their widths in the graphical representation of the detector trace, the chromatogram. Strictly speaking, both $N$ and $H$ mainly include information about the amount of dispersive transport and hence another parameter, resolution *Rs*, is introduced, combining the effects of differential and dispersive transport.

The goal of modern separation technology is to provide fast and efficient results. In practical terms, this corresponds to pushing the sample through the stationary phase as fast as possible (i.e., maximizing the linear velocity, $u$) while maintaining efficient separations (i.e., minimizing the plate height). A plot of plate heights measured vs. the linear velocity typically results in curves with a minimum corresponding to the optimum linear velocity at which the smallest plate heights are achieved. These curves can be described following the van Deemter formalism [7]:

$$H=A+B/u+Cu, \tag{1}$$

where the term $A$ includes contributions to $H$ due to differences in pathlength through packed/porous stationary phases, the term $B$ relates mainly to longitudinal diffusion, and the term $C$ includes contributions based on the mass transfer kinetics related to the establishment of an equilibrium between the analyte concentration in the mobile and in the stationary phase. These plots are most often used as practical tools to investigate separation efficiencies, identify main sources for band broadening, and find the most suited linear velocity.

From the point of view of miniaturization, at first glance it seems obvious that miniaturized separation systems would suffer from a lack of sufficient available separation length, which consequently would lead to small numbers of theoretical plates and, hence, inefficient separations. However, microsystems can make up for this "shortcoming" by offering smaller diffusion lengths than their more conventional counterparts, which reduces, for example, the term $C$ in the van Deemter equation and therefore allows separations to be run at higher speeds without significant losses in efficiency. On the other hand, ways to "compact" longer separation lengths onto the small footprints of microchips have since been developed and successfully tested [8, 9]. Also, due to the possibilities in design, integration, and material, miniaturized separation systems can minimize the contributions from a number of notorious sources for band broadening, such as the injection plug length, Joule heating, and excessive connectors and transfer tubing. Finally, the short separation lengths allow the establishment of high electric fields with relatively small voltages, opening up the possibility to have electrodriven portable separation units running off batteries.

Two excellent and more in-depth discussions on the benefits of miniaturization for separation techniques can be found in the literature [10, 11].

## 2. Examples of Chromatography on Microchips

### *2.1. Gas Chromatography (GC)*

The first report of a microfabricated device for chemical analysis was published by Terry and co-workers at Stanford University in 1979 [12]. They had constructed a planar gas chromatograph by using photolithography and wet-etching techniques, as previously mainly employed in the microelectronics field. The gas chromatography column was made by etching a spiral channel into a 5-cm-diameter silicon wafer, resulting in a 1.5-m-long, 200 μm-wide, and 30-μm-deep column.

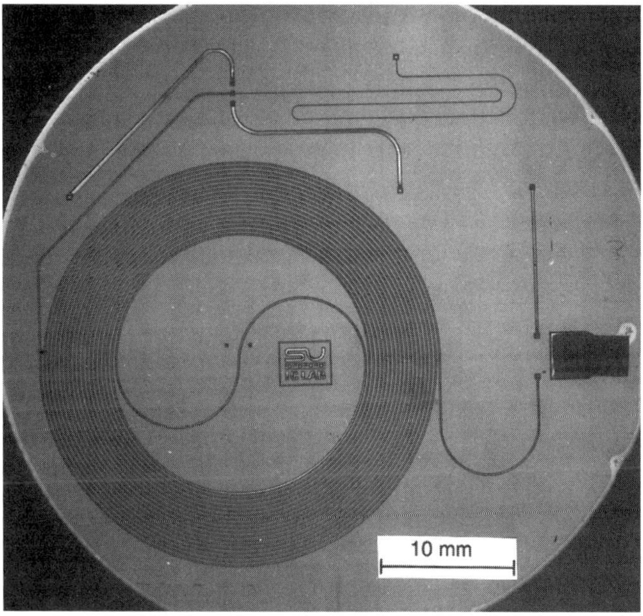


**Fig. 1.** Gas chromatograph with integrated sample injection and thermal conductivity detector. The chromatography column is the spiral channel. The channels on the top of the wafer make up the injection valve and on the right side of the wafer the thermal conductivity detector is mounted [13]. Reprinted with permission

The system had a sample injection valve fabricated onto the wafer, and a thermal conductivity detector that was mounted on the wafer. The whole device is pictured in Fig. 1. The system was sealed by a Pyrex glass plate that was anodically bonded to the silicon wafer. Pressurized gas was then coupled to the chip to introduce carrier gas and a sample gas stream. A viscous liquid stationary phase was introduced into the chip by filling the chip with a mixture of the stationary phase and a volatile solvent. The organic solvent was subsequently blown out of the capillary, leaving a thin layer of the stationary phase on the channel walls.

Using this setup, a separation of eight nonpolar substances was performed in less than 10 s, whereas a comparable separation by conventional column required several minutes. The effective plate numbers achieved in the separation ranged from 385 to 2,300 plates, which is significantly lower than what can routinely be reached in a conventional capillary column with the same length (up to 15,000 plates [7]).

This microfabricated GC device was far ahead of its time and hardly recognized by the analytical chemistry community, and it was not until the early 1990s that the miniaturization of analytical methods and devices really started gathering momentum when Manz and co-workers fabricated a miniaturized liquid chromatograph on a silicon wafer [13].

More recently, Reidy et al. from the University of Michigan reported on a microchip gas chromatographic separation of 24 analytes in around 6 min (see Fig. 2). The sample was a mixture of aromatic and aliphatic hydrocarbons with boiling points ranging from 30°C to 216°C. The peak capacity of the 3-m-long column was over 100 and an efficiency of around 12,500 plates was achieved using air as carrier gas. Figure 2 shows a separation of the mixture [14]. Compounds that co-elute are compounds with very similar boiling points, for example compounds 11, 12, and 13, which are 2,3-dimethylheptane, *m*-xylene, and *p*-xylene and have boiling points that are 141°C, 139°C, and 138.3°C, respectively.

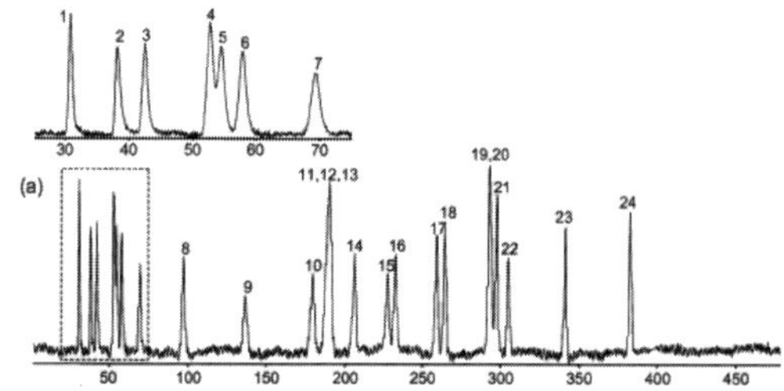


**Fig. 2.** Microchip gas chromatographic separation of 24 aromatic and aliphatic hydrocarbons. The upper image shows an expanded view of the first seven peaks [14]. Reprinted with permission

The same group from the University of Michigan reported in 2005 about using two 3-m-long microfabricated gas chromatography columns in tandem [15]. The two columns contained different stationary phases, one of them having a nonpolar dimethyl polysiloxane stationary phase and the other a moderately polar trifluoropropylmethyl polysiloxane stationary phase. Compounds have different retention factors in the two columns. One drawback of such a system is that some peaks that are separated in the first column can be recombined in the second column if the first compounds to elute are more retained in the second column. To overcome this problem, Lambertus and co-workers designed the system so that the first column could be bypassed for a certain time so that the flow was essentially put "on hold" for a few seconds in the first column. This allowed for greater control of the separations in the column ensemble. Figure 3 shows chromatograms of a 14-component mixture when operating with and without stopping the flow in the first column. Several components that otherwise coelute from the column ensemble can thus be separated very well by stopping the flow through the first column a few times. Because of the short separation columns, all of the 14 compounds are separated in less than 5 min using the stop-flow operation.

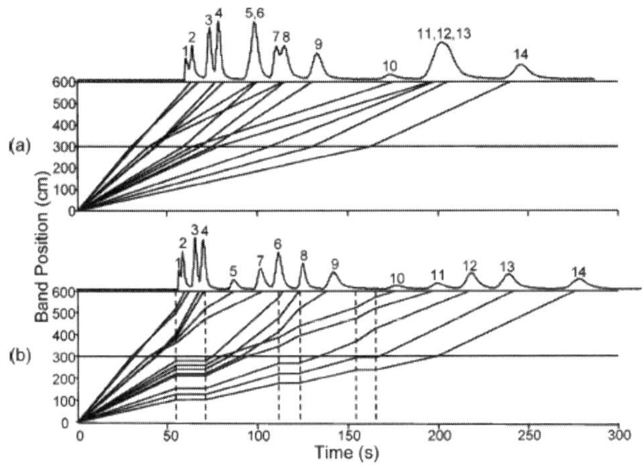


**Fig. 3.** Band positions in the two columns vs. time and the corresponding chromatograms. The band trajectories through the columns are indicated with lines. *Horizontal lines* mean that an analyte has stopped in the column. (**a**) No stop flow pulses. (**b**) Three stop flow pulses in the first column are indicated by *dashed vertical lines*. Analytes: (1) Methanol, (2) ethanol, (3) pentane, (4) isoprene, (5) propanal, (6) cyclopentane, (7) acetone, (8) 1-hexene, (9) isopropyl ether, (10) butanal, (11) benzene, (12) ethyl acetate, (13) cyclohexene, (14) heptane [15]. Reprinted with permission

Work is also being done on integrating several microfabricated elements to make complete gas chromatography systems. One such system was demonstrated in 2005 where a gas chromatograph containing a 3.0-m separation column, a chemiresistor array for detection, a multistage preconcentrator/ focuser, and a calibration vapor source were assembled and tested [16]. The system was able to analyze complex mixtures of common indoor air contaminants in the part-per-billion range in less than 90 s. Figure 4 shows the assembly of the μGC from the microfabricated parts.

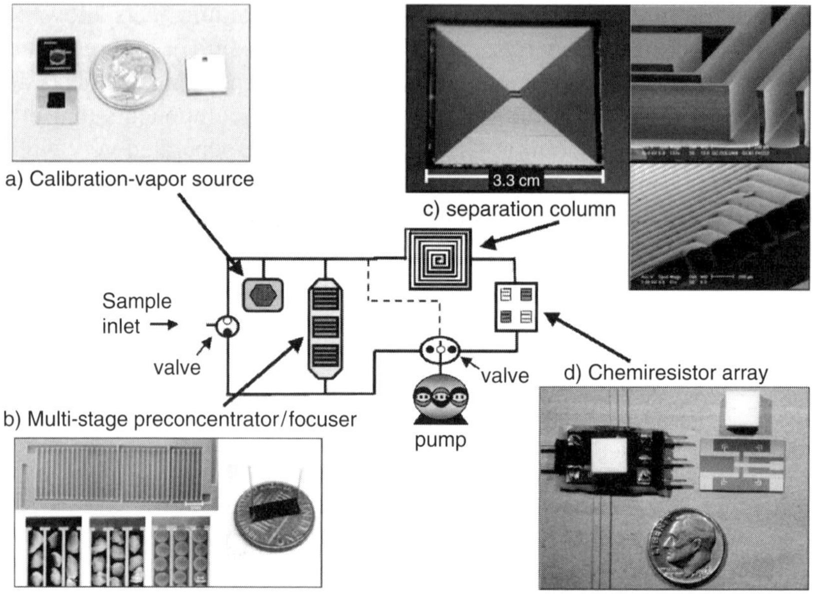


**Fig. 4.** A schematic of a miniaturized gas chromatograph. (**a**) Calibration vapor source, (**b**) Preconcentrator, (**c**) Separation column, (**d**) Chemiresistor detection array [16]. Reprinted with permission

Some groups have opted for using packed microfabricated columns to avoid the problem of uniformly coating the microchannels with a stationary phase, but this greatly increases the analysis time and reduces the efficiency of the separation. A separation presented by Zampolli et al. took 25 min to perform on a 75-cm column [17]. Figure 5 shows a photo of the microfabricated GC column packed with commercial (Carbograph + Carbowax™) stationary phase. The particles are between 150 and 250 μm in diameter.

**Fig. 5.** Encapsulated silicon spiral column packed with a commercial gas chromatography stationary phase [17]. Reprinted with permission

In 2006, Stadermann and co-workers from Lawrence Livermore National Laboratory and the University of Washington published an article about ultrafast gas chromatography on microchips using single-walled carbon nanotubes as a stationary phase. The carbon nanotubes were grown on the bottom of a 50-cm-long separation channel by chemical vapor deposition. The carbon nanotubes have a very high surface to volume ratio and therefore have great potential as a stationary phase. Furthermore, the microchip had integrated heaters that allowed for fast temperature programming. Separation of two 4-compound mixtures was achieved in less than 1 s when using an appropriate temperature gradient [18].

### 2.2. Pressure-Driven Liquid Chromatography (LC)

The first microfabricated liquid chromatography column was described and fabricated by Manz and co-workers in 1990 [19]. Their design was based on a 15-cm open tubular column that was 6 μm wide and 2 μm deep, similar to the "Stanford gas chromatograph" presented earlier, but no separations that were performed on the device were presented. The authors claimed that using microchannels for open tubular LC instead of glass capillaries is attractive because the possibilities offered by microfabrication allow for the reduction of the inner diameter of the chromatography column, reduction of the column length, and the integration of a detection sensor onto the chip. Open tubular LC columns offer relatively short analysis times for a given performance. On the other hand, the main disadvantage of using open tubular system is the low capacity of the column as a result of the low surface-area-to-volume ratio and hence a low phase ratio (volume of the stationary phase to volume of the mobile phase).

The most common chromatography columns in LC to date are packed columns, where the packing consists of small porous particles throughout the entire lumen of the column resulting in a high phase ratio. Realizing packed columns in microfluidic channels is nontrivial, however, both from a purely practical point of view but also because retaining structures (e.g., frits) and nonuniform packings can reduce the separation efficiency [13]. Still, some groups have been successful in making packed microfabricated LC columns. Agilent Technologies currently sell a microfluidic chip for peptide analysis that integrates a packed LC separation column, a preseparation sample enrichment column, and a nanoelectrospray tip for introduction of the eluate into a mass spectrometer [20]. The microchip is fabricated by laser-ablating polyimide films, resulting in a 45-mm-long column with a 75 μm × 50 μm cross-sectional area. The separation and the sample enrichment columns are filled with a 3.5–5 μm stationary phase packing material. External pumps are used to drive the mobile phase, and a rotary switching valve is used to control flow between sample loading and elution/separation on the chip.

A system such as this is very useful for proteomic analysis where the proteins are obtained from 2D gel spots and diluted to low concentrations in microliter amounts of solvent. The samples are then concentrated in the on-chip sample enrichment column before they are separated in the chromatography column. Figure 6 shows base peak chromatograms for a tryptic digest of bovine serum albumin (BSA) separated on the LC microchip using an octadecylsilane stationary phase for five different flow rates. The separation times in these runs are quite long for a microchip separation, up to 40 min. The fluid velocity in the column is quite low because of the increased backpressure when using small particles as a stationary phase support in pressure-driven chromatography.

One way around having to rely on packed columns that is becoming increasingly popular is the use of porous polymer monoliths for a stationary phase. These types of stationary phases are made by mixing monomers with the desired properties (e.g., monomers with octadecyl side chains, cross-linking monomers, porogens, etc.), introducing this low-viscous cocktail into the channels and polymerizing them in situ. By controlling the monomer composition and the solvents, it is possible to control the pore size in the polymer so that the resulting polymer monolith still allows adequate transport of liquid through it. The monoliths have a phase ratio that is similar to packed columns. For added control over the monolith formation, i.e., to make sure that it only forms in the separation channel and not in the rest of the microfluidic channels, a photoinitiator is used to start the polymerization. By masking all the channels except the separation channel, and then exposing the chip with UV light, polymerization will only occur in the separation channel. After polymerization, the unreacted monomers are removed from the channel network, only leaving a porous polymer monolith in the separation channel. Figure 7 shows a SEM image of a porous polymer monolith in a microchannel.

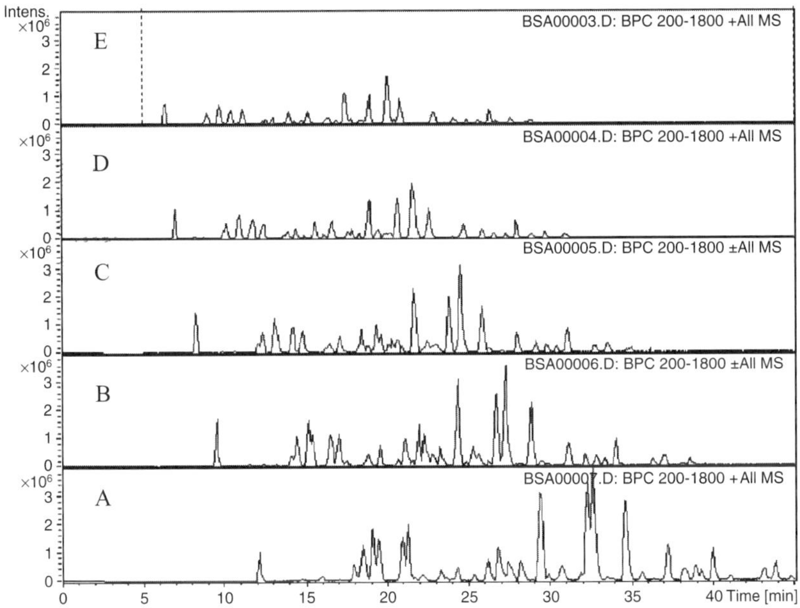


**Fig. 6.** Microchip LC separation of a tryptic BSA digest. Base peak chromatograms from the mass spectrometric data at different flow rates. (**A**) 100, (**B**) 150, (**C**) 200, (**D**) 300, and (**E**) 400 nL min$^{-1}$ [20]. Reprinted with permission

An ion chromatography separation of four proteins on a microchip is shown in Fig. 8, where the stationary phase was a porous polymer monolith. The separation was performed on a 4.5-cm effective length column that was 40 μm wide. All of the analytes eluted within 1 min. The system used an external rotary valve for injection and also controlled the mobile phase gradient externally. Note that the peaks are very sharp due to the zone-sharpening effect of the mobile phase gradient [21].

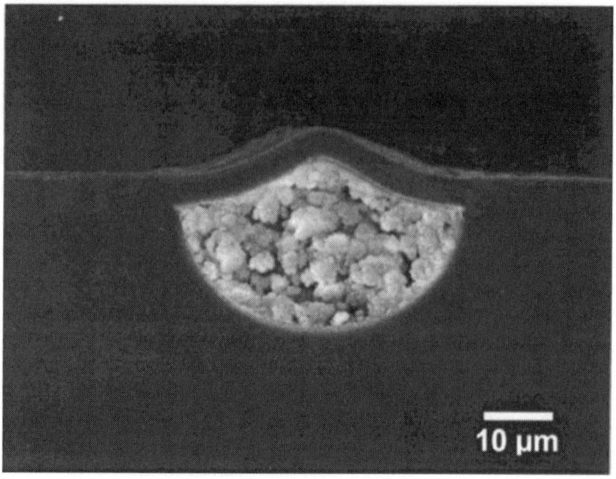


**Fig. 7.** Porous polymer monolith for anion exchange chromatography in a microfluidic channel (cross-sectional view) [21]. Reprinted with permission

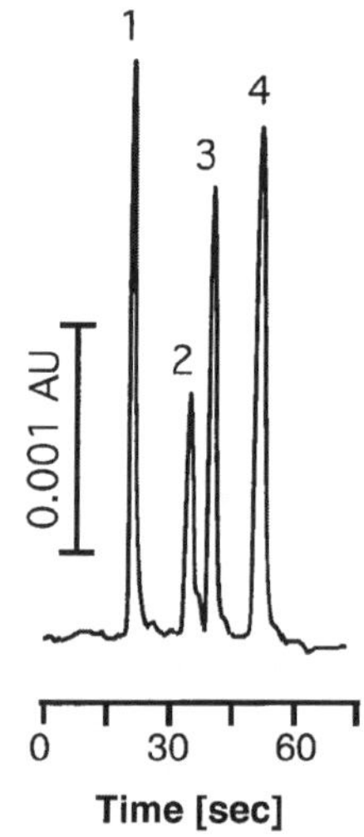


**Fig. 8.** Microchip anion-exchange chromatography of four proteins. The stationary phase is a continuous bed derivatized with alkylammonium moieties. UV-absorption detection at 230 nm. Sample: (**1**) myoglobin, (**2**) conalbumin, (**3**) ovalbumin, (**4**) trypsin inhibitor [21]. Reprinted with permission

In the case described above, the sample injection into the microfabricated LC column was done by using an external rotary valve, like the ones used in regular LC. This reduces the efficiency of the separation as the defined plug will start to disperse even before it reaches the chromatography column, mainly due to extracolumn broadening in the transfer lines. To reduce this dead volume work has been performed on on-chip injection systems. In 2004, Reichmuth and co-workers from Sandia National Laboratories published an article on an on-chip high pressure picoliter injector for microchip applications [22]. Their design consisted of a mobile polymer monolith that blocked the sample inlet, and was then moved to inject sample into the separation column (see Fig. 9). The injection volume was controlled by the time the sample injection channel was open resulting in reproducible injection volumes down to a few hundreds of picoliters. The authors also demonstrated that this system was capable of rapid sample switching without any carryover [23].

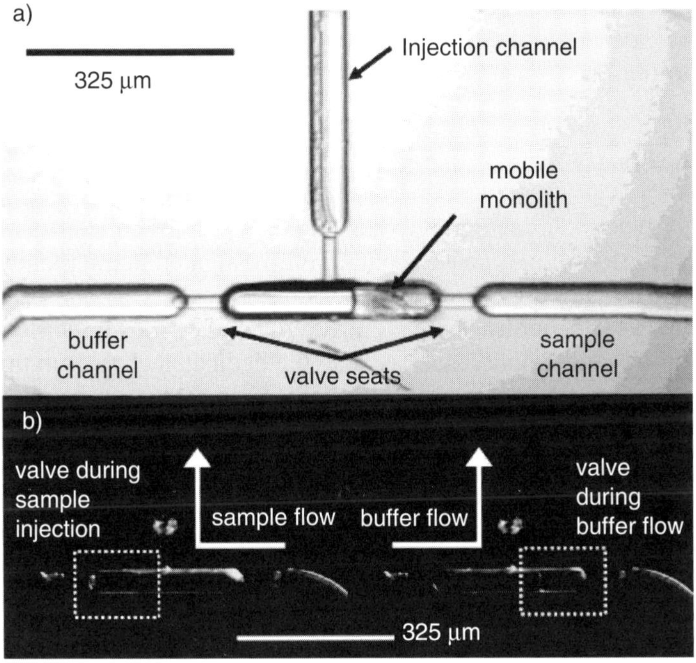


**Fig. 9.** (**a**) Photograph of a microfabricated injection valve. The mobile monolith plug blocks the sample channel, isolating it from the rest of the system. (**b**) Movement of the mobile monolith allows control of the flow into the separation column [22]. Reprinted with permission

While LC is a very well established and robust method, it is somewhat less suited for use in microfluidic systems. As the dimensions of the channels are reduced, the fluidic resistance increases dramatically making it harder and harder to pump liquid through the conduits by using a pressure difference. Even so, in most cases, an external pump needs to be connected to the microchannels, but in order to make small, compact, and even mobile systems, a large pump to achieve high pressures is not feasible. Therefore, external miniaturized pumps, such as an electroosmotically driven pump [24], can be used to accommodate the size requirements. Another pumping method that has been suggested for liquid chromatography is evaporation-driven pumping, where the constant evaporation from the column outlet and the refilling of the channel by capillary forces will "pump" the mobile phase through the channel [25].

However, with these challenges in mind, most of the work on chromatographic separations in liquid phase in microchips has been directed toward capillary electrochomatography (CEC), where, instead of a pressure-driven flow, the electroosmotic flow generated inside the microchannels is used to transport liquids.

### 2.3. Capillary Electrochromatography (CEC)

Capillary electrochomatography has quickly established itself as the most common form of liquid phase separation investigated in microchips. It is a hybrid technique between capillary electrophoresis (CE) and liquid chromatography. The mobile phase is moved by an electroosmotic flow (EOF), as in capillary electrophoresis (see also the dedicated chapter in this book). The main separation principle, on the other hand, is again based on a specific interaction with a stationary phase in the separation column, just as in liquid chromatography. In the case of charged analytes, an electrophoretic separation based on their mass/charge ratio may be superimposed on the chromatographic separation.

One of the great advantages of CEC compared with LC is that complex fluid-handling protocols can be realized fairly easily in CEC for several channels by applying and controlling different voltages at all the inlets and outlets of the chip. Another advantage when using EOF instead of pressurized flow is that there is no increase in backpressure as the channels (or stationary phase packing) get smaller. This is because the EOF originates from the surface of the channels and on the surface of the stationary phase (beads/porous polymer monolith). This offers the possibility of using stationary phases with smaller particles (packed) or smaller pores (porous polymer) and thereby achieving higher separation efficiency.

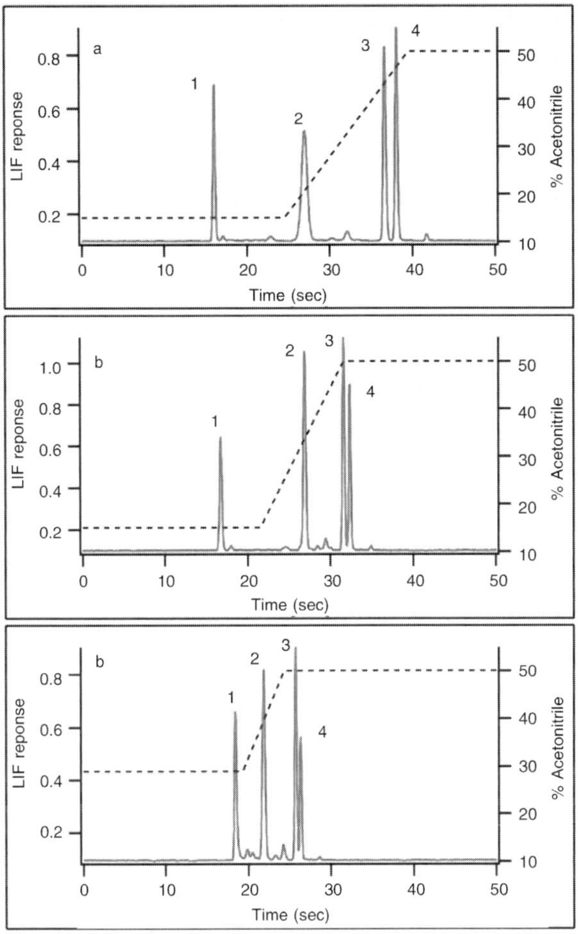


**Fig. 10.** Gradient elution in microchip open channel electrochromatography. Chromatograms (*solid lines*) for the separation of (1) Coumarin 440, (2) Coumarin 450, (3) Coumarin 460, and (4) Coumarin 480 using laser-induced fluorescence (LIF) detection. The LIF response is on the left axis, while the *dotted line* showing the acetonitrile concentration in the mobile phase relates to the right axis (**a**) Gradient from 15 to 50% in 15 s, starting 8 s after injection, (**b**) gradient from 15 to 50% in 10 s, staring 5 s after injection, and (**c**) gradient from 29 to 50% in 5 s, starting 3 s after injection. Adapted from Kutter et al. [27]. Reprinted with permission

Microchip CEC was first reported by Jacobson and co-workers at Oak Ridge National Laboratory in 1994. They used an open channel electrochromatographic approach (OCEC) where an octadecylsilane stationary phase was attached to the surface of a glass channel. They were able to achieve plate heights as low as 5 μm for a retained coumarin dye [26]. The same group has since published many different variations and applications of OCEC. In 1998, Kutter et al. reported on using solvent programming on microchips in combination with OCEC [27]. Solvent programming or gradient elution means that the mobile phase composition is changed during the run to gradually affect the interactions between the analytes, the stationary phase and the mobile phase. This can, for e.g., be used to speed up later eluting analytes by increasing the concentration of an organic modifier (such as methanol or acetonitrile) in the mobile phase.

Figure 10 shows the effect of changing the acetonitrile concentration during a run using different gradients. By increasing the acetonitrile concentration and increasing the slope of the gradient, the peaks become much more closely spaced, while still retaining baseline resolution. This results in a faster separation where all components have eluted in around 25 s in Fig. 4c while in Fig. 4a the last analyte elutes at close to 40 s.

Gottschlich et al. presented a 2D separation on a microchip, where the effluent from a 25-cm spiral OCEC column was directly injected into a 1.2-cm CE column without the need for any interconnects or transfer line, and thus being dead-volume free. Figure 11 shows an image of the microchip used. Note that the spiral is one way of compacting a long separation length onto small footprint, while avoiding detrimental effects stemming from too tight and improperly designed turns [28].

The injections on the chip were controlled so that a TRITC-labeled tryptic digest of β-casein was first injected for 0.5 s into the relatively slow OCEC column (first dimension). Every 3 s a 0.2-s plug of the effluent from the OCEC column was injected into the second dimension (CE separation). This corresponds to about 9% of the total elution volume from the first dimension being injected into the second dimension, and as such does not constitute a comprehensive 2D separation, and loss of some information has to be accepted. An example of a 2D separation performed in the device is shown in Fig. 12.

One of the advantages of using microfabrication is that it is relatively easy to combine and integrate different functional elements on a microchip. This was demonstrated by Broyles et al. in 2003, when they presented a microchip that integrated sample filtration for particles above 1 μm, solid phase extraction that increased the detected signal over 400-fold, and an OCEC gradient separation on a single microchip [29].

Some groups have opted for packed columns for microchip electrochromatography. As mentioned before, the main problem associated with using packed chromatography columns on microchips is the ineffective packing one often gets on account of the polydispersity of the particles, insufficient

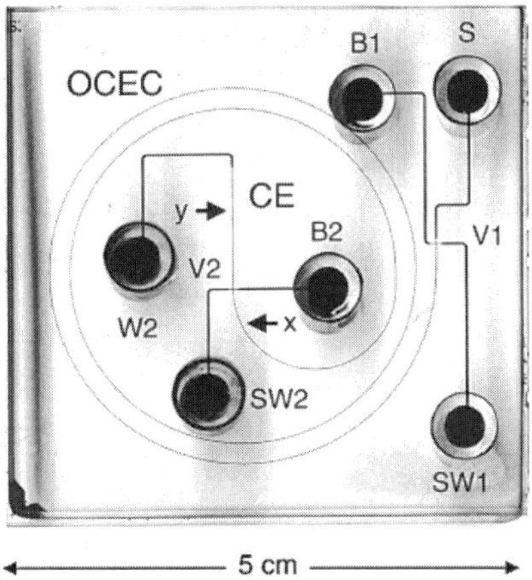


**Fig. 11.** Microchip for 2D separation. Black circles are chip reservoirs for the following: sample (S), buffer 1 and 2 (B1, B2), sample waste 1 and 2 (SW1, SW2), and waste (W). injection valves, V1 and V2, for dimension 1 and 2, respectively. The first dimension (OCEC) extends from V1 to V2 through the spiral channel, while the second dimension (CE) goes from V2 to the detection point y. Detection point x is used for the first dimension only [28]. Reprinted with permission

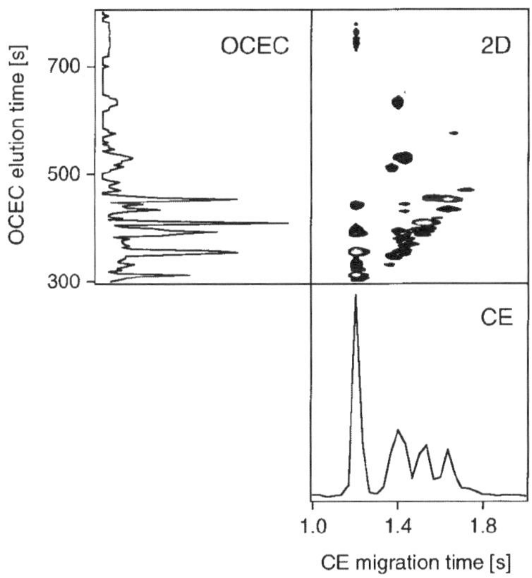


**Fig. 12.** 2D microchip separation of a TRITC-labeled tryptic digest of β-casein. The projections of the first dimension (OCEC) and the second dimension (CE) are shown to the left and below the 2D contour plot, respectively. LIF detection was employed [28]. Reprinted with permission

applicable packing pressures, or difficulties in filling the often noncircular cross sections of microchannels. Trying to improve the packing efficiency, Oleschuk and co-workers at the University of Alberta, Canada, published an article in 2000 where they described a method for trapping beads inside a microsystem using electrokinetic forces. They made a microfluidic channel with weirs at the start and finish of the chromatography column in order to retain the stationary phase beads while allowing mobile phase to pass (see Fig. 13).

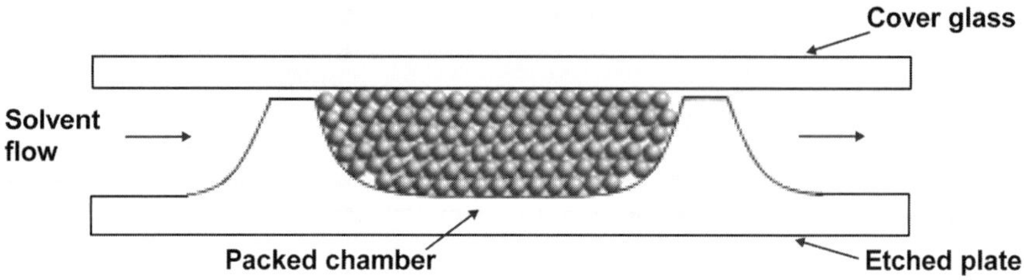


**Fig. 13.** Microchannel with retaining weirs to define a chromatography column [30]. Reprinted with permission

To fill the chamber with the stationary phase packing, an extra channel was fabricated on the microchip that connected to the chromatography column. Electrokinetic pumping was then utilized to direct the stationary phase packing material into the chamber. Figure 14 shows images of the packing process, where beads are "ejected" into the chamber until the entire chamber is filled with beads.

While this method is successful for making short chromatography columns, the authors note that 5-mm-long columns are difficult to prepare routinely [31]. Using these packed columns, a separation of three fluorescent dyes was performed yielding plate heights ranging from 2.8 to 4.1 μm. Other groups have tried different variations of this approach such as using a microchip fabricated in PDMS instead of glass and using pressure to fill the chromatography column with stationary phase beads [32]. This makes the fabrication of the microchip both faster and cheaper, but absorption of the analytes into the PDMS matrix limits the possible use of this polymer for CEC applications. A couple of groups have been studying methods to prevent analyte absorption into the PDMS matrix and, at the same time, increase the magnitude of the EOF by coating the microchannels with polyelectrolyte multilayers [32], using polymer grafting [33], or by mixing transition metal sol–gels into the matrix [34].

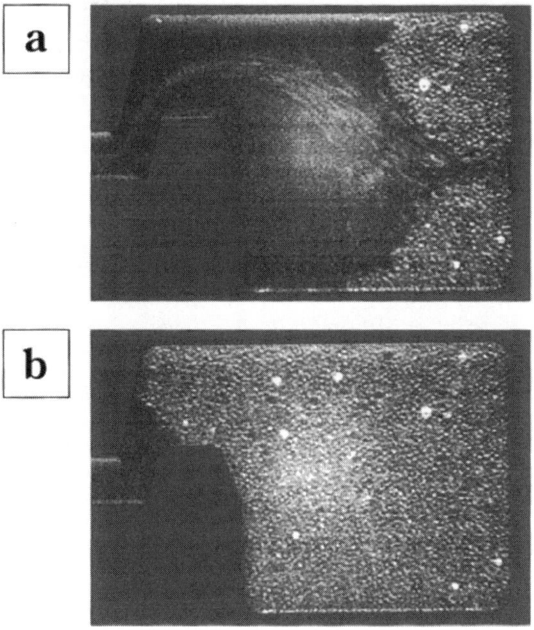


**Fig. 14.** Images of packing the microchip chamber. (**a**) Initial stages of electrokinetic packing. (**b**) Chamber filled with beads. The chamber is about 600 × 600 μm in size [31]. Reprinted with permission

A very different approach to emulate a "packed" stationary phase that can only be realized in microchips is to fabricate microchannels that contain support structures for a stationary phase. This was first introduced by He and co-workers at Purdue University in 1998 when they published an article on the in situ fabrication of stationary phase support pillars that were 5 × 5 × 10 μm separated by 1.5-μm rectangular channels.

Figure 15 shows the design of the inlet structure to a microfabricated chromatography column with microfabricated solid support structures in the separation channel [35].

These structures have the advantage that all dimensions and all geometrical aspects are precisely controlled by the design and the fabrication, thereby potentially reducing band-broadening effects as caused by nonideal packings. In particular, the widths of the channels between the solid support structures can be precisely controlled down to below a micrometer [36], which is expected to reduce the contribution to band broadening by mass transfer allowing faster separations without reduced efficiency. A stationary phase can be attached to the solid support structures through standard silane chemistry for glass and silicon microchips or via polymer grafting for PDMS chips [33]. Separations using these pillar arrays have yielded plate heights around 1.7 μm, which are comparable to those of packed columns with particles less than 1 μm in diameter [37]. Figure 16 shows a CEC separation of a tryptic digest of bovine serum albumine (BSA) on a PDMS microchip with solid support structures in the chromatography column.

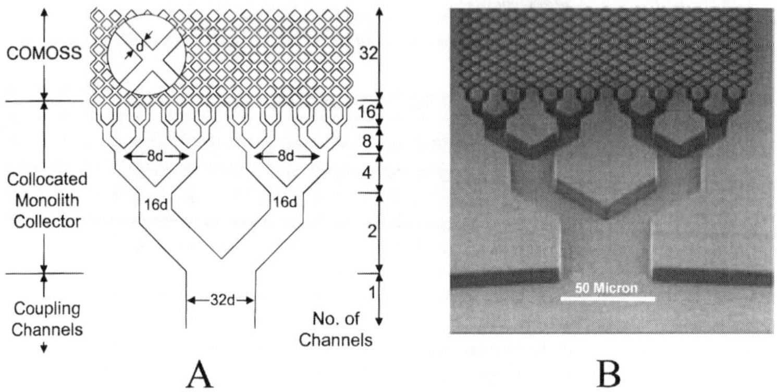


**Fig. 15.** (**a**) Design of the inlet channels to a chromatography column with solid support structures: The microchannels branch out into the chromatography column itself, retaining the cross-sectional area of the initial channel. (**b**) SEM image of the microfabricated structure [35]. Reprinted with permission

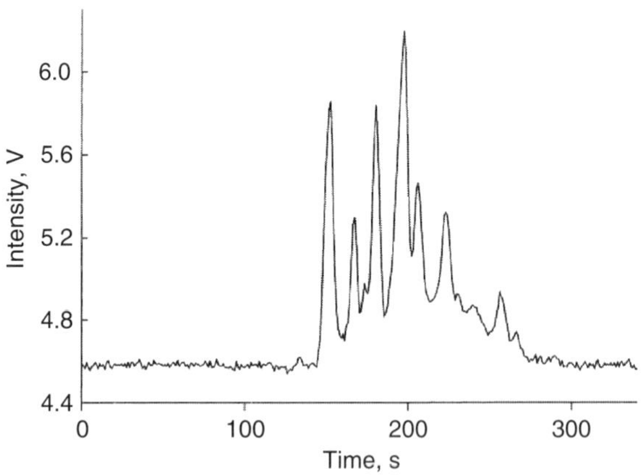


**Fig. 16.** CEC separation of a FITC labeled bovine serum albumine digest using a PDMS microchip with solid support structures in the separation channel. The channels are coated with a charged polymer for EOF support and $C_{18}$ for chromatographic interaction [33]. Reprinted with permission

One of the drawbacks of micromachined solid support structures is the fact that they are, indeed, solid. Even though they help increase the surface-to-volume ratio of such a column compared to open channel chromatography, the phase ratio is still well below that of (classical) packed columns because the beads used for packing are typically highly porous. One recently suggested approach to increase the phase ratio for micromachined support structures is to make porous pillars using a microfabrication process sequence yielding the so-called "black silicon" [38], which resembles a grassy surface with silicon needles instead of grass blades.

Microchip CEC has also been reported with porous monoliths, as described for microchip LC earlier. Several groups have employed porous polymer monoliths in microchips [21, 39, 40]. One of the great advantages of using porous polymer monoliths is that there is a vast selection of monomers commercially available, and these can be mixed and matched to make a polymer monolith with desired and tailor-made properties. A typical monomer mixture for CEC includes cross-linking monomers to link together the polymer chains, a monomer with a strong acidic moiety for EOF generation, and a monomer with a side group that can be used as a stationary phase, such as lauryl acrylate [39]. Again, the monoliths can be prepared restricted to the separation channel by using a photoinitiator to start the polymerization and masking off all other channels during polymerization. This is usually an advantage as it keeps both the injection and the detection regions free of polymeric monolith. CEC separations of 13 unlabelled polycyclic aromatic hydrocarbons were reported by Fintschenko et al. in 2001 on a butyl acrylate-based polymeric monolith using UV laser-induced fluorescence detection. The separation was performed within 15 min, with 10 of the 13 analytes resolved and plate heights of around 5 μm (see Fig. 17).

Other groups have reported even faster microchip CEC separations using porous polymer monoliths. For example, a reversed phase separation of five peptides in 45 s [39] and a separation of uracil, phenol, and benzyl alcohol in less than 20 s [21] have been published.

A slight variant of the porous polymer monoliths are the porous sol–gel monoliths for CEC. These use an in situ polymerized sol–gel instead of organic polymers. The sol–gels are silica-based and have been reported to yield plate heights at around 5 μm [41, 42].

A further variant of using a porous polymeric monolith in a microchip was reported by Zeng and co-workers at the Chinese Academy of Science in 2005 [43]. They immobilized a molecule with chiral recognition abilities (γ-cyclodextrin) in the porous polymer monolith. This allowed the authors to separate enantiomers of two amino acids in less than 2 min (see Fig. 18).

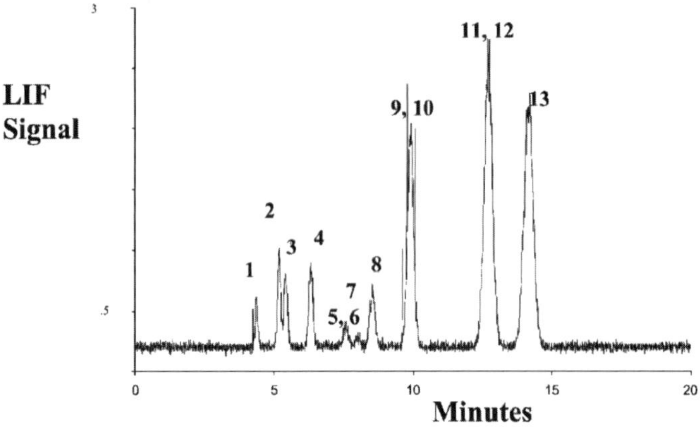


**Fig. 17.** Microchip CEC separation of 13 polycyclic aromatic hydrocarbons. LIF detection was employed with excitation at 257 nm [40]. Reprinted with permission

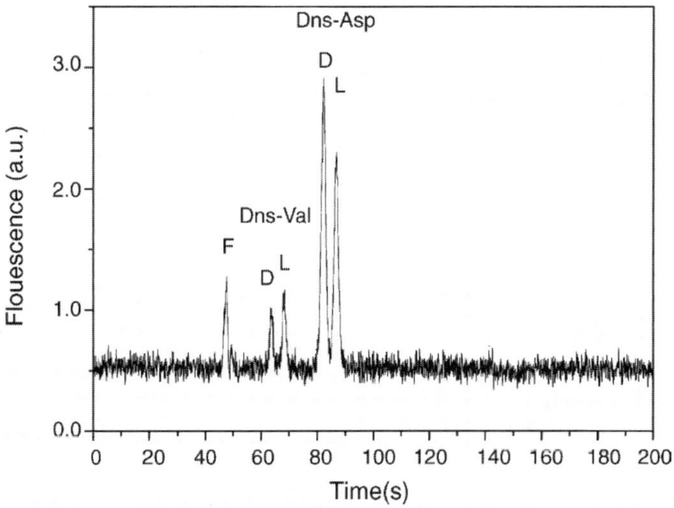


**Fig. 18.** Microchip chiral separations of FITC labeled Dns-valine and Dns-aspartic acid in a $\gamma$-cyclodextrin-modified porous polymer monolith. LIF detection was employed [43]. Reprinted with permission

### 2.4. Other Chromatographic Methods on Microchips

Shear-driven chromatography (SDC) is a variation of chromatography where flow is generated by a moving channel wall and the viscous drag exerted by this moving part on the mobile phase (see Fig. 19).

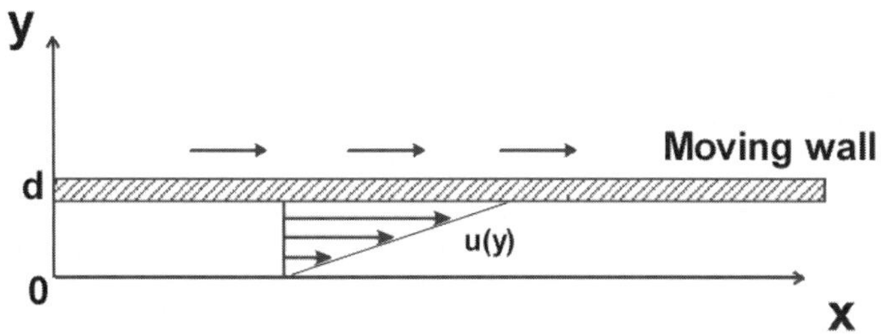


**Fig. 19.** Principle of shear-driven flow: The moving wall drags the mobile phase. The average flow velocity is half the velocity of the moving wall [44]. Reprinted with permission

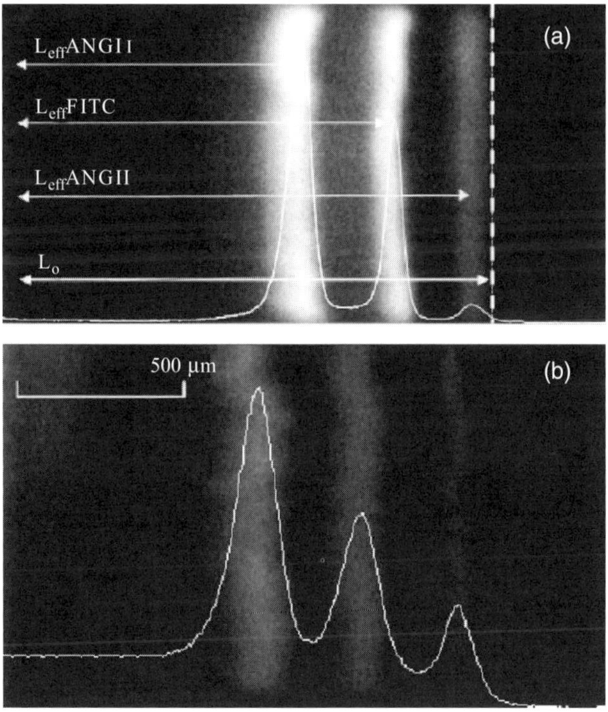


**Fig. 20.** Two SDC separations of FITC-labeled angiotensins. The moving wall velocity is (**a**) 2 mm $s^{-1}$ and (**b**) 35 mm $s^{-1}$. The middle peak is unreacted FITC labeling reagent, while the first and the last peaks are the labeled angiotensins [45]. Reprinted with permission

The average flow velocity is equal to half the velocity of the moving part. Microchip shear-driven chromatography has been reported in channels that were 300–400 nm deep, where only the nonmoving part was coated with a stationary phase [45, 46]. Because of the short diffusion distances in such shallow channels, it is possible to achieve very low plate heights and very fast separations. Vankrukelsven et al. from Vrije Universiteit Brussel published an article in 2006 where they used SDC to separate two FITC-labeled angiotensin peptides with plate heights as low as 0.4 μm [45]. The separation was performed in less than 0.2 s. Figure 20 shows photographs of two SDC separations at different moving wall velocities.

SDC has also been reported in a microchip with a circular channel layout and a rotating moving counterpart. In essence, this yields variable length columns in the same system (and theoretically infinitely long columns), since the progress of the separation can be monitored for each revolution. However, a Fourier transform is necessary to deconvolute the signal for each of the analyte bands as the collected signal trace becomes increasingly complex for each revolution. When the analytes have completed the circular column length several times, lightly retained analytes are likely to have overtaken highly retained analytes [44].

Another chromatographic separation technique realized on microchips is size exclusion chromatography (SEC), which was demonstrated in 2003 by Baba et al. [47]. They used a nanostructured region that acted as a stationary phase, representing an artificial gel. The nanostructured region contains narrow 400-nm "gaps," which molecules can diffuse into and be held back before diffusing out into the flow again (see Fig. 21a, b). Smaller molecules (with larger diffusion coefficients) are more likely to diffuse into the narrow gaps and are therefore retained longer in the separation system. A separation of three DNA fragments with different sizes (10, 5, and 2 kbp) was demonstrated using such a device (see Fig. 21c).

Finally, we will briefly discuss the realization of Hydrodynamic Chromatography (HDC) on microchips. HDC is, despite its name, not strictly a chromatographic technique since it does not have a stationary phase. Nevertheless, it is employed to separate macromolecules or particles based on their size. The principle of HDC is to exploit the parabolic flow profile in pressure-driven flows and the fact that different sized molecules or particles only can "sample" a certain percentage of the flow velocity distribution, depending on how close their centers of mass can get to the channel walls. Additionally, for this to work, the height of the channel must ideally be not too much larger than the particles' dimensions. Then, larger particles, which cannot get as close to the wall with their centers of mass as smaller particles, will on average experience a larger velocity and thus be separated from smaller particles [48–50]. For more details on this particular technique, the reader is referred to Chap. 12 in this book.

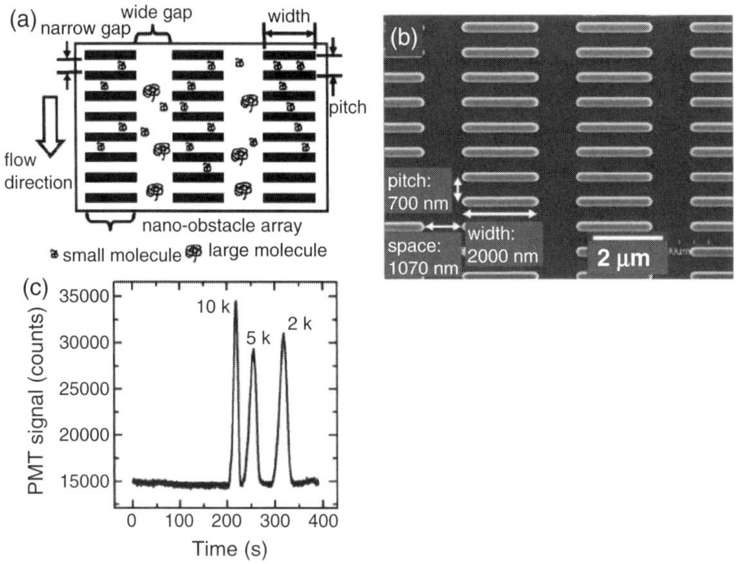


**Fig. 21.** (**a**) Design and principle of a nanostructured SEC column: Small molecules can diffuse into the small gaps and are retained, where as larger molecules flow through the wider gaps only and are hence much less retained. (**b**) SEM image of nano obstacles with a 700-nm pitch, the space between rows is 1,070 nm. (**c**) Microchip SEC separation of three DNA fragments with sizes of 2, 5, and 10 kbp, respectively, using a nanostructured stationary phase [47]. Reprinted with permission

## 3. Conclusions

Miniaturization offers a range of advantages and unique possibilities for realizing chromatographic systems on small scales. Fast and efficient microchip-based chromatography systems can become very powerful competitors or complementary partners to simpler sensor systems, with applications in all aspects of modern life, from food quality monitoring to medical diagnosis. Likewise, in more industrial settings, miniaturized chromatography systems will be employed for in-line process monitoring and high-throughput screening tasks. While the first commercial products using systems as described in this chapter are already appearing, the technical and scientific investigations and developments will continue. Even though there are ultimate limits imposed by the sample (typically the amount of available molecules in a given volume), trends are emerging to further look into the possibilities of nanotechnology for the benefits of improved separation devices.

## References

1. Giddings JC (1991) Unified Separation Science. Wiley, New York
2. Kutter JP (2000) Current developments in electrophoretic and chromatographic separation methods on microfabricated devices. Trends Anal Chem. 19(6): 352–363
3. Kutter JP, Fintschenko Y (eds) (2006) Separation Methods in Microanalytical Systems. CRC Press, Boca Raton
4. Snyder LR, Kirkland JJ (1979) Introduction to Modern Liquid Chromatography. Wiley, New York
5. Poole CF (2002) The Essence of Chromatography. Elsevier, Amsterdam
6. Gústafsson Ó, Kutter JP (2006) Electrokinetic Chromatography on microfluidic devices. In: Pyell U (ed) Electrokinetic.chromatography: Theory, Instrumentation and Applications. Wiley, Chichester, pp 337–349
7. Harris DC (2003) Quantitative Chemical Analysis, 6th edn. W.H. Freeman and Company, New York
8. Culbertson CT, Jacobson SC, Ramsey JM (1998) Dispersion sources for compact geometries on microchips. Analytical Chemistry. 70(18): 3781–3789
9. Griffiths SK, Nilson RH (2001) Low dispersion turns and junctions for microchannel systems. Analytical Chemistry. 73(2): 272–278
10. Jacobson SC, Culbertson CT (2006) Microfluidics: Some basics. In: Kutter JP and Fintschenko Y (eds) Separation Methods in Microanalytical Systems. CRC Press, Boca Raton, pp 19–54
11. Manz A, Harrison DJ, Verpoorte E, Widmer HM (1993) Planar chips technology for miniaturization of separation systems – A developing perspective in chemical monitoring. Advances in Chromatography. 33: 1–66
12. Terry SC, Jerman JH, Angell JB (1979) Gas-chromatographic air analyzer fabricated on a silicon-wafer. IEEE Transactions on Electron Devices. 26(12): 1880–1886
13. de Mello A (2002) On-chip chromatography: The last twenty years. Lab on a Chip. 2(3): 48N–54N
14. Reidy S, Lambertus G, Reece J, Sacks R (2006) High-performance, static-coated silicon microfabricated columns for gas chromatography. Analytical Chemistry. 78(8): 2623–2630
15. Lambertus G, Sacks R (2005) Stop-flow programmable selectivity with a dual-column ensemble of microfabricated etched silicon columns and air as carrier gas. Analytical Chemistry. 77(7): 2078–2084
16. Lu CJ, Steinecker WH, Tian WC, Oborny MC, Nichols JM, Agah M, Potkay JA, Chan HKL, Driscoll J, Sacks RD, Wise KD, Pang SW, Zellers ET

(2005) First-generation hybrid MEMS gas chromatograph. Lab on a Chip. 5(10): 1123–1131
17. Zampolli S, Elmi I, Sturmann J, Nicoletti S, Dori L, Cardinali C (2005) Selectivity enhancement of metal oxide gas sensors using a micromachined gas chromatographic column. Sensors and Actuators B - Chemical. 105(2): 400–406
18. Stadermann M, McBrady AD, Dick B, Reid VR, Noy A, Synovec RE, Bakajin O (2006) Ultrafast gas chromatography on single-wall carbon nanotube stationary phases in microfabricated channels. Analytical Chemistry. 78(16): 5639–5644
19. Manz A, Miyahara Y, Miura J, Watanabe Y, Miyagi H, Sato K (1990) Design of an open-tubular column liquid chromatograph using silicon chip technology. Sensors and Actuators B: Chemical. 1(1–6): 249–255
20. Yin NF, Killeen K, Brennen R, Sobek D, Werlich M, van de Goor TV (2005) Microfluidic chip for peptide analysis with an integrated HPLC column, sample enrichment column, and nanoelectrospray tip. Analytical Chemistry. 77(2): 527–533
21. Ericson C, Holm J, Ericson T, Hjerten S (2000) Electroosmosis- and pressure-driven chromatography in chips using continuous beds. Analytical Chemistry. 72(1): 81–87
22. Reichmuth DS, Shepodd TJ, Kirby BJ (2004) On-chip high-pressure picoliter injector for pressure-driven flow through porous media. Analytical Chemistry. 76(17): 5063–5068
23. Reichmuth DS, Shepodd TJ, Kirby BJ (2005) Microchip HPLC of peptides and proteins. Analytical Chemistry. 77(9): 2997–3000
24. Brask A, Snakenborg D, Kutter JP, Bruus H (2006) AC electroosmotic pump with bubble-free palladium electrodes and rectifying polymer membrane valves. Lab on a Chip. 6(2): 280–288
25. Goedecke N, Eijkel J, Manz A (2002) Evaporation driven pumping for chromatography application. Lab on a Chip. 2(4): 219–223
26. Jacobson SC, Hergenroder R, Koutny LB, Ramsey JM (1994) Open-channel electrochromatography on a microchip. Analytical Chemistry. 66(14): 2369–2373
27. Kutter JP, Jacobson SC, Matsubara N, Ramsey JM (1998) Solvent-programmed microchip open-channel electrochromatography. Analytical Chemistry. 70(15): 3291–3297
28. Gottschlich N, Jacobson SC, Culbertson CT, Ramsey JM (2001) Two-dimensional electrochromatography/capillary electrophoresis on a microchip. Analytical Chemistry. 73(11): 2669–2674
29. Broyles BS, Jacobson SC, Ramsey JM (2003) Sample filtration, concentration, and separation integrated on microfluidic devices. Analytical Chemistry. 75(11): 2761–2767
30. Oleschuk RD, Shultz-Lockyear LL, Ning YB, Harrison DJ (2000) Trapping of bead-based reagents within microfluidic systems: On-chip solid-phase extraction and electrochromatography. Analytical Chemistry. 72(3): 585–590

31. Jemere AB, Oleschuk RD, Harrison DJ (2003) Microchip-based capillary electrochromatography using packed beds. Electrophoresis. 24(17): 3018–3025
32. Ro KW, Chan WJ, Kim H, Koo YM, Hahn JH (2003) Capillary electrochromatography and preconcentration of neutral compounds on poly (dimethylsiloxane) microchips. Electrophoresis. 24(18): 3253–3259
33. Slentz BE, Penner NA, Regnier FE (2002) Capillary electrochromatography of peptides on microfabricated poly(dimethylsiloxane) chips modified by cerium(IV)-catalyzed polymerization. Journal of Chromatography A. 948(1–2): 225–233
34. Roman GT, Culbertson CT (2006) Surface engineering of poly (dimethylsiloxane) microfluidic devices using transition metal sol-gel chemistry. Langmuir. 22(9): 4445–4451
35. He B, Tait N, Regnier F (1998) Fabrication of nanocolumns for liquid chromatography. Analytical Chemistry. 70(18): 3790–3797
36. Mogensen KB, Eriksson F, Gustafsson O, Nikolajsen RPH, Kutter JP (2004) Pure-silica optical waveguides, fiber couplers, and high-aspect ratio submicrometer channels for electrokinetic separation devices. Electrophoresis. 25(21–22): 3788–3795
37. He B, Regnier F (1998) Microfabricated liquid chromatography columns based on collocated monolith support structures. Journal of Pharmaceutical and Biomedical Analysis. 17(6–7): 925–932
38. Costa R, Mogensen KB, Kutter JP (2005) Microfabricated porous glass channels for electrokinetic separation devices. Lab on a Chip. 5(11): 1310–1314
39. Throckmorton DJ, Shepodd TJ, Singh AK (2002) Electrochromatography in microchips: Reversed-phase separation of peptides and amino acids using photopatterned rigid polymer monoliths. Analytical Chemistry. 74(4): 784–789
40. Fintschenko Y, Choi WY, Ngola SM, Shepodd TJ (2001) Chip electrochromatography of polycyclic aromatic hydrocarbons on an acrylate-based UV-initiated porous polymer monolith. Fresenius' Journal of Analytical Chemistry. 371(2): 174–181
41. Morishima K, Bennett BD, Dulay MT, Quirino JP, Zare RN (2002) Toward sol–gel electrochromatographic separations on a chip. Journal of Separation Science. 25(15–17): 1226–1230
42. Dulay MT, Quirino JP, Bennett BD, Zare RN (2002) Bonded-phase photopolymerized sol–gel monoliths for reversed phase capillary electrochromatography. Journal of Separation Science. 25(1–2): 3–9
43. Zeng HL, Li HF, Lin JM (2005) Chiral separation of dansyl amino acids by PDMS microchip gel monolithic column electrochromatography with gamma-cyclodextrin bonded in polyacrylamide. Analytica Chimica Acta. 551(1–2): 1–8
44. Yang X, Jenkins G, Franzke J, Manz A (2005) Shear-driven pumping and Fourier transform detection for on chip circular chromatography applications. Lab on a Chip. 5(7): 764–771

45. Vankrunkelsven S, Clicq D, Cabooter D, De Malsche W, Gardeniers JGE, Desmet G (2006) Ultra-rapid separation of an angiotensin mixture in nanochannels using shear-driven chromatography. Journal of Chromatography A. 1102(1–2): 96–103
46. Clicq D, Vankrunkelsven S, Ranson W, De Tandt C, Barn GV, Desmet G (2004) High-resolution liquid chromatographic separations in 400-nm deep micro-machined silicon channels and fluorescence charge-coupled device camera detection under stopped-flow conditions. Analytica Chimica Acta. 507(1): 79–86
47. Baba M, Sano T, Iguchi N, Iida K, Sakamoto T, Kawaura H (2003) DNA size separation using artificially nanostructured matrix. Applied Physics Letters. 83(7): 1468–1470
48. Blom MT, Chmela E, Gardeniers JGE, Tijssen R, Elwenspoek M, van den Berg A (2002) Design and fabrication of a hydrodynamic chromatography chip. Sensors and Actuators B: Chemical. 82(1): 111–116
49. Chmela E, Tijssen R, Blom M, Gardeniers H, van den Berg A (2002) A chip system for size separation of macromolecules and particles by hydrodynamic chromatography. Analytical Chemistry. 74(14): 3470–3475
50. Blom MT, Chmela E, Oosterbroek RE, Tijssen R, van den Berg A (2003) On-chip hydrodynamic chromatography separation and detection of nanoparticles and biomolecules. Analytical Chemistry. 75(24): 6761–6768

# Chapter 12
# Microscale Field-Flow Fractionation: Theory and Practice

Himanshu J. Sant and Bruce K. Gale

State of Utah Center of Excellence for Biomedical Microfluidics, Department of Bioengineering and Department of Mechanical Engineering, University of Utah, 50 S. Central Campus Drive, Rm. 2110, Salt Lake City UT 84112, USA

## 1. Introduction

The last decade has seen exponential growth in the development of lab-on-a-chip or micro-total-analysis system (μ-TAS) components to create better, faster, and cheaper chemical and biological analysis platforms [1]. Lab-on-a-chip type analysis systems typically include a separation-based sample preparation unit to achieve this objective or to prepare the sample for further interrogation using orthogonal techniques. Researchers have employed a host of sample preparation techniques based on electrophoresis [2, 3], ultrasound [4, 5], flow [6, 7], mechanical ratchets [8, 9], electrokinetics [10, 11], packed bed systems [12], membranes [13] magnetics [14, 15], temperature [16], optics [17], dielectrophoresis [18, 19], and so forth. Microscale field-flow fractionation (FFF) techniques have been an integral part of these efforts. Most of these techniques are simply miniaturized versions of conventional macroscale units with the rationale being that the reduction in physical size of the instrument results in smaller sample volumes and faster analysis times. While, many of these systems work well when miniaturized, this approach proves inadequate for systems that do not scale well. FFF, at least for many subtypes, has been shown to scale very well and FFF meets many of the design challenges for a successful separation module in a μ-TAS including (a) ease of manufacturing, (b) low power, (c) wide range of sample type and size, (d) integration to fluidic components, and (e) material compatibility. Thus, FFF is potentially an important solution to many problems in microfluidic system design.

Field-flow fractionation clearly improves when miniaturized due to the reduction in sample and carrier volumes, analysis times, and more notably an increase in the separation resolution (at least for the electrical and thermal subtypes). Other advantages of miniaturized FFF can include the following: parallel processing with multiple separation channels, batch fabrication with reduced costs, high quality manufacturing, and potentially disposable systems. Additionally, the possibility of on-chip sample injection, detection, and signal processing favors the microfabrication of FFF systems.

Several demonstrations of the effectiveness of FFF systems on the microscale have been made and will be reviewed in the work. Techniques that are often lumped in with FFF include split flow thin cell (SPLITT) fractionation and hydrodynamic chromatography. These related techniques have also been miniaturized and will be discussed later in this work.

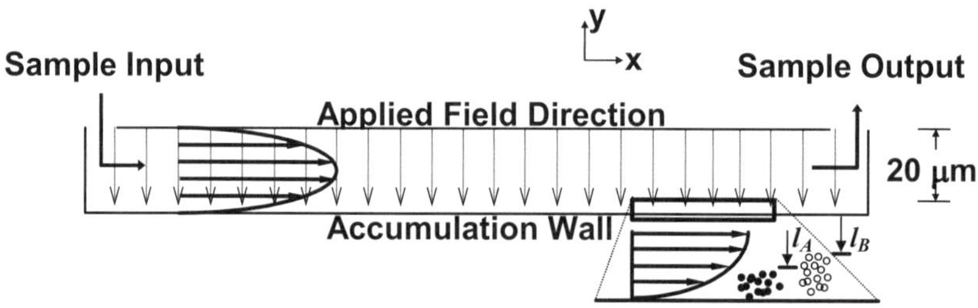


**Fig. 1.** FFF operational principle with two parallel plate type channel walls, laminar flow profile with transverse field direction and location of particle clouds near accumulation wall. The particle clouds depicted by closed circles and open circles in inset figure are particle cloud A with average thickness $l_A$, and particle cloud B with average thickness $l_B$, respectively

## 2. Background and Theory

FFF is a versatile separation technique that relies on the dual effect of the flow behavior and field distribution in a thin, open channel. FFF channels typically consist of a thin spacer enclosed by two parallel plates, modified to impart the external field as shown in Fig. 1. Flow in the channel is laminar resulting in a parabolic fluid velocity profile with differential velocity zones across the height of the channel. The versatility of FFF stems from the numerous types of fields and operating modes that can be employed to separate a wide range of sample types. Researchers over the years have developed different types of FFF systems differentiated primarily by the type of the field employed. Electrical, thermal, magnetic, sedimentation, flow, and dielectrophoretic fields are all commonly used in FFF. In FFF the field is applied perpendicular to the flow of the carrier. Like in chromatography, an impulse injection of sample is made into a continuously flowing carrier solution. Under the influence of the applied field and possibly other hydrodynamic or gravitational forces, the injected sample migrates to an equilibrium position between the two walls of the channel, as shown in Fig. 1. The location of the equilibrium depends on the operating mode of the FFF channel, as will be discussed shortly. Sample particles then travel down the channel at the velocity associated with the flow at the equilibrium distance from the wall. Selectivity in FFF separations is determined by the system's ability to differentially retain the samples based on their physiochemical properties. FFF operational parameters like field and flow rate can be varied to allow the user to tune resolution and analysis times for a given set of sample particles. FFF channels are also naturally gentle and can be used with delicate samples such as cells and liposomes since there is no stationary phase and the shear rates are low. In addition, a single channel can be used to separate a large range of sample sizes, thus enhancing the utility of FFF instruments when compared to many chromatography techniques.

### 2.1. FFF Operating Modes and SPLITT Fractionation

FFF can be classified into five broad modes of operation based on the separation mechanism: (a) normal, (b) steric, (c) focusing, (d) cyclical, and (e) zero-field or hydrodynamic FFF.

In normal or classical FFF, the sample particles are forced towards the accumulation wall by the applied field. At the accumulation wall, diffusive forces associated with Brownian motion cause particles to move away from the accumulation wall. At equilibrium, the field-induced migrative forces and diffusive forces balance each other and generate an exponential concentration profile of the particle cloud. The average distance $l$ of the particle cloud from the wall depends on the extent of interaction between the particles and the field and determines the average rate of travel for a particle down the length of the separation channel. For a mixture of particles "A" and "B" in a FFF channel as shown in Fig. 1, if $l_A < l_B$, then the "B" particles will spend relatively more time in the high velocity zone and move faster down the length of the channel compared to the "A" particles. Thus retention and separation can be generated by manipulating the average distance a particle spends away from the wall. For particles with similar mobilities, larger particles tend to be forced closer to the wall due to their slower diffusion and so smaller particles typically elute from the FFF channel first.

The limit to normal FFF occurs when high field strengths force the sample particles to contact the wall. The distance particles are away from the wall is then controlled by the diameter of the particle or steric effects, and this is referred to as the steric mode of FFF. In steric mode, larger particles protrude farther into the flow stream than do smaller particles and thus larger particles elute first. For microscale systems that generate relatively high fields, the on-set of steric FFF is an important factor in microscale system evaluation. In steric FFF, the elution sequence is reversed in comparison to normal FFF with larger particles eluting earlier.

Another FFF mode related to steric FFF is focusing FFF or hyperlayer FFF. This mode is realized by controlling the location of the equilibrium concentration distribution inside the channel. In the case of focusing FFF, a Gaussian-type concentration profile is generated within the separation channel by using a balance of dispersive flux and migrative flux. Retention can be induced by differentially controlling the lift for different sets of particles away from the accumulation wall. Since lift is generally more significant for larger particles, they tend to move away from the accumulation wall and towards the center of the channel, which causes them to elute before smaller particles. Thus elution patterns are similar to those for steric mode of FFF and it can often be challenging to determine which mode is in operation using only experimental results. Generally, higher carrier velocities are associated with this mode and that can result in shorter elution times.

A recently developed microscale FFF mode involves the use of cyclical fields instead of a steady, uniform field. In this case, particles move either back and forth between the walls or oscillate against one wall of the channel. The retention time for the particle is determined by whether the particle spends more time in the fast flow lines or in the slower flow areas. The amount of time spent in the different flow areas can be tuned by adjusting the field strength and the frequency of the applied field. Cyclical methods have primarily been demonstrated with electrical systems and have the advantage that retention is dependent only upon the susceptibility of the particle to the applied field. Equilibrium processes are not involved and diffusion processes are essentially eliminated from the retention process, so very high speed separations can be generated.

Another technique very closely related to FFF is the SPLITT technique, which generates a continuous separation process. SPLITT has two separate inlets for the sample mixture and a carrier and two outlets for the bifurcated/separated samples. The carrier stream compresses the sample particles against one wall and the field perpendicular to the flow drives sample particles with some minimum interaction across the carrier stream interface where they elute from the opposite outlet. Particles that do not exhibit this minimum interaction with the field continue in the original sample stream and elute from the outlet on the same side as the sample inlet port. SPLITT typically induces very fast fractionation and can be used in serial and parallel fashion to separate complex mixtures with high resolution and high throughput.

A technique called hydrodynamic chromatography is also related to FFF and is generated when the field in the FFF channel is zero such that there is no transverse flux of the particles due to the field and particles are dispersed randomly in the channel. When the particle sizes are comparable to the channel thickness, larger particles will be located in the higher velocity zones as they can not approach as close to the channel walls as smaller particles might due to their large size. In contrast, smaller particles can, on average, approach closer to the channel walls and spend more time in slower velocity zones. This zero-field separation mode has an elution sequence similar to steric FFF. Typically the size selectivity of this technique is poor but can prove to be an efficient tool to separate larger macromolecules.

## 2.2. FFF Retention Theory

In general, the theory behind FFF systems is well developed [20–22] and in principle the theory can be applied to all the FFF subtypes, including microscale FFF systems. The FFF channel, as shown in Fig. 1, is a thin open ribbon-like channel of rectangular cross section with an aspect ratio (the ratio of width to height) over 80 so that channel walls can be closely approximated as two infinite, parallel plates [23, 24]. Flow between parallel plates separated by small distances is laminar for the flow velocities of interest and is described by

$$v = \frac{wy - y^2}{2\eta}\frac{\mathrm{d}p}{\mathrm{d}x} = 6\langle v\rangle\left(\frac{y}{w} - \left(\frac{y}{w}\right)^2\right), \tag{1}$$

where, $v$ is the flow velocity at a distance $y$ from one of the plates, $\eta$ is the viscosity of the fluid, $w$ is the plate separation or channel height, $\langle v\rangle$ is the average flow velocity across the channel, and $\mathrm{d}p/\mathrm{d}x$ is the pressure gradient along the flow axis. As the parabolic distribution given in equation (1) implies, the fluid velocity at the surface of the channel walls is zero (non-slip flow) while at a maximum in the center of the channel. Thus, if a particle or cloud of particles were to maintain an average distance $y$ different from another particle or cloud of particles, their average velocities through the channel would be different and they would exit the channel at distinct times.

Retention in FFF is the measure of the ability of the system to retain or retard the travel of a particle through the channel compared to a particle unaffected by the applied field. Experimentally, the retention ratio $R$ is found by

$$R = \frac{t_0}{t_r} = \frac{V_0}{V_e}, \quad (2)$$

where $t_0$ is the time required for an unretained particle to exit the channel, $t_r$ is the time for the retained sample to exit, $V_0$ is the void volume of the channel, and $V_e$ is the elution volume of the sample. The elution or retention time in FFF is directly related to the properties of the sample and the sample's response to the applied field according to the equation [20]

$$R = 6\lambda\left[\coth\left(\frac{1}{2\lambda}\right) - 2\lambda\right], \quad (3)$$

where $\lambda$ is a nondimensional parameter given by

$$\lambda = \frac{l}{w}. \quad (4)$$

The $l$ in (4) is the average distance of a sample particle from the accumulation wall as described earlier and is related to experimental conditions by

$$l = \frac{D}{U}, \quad (5)$$

where $D$ is the particle diffusion coefficient and $U$ is the field-induced drift velocity, which depends on the applied field strength according to

$$U = \frac{S'\phi}{f'}, \quad (6)$$

where $S'$ is the applied field strength, $\phi$ is the field susceptibility of the particles, and $f'$ is the sample friction coefficient. Note that the form of (6) will vary somewhat depending on the type of field used in the particular FFF system. The diffusivity, $D$, can be calculated using the Einstein equation

$$D = \frac{\kappa T}{3\pi\eta d}, \quad (7)$$

where $\kappa$ is Boltzmann's constant, $T$ is the absolute temperature, and $d$ is the particle diameter.

The balance between field-induced migration and diffusion away from the accumulation wall leads to an exponentially defined particle distribution

$$c(y) = c_0 e^{-\left(y\frac{U}{D}\right)}, \tag{8}$$

where $c(y)$ is the concentration of particles at a distance $y$ from the accumulation wall and $c_0$ is the concentration of particles at the wall. Note that this particle distribution is only true for normal mode FFF and can be modified significantly if the retention is occurring in another mode of FFF.

Other effects that may cause deviations from this theory are typically referred to as repulsion effects. Studies on repulsion effects by Tri et al. have been reported for macroscale FFF systems [25]. Miniaturization of FFF channels may be limited by such particle–wall repulsion effects as these interactions typically result in the exclusion or the repulsion of the particles away from the accumulation wall and, hence, reduce the effective retention in FFF channels. These particle–wall interactions include electrostatic forces, hydrodynamic lift, and van der Waal's attractive forces. The wall repulsion layer increases the average particle cloud thickness $l$ and can lead to incorrect measurements of sample properties. To fully understand scaling in FFF and the potential of FFF channel miniaturization, a thorough investigation in wall–particle and particle–particle interactions is needed.

### 2.3. Plate Height

In considering the usefulness and effectiveness of microscale FFF systems, figures of merit for comparing different FFF instrument designs and for comparing FFF instruments to other instruments are required. These figures of merit are generally based on the chromatographic concept of plate height. Thus a brief review of chromatographic plate theory follows.

The length $L$ of a separation column can be broken down in to $N$ theoretical plates of height $H$

$$H = \frac{L}{N}, \tag{9}$$

where the plate height, $H$, is a measure of variance ($\sigma^2$) or spreading that has been created as the band of particles being separated moves through the separation channel. The plate number, $N$, is a measure of the separation efficiency of a system and indicates the number of times a certain separation level is accomplished in a channel. $H$ can be closely approximated by the ratio of variance to the length of the channel, $L$, [26] according to

$$H = \frac{\sigma^2}{L}. \tag{10}$$

The plate height generally represents the length of the separation column required to generate a defined level of separation between two particles. Ideally, $H$ should be as small as possible to maximize the level of separation between two samples generated in a given instrument. In chromatography systems, $H$ and $N$ are used as figures of merit for comparison with various instruments, with the goal being to minimize $H$ and maximize $N$. Systems with a large plate height will have widely dispersed sample bands and will be unable to separate as many different samples simultaneously as an instrument with a small plate height. The goal in microscale systems is to minimize the plate height.

Plate heights in FFF are generated by a combination of factors such as nonequilibrium effects ($H_n$), longitudinal diffusion ($H_d$), sample relaxation ($H_r$), sample polydispersity ($H_p$), sample volume ($H_s$), and instrumental effects ($H_i$) [26]. These factors can be classified in two groups based on their origin. The first group contains effects that give rise to diffusion-based dispersion such as nonequilibrium effects and longitudinal or axial diffusion. The second category encompasses all other band-broadening factors that include sample and instrumental related effects. Sample relaxation, injection volume, polydispersity, and instrument-related plate height are included in this second category. Overall plate height, $H$, can be formulated as a combination of all of these factors [26] and written as

$$H = H_d + H_n + H_r + H_p + H_i + H_s. \tag{11}$$

Polydispersity of a sample is an inherent property of the sample being processed, not a system property, and can be ignored when optimizing an instrument. As diffusion coefficients are relatively low compared to the length of a FFF channel, the contribution of diffusion to plate height is negligible unless very low flow velocities are used. Thus, only the contributions due to geometrically dependent nonequilibrium and instrumental effects require consideration during microscale instrument design and optimization efforts.

#### 2.3.1. Nonequilibrium Plate Height

In FFF, the nonequilibrium component of plate height, $H_n$, is heavily dependent on channel thickness, diffusion, $D$, and average flow velocity, $\langle v \rangle$ and is given by

$$H_n = \frac{\chi(\lambda) w^2 \langle v \rangle}{D}. \tag{12}$$

The function $\chi(\lambda)$ is highly complex and an exact derivation was found by Giddings [27] and so an approximation proposed by Giddings is mentioned here

$$\chi(\lambda) = 24\lambda^3 \left(1 - 8\lambda + 12\lambda^2\right). \tag{13}$$

$H_n$ is a complex function of the channel dimensions and the effect of miniaturization on it can not be inferred directly. A closer look at the scaling of $H_n$ is required to estimate the effects of miniaturization on plate height.

#### 2.3.2. Instrumental Plate Height

In FFF systems, the instrumental component of plate height depends on the instrument setup, channel geometry, the fluidic connections, postcolumn volumes, and the sample injection size and method. These elements that contribute to instrumental band broadening are not easily expressed in a comprehensive theory and so havc been ignored when examining these systems mathematically and conceptually. Thus, no comprehensive theory of instrumental effects exists and the effect of geometry on instrumental plate height is only known conceptually.

### 2.4. Resolution

The resolution of a chromatography system, $R_s$, is a measure of the relative separation ability of a system and can be represented by [21]

$$R_s = \frac{\Delta R}{4\bar{R}} \sqrt{\frac{L}{H}} \tag{14}$$

where $\Delta R$ is the difference in retention ratio for two distinct particles and $\bar{R}$ is the average retention ratio of the two particles being considered.

## 3. Miniaturization Effects in FFF

FFF can be classified in two broad classes based on the type of field involved. Typically, general FFF systems (e.g. sedimentation, flow, gravitational) fall in to the "nongradient based FFF" category where the field does not depend on the channel height, as is evident in the definition of the retention parameter $\lambda$. The miniaturization of such systems was not favored in the early FFF literature as resolution is expected to drop with miniaturization [28, 29]. It was generally believed that field strength manipulation could be used to increase the separation efficiency of the system by modifying the average particle cloud thickness in the channel, so miniaturization was unnecessary. A closer examination of the geometric scaling effects in recent times has shown that miniaturization of nongradient-based FFF systems can lead to a minimal loss in performance with possible improvement in certain situations for these FFF systems [29]. But in certain FFF subtypes (e.g. electrical and thermal), field strengths can not be increased indefinitely. Interestingly, the field experienced by particles in these systems is highly dependent on the channel thickness $w$. These FFF systems are classified as "gradient-based FFF" where the field scales with the channel height and an increase in field strength is expected with miniaturization. Electrical FFF (ElFFF) and thermal FFF have both shown improvement in retention with scaling. Miniaturization of FFF instruments has resulted in a reduction of the instrument size, sample and carrier volumes, and power consumption along with a reduction in analysis time, but the effect of miniaturization on FFF performance can only be determined by examining scaling behavior of plate height and resolution. Both nongradient and gradient-based FFF systems behave differently when miniaturized due to the contrasting dependence of plate height and resolution on channel dimensions.

### *3.1. Instrumental Plate Height*

The packaging or interfacing of microscale FFF systems with the real world requires far more consideration than is needed for macroscale FFF channels. Typically, macroscale FFF systems use large flow cells for detectors, long lengths of extracolumn tubing, and large sample injection volumes. While thc performance of macroscale systems does not get deteriorate considerably in these situations, it can severely affect microsystems. The effect of these large extracolumn volumes and sample sizes is referred to as instrumental effects. Such instrumental effects can play a large role in increased plate heights and a subsequent loss in resolution for microscale FFF systems. As the FFF channels are miniaturized, the importance of instrumental plate height increases further. Typically, ultralow volume sample injections and on-chip detectors are preferred in conjunction with microsystems to reduce plate heights and sustain miniaturization related advantages.

The effect of miniaturization on instrumental plate height can be measured using experimental plate height data collected from elution peaks obtained from similar experiments conducted in FFF channels of various sizes [29]. Instrumental plate height data collected from a variety of FFF instruments of different sizes indicates an empirical correlation between instrumental plate height and channel height as

$$H_i = 3w. \tag{15}$$

It can clearly be seen that the instrumental plate height drops with a dncrease in channel height and the associated improvement in sample injection, channel fabrication, and detector arrangement [30]. With miniaturization, instrumental plate height drops linearly and can be a critical factor in the overall reduction in plate heights.

Accordingly, there appears to be a clear advantage to miniaturization of FFF systems with regard to instrumental plate heights.

### *3.2. Gradient-Based Systems*

Gradient-based systems are the FFF subtypes in which the applied field scales with channel height. Electrical FFF and thermal FFF systems fall into this category. These types of FFF systems are believed to benefit from miniaturization the most.

#### 3.2.1. Plate Height Scaling

The total plate height that can be measured or calculated theoretically is a sum of nonequilibrium and instrumental plate heights. Comparative scaling analysis of these three plate heights: total, nonequilibrium, and instrumental give us a gauge of relative importance. Figure 2a shows the estimates for ElFFF systems, which indicate that plate heights are dominated by nonequilibrium effects, which generate an exponential increase in plate height as $w$ increases. It should be noted that in FFF systems nonequilibrium plate height is a strong function of the applied field also. For example, in the case of ElFFF with a 0.25% effective field strength, the value of $H_n$ is very high, almost 10 times higher than $H_i$, but as field strength is increased to 1.25%, the relative magnitudes of the instrumental and non-equilibrium contributions become similar. Thus, even at relatively moderate field strengths in ElFFF, particularly for microscale ElFFF, a tight control over $H_i$ is very important, which can be achieved only with proper instrument design and operation.

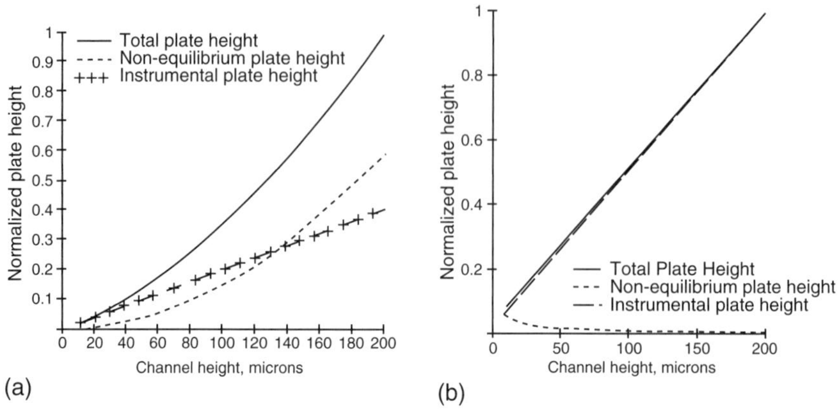


**Fig. 2.** Plots showing the variation of $H_n$ and $H_i$ with channel height for (**a**) Gradient-based systems (where field strength varies with $w$) and (**b**) general FFF systems (field is independent of $w$). Normalization was based on the highest value of the plate height in the dataset. Reprinted from Sant and Gale [29], Copyright (2006), with permission from Elsevier

### 3.2.2. Resolution Scaling

A general expression for resolution with a dependence on only a single geometric dimension can be obtained by making the length a function of channel height ($L = 3{,}000\ w$) and by substituting (3, 12, 13, and 15) into (14) (with total plate height as a combination of nonequilibrium and instrumental plate height) to obtain

$$R_s = \frac{\Delta R}{4\bar{R}} \sqrt{\frac{3000\bar{D}}{\chi\left(\bar{\lambda}\right) w \langle v \rangle + 3\bar{D}}}. \tag{16}$$

Note that (16) has the function $\chi(\lambda)$ embedded in it still (13) and that $R$ and $\lambda$ are also functions of $w$. The bar over variables in (16) indicates the mean values of the two particle clouds. These equations were used to provide a basic framework around which the various scaling effects associated with FFF systems can be compared.

Figure 3 shows the typical dependence of resolution on channel height $w$, for an ElFFF system. It is clear that resolution increases with a decrease in $w$, which is the motivating factor for miniaturizing ElFFF systems [29].

For typical experimental conditions, the resolution is nine times higher with a 10-fold reduction in channel height.

It should be noted that a reduction in channel height is always accompanied by a reduction in length when small analysis times are envisioned (the lower curve in Fig. 3). When the channel length scales proportionally to width, we expect a drop in resolution, but a large reduction in overall analysis time is achieved. Also, there is close to a 20% error when resolution [29] is computed with $\chi(\lambda)$ values obtained from (13) instead of the exact equation derived by Giddings [27].

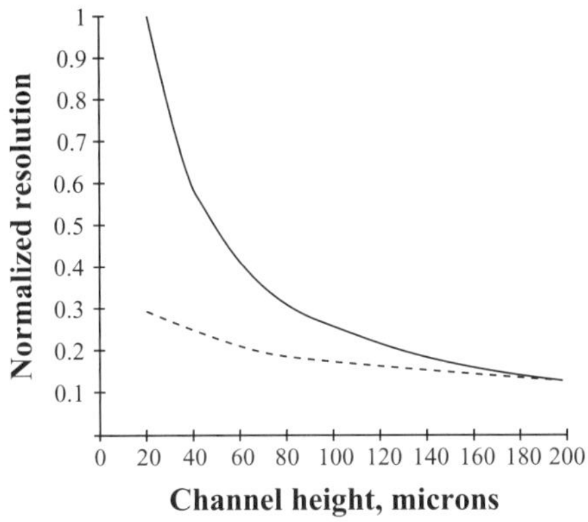


**Fig. 3.** Geometric scaling models for ElFFF system using a constant length of 60 cm. The assumed diameter and electrophoretic mobility of particles for the simulations are 50 nm and $-1.75 \times 10^{-11}$ $m^2$ $V^{-1}s^{-1}$. (*solid line*) Equation (16) with constant *L* and zero $H_i$, (*Dotted line*) Equation (16) with scaling *L* and scaling $H_i$. Reprinted from [29], Copyright (2006), with permission from Elsevier

### *3.3. Nongradient-Based Systems*

#### 3.3.1. Plate Height Scaling

Fig. 2b shows how plate height changes as the channel size is reduced for general FFF/nongradient-based systems and demonstrates how the relative importance of nonequilibrium and instrumental effects has switched and why there is little motivation to miniaturize these systems if only nonequilibrium plate heights are considered.

In this case the nonequilibrium effect (which increases with a reduction in channel height) is almost negligible and the plate height is dominated by instrumental effects.

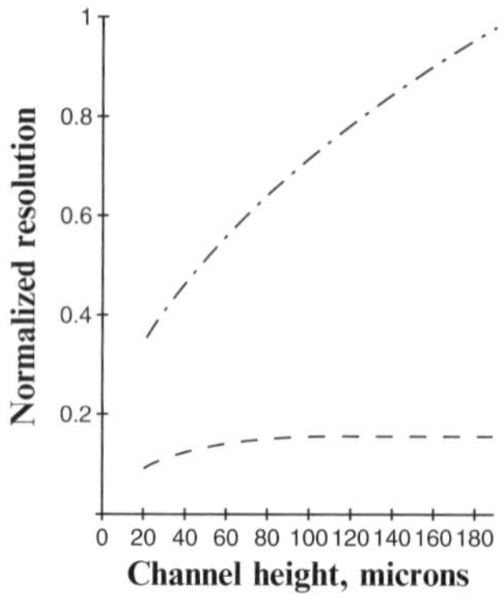


**Fig. 4.** Plot showing the effect of length and instrumental plate height on the resolution of general FFF systems as the channel height is reduced. (*Dashed line*) Equation (16) with constant $L$ and zero $H_i$, (*Dashed-dotted line*) Equation (16) with scaling $L$ and scaling $H_i$. Reprinted from [29], Copyright (2006), with permission from Elsevier

Unlike gradient-based FFF systems, mathematical models of general FFF systems predict a loss in resolution with miniaturization. The inclusion of instrumental plate height scaling, though, points to the possibility of an effective miniaturized general FFF channel. The top trace in Fig. 4 is the simulation result for normalized resolution (16) where $L$ and $H_i$ are kept constant at 60 cm and zero μm respectively, a typical geometry for general FFF systems. As expected, there is a considerable loss in resolution (~70%) when $w$ is reduced to 20 μm, but there is also a 10-fold reduction in retention time – an advantage at a heavy price.

The bottom trace from Fig. 4 shows a nearly constant resolution if $L$ and $H_i$ scale with $w$, with only an 8% loss in resolution when $w$ is reduced from 200 μm to 10 μm, while the 100 times reduction in retention time still occurs. This situation is the most likely one to be experienced in a practical situation, and provides evidence that miniaturization could be practical for general FFF systems [31].

Thus, a well designed general FFF should show improvement in resolution with miniaturization due to the major improvements related to instrumental effects. To gain all the advantages associated with miniaturization, though, general FFF systems may be required to operate under low retention conditions. The lower retention times associated with the high retention ratio will result in the reduced overall analysis time, while only sacrificing a small percentage of the potential resolution. Table 1 summarizes the scaling behavior of important FFF parameters and whether it is an advantage or disadvantage for both gradient and nongradient-based (general) FFF systems [32].

## 4. Microscale Electrical FFF

Electrical FFF was the first FFF subtype to be miniaturized using MEMS techniques [33]. In ElFFF, a voltage is applied across the two channel walls bounding the FFF channel [25, 34]. The separation criteria is based on the $\zeta$-potential or electrophoretic mobility possessed by the particles suspended in the carrier solution, which is typically DI water or a low ionic strength buffer, a potential challenge when analyzing certain biological materials. With the ability to measure the electrophoretic mobility of sample particles with known sizes, ElFFF can be used both as a separation unit or a diagnostic instrument.

Applications of ElFFF include the following: separation of cells and organelles, bacteria and viral separations, characterization of emulsions, liposomes, and other particulate biological vehicles, separation of macromolecules, environmental monitoring, and biomaterial studies. ElFFF has been used to study protein adsorption by analyzing surface-modified particles for biomaterial applications. In addition to many of these appli-

**Table 1.** Nongradient FFF and gradient-based FFF parameters affected by miniaturization [32]

| Parameter | General FFF scale factor | Advantage or disadvantage | ElFFF scale factor | Advantage or disadvantage |
|---|---|---|---|---|
| Retention ratio (*R*) | 1/s | Disadvantage | 1 | Potential Advantage |
| Analysis time | s$^2$ | Limited Advantage | s | Advantage |
| Drift velocity (*U*) | 1 | Neither | 1/s | Advantage |
| Plate height (*H*) | 1/s | Disadvantage | s$^2$ | Advantage |
| Resolution ($R_s$) | s | Disadvantage | 1/√s | Advantage |
| Steric transition ($d_i$) | 1 | Neither | √s | Potential Advantage |
| Equilibration time ($\tau_e$) | s | Advantage | s$^2$ | Advantage |
| Field time constant ($\tau$) | N/A | Subtype Specific | s | Advantage |
| Required sample size | s$^3$ | Advantage | s$^3$ | Advantage |
| Solvent consumption | s$^3$ | Advantage | s$^3$ | Advantage |
| Instrument size | s | Advantage | s | Advantage |
| Separable particle size | s | Relative | s | Relative |

cations, Gale et al. have employed μ-ElFFF in whole blood separations for medical diagnostics [35]. ElFFF also finds application as a sample pretreatment system by performing an initial separation on a sample that can later be collected for further testing by another analysis system. For example, ElFFF can be used as a sample preparation unit prior to a PCR step in a total analysis system.

A major advantage ElFFF enjoys over similar separation systems is low power and voltage requirements. In comparison to electrophoresis systems, which typically require several thousand volts, ElFFF operates below 3V. Even such a small amount of applied voltage across thin ElFFF channel results in a voltage gradient similar to that in electrophoresis systems that operate at about 1,000 times higher voltage. Thus, miniaturization proves beneficial in reducing power requirements and raises the possibility of a portable instrument with small batteries as power source.

### 4.1. Theory

Most of the general FFF equations can be applied directly to ElFFF by replacing $U$ in the particular (5). For example, $\lambda$, the nondimensional parameter relating experimental parameters to $R$, is represented by

$$\lambda = \frac{D}{\mu E w} = \frac{\kappa T}{3\pi\eta\mu d E w}, \tag{17}$$

where $\mu$ is the electrophoretic mobility of the sample and product of $E$ and $w$ is effective voltage $V_{\text{eff}}$. Equation (17) shows that retention in ElFFF systems is still inversely proportional to $w$, but since the effective field $E$ is also a function of channel height, there is no effect on retention as the channel is miniaturized. While this conclusion may seem to indicate that there is no net benefit in terms of retention from miniaturization, the fact that there is no disadvantage allows for the system as a whole to derive a significant advantage from miniaturization.

The steric transition point in FFF systems indicates a change in mode from normal FFF to steric FFF as defined earlier. The steric transition point in ElFFF systems can be determined using

$$d_{\text{i}} = \sqrt{\frac{2\kappa T w}{3 V_{\text{eff}}}}. \tag{18}$$

Examination of (18) suggests that the steric transition point for ElFFF systems is dependent upon channel height, a property significantly different from those of general FFF systems. Thus, by miniaturizing the system, it becomes possible to significantly reduce the steric transition point and make available the high-speed separations possible in steric mode. The separation mode, though, cannot be changed to steric mode simply by increasing the applied field, as can be done in general systems, due to the electrolysis of water at even moderate applied voltages. Thus, for ElFFF systems, steric separations of smaller particles might be impossible unless channels with smaller dimensions are fabricated.

The mechanics of ElFFF are different from most other FFF systems due to the presence of the electrical field and its interaction with the aqueous carrier. An electrical double layer is created at the interface of the polarized electrode and carrier solution as shown in Fig. 5. A major portion of the applied field drops across this double layer resulting in an effective field in the bulk of the channel that is only a fraction of the applied voltage available for retention of the sample. Effective voltages on the order of 1% have been reported in case of μ-ElFFF systems [36, 37]. This loss of effective field is caused by two related electrochemical phenomena. First, a significant portion of the voltage drops at the electrode/carrier interface, which may be attributed primarily to the electrode material properties. Second, the applied voltage has to overcome a potential barrier before any significant charge-transfer starts between the electrode and the carrier solution. The severity of the effective field reduction depends largely on the thickness of the double layer or the ionic strength of the carrier solution. A compact double layer, as occurs with a high ionic strength carrier, may result in very low field in the bulk with little or no retention in the channel. A critical concern regarding miniaturization of ElFFF is the creation of a more compact double layer and relatively low ionic strength solutions that can be used in the systems.

If the effective field is very low, the first solution to solving the problem would be to raise the applied voltage. Unfortunately, since the electrodes are in direct contact with an aqueous carrier, at voltages over about 2 V, electrolysis occurs and bubbles are rapidly generated that destroy the flow profile and cause severe mixing that makes the system nonfunctional. Thus, applied voltages are generally proscribed to a value where electrolysis does not occur.

One of the major challenges in ElFFF is the determination of the effective field and its associated retention of sample in the ElFFF channel. Unfortunately, this problem is highly complex and involves a number of operational and instrument variables such as voltage, sample, carrier composition, pH and ionic strength, electrode material and history, and so on. Only a rudimentary model for simulating transport properties using the convection-diffusion equation has been presented by Chen et al. [38]. The convection-diffusion equation was used to mimic ion and particle transport (DNA with an anisotropic diffusion coefficient) in ElFFF with an arbitrary value for the effective field (0.5% of applied voltage). The standard theory of ElFFF, though, embeds the solution to the convection-diffusion equation as illustrated by Palkar et al. [39] and numerical solution is not generally required if an effective field value is assumed, even for a sample with anisotropic diffusion. A more complete understanding of the inner workings of an ElFFF channel has not yet been presented.

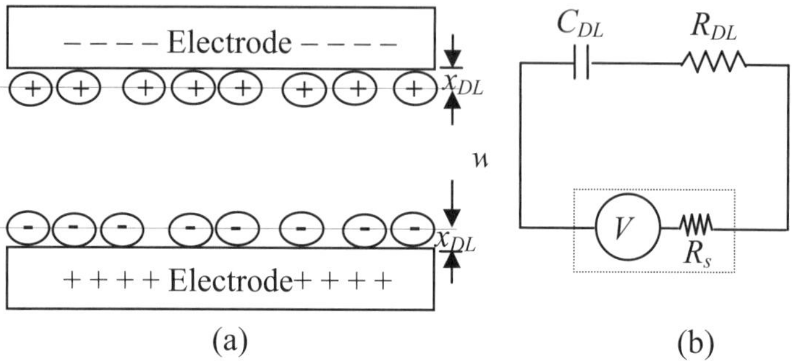


**Fig. 5.** (**a**) Electrical double layer in ElFFF system and (**b**) electrical circuit equivalent of ElFFF system showing double layer capacitance, source and bulk channel resistance, and voltage source

To get around this difficulty in understanding all of the processes taking place in the ElFFF channel, a way to predict ElFFF behavior based on an electrical circuit parameter model was developed by Kantak et al. [37, 40]. The ElFFF system can essentially be represented by electrical circuit components as shown in Fig. 5. In this model, the electrical double layer at each electrode can be represented by a parallel plate capacitor $C_{\mathrm{DL}}$ in parallel with the electrode–solution interface resistance $R_{\mathrm{DL}}$. The effective potential responsible for separation in the channel is identified as the potential drop across the bulk resistance $R_{\mathrm{B}}$. It should be noted that the bulk capacitance due to the channel itself will be very small at the low frequencies used in ElFFF (typically DC). $R_{\mathrm{s}}$, the source resistance can play an important role in voltage distribution as will be described in more detail in the section on cyclical ElFFF. The value of each of these circuit elements can be measured experimentally with little difficulty and predictions of effective field and elution times made that are highly accurate.

Using the circuit in Fig. 5, an interesting observation regarding the speed of various configurations of ElFFF systems can be made. From basic electrical circuit analysis, the time constant for electric field stabilization, $\tau$, of the system can be given as

$$\tau = \frac{(R_B + R_S)R_{DL}C_{DL}}{R_B + R_S + R_{DL}}. \tag{19}$$

Scaling analysis of the time constant reveals that a time constant of 40 s for a macroscale system drops to ~3.6 s for a similar microscale system, allowing a significant improvement in overall analysis time.

### 4.2. Fabrication and Packaging

Conventional semiconductor fabrication processes were applied towards fabrication of the earliest μ-ElFFF systems as outlined in Fig. 6 [33]. KOH etching was used to realize input and output ports on a silicon wafer. Titanium and gold layers were sputtered on the silicon as well as a glass substrate used as the second wall of the FFF channel. Platinum has also been used as an electrode material [37]. Thick photosensitive polyimide/SU-8 was photolithographically patterned to realize the microfluidic channels and provide a spacer between the electrodes. The two substrates were then bonded together to make an enclosed ElFFF channel. In several versions of the system, an adhesive trough was provided around the channel to facilitate adhesive bonding of the silicon wafer with polymer channel to a glass substrate with identical channel electrodes. For fluidic connections, PEEK tubing was attached to the silicon wafer over the ports using a ferrule glued to the substrate. Electrical connections were made by bonding wires to extensions of the electrodes. Another microfabricated system was reported by Lao et al., which used indium tin oxide (ITO) as electrodes [41]. ITO

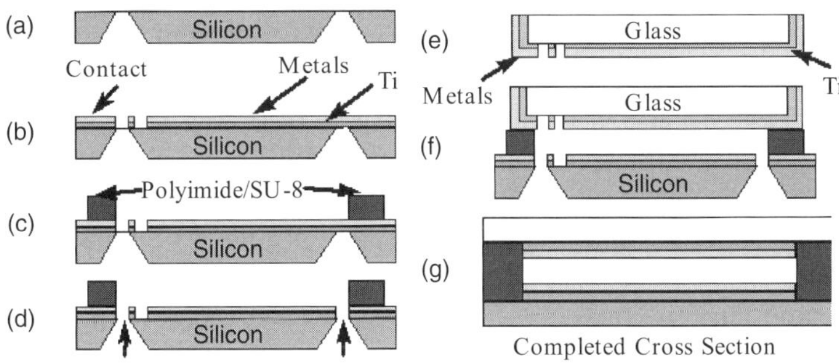


**Fig. 6.** Fabrication flow chart for the μ-ElFFF system. (**a**) Etching of input and output ports in silicon. (**b**) Deposition and patterning of titanium as adhesion layer and gold as channel electrode an. (**c**) Spinning and patterning of polyimide/SU-8 as channel walls. (**d**) Removal of $Si_3N_4$ membranes. (**e**) Deposition and patterning of titanium as adhesion layer and gold as channel electrode on glass. (**f**) Bonding of glass and silicon substrate using UV-curable adhesive. (**g**) Cross section of completed μ-ElFFF system

(transparent ceramic)-coated glass of 3 mm thickness was patterned to obtain electrodes with sheet resistance of 14 Ω $cm^{-1}$. This manufacturing process also used SU-8 as channel walls.

For more recent microscale ElFFF systems, the fabrication process was modified so as not to include any special micromachining processes and yet still achieve the advantages related to the miniaturized systems. In this design, polished graphite plates were used as both channel electrodes and a microfluidic channel was cut in a 25-μm thick double side adhesive tape, using xurography, an inexpensive rapid prototyping tool based on knife plotting [42, 43]. This system provided more reproducible fabrication results in a cost effective manner and proved efficient in producing prototypes for research purposes.

Typical μ-ElFFF system geometrical dimensions are 6 cm length, 2 mm width, and 25 μm height in comparison to a macroscale channel of 64 cm length, 2 cm width, and 176 μm height. The sample injection for microsystem is reduced to 0.1 μL from a 1-5 μL for macroscale system.

### 4.3. System Characteristics

With the fabrication of the first microscale ElFFF systems an effort was made to understand the operation of these systems and compare the results to macroscale systems. These comparisons included basic electrical operation of the systems followed by retention and separation experiments.

The most basic characteristic for ElFFF systems is the current–voltage relationship. Typically, the currents are relatively small at low voltage values (<1.2 V for platinum electrode systems). Such low currents do not generate enough effective field to induce any retention of the sample. In the vicinity of 1.5 V, the current increase is approximately linear in proportion to voltage. This voltage is loosely termed the "turn-on" voltage and is a function of the overvoltage for electrode material [44]. This turn-on voltage changes with electrode system and the order of overvoltage for the systems we have tested are graphite < platinum < gold. At higher voltages, current growth slows indicating limiting charge transfer reactions.

Other important parameters for an ElFFF system include the electrical time constant and the required stop-flow/sample relaxation time. The measured time constants for the early microsystems measured from 1 to 4 s in comparison to 40 s for its macroscale counterpart. More recently, microscale systems have been shown to have time constants of less than 10 ms depending on testing method and the materials involved [45, 41]. The stopflow time in FFF is designed to allow the injected particles to move to their equilibrium positions, which reduces band broadening and improves the retention characteristics. The equilibration time, $t_e$ can be determined by [20]

$$t_e = \frac{w^2}{\mu V_{eff}}. \tag{20}$$

For macroscale systems, the stop flow time can be 5 min, but for microscale systems a stop flow time of only 2 to 3 s is sufficient.

### 4.3.1. Retention

Retention is directly related to current in ElFFF and a plot of the inverse of current with retention ratio $\lambda$, shows that retention in microscale systems follows the theoretical predictions and that a straight line through the data goes through the origin as shown in Fig. 7a [37]. Microscale ElFFF systems have been demonstrated for a range of samples, conditions, and configurations. An example of differential retention of particles based on a difference in electrophoretic mobility is shown in Fig. 7b.

A retention related ElFFF characteristic is size selectivity, which is a measure of a systems ability to separate based on size. Typically, macrosystems possess size selectivity between 0.67 and 1 [34], but size selectivity in microscale ElFFF has been shown to be close to unity [37].

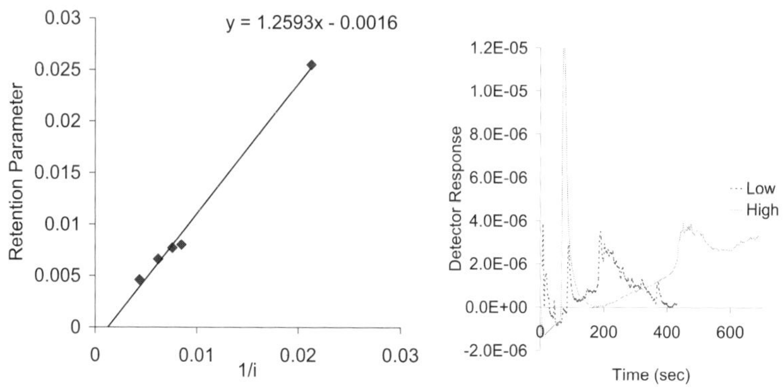


(a) (b)

**Fig. 7.** Retention results in ElFFF [37]. (**a**) Graph of retention parameter, $\lambda$, compared to the inverse of the measured current, *1/i*. The experiments were performed using 94 nm PS particles in DI water with a flow rate of 0.3 mL/h in a system with platinum electrodes. (**b**) Fractograms of particles with the same diameter, but differing levels of carboxylation. The run labeled "low" had a lower density of COOH groups on the surface (67 μequiv $g^{-1}$) compared to the sample labeled "high" (510 μequiv $g^{-1}$). The carrier was DI water with a voltage of 1.392 V (10 μA) and a flow velocity of 1.48 mm $s^{-1}$. Reprinted with permission from Gale et al. [37]. Copyright (2002) American Chemical Society

The theory for FFF predicts that the steric transition point should fall as the channel is miniaturized, but this was found not to be the case. For a 28 μm channel μ-ElFFF system, the steric transition point was found to be 450 nm, which is similar to the limit for macroscale systems. While there may be potential for the steric transition point to be pushed to lower diameters, the likely reason for similar results is particle/wall repulsion effects [25], which would likely operate in a similar manner for both macroscale and microscale systems. While, the origins of the exclusionary effects are not well understood, they may be linked to particle–wall repulsion, frictional drag, and lift forces. It will be interesting to see how this exclusion zone scales and whether it becomes the limiting factor in the performance of microsystems.

### 4.3.2. Separations

A goal for μ-ElFFF is high-speed separations that can be detected using an on-chip detector [37], which should significantly reduce band broadening and maximize resolution. Figure 8a shows such a separation of a multi-component polystyrene particle mixture in under 120 s. A separation of identical resolution in a macroscale system would require 2 h. As shown in Fig. 7b μ-ElFFF can also generate charge-based separation, an attractive feature for biological separations. Figure 8b demonstrates how identical particles with and without proteins attached to the surface can be differentially retained and separated in a μ-ElFFF system. Separations of blood components both before and after homogenization have been demonstrated using these same systems [35].

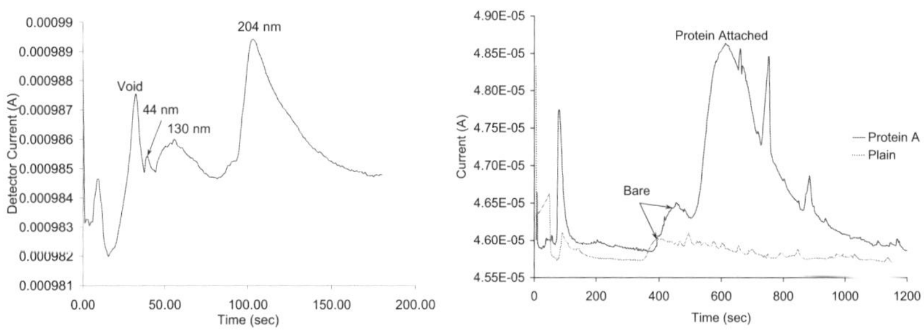


(a) (b)

**Fig. 8.** Separations performed in microscale ElFFF system. (**a**) Separation of a mixture of polystyrene particles detected using an on-chip conductivity detector. The operating conditions are 1.6 V, 0.3 ml $h^{-1}$ flowrate and 27 μA current in a 2-mm wide channel with platinum electrode with DI water carrier. Reprinted with permission from Gale et al. [37]. Copyright (2002) American Chemical Society. (**b**) Fractograms showing differential retention and separation between bare particles and particles with attached protein A [32].

The retention and separation data for μ-ElFFF was obtained with an on-chip conductivity detector [46, 47]. The on-chip conductivity detector was used to minimize the band broadening due to postcolumn volumes and to maximize the miniaturization related advantages of the μ-ElFFF channel. Another type of on-chip detection scheme employed with μ-ElFFF was based on the resonance light scattering (RLS) principle [48]. On-chip RLS is based on the shift in absorption and scattering spectral profiles for different particles, i.e., different sized particles yield different colored scattered light. The main advantages of this technique include the following: its noninvasive nature, high sensitivity for particle detection, and low cost.

### 4.3.3. Effective Field Scaling

The most important performance factor in ElFFF is the effective field. Since the effective field in the channel is difficult to determine using transport and electrochemical models, the effective field is usually calculated from retention data of standardized particles with known electrophoretic mobility. The effective electric field, $E_{\text{eff}}$, responsible for the given retention is then calculated using

$$E_{\text{eff}} = \frac{6Dt_{\text{r}}}{\mu w t_0} = \frac{2\kappa T t_{\text{r}}}{\mu w \pi \eta d t_0}. \tag{21}$$

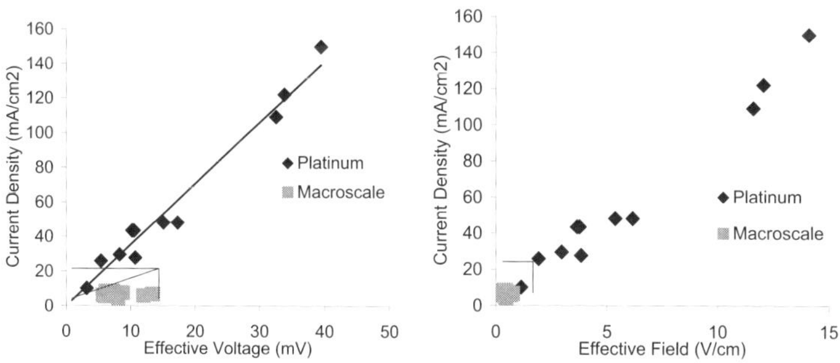


(a) (b)

**Fig. 9.** Comparison of electrical properties of microscale and macroscale ElFFF system. The boxes in both figures represent the limits of the macroscale system. Plots show (**a**) effective voltage and (**b**) effective field as a function of current density for both systems. Platinum refers to data from a platinum-based microscale system. Reprinted with permission from Gale et al. [37]. Copyright (2002) American Chemical Society

The effective field calculated using (21) is generally only ~1% of the applied field. Figure 9 shows the measured effective field for a μ-ElFFF system as a function of current density and shows that microscale systems significantly outperform macroscale systems, which are constrained to the boxes drawn in each figure. It can be seen that macroscale system is more efficient in terms of generating more effective field in comparison to microscale system for a given amount of current, but the amount of effective field that can be generated using a microsystem far exceeds that of a macroscale system. The limits in this comparison may be related to the electrode composition (platinum for microscale, graphite for macroscale) and similar data generated for graphite microsystems may show even better results.

## 5. Microscale Cyclical Electrical FFF

Once μ-ElFFF channels had been developed, efforts were made to determine methods to overcome the electrochemical limits associated with ElFFF. Cyclical fields were proposed as a way to overcome some of the capacitance associated with the double layer and drive effective fields up substantially. This method has proven very effective and has turned into a useful method worth considering on its own. Note that Cyclical Field-Flow Fractionation (CyFFF) was first proposed as a method by Giddings in 1986 [49] and demonstrated using a gravitational FFF system [50], but the only microscale systems have been cyclical electrical (CyElFFF). Thus, CyElFFF will be the focus of this text, though much of the analysis also applies to other CyFFF techniques.

Physically, a CyElFFF system is identical to an ElFFF system and both techniques can be performed using the same channel and electrode setup [51, 52]. As shown in Fig. 10, the mechanics of CyFFF are quite different from normal FFF due to the presence of the oscillating field.

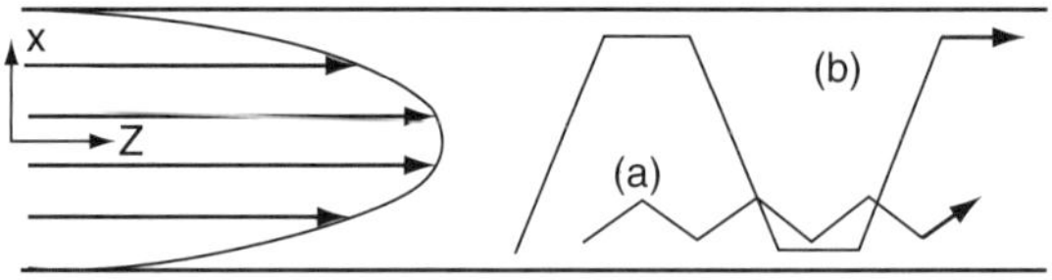


**Fig. 10.** Schematic diagram of the particle mechanics for (**a**) CyFFF mode I and (**b**) CyFFF mode III

Unlike normal ElFFF, the particles interacting with the field do not reside near the accumulation wall or form a static exponential concentration profile (Fig. 10a) but move back and forth between the parallel electrodes under the influence of cyclical field. A number of operational and physical parameters (applied voltage and its frequency, electrophoretic mobility and size of the particles, flow rate, pH, and ionic strength of the carrier solution) determine how fast and how far the sample particles move between the electrodes and the average location of the particles in the channel. Depending on the motion of the particles, a particular group of particles will spend more or less time in the faster velocity zone (away from the channel electrodes) compared to other particles, generating differential retention and separation.

Based on the magnitude of the particle movement between the channel electrodes, CyElFFF can be classified into three different modes of operation. If particles oscillate only against one wall/electrode and do not completely cross the channel and reach the other wall, the particles are operating in Mode I. If the particles completely cross in one half cycle and rest for some period against the opposite wall, the particles are operating in mode III. Mode II occurs when particles reach the opposite wall just as the field is reversed.

### 5.1. Theory

In CyElFFF the retention ratio, $R$ can be related to a nondimensional parameter $\lambda_0$, which for an applied square wave electrical field is given by

$$\lambda_0 = \frac{\mu E_{\text{eff}}}{2 f w} = C \frac{\mu E_{\text{eff}}}{f w^2}, \tag{22}$$

where $f$ is the frcquency of the field oscillation in Hertz. $\lambda_0$ values can be used to determine the mode of CyElFFF operation and retention ratio $R$. For mode I, $R$ is given by

$$R = 3\lambda_0 \left(1 - \frac{2\lambda_0}{3}\right) \text{ for } \lambda_0 \leq 1; \tag{23}$$

whereas, for Mode III $R$ is given by

$$R = \frac{1}{\lambda_0} \text{ for } \lambda_0 \leq 1. \tag{24}$$

Mode II can be represented by either model as the unity value of $\lambda_0$ holds true for both (23) and (24).

Equations (22) to (24) were first developed by Giddings [49]. Gidding's model does not account for three factors related to FFF, which can occasionally be significant: steric effects due to finite particle size, diffusion effects, and, in CyElFFF, the reduced effective field due to the electrical double layer induced shielding. In reality, when a spherical particle approaches an accumulation wall, it cannot get any closer than its radius where the center of mass is located. Because of this steric effect, particles never have a truly zero velocity, and hence, retention times are less than predicted. Particle diffusion across the channel ($x$-direction in Fig. 10) can change the resultant average particle location in the channel and hence, the elution time. Finally, the presence of the electrical double layer should reduce the effective field for CyElFFF, creating a host of challenging problems.

These three major modifications to the CyElFFF theory were made by Kantak et al. [40] with a few assumptions and using a single model particle in the improved CyElFFF model. The assumptions made were (a) no overlap of electrical double layers for particles and electrodes, (b) no frequency dependence of particle double layer, (c) constant particle electrophoretic mobility, (d) except for the diffusion modeling, coherent particle motion that can be represented by a single particle, and (e) a relaxed sample with motion starting at one of the accumulation walls. This model was found to closely predict the elution times of particles, as will be shown later.

#### 5.1.1. Effective Field Model

There are several ways that the effective field distribution in ElFFF or CyElFFF channel can be computed. Biernacki et al. [53] and Chen et al. [54] used a first principles approach to compute effective field from transport models of ions and particles. A combination of Poisson's and the Nernst-Planck [53] or the convection-diffusion equations [54] were used to model the particle and ion movement in the channel. While, this approach should result in a very generic model that can be applied to any situation concerning CyElFFF operation, the field values used in both these models were arbitrary and did not take into account the electrochemical processes occurring at electrodes and electrode-carrier interface. A complete model of CyElFFF should actually account for all electrochemical kinetics and species transport without any empirical quantities.

Another approach to compute effective field is by using the electrical analog of CyElFFF (Fig. 5) [40]. Using basic electronics principles the effective field across the bulk of the channel can be predicted by

$$E_{\text{eff}} = \frac{VR_{\text{B}}}{w}\sqrt{\frac{1+(2\pi fR_{\text{DL}}C_{\text{DL}})^2}{(R_{\text{B}}+R_{\text{S}}+R_{\text{DL}})^2+(2\pi f(R_{\text{B}}+R_{\text{S}})R_{\text{DL}}C_{\text{DL}})^2}}. \tag{25}$$

The circuit parameters required to calculate effective field can be measured using a simple series of experiments [40]. Thus a semiempirical model of CyElFFF can be generated by using the experimentally derived electrical circuit parameters towards predicting effective field in the system using (25).

### 5.1.2. Steric Effects in CyFFF

The main contribution of steric effects is in limiting the approach of particles to the accumulation wall. To account for the steric effects in CyFFF, one has to include the size of the particle while calculating the particle displacement in the channel. Making the adjustment for steric effects, the retention in CyFFF systems for Mode I and III are then given by

$$R_{\mathrm{I}} = 3\lambda_o\left(1-\frac{2\lambda_0}{3}\right)-6\frac{d}{2w}\left(1-\lambda_0-\frac{d}{2w}\right) \text{ and} \tag{26}$$

$$R_{\mathrm{III}} = \frac{(w-d)}{\lambda_0 w}\left(1+\frac{d}{w}\left(1-\frac{d}{2w}\right)\right)+6\gamma\frac{d}{w}\left(\frac{1}{2}-\frac{(w-d)}{2w\lambda_0}\right) \tag{27}$$

respectively, where $\gamma$ is a dimensionless correction factor accounting for particle migration under steric conditions [55] and is between 0.5 and 1.0.

### 5.1.3. Particle Diffusion Effects

For mode I where particles just oscillate near one of the channel walls, diffusion in the $x$-direction can be critical as it significantly affects the location of the particle clouds and in turn elution times. For coherent motion of particles, linear unidirectional diffusion length of a single particle can be obtained from random walk theory as

$$l_{\mathrm{D}} = \sqrt{2Dt}\,. \tag{28}$$

The diffusion displacement in the $x$-direction, $x_{\mathrm{D}}$, with a geometric factor for the centroid of the particle cloud is given by

$$x_{\mathrm{D}} = \frac{4\sqrt{2Dt}}{3\pi}. \tag{29}$$

For mode III CyElFFF begins to approach normal FFF at high values of $\lambda_0$, and diffusion can begin to play a role. If steric effects are not controlling, the distance traveled by the particle ($z_{\mathrm{III}}$) along the channel ($z$-direction) is given by

$$z_{\mathrm{III}} = 6<v>\left(\frac{(w-2l)}{2\mu V_{\mathrm{B}}}+\frac{(w-2l)l}{\mu V_{\mathrm{B}}}-\frac{(w-2l)}{3w\mu V_{\mathrm{B}}}-\frac{(w-2l)l}{w\mu V_{\mathrm{B}}}-\frac{(w-2l)\;l}{w\mu V_{\mathrm{B}}}\right)$$
$$+6<v>\left[\frac{1}{2w}-\left(\frac{1}{2w}\right)^2\right]\left(\frac{1}{2f}-\frac{(w-2l)w}{\mu V_{\mathrm{B}}}\right). \tag{30}$$

It should be noted this lumped electrical parameter model does not account for all potential physical effects and for this reason, the scope of this model can be limited and should be used with care [40, 45].

## *5.2. Experimental Results*

CyElFFF has shown the ability to significantly retain nanoparticles and to perform separations on nanoparticles, especially using low ionic strength carriers. Examples of some of these experiments are summarized in the following sections.

### 5.2.1. Comparison of Theory with Experimental Data

Figure 11 shows the typical elution characteristics of the μ-CyElFFF as a function of frequency of the applied field [40]. It can be deduced from Fig. 11 that Giddings's model clearly does not match with the experimental data and deviates even with the inclusion of steric and diffusion effects. Elution times computed using the estimated effective field instead of the nominal field yield a better match with the experimental results and show that electrical double layer related effects are of prime importance in CyElFFF. Also, mode transition can be predicted correctly with the use of the lumped electrical parameter model for the evaluation of $\lambda_0$.

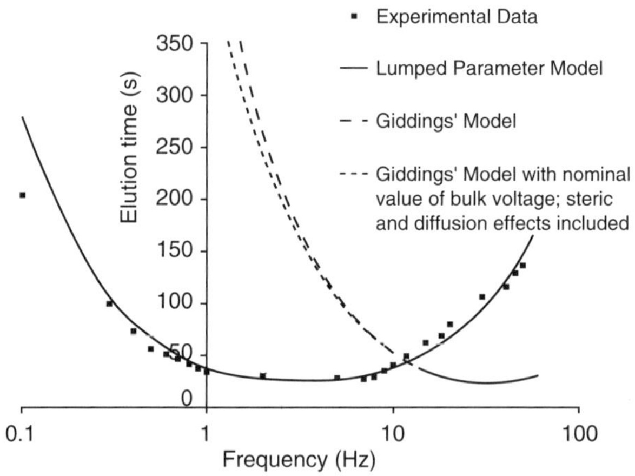


**Fig. 11.** Comparison between model and experimental model for μ-CyElFFF system. Reprinted with permission from Kantak et al. [40]. Copyright (2006) Wiley nanoparticles in a 50 μM ammonium carbonate carrier

Lao et al. showed that the increased effective field with the pulsed field resulted in 50-fold increase in current and that there was a strong influence of pulse frequency on retention time [41]. The retention dependence on voltage is straightforward and an increase in applied voltage results in increased retention when in Mode III. For example, the elution time quadruples when the peak to peak voltage (square wave at 1 Hz) is increased from 1 to 8 V for retention of 100 nm amino-coated polystyrene nanoparticles in a 50 μM ammonium carbonate carrier.

### 5.2.2. Separations

Separations in μ-CyElFFF are dependent on a difference in the electrophoretic mobility of the samples. In the earliest separation results published by Lao et al., pulsed ElFFF, a variation of CyElFFF, was able to resolve 0.105 μm and 0.405 μm particles [41].

Figure 12 shows high-speed separations (<3 min) of nanoparticles using μ-CyElFFF [52]. It can be seen that resolution is highly dependent on frequency and in this case separation resolution increases with the frequency of the applied field. It should be noted that plate heights also vary with frequency and tend to follow a pattern similar to the elution times (Fig. 11) with a minimum plate height obtained near the mode transition point. This translates to limits on separation power in μ-CyElFFF with reduced peak capacities in comparison to μ-ElFFF. An offset voltage is used to force the particles towards one accumulation wall while operating in mode I to relax the sample and avoid initial random distribution of the particles across the channel height.

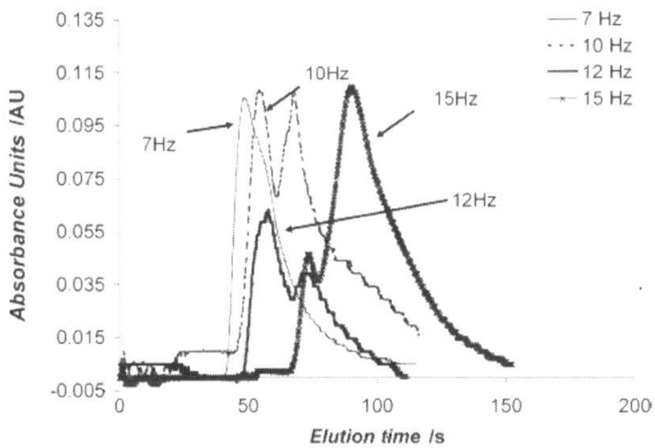


**Fig. 12.** Separations of 100 nm silica nanoparticles (first peak) and polystyrene (PS) amine group nanoparticles (second peak). The operating conditions are a square wave of 3.0 VPP with 0.3 VDC offset, DI water as carrier with 1.0 mL $h^{-1}$ flow rate [52]. Reproduced by permission of The Royal Society of Chemistry

### 5.2.3. Effects of Carrier pH and Ionic Strength

The ionic strength of the carrier affects the retention characteristics and effective field of the CyElFFF to a great extent. Bulk channel conductivity, double layer impedance, current across the channel, electrophoretic mobility of the particles, and source resistance all are affected at various levels across the frequency spectrum of the applied field [45].

In general an increase in current associated with an increased ionic strength correlates to a drop in available effective field, while an increase in current for constant carrier conditions results in an increase in effective field. Similar logic applies to an increase in pH. It should be emphasized that in order to obtain reliable operation using either ElFFF or CyElFFF, the carrier conditions should be maintained carefully. Also, the electrophoretic mobility increases with an increase in ionic strength [56]. The overall effect of increased electrophoretic mobility and reduced effective field results in very convoluted CyElFFF retention results with increase in ionic strength as shown in Fig. 13. Not only does retention drop, but the band broadening of the elution peaks also increases with an increase in the ionic strength of the carrier, leading to very poor results.

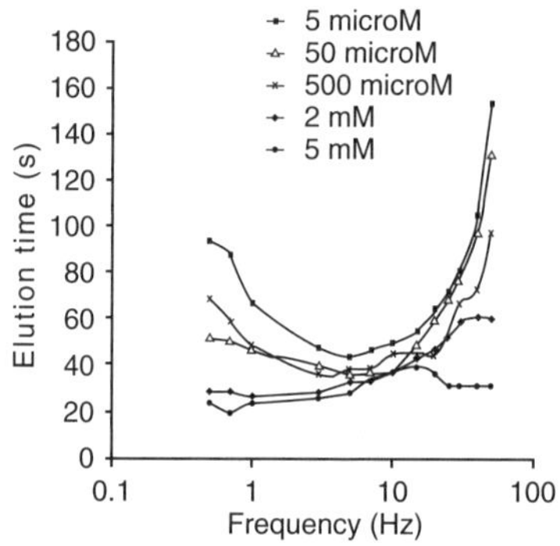


**Fig. 13.** Retention characteristics of μ-CyElFFF as a function of ionic strength of carrier and frequency of the applied oscillating field. Reprinted with permission from Kantak et al. [45]. Copyright (2006) American Chemical Society

Microscale CyElFFF has shown promise as a nanoparticle separation and analysis technique with the ability to generate significant retention and high electric fields in both Mode I and Mode III. The primary limitation of the technique is that it does not work well with high ionic strength carriers.

## 6. Microscale Dielectrophoretic FFF

Pioneering work on microscale dielectrophoretic FFF (DiFFF) was done by Gascoyne's group with applications in cancer research and cell separations [57–63]. These systems are technically Dielectrophoresis-Gravitational FFF systems in which dielectrophoretic (DEP) fields are used to drive particles away from the wall and gravitational forces drive the same particles towards the wall, similar to hyperlayer FFF. Thus a balance between these competing fields determines the average particle location in the channel and the elution time.

DiFFF devices were created using a combination of microfabrication and conventional FFF assembly techniques. Fabrication of the electrodes to impart the dielectrophoretic field requires the use of microfabrication techniques. Standard photolithography techniques were used to manufacture interdigitated electrodes of 50 μm in both width and gap on a glass substrate [59]. A Teflon spacer of 400 μm thickness was cut to realize an open channel of 50 mm width with 288-mm tip to tip length. A set of 36 nylon screw-clamps were used to assemble the DiFFF chamber with the Teflon spacer being sandwiched between two glass substrates. The glass substrate allowed optical interrogation of the sample as it passes through the FFF channel. For cellular separations, use of a fraction collector is a common practice enabling the user to perform more diagnostic tests on the individual fractions.

### 6.1. Theory

Figure 14 shows the operational principle of DiFFF and the forces involved in this separation technique. Dielectrophoretic forces are generated

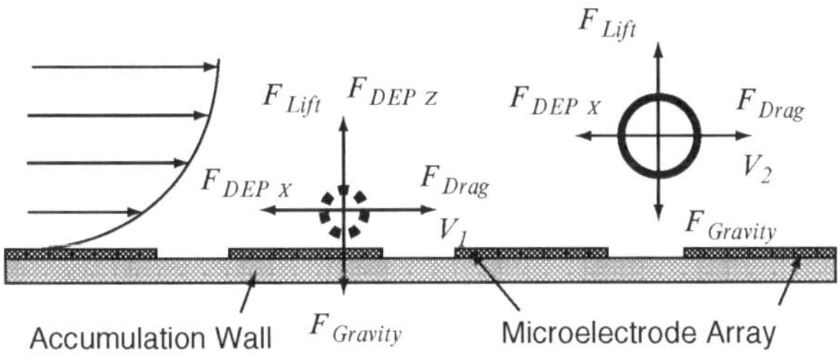


**Fig. 14.** The DiFFF principle. A close look at the accumulation wall shows the parabolic velocity profile, the microelectrode array and two particles with different dielectric properties at different locations away from the accumulation wall determined by a force balance of DEP levitation, hydrodynamic lift, gravity, and fluid flow-induced drag

by a nonuniform AC field imparted by electrodes on one wall of the channel [60]. For dielectrophoretic separations, polarizability of the particle plays an important role as the sample particles are suspended in an aqueous medium in the presence of a high frequency electric field. Electrostatic interactions between polarized particles and the field induce movement of the particles. The degree of polarizability and the particle's size cause a particle to experience different relative forces away or towards the channel walls, which induces a unique velocity along the channel, as is typical for FFF. In addition to the particle properties, the carrier properties (aqueous medium), and the frequency and magnitude of the applied field help determine the forces on a particle. If the particle polarizability is less than that of the medium, particles experience negative DEP forces and are moved away from the high field strength area. In positive DEP particles are retained near the electrodes where high field strength zones are located. Both of these types of DEP are very useful and can be employed to separate particles if proper elution modes are used. Depending on the particle properties, two different types of DEP modes: migration and retention can be used. In DEP migration mode, different types of sample particles with polarizability on either side of medium are separated. Positive DEP causes retention of one type of particle at the electrodes and the particles experiencing negative DEP are forced away from the electrode region (into faster flow lines typically) to achieve spatial separation across the channel. DEP retention mode utilizes the balance between DEP and hydrodynamic forces to realize discreet retention of the sample particles. Particles that experience positive DEP are retained strongly at the electrode surface and particles with weak interaction forces (small positive DEP or even some negative DEP) elute earlier. This mode of DEP maintains the particles at different heights and is the most common mode used in DiFFF [59]. Retention mode DiFFF has the potential to obtain higher resolution than migration mode DiFFF due to inherently higher selectivity as it requires smaller differences between the dielectric properties of cells to generate a significant retention difference.

The electrodes used to impart the dielectrophoretic force field are typically interdigitated to generate nonhomogeneous field. Particles with different polarizabilities levitate in distinct velocity zones under the influence of the collective electrical and hydrodynamic/gravitational forces. Typically, the resultant $F_{\text{DEP-X}}$, dielectric force component opposing the carrier flow is very small compared to the $F_{\text{DEP-Z}}$ acting in the perpendicular direction, especially since the horizontal force averages out due to the interdigitated nature of the electrodes. For the particle sizes typically involved in DiFFF, gravitational and lift forces are also significant. Forces describing the physics behind dielectrophoresis can be summarized by equations (31) to (35), where $V_{\text{RMS}}$ is

$$F_{\mathrm{DEP_Z}} = 2\pi\varepsilon_{\mathrm{m}} r^3 p\,\mathrm{Re}\left(f_{\mathrm{CM}}\right) q\left(h\right) V_{\mathrm{RMS}}^2, \tag{31}$$

$$F_{\mathrm{gravity}} = \frac{4}{3}\pi r^3 \left(\rho_{\mathrm{p}} - \rho_{\mathrm{m}}\right) g, \tag{32}$$

$$F_{\mathrm{Lift}} = C\frac{6\eta r^3 \langle v \rangle}{w\left(h-r\right)}, \tag{33}$$

$$\mathrm{Re}\left(f_{\mathrm{CM}}\right) = \frac{f^2 - f_0^2}{f^2 + 2f_0^2}, \tag{34}$$

$$f_0 = \frac{\sigma_{\mathrm{m}}}{\pi r C_{\mathrm{mem}}}, \tag{35}$$

the applied RMS voltage, $\varepsilon_{\mathrm{p}}$ and $\varepsilon_{\mathrm{m}}$ are the particle and medium dielectric permittivities, $\rho_{\mathrm{p}}$ and $\rho_{\mathrm{m}}$ are the particle and medium densities respectively. $\mathrm{Re}_{(f\mathrm{CM})}$ is the real part of the Clausius-Mossotti factor $f_{\mathrm{CM}}$, which is a function of the frequency-dependent complex dielectric permittivities of the particle and the medium and essentially reflects the frequency dependent magnitude and direction of the field-induced polarization $\alpha_{\mathrm{DEP(f)}}$. $f_0$ is the crossover frequency at which DEP forces acting on the cells are zero, $C_{\mathrm{mem}}$ is the membrane capacitance, and $\sigma_{\mathrm{m}}$ is the electrical conductivity of the cell-suspending medium. The capacitance difference due to membrane surface area has more effect on cell separation mechanics than the membrane composition or morphological differences. $q(h)$ is the measure of height dependence of $F_{\mathrm{DEPZ}}$, $p(f)$ is a frequency-dependent factor to account for the electric double layer induced voltage drop, $h$ is the position of the particle, $r$ is the particle radius, $<v>$ is the average velocity of the fluid, and $C$ is a constant (0.172). An cxpression for the levitation height of the sample can be derived by combining and simplifying (31)–(33) as

$$h_{\mathrm{eq}} = \frac{1}{q(h)}\left(\frac{2\left(\rho_{\mathrm{p}} - \rho_{\mathrm{m}}\right)g - {}^{3F_{\mathrm{Lift}}}\!/_{2\pi r^3}}{3\varepsilon_{\mathrm{m}} p(f)\,\mathrm{Re}(f_{\mathrm{CM}}) V_{\mathrm{RMS}}^2}\right). \tag{36}$$

Typically, the geometrical dimensions of the dielectrophoretic chamber are limited by the size of sample being processed. As DiFFF was developed to analyze cells, channel thickness is generally 250–400 μm. Since retention in DiFFF is partially a function of the channel height, work on optimization of channel dimensions would be of interest for future miniaturization efforts.

### 6.2. Experimental Results

DiFFF was used to separate human leukemia (HL-60) cells from normal human mononuclear cells, polystyrene beads with different surface properties, and human breast cancer cells from whole blood. Excellent separation results were obtained for a variety of cell types that closely match with the theoretical models proposed for them [57, 58, 63].

For a typical experiment, the DiFFF channel was loaded with sucrose buffer and leukocyte samples were loaded using an injection valve [63]. A 10 kHz DEP voltage (4 V peak to peak) was applied to the electrodes during loading of the cells to prevent adhesion of the cells to the electrode surface. Cells were allowed to sediment towards the DEP electrodes in the chamber for 5 min. Table 2 is the summary of leukocyte analysis results using DiFFF.

Field programming in terms of the use of a combination of swept frequency and fixed frequency improves the discrimination of the different cell types. Details of these experiments are available [57, 58, 63].

Similar separations of human breast cancer cells (MDA-435) from normal T-lymphocytes resulted in 99.2% purity after separation within 11 min with a close to 69% total cell recovery [58].

It can be seen from Fig. 15 that most of the breast cancer cells were retained near the electrode channel under positive DEP at 20 kHz as only 35% of cells elute from the channel. These results suggest that designing the experimental conditions (mainly frequency) for different types of cells is an important issue. The effect of voltage on each cell type was similar, so voltage cannot be used to aid the selectivity of the system.

**Table 2.** Summary of leukocyte separation by DiFFF

| Experimental parameters | Cell types | % Purity | Separation time (min) |
|---|---|---|---|
| Monocyte: T-lymphocytes (20–50 kHz, 10 min: 5 kHz, 6 min) | Monocytes | 98 | 16 |
| | T-lymphocytes | 92 | |
| Monocyte: B-lymphocytes (20–40 kHz, 10 min: 5 kHz, 6 min) | Monocytes | 94 | 16 |
| | B-lymphocytes | 92 | |
| Granulocytes: T-lymphocytes (40–50 kHz, 8 min: 5 kHz, 5 min) | Granulocytes | 94 | 13 |
| | T-lymphocytes | 87 | |
| Monocytes: Granulocytes (30–35 kHz, 8 min: 5 kHz, 5 min) | Monocytes | 97 | 13 |
| | Granulocytes | 91 | |

(Adapted with permission from Duhr et al. [63]. Copyright (2000) Biophysical Society). The DEP field consists of a swept frequency followed by a fixed frequency

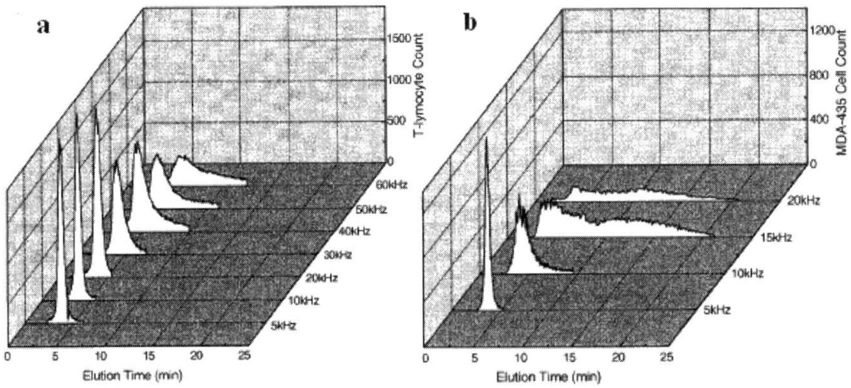


**Fig. 15.** Frequency-dependence of DiFFF elution profile for (**a**) T-lymphocytes and (**b**) human breast cancer MDA-435 cells. Rapid band broadening of elution fractograms for MDA-435 cells is clearly visible above 10 kHz. Isotonic sucrose/dextrose buffer was used as carrier with 4 V peak to peak DEP field. Reprinted with permission from Wang et al. [58]. Copyright (2000) American Chemical Society

An important FFF related observation made using the DiFFF systems was that the hydrodynamic lift effect was relatively small compared to conventional gravitational FFF systems. Cell elution times were found to be inversely proportional to the flowrate indicating insensitivity of cell equilibrium heights to flow rate. The reason for this result is that hydrodynamic lift forces are very weak (less than 2%) in comparison to the DEP fields [57].

DiFFF is one of the most successful subtypes of microscale FFF and has the potential to be a very important diagnostic tool in cell separation and cancer research. With the ability to sort thousands of cells per minute, DiFFF can be used in conventional labs for sample preparation and other analyses.

## 7. Microscale Thermal FFF

Thermal field-flow fractionation (ThFFF), one of the oldest FFF subtypes, has primary application in the separation and analysis of dissolved and suspended polymer samples. ThFFF utilizes a temperature gradient across the channel walls to induce separation based on a particle's thermal diffusion coefficient. Thermophoresis in liquid solutions is not well understood and a comprehensive model to predict thermal diffusion of polymers is not available [64]. ThFFF is similar to ElFFF in that the gradient of temperature generates the field, just as the gradient of voltage generates retention in electrical FFF. To generate the temperature gradient, thermally conductive channel walls are maintained at different temperatures. Polymers typically migrate towards the cold wall under the influence of thermal diffusion and accumulate at the cold wall in what is thought to be an entropic process [65]. A balance between transport and normal-diffusion determines particle retention. A range of solvents such as methanol, THF, acetonitrile, DMSO, toluene, and aqueous solutions with different types of detergents have been employed as carriers in ThFFF.

Applications of ThFFF include the following: molecular weight and size distribution determination, particle size measurement, thermal diffusion coefficient measurements, physiochemical and surface property studies, and separation of particle colloidal mixtures.

Retention in ThFFF system depends on the Soret coefficient, a ratio of thermal diffusion and Fickian diffusion coefficients. The retention parameter can be related to these sample physiochemical properties by

$$\lambda = \frac{D}{\left(D_{\mathrm{T}}\Delta T\right)} = \frac{\kappa T}{3\pi\eta}\left[\frac{10\pi N_A}{3[\eta]MW}\right]^{1/3}\frac{1}{\left(D_{\mathrm{T}}\Delta T\right)}, \tag{37}$$

where, $D_{\mathrm{T}}$ is the thermal diffusion coefficient, $\Delta T$ is the temperature drop across the channel height, $\kappa$ is the Boltzmann's constant, $T$ is the average temperature, $\eta$ is the average carrier viscosity, $N_{\mathrm{A}}$ is the Avogadro's constant, $[\eta]$ is the intrinsic viscosity of the dissolved sample, and MW is the average molecular weight of a dissolved sample.

The first successful microfabricated ThFFF system was demonstrated by Edwards et al. [66] and was followed by several reports on mesoscale ThFFF with reduced geometrical dimensions [67, 68] and a microfabricated ThFFF without showing any particle separation [69]. The early ThFFF microsystems were fabricated using techniques similar to the μ-ElFFF fabrication (Fig. 6) with a 27 μm channel thickness, 2–4 mm breadth and 4–6 cm length. These microsystems made of silicon and glass with SU-8 to define the channel walls were fabricated using conventional microfabrication techniques. A thin-film titanium heater on silicon or boron-doped silicon heaters were used to generate the hot wall whereas a glass slide was used as the cold or accumulation wall. Results obtained this early microsystem presented showed a poor temperature drop across the channel due to a poor heat transfer setup. Essentially, silicon with its very high thermal conductivity was found to waste a lot of input energy that was transferred to the environment rather than across the channel. In addition, glass has a high heat capacity and acts as an insulator making it difficult to maintain a good cold wall temperature. The overall result of these efforts was a very low effective temperature drop available for retention and separation. Later communications [70] from our group showed how the efficiency of heat transfer in ThFFF can be improved by just switching the roles of silicon and glass to cold wall and hot wall respectively. Even with a low temperature drop of ~5.3°C, retention ratios of 0.46 and 0.33 for particles of 204 nm and 272 nm were obtained in a clear separation. Power consumption was reduced 300-fold for a one order of magnitude smaller temperature drop generating similar results when compared to a macroscale system.

## 8. Miniaturized Flow FFF

Flow field-flow fractionation (FlFFF) is an FFF subtype that relies on cross-flow to compress the sample particles towards the accumulation wall. There are a number of variations of this FFF subtype: conventional (symmetrical), asymmetrical, frit inlet and frit outlet, frit inlet asymmetrical, and hollow fiber FlFFF. Asymmetrical channels have come to dominate in the past few years, since they do not require a perpendicularly applied flow, but rely on flow out of one of the channel walls to generate the gradient. Unlike conventional FFF, the width of the channel progressively shrinks for asymmetrical FFF to maintain a constant field gradient along the length of the channel. For frit inlet asymmetrical FlFFF, the sample is hydrodynamically compressed to the accumulation wall using cross-flow of the carrier solution entering through a small inlet frit as shown in Fig. 16.

A mesoscale frit inlet asymmetrical flow FFF system was fabricated using conventional FFF manufacturing technique [71]. This paper characterized a flow FFF system with only the width and length dimensions being reduced three fold. The channel height was kept at 250 μm as in most conventional FFF systems. This reduced volume system was used to separate a mixture of protein standards with size range between 29 and 700 kD. The reduction in channel width and overall channel volume allowed the use of 10 times smaller flowrates and sample injection volumes. The analysis time was reduced by half (~ 20 min) in comparison to a full-size system for comparable resolution.

While the mesoscale system allowed characterization of a smaller flow-FFF system, it will be interesting to test the performance of a true microsystem with channel heights less than 100 μm. Comprehensive experiments will have to be carried out to check if crossflow can generate a high enough field gradient to induce high resolution separations.

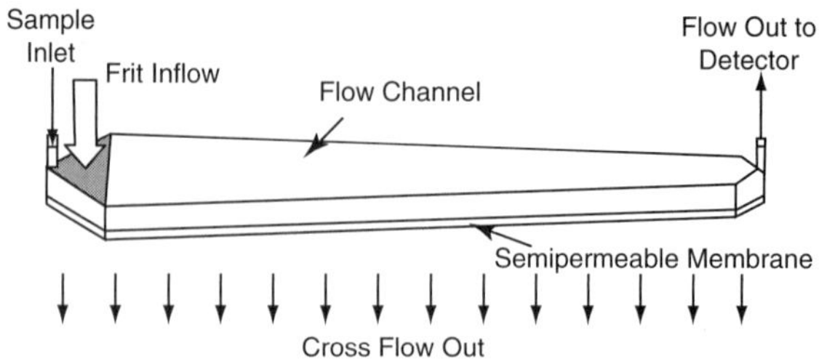


**Fig. 16.** The asymmetrical-flow field-flow fractionation system with inlet frit and sheet membrane at the accumulation wall

## 9. Microscale Acoustic FFF

Acoustic/ultrasound-based field-flow fractionation (AcFFF) was first demonstrated by Semenov et al. [72] on the macroscale and fractionation of particles was shown by Beckett's group in several communications for acoustic FFF and SPLITT systems [73–75]. AcFFF is able to separate samples based on density or compressibility of the sample in addition to its size. While most of FFF techniques generate a constant field in the direction normal to the separation direction, the field across the channel varies in AcFFF according to

$$F_{Ac} = \left(-\frac{4}{3}\pi r^3\right) k \overline{E} f(\rho,\gamma) \sin(2ky), \tag{38}$$

where $F_{Ac}$ is the acoustic force in direction of the field, $r$ is the particle radius, $k$ is the acoustic wave number based on the fundamental wavelength of the acoustic wave, and $\overline{E}$ is the acoustic energy density (function of carrier fluid density and acoustic wave velocity). Equation (38) also indicates a strong dependence of acoustic force on particle size. Factor $f_{(\rho,\gamma)}$ is a function of density and the compressibility of the medium and the particles.

The retention parameter for high retention mode in AcFFF is [76]

$$\lambda = \frac{6\kappa T}{\pi^2 \rho u^2 r^3 \left| f' \right|}, \tag{39}$$

where $u$ is the acoustic wave velocity and $|f'|$ is a function similar to $f_{(\rho,\gamma)}$.

The design of a microsystem with the ability to impart acoustic fields was presented by Edwards et al. [77, 78] and consisted of three layers: an acoustic transducer, a microchannel, and a reflector. Thicknesses for each layer were determined based on the fundamental frequency of the system. Lithium niobate was used as the transducer material and a silicon wafer with input and output ports was used as the reflector plate. A $CO_2$ laser was used to pattern a thin layer of PDMS, which when sandwiched between the transducer and reflector layers resulted in a μ-AcFFF system [78]. This microsystem showed very poor retention and the retention characteristics did not match with the normal mode or steric mode theory. Nevertheless, it showed a dependence of retention on the applied field and better results were predicted if longer channels were used [78].

## 10. Other Microscale FFF Efforts

A few other nonconventional FFF techniques have been developed at the microscale, but only a little information has been published on them. For example, the FFF-like SPLITT system, zero-field hydrodynamic chromatography, and a dual field thermal–electrical FFF system have been demonstrated on the microscale. These techniques hold a lot of promise as they expand the FFF applications and provide researchers with more versatile instrumentation.

A microfabricated thermal–electrical system that is capable of imparting both thermal and electrical fields simultaneously using a single instrument was demonstrated on nanoparticles. This system improved retention by 20% with the use of a thermal field (15°C temperature difference) in conjunction with the electrical field [70]. This device design also demonstrated better channel wall material selection can increase the temperature drop across the μ-ThFFF channel.

A technique for minimizing end effects in FFF channels was uniquely demonstrated in a microscale FFF system [79]. For this work, microstructures in the triangular inlet end-pieces were used to distribute sample uniformly across the breadth of the microchannel. A comprehensive fluid flow analysis was used to optimize the size, shape, and locations of the microstructures. Particle dispersion analysis and experimental plate height measurements showed close to a 50% reduction in total plate height for common FFF operating conditions.

### *10.1. Microscale Split-Flow Thin Fractionation*

Split flow thin fractionation (SPLITT) cell is a separation technique very similar to FFF in principle and geometry [80]. Unlike FFF, SPLITT operates continuously and separates the sample stream into two fractions. Since SPLITT can operate continuously, it is ideal for use in applications where high throughput is required.

SPLITT channels have two inlets to introduce the sample and carrier solutions respectively, and two outlets to elute the separated fractions as shown in Fig. 17. The inlet flow splitter prevents unwanted mixing of the two inlet streams and forces the sample input line into a thin stream along one wall of the channel and thus forming the inlet splitting plane (ISP). The applied field perpendicular to the flow of the solution forces susceptible particles across the ISP towards the opposite wall, inducing a binary separation based on the particles' susceptibility to the applied field. The system in Fig. 17 is an example of transport mode SPLITT where the higher transport rates of the black particles allows them to cross the ISP and elute through the lower outlet. For those familiar with microfluidics, this system is similar to the H-filter in principle [81], but adds an external field instead of a concentration gradient as the driving force, which gives SPLITT the possibility of high power separations with higher resolution.

The inherent capability of the system to divide a sample based on a property of the sample particles provides the impetus for sequential or serial separation to yield high volume separation of a complex multicomponent sample. As a first step towards achieving this goal, a microfabricated electrical SPLITT system was fabricated and characterized using polystyrene nanoparticles [80]. Micromachining techniques were used to fabricate a 40 μm thick (channel height) and 2 cm long channel on a glass substrate with gold electrodes in a method similar to that for μ-ElFFF. The microchannel was realized using SU-8 and bonded to another glass substrate using UV curable adhesive.

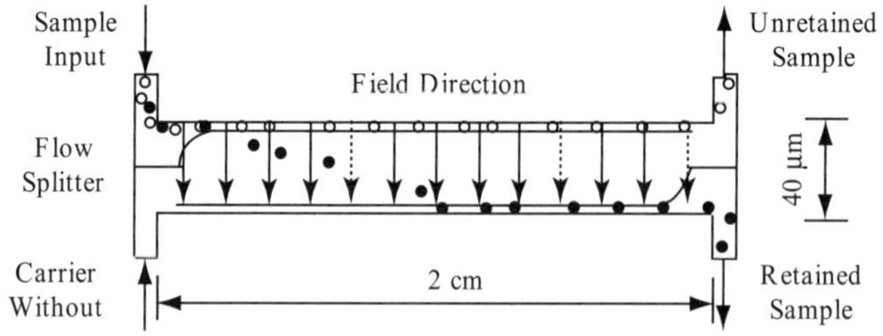


**Fig. 17.** Schematic diagram of SPLITT system with input and output connections for sample and carrier flow with field direction and illustration of the separation mechanism

Characterization of the system was done using a mixture of 108 nm ($\mu$ of $2.47 \times 10^{-4}$ cm$^2$ V$^{-1}$ s$^{-1}$) and 220 nm ($\mu$ of $4.48\times10^{-4}$ cm$^2$ V$^{-1}$ s$^{-1}$) diameter polystyrene particles with amine surface groups. SEM images of the eluted particles showed 94% 220 nm particles in one outlet and 70% 108 nm particles in other outlet, a high number selectivity separation at only 1.2 V across the channel. A logical solution to improve the resolution of 108 nm particle sample would be to pass it through another SPLITT channel and adjust the voltage to further separate out 220 nm particles from this stream.

The microfabricated electrical SPLITT system showed potential as a powerful separation technique even with only a 1% effective field due to double layer effects. Serial and parallel combinations of SPLITT channels should result in high resolution and high throughput separations of nanoparticles. In electrical SPLITT system it is relatively easy to control the electrical field in designated areas when compared to thermal or magnetic systems. The ability to program each electrical SPLITT channel separately should result in a tunable resolution and should increase the robustness of the system. There is still significant room for determining optimized channel dimensions for a variety of sample sizes and types in microscale electrical SPLITT channels.

### 10.2. Microscale Hydrodynamic Chromatography

Hydrodynamic chromatography (HDC) is similar to FFF and is carried out in a channel identical to that used in FFF, but no field is applied. The channel thickness is comparable to the analyte size. A particle size based separation is induced due to the exclusion of particles from areas near the channel walls, just as in steric mode separations in FFF. Larger analytes experience higher velocity flow lines compared to the smaller analytes that can move closer to the wall where flow velocity is a minimum. Figure 18 shows how particles in a hydrodynamic chromatography channel are limited in their motion.

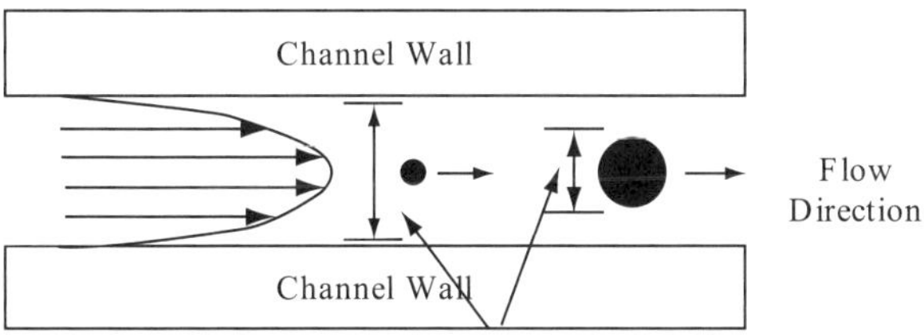


**Fig. 18.** Hydrodynamic chromatography channel showing the parabolic velocity profile and the regions where a particle can be located

Retention in hydrodynamic chromatography is discussed in several papers [82, 83]. A simple relation between residence time and analyte size is

$$\tau = (1 + B\lambda - C\lambda^2)^{-1}, \tag{40}$$

where $\tau$ is the retention time and $\lambda$ is the nondimensional size of the sample with respect to channel height. $B$ and $C$ are geometry-dependent constants. Several correction factors have been suggested to this basic theory to accommodate for effects such as hydrodynamic interactions with the wall, high velocity-induced inertia effects, and electrostatic or electrokinetic lift forces.

Channel dimensions are predominantly determined by the size of the sample being analyzed. For efficient separation, the channel height should be comparable to the analyte size thus restricting the geometrical dimensions of the HDC channel. This requires thin channels to interrogate smaller samples as is shown in a demonstration of on-chip HDC.

Silicon micromachining was used to realize 1 μm deep, 1,000 μm wide and 8 cm long channels in a 100 mm Si wafer with a 1 μm thick, thermally grown silicon dioxide. Fusion bonding was used to bond the silicon surface with polished Pyrex glass to enclose the channels. Anodic bonding of a Kovar piece with soldered HPLC tubing was used to connect to the input and output ports obtained by powder blasting through the Pyrex. The very small channel volume required on-chip sample injection and UV detection to prevent any extracolumn band broadening that otherwise would have severely affected the separation efficiency.

The distinct separation of particle plugs can be seen in Fig. 19 as the sample travels along the length of the channel [83]. A narrower channel results in better resolution for the smaller sized nanoparticles, likely since band broadening can be associated with the transition from a point injection to a wide channel. Other work showed the separation of proteins and dextran to show biomedical applications of the HDC chip [82].

It is clear that HDC is capable of inducing separation at microscale dimensions without the presence of any external field. Since there is no reliance on an external field, this technique is insensitive to any physiochemical property of the sample and a purely size-based separation can be achieved. Thus, analysis of different sample types using the same instrument should be feasible. Drawbacks of this technique include the requirement for a long separation channel and the limited particle size range that can be separated by a given channel. In addition, a mixture of particles a large size range will result in poor resolution for the smaller particles if the channel is designed to match the larger particles.

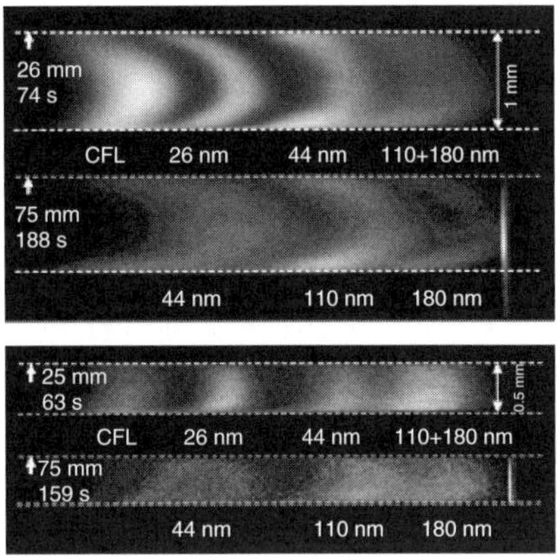


**Fig. 19.** Top view of the HDC separation of polystyrene nanoparticles and fluorescent marker in different sized microchannels in a silicon-pyrex chip. Reprinted with permission from Blom et al. [83]. Copyright (2003) American Chemical Society

## 11. Nanoscale FFF

With the recent emphasis on nanotechnology and the development of techniques for nanoscale manufacturing, the question of how FFF might scale to nanometer dimensions comes up. A few researchers have already begun to answer this question, but the fields applied in these systems are usually self generating and not applied externally. Also, electroosmotic flow, instead of pressure-driven flow, tends to be used at nanoscale dimensions, which changes some of the physical processes generating separation.

In a typical electrokinetic separation, such as capillary electrophoresis, the electric field is applied in the direction of the separation in a channel with a significant surface charge on the walls and without any pressure-driven flow. When the dimensions of a channel become comparable to the thickness of the electrical double layer, a particle is subjected to a transverse electrostatic field and a nonuniform flow profile when compared to the plug flow typically associated with electro-osmotic flow. This setup in a nanoscale electrokinetic channel rapidly begins to approach that in a typical FFF channel. Because of the surface charge on the walls, the electrical double layer spreads across the entire height of the electrokinetic nanochannel with most cations approaching the negatively charged glass walls and anions overlapping in the middle of the channel. This charge distribution across the height of the nanochannel results in nonuniform (approaching parabolic) flow velocity distribution while also exerting a lateral electro-migrative force on the sample. Particles in the channel will be distributed by physics similar to those in normal electrical FFF. The axial migration of the sample is assisted by the main potential gradient along the length of the channel. The main advantages of nanoscale channels over normal electrical FFF is the smaller zone broadening and the ability to operate with high ionic strength carriers. These nanochannels also have the same charge on every wall and so the gradient is between the walls and the center of the channel rather than between the two walls. Thus, every wall could be considered an accumulation wall.

Figure 20 shows images of experimental results showing the separation of fluorescein and bodipy molecules with valence charge ($z_s$) of –2 and –1 respectively. The FFF-like separation mechanism can be observed as the charge difference of the species results in a distinct retardation in the electromigration of the particles. One important fact to be noticed is the better separation in case of 2-μm deep channel than a 40 nm channel [84, 85]. Extensive numerical modeling of FFF in these types of nanochannels has also been performed [86].

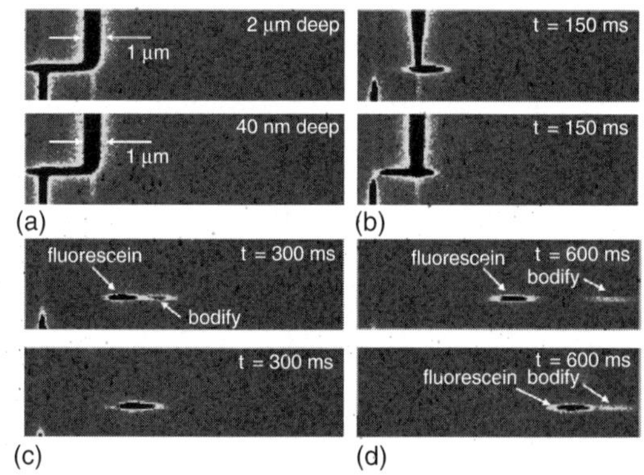


**Fig. 20.** Separation of fluorescein ($z_s$ = –2) and bodipy ($z_s$= –1) in channels with depths of 2 μm and 40 nm. The top image in each pair is for a channel of 1 μm width × 2 μm depth with an average liquid velocity of 25 μm s$^{-1}$. The bottom image in each pair is for a channel of 1 μm width × 40 nm depth with an average liquid velocity of 25 μm s$^{-1}$. Images are taken at (**a**) injection (time, $t$=0), (**b**) $t$=150 ms, (**c**) $t$=300 ms and (**d**) $t$=600 ms. Reprinted with permission from pennathur and Santiago [85]. Copyright (2005) American Chemical Society

## 12. Conclusion

Microscale FFF has shown significant progress since it was first published in early 1997 [87]. Various microscale FFF systems have been used to analyze nanoparticles, DNA, proteins, cells, viruses, polymers, and other materials. A summary of applications is provided in Table 3. Several research groups are now actively exploring applications for these technologies. Since many formats of FFF show significant scaling advantages, there is an opportunity for improvement in these technologies as they are miniaturized. With the advent of the first nanoscale systems, the final potential of these systems is just becoming evident. These FFF channels have also been integrated into more complex analysis systems, and have the potential to be integral components of a lab-on-a-chip system because of their simple operation and easy tuning for specific applications. There are also opportunities to investigate subtechniques in microscale FFF such as magnetic, flow, or sedimentation FFF, which have not yet been explored.

**Table 3.** Summary table for microscale field-flow fractionation systems

| FFF subtype | Physiochemical properties | Applications |
|---|---|---|
| Electrical | Size, electrophoretic mobility | Cells and organelles, bacteria and viral separations, characterization of emulsions, liposomes, protein adsorption |
| Thermal | Size, thermal diffusion coefficient | Separation of dissolved and suspended polymers, polymer and silica nanoparticle analysis |
| Cyclical electrical | Electrophoretic mobility | Biopolymer separations and zeta potential measurements |
| Dielectrophoresis | Dielectric permittivity, size | Cell Separation and dielectric property measurements and cancer cell separation |
| Electrical SPLITT | Size, electrophoretic mobility | High-throughput nanoparticle purification, proteins and starch, clay, viruses, spores, bacteria |
| Asymmetrical flow | Density, size | Proteins, DNA, polymers, cells, micro and nanoparticles |
| Hydrodynamic chromatography | Size | Large macromolecule without any charge requirement |
| Acoustic | Size, density, or compressibility | Macromolecules and nanoparticle separations |

Overall, significant promise has been shown, but substantial work needs to be completed before these techniques will be applied broadly.

## Acknowledgements

The authors acknowledge funding from the State of Utah Center of Excellence program for some of this work.

## References

1. Dittrich P, Tachikawa K, Manz A (2006) Micro total analysis systems. Latest advancements and trends. Anal. Chem. 78: 3887–3907
2. Wang Y, Choi M, Han J (2004) Two-dimensional protein separation with advanced sample and buffer isolation using microfluidic valves. Anal. Chem. 76: 4426–4431
3. Xiao D, Van Le T, Wirth M (2004) Surface modification of the channels of poly(dimethylsiloxane) microfluidic chips with polyacrylamide for fast electrophoretic separations of proteins. Anal. Chem. 76: 2055–2061
4. Araz M, Lee C, Lal A (2003) Ultrasonic separation in microfluidic capillaries. Proc. IEEE Ultrason. Symp. 1: 1066–1069
5. Harris N, Hill M, Beeby S, Shen Y, White N, Hawkes J, Coakley W (2003) Silicon microfluidic ultrasonic separator. Sens. Actuator. B Chem. 95: 425–434
6. Yang S, Zahn J (2004) Particle separation in microfluidic channels using flow rate control. In Proceedings of the ASME Fluids Engineering Division 2004, 127–132
7. Yamada M, Seki M (2006) Microfluidic particle sorter employing flow splitting and recombining. Anal. Chem. 78: 1357–1362
8. Tessier F, Slater G (2002) Strategies for the separation of polyelectrolytes based on non-linear dynamics and entropic ratchets in a simple microfluidic device Appl. Phys. Mat. Sci. Process. 75: 285–291
9. Han J, Craighead H (2000) From microfluidics to nanofluidics: DNA separation using nanofluidic entropic trap array device. Proc. SPIE 4177: 42–48
10. Skulan A, Barrett L, Singh A, Cummings E, Fiechtner G (2005) Fabrication and analysis of spatially uniform field electrokinetic flow devices: Theory and experiment. Anal. Chem. 77: 6790–6797
11. Bousse L (1999) Electrokinetic microfluidic systems. Proc. SPIE 3875: 2–8
12. Ramsey J, Collins G (2005) Integrated microfluidic device for solid-phase extraction coupled to micellar electrokinetic chromatography separation. Anal. Chem. 77: 6664–6670

13. Wang P, Gao J, Lee C (2002) High-resolution chiral separation using microfluidics-based membrane chromatography J. Chromatogr. A 942: 115–122
14. Inglis D, Riehn R, Austin R, Sturm J (2004) Continuous microfluidic immunomagnetic cell separation. Appl. Phys. Lett. 85: 5093–5095
15. Pekas N, Granger M, Tondra M, Popple A, Porter M (2005) Magnetic particle diverter in an integrated microfluidic format. J. Magn. Magn. Mater. 293: 584–588
16. Buch J, Kimball C, Rosenberger F, Highsmith Jr. W, DeVoe D, Lee C (2004) DNA mutation detection in a polymer microfluidic network using temperature gradient gel electrophoresis. Anal. Chem. 76: 874–881
17. Terray A, Hart S, Kuhn K, Arnold J (2004) Optical chromatography in a PDMS microfluidic environment. Proc. SPIE 5514: 695–703
18. Kanagasabapathi T, Dalton C, Kaler K (2005) An integrated PDMS microfluidic device for dielectrophoretic separation of malignant cells. Proc. 3rd Int. Conf. Microchannels Minichannels PART B: 411–418
19. Li Y, Dalton C, Said H, Kaler K (2005) An integrated microfluidic dielectrophoretic (DEP) cell fractionation system. Proc. 3rd Int. Conf. on Microchannels and Minichannels PART B: 403-410
20. Giddings J (1968) Non-equilibrium theory of field-flow fractionation J. Chem. Phys. 81–85:49
21. Giddings J, Caldwell K (1989) Field-flow fractionation. In Rossiter B, Hamilton J (eds) Physical Methods of Chemistry Vol IIIB. Wiley, New York
22. Schimpf M (2000) Resolution and fractionating power. In Schimpf M, Caldwell K, Giddings J (eds) Field-flow fractionation handbook. Wiley-Interscience, New York
23. Giddings J, Schure M (1987) Theoretical analysis of edge effects in field-flow fractionation. Chem. Eng. Sci. 42: 1471–1479
24. Hovingh M, Thompson G, Giddings J (1970) Column parameters in thermal field-flow fractionation. Anal. Chem. 42: 195–203
25. Tri N, Caldwell K, Beckett R (2000) Development of electrical field-flow fractionation. Anal. Chem. 72:1823–1829
26. Davis J (2000) Band broadening and plate height. In Schimpf M, Caldwell K, Giddings J (eds) Field-Flow fractionation handbook. Wiley-Interscience, New York
27. Giddings J (1968) Nonequilibrium theory of field-flow fractionation. J. Chem. Phys. 49: 81
28. Giddings J (1993) Micro-FFF: Theoretical and practical aspects of reducing the dimensions of field-flow fractionation channels. J. Microcolumn. Sep. 5: 497–503
29. Sant H, Gale B (2006) Geometric scaling effects on instrumental plate height in field-flow fractionation. J. Chromatogr. 1104: 282–290.
30. Gale B, Besser R, Papautsky I, Brazzle J, Frazier A (2000) Packaging of biomedical analysis systems. In Proceedings of the Advanced Technology Workshop (ATW) for MEMS and Microsystem Packaging and Integration, Orlando, FL, November 10–12, 2000

31. Kang D, Moon M (2004) Miniaturization of frit inlet asymmetrical flow field-flow fractionation. Anal. Chem. 76: 3851–3855
32. Gale B (1999) Scaling effects in a microfabricated electrical field-flow fractionation system with integrated detector, Ph.D. Dissertation, University of Utah, Utah
33. Gale B, Caldwell K, Frazier A (1998) A micromachined electrical field-flow fractionation (μ-EFFF) system. IEEE Trans. Biomed. Eng. 45:1459–1469
34. Caldwell K, Gao Y (1993) Electrical field-flow fractionation in particle separation. 1. Monodisperse standards. Anal. Chem. 65: 1784–1772
35. Gale B, Caldwell K, Frazier A (2000) Blood and protein separations using a micromachined electrical field-flow fractionation system. In Proceedings of the MicroTAS 2000, Enschede, Netherlands, 399–402
36. Gale B, Caldwell K, Frazier A (2001) Geometric scaling effects in electrical field-flow fractionation. 1. Theoretical analysis. Anal. Chem. 73: 2345–2352
37. Gale B, Caldwell K, Frazier A (2002) Geometric scaling effects in electrical field-flow fractionation. 2. Experimental results. Anal. Chem. 74: 1024–1030
38. Chen Z, Chauhan A (2005) DNA separations by EFFF in a microchannel. J. Colloid Interface Sci. 285: 834–844
39. Palkar S, Schure M (1997) Mechanistic study of electrical field-flow fractionation. II: The effect of sample conductivity on retention. Anal. Chem. 69: 3230–3238
40. Kantak A, Merugu S, Gale B (2006) Particle size and electric field effects in cyclical electrical field-flow fractionation. Electrophoresis 27: 2833–2843
41. Lao A, Trau D, Hsing I (2002) Miniaturized flow fractionation device assisted by a pulsed electric field for nanoparticle separation. Anal. Chem. 74: 5364–5369
42. Sant H, Gale B (2004) Flexible coupling of a waveguide detector with a microscale field-flow fractionation device. Proc. SPIE 5345: 250–257
43. Bartholomeusz D, Boutte R, Andrade J (2005) Xurography: Rapid prototyping of microstructures using a cutting plotter. JMEMS 14: 1364–1374
44. Bard A, Faulkner L (2001) Electrochemical methods: Fundamentals and applications. Wiley, New York
45. Kantak A, Merugu S, Gale B (2006) Carrier ionic strength effects in cyclical electrical field flow fractionation. Anal. Chem. 78: 2557–2564
46. Gale B, Caldwell K, Frazier A (1998) Electrical conductivity particle detector for use in biological and chemical micro-analysis systems. Proc. SPIE 3515: 230–242
47. Gale B, Frazier A (1999) Electrical impedance spectroscopy particle detector for use in microanalysis systems. Proc. SPIE 3877: 190–201
48. Graff M, Frazier A (2006) Resonance light scattering (RLS) detection of nanoparticle separations in a microelectrical field-flow fractionation system. IEEE Trans. Nanotech. 5: 8–13
49. Giddings, JC (1986) Cyclical field field-flow fractionation: A new method based on transport rates. Anal. Chem. 58: 2052–2056

50. Lee S, Myers M, Beckett R, Giddings J (1988) Experimental observation of steric transition phenomena in sedimentation field-flow fractionation. Anal. Chem. 60: 1129–1135
51. Gale B, Merugu S (2005) Cyclical electrical field flow fractionation. Electrophoresis 26: 1623–1632
52. Kantak A, Merugu S, Gale B (2006) Characterization of a microscale cyclical electrical field flow fractionation system. Lab Chip 6: 645–654
53. Biernacki J, Vyas N (2005) A one-dimensional transient model of electrical field flow fractionation. Electrophoresis 26: 18–27
54. Chen Z, Chauhan A (2005) Separation of charged colloids by a combination of pulsating lateral electric fields and poiseuille flow in a 2D channel J. Colloid Interface Sci. 282: 212–222
55. Peterrson M, Giddings J (1984) Sep. Sci. Technol. 19: 307
56 Lyklema J (1995) Fundamentals of Interface and Colloid Science: Vol. II: Solid-Liquid Interfaces. Academic Press Inc, San Diego
57. Yang J, Huang Y, Wang X, Becker F, Gascoyne P (1999) Cell separation on microfabricated electrodes using dielectrophoretic/gravitational field-flow fractionation. Anal. Chem. 71: 911–918
58. Wang X, Yang J, Huang Y, Vykoukai J, Becker F, Gascoyne P (2000) Cell separation by dielectrophoretic field-flow fractionation. Anal. Chem. 72: 832–839
59. Wang X, Vykoukai J, Becker F, Gascoyne P (1998) Separation of polystyrene microbeads using dielectrophoretic/gravitational field-flow fractionation. Biophys. J. 74: 2689–2701
60. Huang Y, Wang X, Becker F, Gascoyne P (1997) Introducing dielectrophoresis as a new force field for field-flow fractionation. Biophys. J. 73: 1118–1129
61. Wang X, Vykoukal J, Becker F, Gascoyne P (1998) Separation of polystyrene microbeads using dielectrophoretic/gravitational field-flow fractionation. Biophys. J. 74: 2689–2701
62. Becker F, Wang X, Huang Y, Pethig R, Vykoukal J, Gascoyne P (1995) Separation of human breast cancer cells from blood by differential dielectric affinity. Proc. Nat. Acad. Sci. 92: 860–864
63. Yang J, Huang Y, Wang X, Becker F, Gascoyne P (2000) Differential analysis of human leukocytes by dielectrophoretic field-flow fractionation. Biophys. J. 78: 2680–2689
64 Duhr S, Arduini S, Brauna D (2004) Thermophoresis of DNA determined by microfuidic fluorescence. Eur. Phys. J. E 15: 277–286
65 Janca J (1988) Field-flow fractionation: Analysis of macromolecules and particles, Marcel Dekker, New York
66. Edwards T, Gale B, Frazier A (2002) A microfabricated thermal field flow fractionation system. Anal. Chem. 74: 1211–1216
67. Janca J (2002) Micro-channel thermal field-flow fractionation: New challenges in analysis of macromolecules and particles. J. Liq. Chrom. Relat. Techn. 25: 683–704

68. Janca J, Ananieva I, Menshikova A, Evseeva T, Dupak J (2004) Effect of channel width on the retention of colloidal particles in polarization, steric, and focussing micro-thermal field-flow fractionation. J. Chromatogr. A 1046: 167–173
69. Bargiel S, Dziuban J, Gorecka-Drzazga A (2004) A micromachined system for the separation of molecules using thermal field-flow fractionation method. Sensor. Actuator. Phys. 110: 328–335
70. Sant H, Gale B (2001) A microfabricated thermal electric field flow fractionation system. Proc MicroTAS 2001 563–564
71. Kang D, Moon M (2004) Miniaturization of frit inlet asymmetrical flow field-flow fractionation. Anal. Chem. 76: 3851–3855
72. Semenov S, Maslow K (1988) Acoustic field-flow fractionation. J. Chromatogr. 446: 151–156
73. Tri N, Beckett R (1999) Ultrasonic forces for particle separations. In Proceedings of the Eighth International Symposium of Field-Flow Fractionation, Paris, France, September 6–8, 1999
74. Beckett R (2001) The use of acoustic forces of thin channel separations. In Proceedings of the Ninth International Symposium on Field-Flow Fractionation, Golden, CO, June 26–29, 2001
75. Tri N, Beckett R (2001) Acoustic field-flow fractionation for particle separation. In Cazes J (eds) Encyclopedia of Chromatography. Marcel Dekker, New York
76. Martin M, Williams P (1992) Theoretical basis of field-flow fractionation. In Dondi F and Guiochort G (eds.) Theoretical advancement in chromatography and related separation techniques. NATO ASI Series C: Mathematical and Physical Sciences Kluwer, Dordrecht, The Netherlands, 383: 534
77. Edwards T, Frazier B (2004) A micro acoustic field-flow fractionation system (μ-AcFFF) for nano-scale separations. In Proceedings of the MicroTAS 2004
78. Edward T (2005) Microfabricated acoustic and thermal field-flow fractionation systems, Ph.D. Dissertation, Georgia Institute of Technology, May 2005
79. Sant H, Kim J, Gale B (2006) Reduction of end-effect induced zone broadening in field-flow fractionation channels. Anal. Chem. 78: 7978–7985
80. Narayanan N, Saldanha A, Gale B (2005) A microfabricated electrical SPLITT system. Lab Chip 6: 105–114
81. Brody J, Yager P (1997) Diffusion-based extraction in a microfabricated device. Sens. Actuator. Phys. 58: 13–18
82. Chmela E, Tijssen R, Blom M, Gardeniers H, Van den berg A (2002) A chip system for size separation of macromolecules and particles by hydrodynamic chromatography. Anal. Chem. 74: 3470–3475
83. Blom M, Chmela E, Oosterbroek R, Tijssen R, Van den berg A (2003) On-chip hydrodynamic chromatography separation and detection of nanoparticles and biomolecules. Anal. Chem. 75: 6781–6768
84. Pennathur S, Santiago J (2005) Electrokinetic transport in nanochannels. 1. Theory. Anal. Chem. 77: 6772–6781

85. Pennathur S, Santiago J (2005) Electrokinetic transport in nanochannels. 2. Experiments. Anal. Chem. 77: 6778–6789
86. Griffiths SK, Nilson RH (2006) Charged species transport, separation and dispersion in nano-scale channels: Autogenous electric field-flow fractionation. Anal. Chem. 78: 8134–8141
87. Gale B, Frazier A, Caldwell K (1997) A micromachined electrical field-flow fractionation system. In Proceedings of the tenth IEEE International Workshop on Micro Electro Mechanical Systems (MEMS '97), Nagoya, Japan 317–322

# Chapter 13
# Nucleic Acid Amplification in Microsystems

Götz Münchow and Klaus-Stefan Drese

Institut für Mikrotechnik Mainz GmbH
Fluidics and Simulation Department
Carl-Zeiss-Straße 18-20, D-55129 Mainz
Germany

## 1. General Elements of Amplification

One of the most basic sources of information in biology is nucleic acid. While the nature of DNA (desoxyribonucleic acid) is comparable to a library where all details are stored, RNA (ribonucleic acid) is more involved in information transfer. In the century of information technology, it is only natural to analyze both kinds of nucleic acids for identifying predispositions (reading DNA) as well as the actual status of a biological system (reading RNA). However, the usually very low number of nucleic acid molecules, e.g., only two DNA strands of a single gene within a mammalian cell, lies beyond detection limits as long as detection techniques do not reach single molecule sensitivity. Moreover, the effort needed for electrical, optical, or chemical signal amplification is so high that such a technique cannot be applied to numerous applications, especially in the field of point of care testing. Hence, the solution is not improving detection limits, but performing target amplification.

For the nucleic acid amplification to occur, some type of copying machine is required, because the sequence of a gene is a priori not known and currently there exists no sufficiently cheap method to read the information that would enable a subsequent synthesization in a more or less chemical approach. At this point, biology helps by providing different kinds of biological copying mechanisms. As a consequence, most amplification strategies imitate microbiological processes, e.g., TMA (transcription mediated amplification) has its equivalent in reverse transcription, HDA (helicase dependent amplification) in DNA unwinding, SDA (strand displacement amplification) in excision DNA repair, Q-beta-replicase amplification in bacteriophage replication, etc. [1].

In this sense, NASBA (nucleic acid sequence based amplification) is the in vitro version of the natural replication of retroviral RNA, formerly known as 3SR (self-sustaining sequence replication) [2, 3]. The reaction takes place at 42°C and is initialized by binding a primer, a short oligonucleotide strand that serves as a starting point, to the RNA-template. The enzyme reverse transcriptase translates the RNA into a complementary DNA (cDNA) and the RNA of the hybrid is digested by RNAse H. Afterwards, a second primer binds to the previously generated single-stranded cDNA and is extended by the polymerase activity of the reverse transcriptase. In this step, a new promoter is also generated at which the T7 polymerase binds and synthesizes about 40–100 new RNA strands per double-stranded DNA. From this point on, the above steps continue in a cyclic manner leading to an amplification of up to 12 orders of magnitude within 90 min. Since the product of one cycle provides the starting templates for the following cycle, the amplification is in principle exponentially. However, because of depletion effects and generation of side products, this is only true in the beginning. Later, a linear behavior is found that finally saturates [4].

In contrast to isothermal protocols like NASBA, there exists another group of thermal cycling methods with its most prominent representative named PCR (polymerase chain reaction). During PCR, specific DNA sequences are amplified. In the first step, the DNA double helix is denatured at about 95°C, meaning that the DNA is separated into two single strands. In the second step, the temperature is lowered to 40–65°C and two primers, about 17–28 bp long, specifically bind to the single-stranded DNA. During the last step of the cycle, these primers are extended by the enzyme Taq polymerase, cf. Fig. 1. The Taq polymerase is a thermostable polymerase first isolated from *Thermus aquaticus*, a bacterium that lives in hot springs, and is able to generate a double strand from initially one single strand. Hence, the number of DNA strands increases exponentially with the number of cycles. As for NASBA, the exponential behavior finally saturates leading to an amplification of up to 12 orders of magnitude.

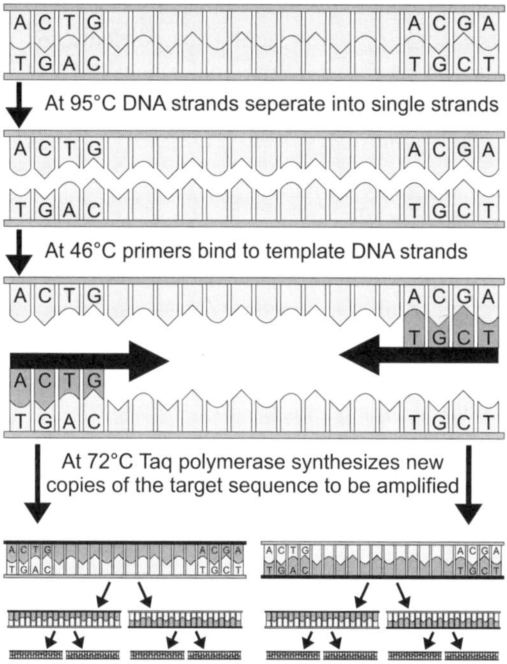


**Fig. 1.** Principle of the polymerase chain reaction

A possible classification is described by Zhao who identified three different types of amplification processes [1]. The first one is target based, meaning that the target gene of interest is amplified. The second one is probe based, relying on the amplification of a probe analog to a gene sequence. The third type is based on signal amplification where the specificity is not achieved during amplification but prior to it, e.g., by antibody-antigen interaction as a specific binding event known from immuno-PCR. A fourth set of amplification methods, which was not mentioned by Zhao, are unspecific amplifications like the whole genome amplification. Zhao decided not to include this class because unspecific amplification is mainly used as a preparative and not as a diagnostic step unless it is followed by specific assay steps.

Some of these different amplification methods are described in more detail later in this chapter. Nevertheless, it should be noted that the most widely used chemical procedures are too expensive for most amplification methods to be considered economically responsible, except for the fact that they may offer very sensitive analytical assays.

For the detection of amplified nucleic acids, principally most of the microtechnological detection methods can be applied, e.g., optical (fluorescence), electrochemical, gravimetrical (mechanical), or magnetic. These detection methods can be divided into two classes. The first class is based on labels, e.g., fluorescence or magnetic detection. The second class works label-free, e.g., electrochemistry and gravimetry. However, even though detection without labels is possible, labels are primarily used as they provide an increased specificity. One of the best examples is real-time amplification using fluorescent dyes, which indicate the total amount of DNA. The counterpart to this class of detection methods is the endpoint detection. The main representatives here are electrophoresis, including slab gel and capillary electrophoresis, and microarrays. During electrophoresis, the end-point product is separated by size followed by a detection using one of the above-mentioned sensor methods. In the case of microarrays, immobilized oligonucleotides are prepared while the amplified single-stranded DNA is brought into contact for hybridization and final detection. However, in the case of unspecific amplifications, a specific response reaction, e.g., hybridization on an array, is carried out subsequently.

## 2. Micro–Macro Comparison

By definition, the only element characterizing micro- or nanosystems is their typical length scale. Therefore, direct effects due to downscaling are discussed at first, which, compared to macroscopic systems, lead to generic differences. Further but indirect effects also occur in microfluidic designs and have to be considered, e.g., enhancement of unspecific binding. The third type of differences represents new effects unknown in macroscale, e.g., quantum effects. Since only a macroscopically established process is downscaled, these effects are irrelevant here. Below, typical length scales are identified and followed by a discussion relating to volumetric, two-dimensional, and one-dimensional effects.

### *2.1. Typical Length Scales*

For the direct comparison, a 96-well microtiter plate is chosen as a macroscopic reference system. Here the well-to-well distance is 9 mm. Since the walls have a finite thickness and there is usually additional spacing between neighboring walls, a typical length of 4 mm is defined. This leads to a typical volume of 64 μL. To keep the discussion mathematically simple, for the microsystem a typical length scale of 400 μm is set, which corresponds to a volume of 64 nL. If not mentioned otherwise, this comparison between the 4 mm and the 400 μm is taken as the standard example for the dissections below. The reader should notice that 400 μm is not the lower limit for miniaturized biosystems at all and might find microfluidic systems with typical length scales that are one or two orders of magnitude smaller, i.e., 40 μm or 4 μm resulting in 64 pL or 64 fL. However, the limit is given by the biology itself, e.g., cell sizes are in the order of 2–10 μm, and because of the small size of enzymes and DNA, approximately a few nanometers, nucleic acid amplification processes also allow nanometer-sized systems.

### 2.2. Volumetric Effects

Volumetric effects show cubic dependence on system dimensions. Supposing that the concentrations of the used reagents could be kept constant, which is usually not possible, the amount of reagents for a single assay has to be reduced by a factor of 1,000 with respect to the above example. However, using an optical real-time assay, the fluorescence signal intensity is also reduced by three orders of magnitude. But microsystems are usually integrated solutions in contrast to our macroscopic example. This leads to none generic differences that are introduced by special microsystem designs. Components reducing the loss of fluorescent light can be easily integrated and provide a signal enhancement of about one order of magnitude. On the other hand, only by an improved focusing of the excitation light source, the amount of excitation light emitted from fluorescence labels can be additionally increased by a factor of 100. This means that, although the signal intensity is probably reduced in microsystems, integrated components and modified setups can help to reduce the negative light intensity effect.

Assuming that the ratio between active fluidic volume and thermally controlled surrounding material is kept fixed, the volumetric downscaling would result in a reduction of power consumption by the same factor of 1,000. This kind of difference is again not generic, however. As a result of manufacturability and structural stability requirements, the wall thicknesses do not decrease that much during miniaturization. Consequently, the above argumentation cannot be followed. It is obvious that the power consumption will decrease by miniaturization, even if it is in a linear or quadratic rather than in a cubic way. Furthermore, some microsystems have taken the approach of stationary temperature fields, where the fluid itself is transported to individual temperature zones. Because of such a different technological approach, the scaling cannot be applied for estimating power consumption or material changes during the course of miniaturization.

We have assumed above that the concentrations of reagents stay constant, but in many applications the total number of molecules becomes relevant, e.g., in single cell analysis, only two DNA copies for each gene are present. In return, the concentration of such reagents increases by a factor of 1,000 during miniaturization. With regard to the above mentioned example of a single cell, the calculated concentration of the target-DNA, or any other two molecules, is 0.1 nM when choosing a well of 4 μm in size. Since detection limits are often said to be in femto- or ato-Molar range, it could be assumed that detection does not seem to be a critical factor in miniaturized systems. But in this case, not the reagent concentration itself but the total number of molecules defines the detection limits. So, the consideration of the total number of molecules helps to avoid false routes of developments, which may occur when only reagent concentrations are taken into account.

### 2.3. Surface Effects

Surface effects scale quadratically with the characteristic system size and the most obvious example of this is that the number of spots on an array decreases by a factor of 100 if the spot pitch is kept the same and only the total surface area of the array is reduced. Another example is the signal intensity of a "surface sensor," e.g., an electrochemical sensor, which decreases by two orders of magnitude once its size is reduced by only one order of magnitude. However, the signal intensity loss of an electrochemical microsensor due to downscaling is much less if the corresponding current is not only dependent on the total surface area of the sensor, but also on the thickness of the diffusion layer. In those cases, the above scaling argument becomes invalid because the signal intensity is no longer dependent on the entire area. The two initial examples have been chosen because in most cases the amplification is followed by detection. Again, picking up the point of reagent consumption, the miniaturization allows a saving of a factor of 100 relating to the surface chemistry compared to a factor of 1,000 relating to the volume chemistry. As a consequence, the problem of unspecific binding also becomes more important. Since this is one of the biggest challenges for miniaturized amplification assays, it is scrutinized here in more detail.

Let us assume a molecule of 40 nm in size that finds a binding site every 400 nm on average. A further assumption shall be that equilibrium is completely on the binding site, thus all molecules remain bound once they have reached one. In the macroscopic example, $6 \times 10^8$ molecules can bind to the surface that leads to a concentration of 16 pM. In the microsystem example, there exists only $6 \times 10^6$ binding sites, but the corresponding concentration would be 160 pM. One may argue that a change by one order of magnitude is not all that dramatic. But this one order of magnitude is due to scaling only. In terms of the macroscopic example, the reagents are usually transported by pipetting robots and, therefore, all long distances are covered by mechanical and not by fluidic movement. Also, the amplification is carried out without any fluid movement. Hence, the surface that is exposed to the reagents for unspecific binding remains virtually constant, within a factor of 2 or 3 relating to the total surface area of a single well. In contrast, in nearly all microamplification systems, the reagents are mixed and brought to the temperature control zones via fluidic movement [5–8]. Thus, the surface presented to the reagents is increased by two orders of magnitude only by transport processes. In this case, the amount of absorbed molecules is the same as in macroscopic devices, which leads to a concentration of 16 nM compared to 16 pM as in the case of the macroscopic example. For even smaller systems (4 μm), the concentration can further increase reaching a concentration of 16 mM.

### 2.4. Linear, Timescale and Other Effects

In contrast to amplification, transport and separation processes show a linear dependence on the characteristic system size. Furthermore, during miniaturization, dissipation processes usually increase quadratically divided by a typical length that, on its part, decreases linearly. Thereby, dissipation processes, e.g., diffusion of mass, diffusion of momentum, better known as liquid friction, and heat conduction, become more important with decreasing length and finally dominate. Good examples are the drastically reduced power consumption for heating and the existence of laminar flow instead of turbulent flow. In contrast, inertia effects are volume- dependent and become less important during miniaturization.

Looking at the increase of dissipative processes, an example of timescale reduction due to miniaturization is the fast heat exchange. For the scaling case specified above, a speedup of a factor of 100 is reached if the heat transfer is conduction limited. In the case of power limitation, a reduced heating time by a factor of 1,000 is achieved. However, the first case is more realistic. Using these properties, Kopp et al. presented an ultrafast amplification and even reached the point where the reaction rate of the polymerase became the limiting factor [5]. A second example for this accelerated dissipation is the increased diffusion of molecules from the centre of the fluid toward the walls and vice-versa. In the case of thermal transport, the process is accelerated by a factor of 100 relating to the initial scaling example. A typical diffusion constant for a midsize molecule in water is $1 \times 10^{-10}\ m^2 s^{-1}$, which leads to a typical diffusion time of about 11 h in the macroscopic system. This means that in the case of an absorbing wall, unspecific binding can be neglected because amplification is generally finished within 1h. Compared to the microsystem, the diffusion time is already reduced to approximately 7 min. Thus, even in fast micro-PCR systems, molecules will certainly meet the walls. In addition, this process is accelerated by the convective exchange close to the wall. In smaller systems, like cavities of 40 μm, which are interesting for single cell analysis, the timescale for the diffusion is further reduced to 4 s. Accordingly, supported by higher surface-to-volume ratios and decreased diffusion times, unspecific binding remains a big challenge in microscopic approaches.

At the end of this micro–macro comparison it is noteworthy that the amplification reactions themselves, and in particular their rates, remain unchanged via miniaturization as long as the concentration and the quality of temperature control stay constant. As a consequence, all isothermal amplification processes that do not rely on surface reactions cannot be accelerated. For example, a NASBA microchip is as fast as its macroscopic counterpart. This means that, very likely, only PCR can reach a strong acceleration via miniaturization toward ultrafast amplification.

The comparison between microscopic and macroscopic approaches can be summarized in four points:

1. Microsystems have the potential for faster operations if the macroscopic amplification is transport limited.
2. Unspecific binding is the biggest additional challenge introduced by miniaturization.
3. The signal intensity in real time approaches is reduced.
4. The amount of reagents needed is also reduced.

## 3. Microfluidic Realization Methods

The most commonly used and automated PCR systems have been realized in macrosystems and there are several attempts to fulfill the demands of PCR protocols. Generally, those so-called PCR thermal cyclers consist of a specific number of well adapters, a heating block designed for rapid and accurate temperature change, and electronic controllers. The wells can either be tubes or well plates, which are commonly comprised of 96 or 384 wells. In recent years, 1,536-well plates have also been increasingly used. These were already introduced by Sasaki and co-workers in 1997, and each well maintained a reaction volume below 1 μL [9]. To heat and cool the heating block Peltier cells, which can reach heating rates up to

6°K $s^{-1}$ and lower cooling rates of approximately 1.5–4.0°K $s^{-1}$, are generally used. Using tempered air circulating around capillaries, in which the samples are placed, those heating and cooling rates can also be increased up to 20°K $s^{-1}$ (LightCycler®, Roche Diagnostics). The main advantage of macro-PCR thermal cyclers is the parallelization – meaning the number of amplifications performed at the same time. But they are time-consuming and less applicable for point of care diagnostics or integrated systems.

Because of the price of enzymes, the costs associated with each reaction are still quite high. Accordingly, one main intention to realize microfluidic PCR systems is the reduction of sample amounts. Furthermore, because of their smaller mass compared to macroscopic systems, miniaturized systems show less thermal inertia. As discussed before, this property allows increased heating and cooling rates, which lead to drastically reduced total amplification times.

### 3.1. Substrates

In the first half of the 1990s, miniaturized biomedical and microfluidic systems, besides intelligent sensors for automotive applications, developed as one of the most concrete and vivid application fields of microsystem technologies. At this time, most microsystem solutions consisted of silicon and the manufacturing was adopted from established processes of the microelectronics industry. This is the reason that the first classification of miniaturized PCR systems was made according to the chosen material, silicon or glass [5, 10]. Later, polymers like polymethylmethacrylate (PMMA), polydimethylsiloxane (PDMS), and polycarbonate (PC) as alternative substrates continued to increase in popularity. Today, there is still an ongoing search for further materials, mainly polymers, which can be used for PCR microsystems. An overview of typically used materials is given in Table 1.

In most silicon-based PCR systems, the temperature control system of a chamber consists of integrated thin film heaters and temperature sensors, which are controlled by a personal computer or an electronic control unit [11]. The integrated setup implies that after every use these chambers have to be cleaned to avoid crosscontamination.

**Table 1.** Material used for miniaturized PCR systems

| Material | References |
|---|---|
| Silicon | 11–17 |
| Glass | 5, 18–32 |
| Silicon + glass | 8, 10, 13, 32–54 |
| Polyimide | 55 |
| PTFE (polytetrafluoroethylene, Teflon) | 56–60 |
| PC (polycarbonate) | 61–68 |
| PMMA (polymethylmethacrylate) | 7, 69–71 |
| PDMS (polydimethylsiloxane) | 72–82 |
| COC (cyclic-olefin-copolymer) | 7, 83, 84 |
| PET (polyethylene-terephthalate) | 85 |
| PP (polypropylene) | 86, 87 |
| SU-8 | 88, 89 |
| Fiber optic face plates | 90 |

Because of high fabrication costs of silicon, a general trend toward low-cost polymeric disposables for micrototal analysis systems (μTASs) is observed, and therefore toward setups where the temperature control system and the PCR chamber itself are not monolithically integrated. The advantages over silicon or glass are lower material costs, flexibility, transparence to visible and UV light, easy molding, and improved biocompatibility due to possible surface modifications. Rapid prototyping can be performed by micromilling (down to 50 μm) and laser ablation (down to 5 μm), which permit frequent design changes. After a design freeze, microinjection molding and hot embossing (for thermoplastics) or casting (for elastomers like PDMS) are good choices for inexpensive mass fabrication. On the other hand, the use of polymers also implies some disadvantages, such as low thermal and electrical conductivity.

### *3.2. Types of Setup*

The first realizations of chip-based PCR systems were presented by Northrup et al. in 1993 and Wilding et al. in 1994 [10, 39]. In the sense of current integrated systems, these microfabricated and silicon-based reaction wells were not fully developed microfluidic systems, but already combined the advantages of miniaturized devices.

The different setup types can be divided into three main groups: the amplification in wells where a sample is injected into a chamber and remains at this position after the amplification process is completed; the amplification by continuous flow-through devices where one or multiple samples are transported through different temperature zones and collected at the channel outlet; and special realization methods that utilize, for example, buoyancy forces or electrowetting for sample transport.

### 3.2.1. Amplification in Wells

The basic microstructure of PCR wells consists of a heating element, a well in which the sample is injected, and a cover plate, cf. Fig. 2. To fabricate wells with integrated heaters and sensors, silicon and the long experience in using this material offer great possibilities and has been first utilized by Northrup et. al. [10]. Their device consisted of a reaction chamber, which was embedded in a silicon chip and a silicon nitride membrane serving as the chamber bottom, onto which a thin polysilicon heater was deposited. Finally, the complete chamber was covered by a glass slide and the sample was injected through polyethylene tubing. Up to the present, this basic structure and principle have been replicated and modified by numerous groups [11, 13, 35, 38].

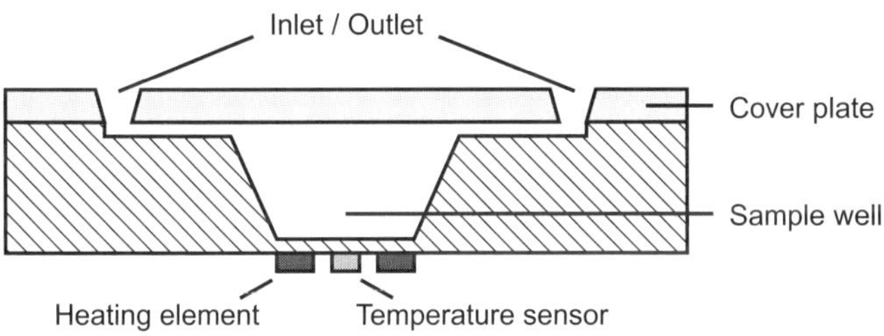


**Fig. 2.** DNA amplification in wells. First, the sample is injected though the inlet port and positioned inside the well. After the ports are sealed, the chip is heated and cooled according to the PCR protocol. The temperature is monitored by an integrated temperature sensor

These systems differ in the used material and layout. The total volume of realized chambers lies between 40 pL [90] and 50 μL [10] but most designs use wells between 1 and 10 μL [13, 34, 39, 55, 72], volumes that can still be handled with conventional laboratory equipment. In addition, the heat transfer and the temperature distribution of the devices during PCR process have been analyzed by FEM (finite element method) simulations in order to reach higher yields by optimizing the reaction conditions or to reduce the total time needed for amplifications [38, 43, 47, 69, 72, 88]. A fast amplification performed in a well has been presented by Yoon et al. [38]. By using additional cooling fins and a fan, they reached heating and cooling rates of about 36° K $s^{-1}$ and 22° K $s^{-1}$ and achieved 30 cycles within 3 min. Poser et al.

even reached heating and cooling rates of 80° K $s^{-1}$ and 40° K $s^{-1}$, respectively [13]. By such an acceleration of temperature changes, different groups reached a successful DNA amplification within 10 min or less [11, 69, 73]. It should be pointed out that nearly all presented microdevices require less than 60 min and that the typical temperature precision stated is below ± 0.2°C [13, 38, 44, 46, 49, 77].

Apart from bulky copper heating blocks, most microchambers are equipped with thin-film heaters and temperature sensors made of platinum [34], indium-tin-oxide [25], polysilicon [10, 12, 87], or tungsten [74]. The structures are in direct contact to the chamber and heat can be easily transferred due to thermal conductions. As a result of the low power consumption, a few systems can be operated using commercially available batteries [49, 73]. Additionally, systems based on noncontacting heat transfer, e.g., infrared [26, 55, 91] and induction heating [49], have been realized. In case of an infrared-mediated temperature control, Giordano et al. used a 50 W tungsten lamp, which was focused on the microchamber through a convex lens. A small T-type thermocouple was placed inside the chamber to measure the temperature directly. The total reagent volume was 1.7 μL and 15 cycles were completed within 240 s [55]. Fermér and co-workers performed PCR in tubes using microwaves demonstrating a further possible noncontacting heating source for miniaturized PCR systems [92].

Decreasing the reaction volume required for amplifications and increasing the number of simultaneously performed amplifications are the main efforts to maximize time and cost efficiencies. Zou et al. designed a 16-well system [85]. The reaction chip was made of low-cost polymer and mounted on a thermally nonconductive substrate for thermal insulation between the wells. The substrate itself accommodated independent silicon heating blocks that allowed thermal multiplexing, i.e., different temperature settings for each well at the same time, showing a thermal cross talk as low as 0.2%. Nagai et al. presented parallel amplifications of high density [14–16]. A single test fragment was amplified in 10,000 reaction wells, which were 86 pL in volume and etched in a single silicon wafer. One of the most impressive combinations of sample reduction and massive parallelization has been achieved by Leamon et al. [90]. Their well plate permitted a simultaneous amplification of 300,000 discrete PCR reactions in wells of 40 pL in volume. To provide the PCR product for subsequent processing, it was immobilized to a DNA capture bead. In the case of a uniform temperature protocol for all wells, the achieved temperature uniformity of microarrays can be measured by liquid crystals thermometry, also applicable for characterizing any other heated microfluidic devices [93].

Because PCR is a temperature-controlled and enzyme-catalyzed process, which requires temperatures up to 95°C, the chamber sealing is essential to avoid evaporation of reagents. For well-based systems, the chambers are often sealed by tape [34, 36, 72], rubber [38], or oil droplets [11, 26, 31, 49, 69, 94]. An advanced yet feasible system was presented by Oh and coworkers [43]. It included four PCR chambers and a world-to-chip microfluidic interface with built-in valves. The sample was injected by a pipette supported by pipette tip guides. Once the samples were loaded on to the microchambers, plastic fittings with rubber sheets were slid to the "sealing mode" position, closing the inlets without dead volume. Phase change valves, another technique for PCR chamber sealing, have been presented by Chen et al. [95]. For this purpose, the used liquid plug can be thermally actuated and either be frozen or melted.

Another approach to perform PCR in microscale is the use of virtual reaction chambers as reported by Neuzil et al. [18]. Here, a 1 µL sample was placed on a glass plate and encapsulated with mineral oil to prevent evaporation of water, cf. Fig. 3. The heating and cooling rates were about $40\,K\,s^{-1}$ and $20\,K\,s^{-1}$, respectively. The surface of such a system has to be hydrophobic as well as oleophobic. Thereby, a small amount of liquid, aqueous or oily, will form a droplet that acts like a closed compartment. For additional transport of those droplets, Guttenberg et al. used surface acoustic waves generated by piezoelectric materials [96]. They first placed the PCR solution and the oil droplet separately on the chip and brought them into contact by moving them toward the same heater element. After amplification, the oil-covered sample droplet was joined with another additional oil droplet encapsulating a sample of concentrated hybridization buffer. As soon as the two aqueous solutions made contact inside the oil, a heater was set on temperature and the hybridization process started.

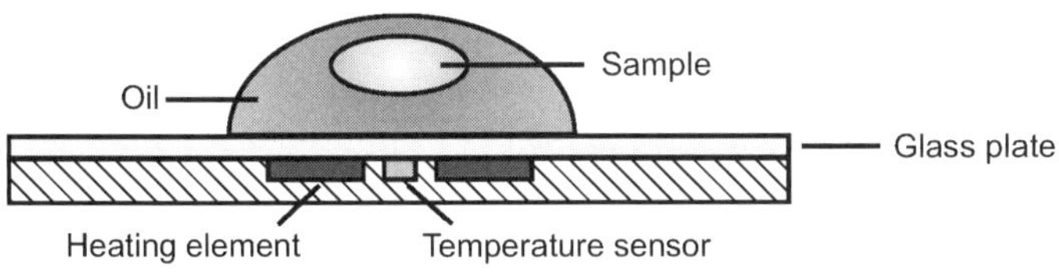


**Fig. 3.** Virtual reaction chamber. The PCR sample is covered with mineral oil and positioned above chip heater and temperature sensor. The glass plate [18] can also be replaced by other materials or setups [44]

### 3.2.2. Amplification by Continuous Flow-Through Devices

The second main group of miniaturized PCR systems relies on a continuous flow of reagents through three thermostatic zones representing the three steps of a PCR: denaturation, annealing, and extension. A simple system consisting of a 5 m long, thin capillary tube and a pump was already presented by Nakano et al. in 1994 establishing a basis for chip-based devices [59]. A 50 μL sample was introduced into the capillary and pumped through three different tempered oil baths in a loop. The duration of the three temperature steps was determined by the length of the capillary in each bath, the cross-section, and the flow rate, respectively. Additionally, a constant continuous flow was essential for the reproducibility of the residence time in each cycle. A similar system using a Teflon tube and immiscible organic liquid to separate the sample plugs was also performed by Curcio et al. [58]. Instead of using temperature baths, Park et al. presented a cylindrical assembly consisting of three equally divided thermostating copper blocks and a fused silica capillary, which wound around the block 33 times [24].

The first continuous flow-through PCR integrated on a chip was introduced by Kopp and his co-workers [5]. It consisted of a glass plate comprising etched microchannels in the form of 20 meanders representing the number of cycles. The chip was mounted on three copper heating blocks which were kept at constant temperature, cf. Fig. 4. Compared to amplifications in microwells, the heat inertia is reduced to the thermal mass of the sample itself, thus allowing high heating and cooling rates which reduce the time of amplification. The calculated time for achieving temperature equilibrium inside a channel of a cross section of 40 × 90 μm is less than

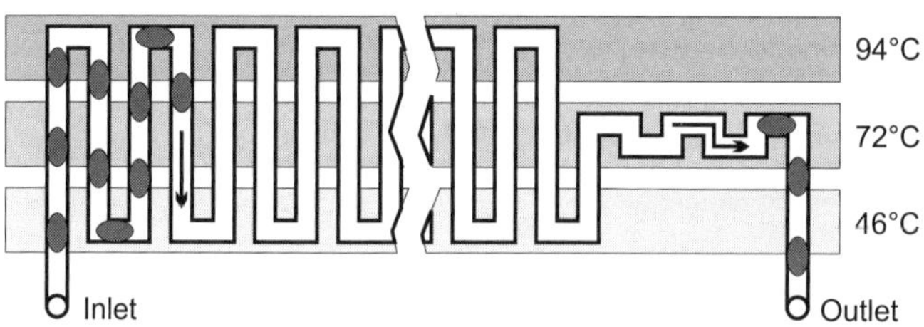


**Fig. 4.** Basic principle of DNA amplification within a continuous flow device on chip. The samples, denoted by dots, are introduced at the inlet and separated by, e.g., mineral oil for serial flow mode. Three individual heaters are maintained at constant temperatures providing adequate thermocycling conditions. The number of cycles is given by the number of windings and the final cycle is followed by an additional extension step

100 ms [5]. The total amplification time of different devices varies from 30 min [33, 41] down to a few minutes [5, 19, 63] depending on the number of cycles and flow rates.

In contrast to continuous flow, the serial flow mode offers the possibility to enhance the number of samples drastically. To create a serial flow, the samples can be separated by either gas or liquid. Gas–liquid systems pose more challenges concerning the realization of reproducible flow conditions [6]. Segmenting the samples by liquid, the interactions between the liquids themselves and between liquid and wall have to be considered. Once a sample passes a wettable surface, a thin film of it will be left at the wall. The remainder could crosscontaminate the next sample segment after reaching this area. Using a channel wall material with low wettability, the problem of crosscontamination can be reduced or almost eliminated.

The best way to prevent surface contacts of the sample is the use of a two-phase system consisting of immiscible liquids. Since the PCR mixture is mainly based on water, nonvolatile mineral oil, as mentioned above, or other types of organic liquids [97] have been used as a carrier liquid for sample droplets. When using a hydrophobized channel wall surface, the oil will form a thin layer. Thus the direct sample–wall contact is omitted and crosscontamination between individual sample droplets can be virtually excluded. A detailed examination concerning the problem of cross-contamination of droplet microreactors and plug-shape reactors has been presented by Barrett et al. [98]. Creating a two-phase and air bubble-free flow, a typical setup consists of two syringe pumps, one of which constantly provides the carrier liquid and one for the actual sample injection [33]. The final amplification products can either be analyzed by, for example, laser detection [58] or slab gel electrophoresis [5, 20, 33].

Further improvements of continuous flow-through PCR have been presented by Li and co-workers [23]. They developed a microchip including a local velocity control and a heating configuration for a more efficient DNA amplification. By varying the widths of the channel, the flow velocities of the samples can be regulated. Because of the modified channel shape, the transport time of a sample between two temperature zones as well as the time needed for temperature transition can be reduced. A detailed temperature analysis has been performed by Zhang et al. [32]. They compared a glass–glass bonding chip with a silicon–glass bonding chip using FEM simulations. Obeid et al. designed a modified chip adapted for both DNA amplification and reverse transcriptase-PCR analyzing RNA templates [19]. Besides silicon and glass, Schneegaß et al. also introduced a flow-through system made of the transparent elastomer PDMS with the benefit of a disposable unit [33]. To stabilize the flow and to prevent air bubble generation during the initial injection of the sample volume into the microchannel,

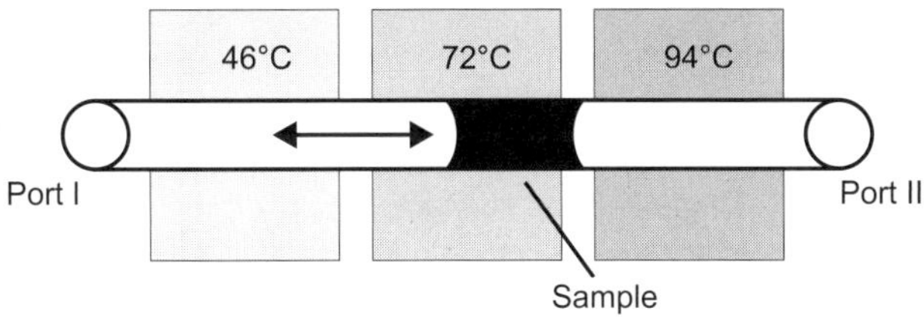


**Fig. 5.** Transport scheme of an oscillating PCR system. The sample is introduced into the channel and pumped back and forth. Three heaters are maintained at constant temperatures regarding the PCR protocol

On the basis of this concept, a microfluidic chip with an integrated peristaltic micropump has been introduced by Bu et al. [37]. Besides a theoretical model of the pump, a thermal analysis by FEM regarding the temperature uniformity and the location of the heaters has been provided. An oscillating-flow PCR chip based on silicon was presented by Wang and co-workers [8]. They also have theoretically analyzed and simulated the thermal performance of the chip. Chiou and co-workers presented a further actuation principle [60]. By applying different pressure values to the capillary outlets a 1 µl sample was transported between different heating zones. The capillary in which the sample was placed was completely filled with oil. Once the sample was loaded, the capillary was pressurized to 5 psi to prevent sample degassing. By applying a pressure increase of 2.5 psi to one capillary outlet, the sample plug was set in motion. The sample position was identified by additional photodiodes in combination with a laser beam that was coupled into the channel. Instead of three separated heating zones at constant temperatures, Cheng et al. introduced a sample plug that was shuttled along a temperature gradient [84]. The circular microfluidic chip comprising the channel structure was made of PMMA and clamped to a glass heater chip. The heater chip itself consisted of an outer and inner heater ring, which generated a radial temperature gradient of 2° K $mm^{-1}$. The denaturation, extension, and annealing segments of the microchannel were specifically placed in relation to the temperature gradient to provide a homogenous temperature distribution, cf. Fig. 6a.

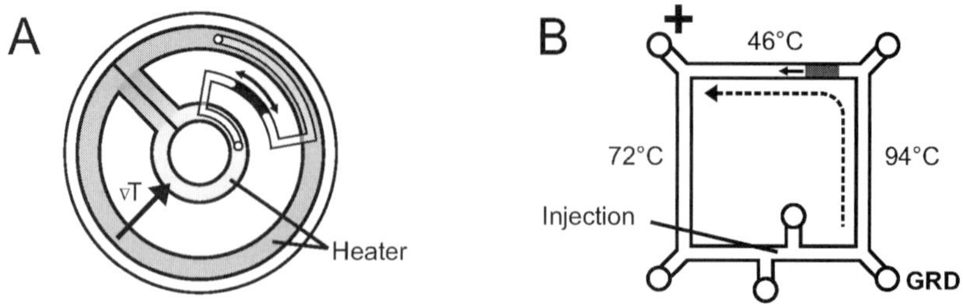


**Fig. 6.** (a) PCR chip with temperature gradient provided by one heating element. The sample is shuttled back and forth by a connected pump performing amplification. (b) Electrokinetically driven sample plug. After injection, the sample is transported by switching the voltage over to specific reservoirs positioned around the loop

Instead of pressure-driven flow for continuous flow-through or oscillating PCR systems, Chen et al. introduced a polycarbonate microchip using electronically driven pumping [68]. The chip consisted of one circular microchannel, 0.5 μL in volume, which was completely filled with a PCR reaction mix. The DNA templates were injected through a double T-injector by applying voltage to connected reservoirs. By switching the voltage to additional reservoirs, the plug that comprises the DNA target was transported around the loop through three independent temperature zones, cf. Fig. 6b. Completing 27 cycles required approximately 18 minutes at an electric field strength of 300 V $cm^{-1}$. A PCR system renouncing moving parts or external pumps has been presented by West and co-workers [89]. Here, AC magnetohydrodynamic actuation is used to circulate the PCR mixture through different temperature zones. Such a setup can be advantageous for pumping strong electrolytes or ionic liquids.

After Pollack et al. had demonstrated that droplets containing PCR reagents can be actuated by electrowetting without having a negative effect on a subsequent DNA amplification [100], Griffith et al. presented a computational approach to design a digital microfluidic system in the form of a PCR analysis array [101, 102]. The algorithm is supposed to coordinate the motions of hundreds of droplets simultaneously and chemical reactions accordingly. Furthermore, Zhang et al. presented a performance comparison with SystemC, a software program suitable for high level modeling of digital systems, between continuous flow systems (pressurized flow) and droplet-based systems based on electronically driven liquid handling, e.g., electrowetting [103]. The comparison, which takes into account the system throughput, system correction capacity, and system design complexity, results in higher performance and less setup complexity of the droplet-based microfluidic system.

Buoyancy forces allow fluid transport without any moving parts and can be induced by only one heating source placed inside a chamber. In this case, a constant temperature gradient creates laminar thermal convection circulating reagents along stationary paths, cf. Fig. 7a. The simplest way of convective polymerase chain reaction has been presented by Henning and Braun [94]. They preheated a chamber to annealing temperature and dipped an immersion heater into 20 µL of the reaction volume. The temperature of the heater stayed constant and generated a continuing recirculation of the reagents between the hot and the cold region, thereby amplifying DNA strands. Previously, Braun and co-workers introduced a similar PCR system [104]. Instead of using a simple heated wire, the thermal convection was induced by an infrared beam that was focused on the center of the chamber. The DNA was replicated 100,000-fold within 25 min.

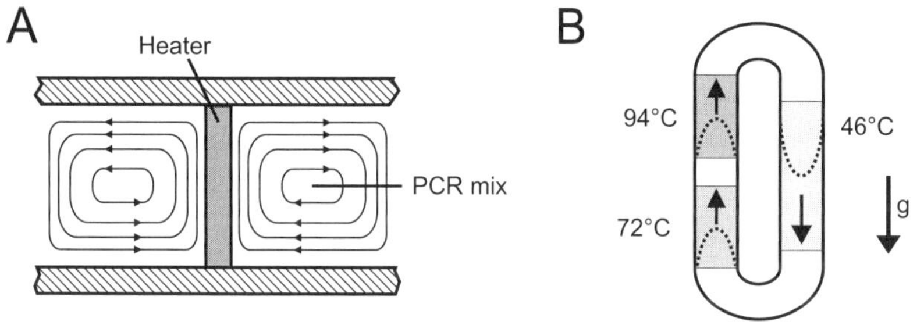


**Fig. 7.** (**a**) Convective flow within PCR chamber. Only one heating element is necessary to perform DNA amplification. (**b**) PCR chamber using buoyancy forces and gravity (g) to circulate reagents in a closed loop continuously

Krishnan et al. developed a similar system based on Rayleigh-Bérnard convection [70]. It consisted of two heating plates that were constantly heated up to 61°C (top plate) and 97°C (bottom plate). In between a 35-µL Rayleigh-Bérnard cell was placed. As soon as the fluid layer is heated from below, the density of the fluid near the bottom plate decreases compared to the fluid density in the upper part of the chamber. As a result, the top region of the fluid becomes unstable and convective motion occurs. This leads to circulating flow forming convection cells. Thus, amplification conditions were provided and also different chamber designs have been examined by simulation and flow visualization [105]. A different development was introduced by Wheeler et al. [86]. Their transportable convection-based thermocycler utilized natural convective forces to circulate PCR reagents in a closed loop. It consisted of two heating zones set at

94°C and 57°C, respectively. This device was later modified by Chen and coworkers [56]. They added a third heater responsible for the primer extension process, cf. Fig. 7b. Additionally, they presented some modeling in order to predict the fluid velocity and the temperature distribution around the loop using a one- and three-dimensional model. A comparable droplet-based set-up was presented by Walsh et al. [106] and Daly et al. [57]. It consisted of nanoliter sample droplets, which were filled into a loop, embedded in an immiscible carrier liquid. The heated carrier liquid was transported by active pumping and buoyancy forces dragged the sample droplets along the loop.

Further approaches toward miniaturized PCR systems are microreactors in the shape of water-in-oil emulsion [107] which have been used for, e.g., amplification of single DNA molecules by Nakano et al. [108]. Also, amplifications inside single cells have been reported for which the microfluidic technology could provide numerous application areas [109–112]. So far, most miniaturized systems have utilized polymerase chain reaction for target amplification. But an isothermal amplification, called NASBA, has also been presented in the form of a microfluidic system for real-time cancer marker detection [50, 83].

An overview of different realization methods is given in Table 2. Further review articles with regard to continuous flow-through systems [6] or PCR microsystems in general [121–127] can be found in the literature.

**Table 2.** Overview of different realization methods

| | |
|---|---|
| Microchamber PCR | 10–13, 17, 22, 25, 26, 28–31, 34–36, 38, 39, 42–44, 49, 53–55, 62, 64–67, 69, 71–73, 77–80, 82, 87, 88, 100, 113–117 |
| Open droplet PCR | 18, 96 |
| Continuous flow-through PCR | 5, 6, 19, 20, 23, 24, 27, 33, 40, 41, 45–47, 57, 58, 63, 74, 75, 118 |
| Oscillating PCR | 7, 8, 21, 37, 60, 84 |
| Convective PCR | 56, 70, 86, 94, 104, 106, 119 |
| Microarray PCR | 14, 15, 16, 61, 76, 85, 90, 120 |
| Electrokinetically driven PCR | 68 |

### 3.3. Surface Treatments

Although microfluidic systems for biochemical analysis offer the great possibility of reducing the sample amount and reagent consumption, the control of surface chemistry and protein adsorption becomes much more important, as already mentioned above. By decreasing the dimension of the device, the surface-to-volume ratio increases inversely proportional to its size (square-cube law). A fluidic system possessing a characteristic length of 1 m results in a surface-to-volume ratio of 1 $m^{-1}$. In the case of a length scale of 1 μm, the surface-to-volume ratio increases to $10^6$ $m^{-1}$. That means that the molecules in a microfluidic system are influenced by the wall surface nearly all the time. Target molecules or analytes may adsorb on-the-wall surfaces completely and nonspecifically [128, 129]. These bindings are mainly noncovalent and based on H-bonding, electrostatic, and hydrophobic interaction [130]. Accordingly, different contact angles of the used substrate material and sample can influence the protein adsorption [131]. Because of the size and three-dimensional structure of a protein, the wall absorbance can lead to physical or chemical transformations. These transformations can cause protein modifications that again may result in inactivity or reduced activity of the protein.

Protein adsorption and its prevention are particularly critical problems for DNA amplification in microsystems. Here, nonspecific adsorption can either lead to loss of reaction efficiency resulting in decreased yield or to no amplification at all. Different groups observed that especially Taq polymerase immobilized on the channel surfaces and denatured [7,52]. In this context, Erill et al. presented an analysis of different glass–silicon chips with respect to reagents binding [52] and Felbel et al. presented the inhibit behavior of silicon powder [132]. The loss of separation efficiencies, especially when using polymeric material, is also known from capillary electrophoresis devices [133].

To prevent protein adsorption, a distinction is drawn between static and dynamic wall coatings. The first type often consists of covalently linked coatings or coatings that are attached to the channel wall by covalent chemical bonds. But pretreatments based on different deposition techniques also have the character of static passivation. The static coatings are usually more time-consuming and complex. Often these types of coatings require several chemical reaction steps but provide a higher potential and better performance. The second type is dynamic coatings at which polymers or molecules, e.g., enzymes, adhere to the wall surface only by adsorption [134]. Dynamic coatings are attractive and easy to create. In most cases, a specific amount of coating material is dissolved in a fluid that is introduced into a chamber or flushed through a microchannel for some minutes.

Because of the weak binding energy of the adsorbed material, parts of the coating material can go again into solution during the PCR process. Dynamic coating can be also performed during DNA amplification itself by adding additional chemical substances to the PCR mixture. These additives can be used as a complementary protection against nonspecific adsorption of proteins without any former wall coating process. Additionally, definite types of additives can also stabilize the activity of Taq polymerase enzymes.

A well-known covalently linked static coating is the silanization of surfaces. Mostly silicon or glass-based miniaturized PCR systems have been coated by silanization. The process was first introduced by Hjertén and makes surfaces hydrophobic [135]. Silanization is performed at moderate temperature and pressure and is achieved by a self-assembly process from an organic solution of silane molecules. By wet chemical treatment, hydroxyl groups are formed on oxidized surfaces that act as covalent binding sites for alkylsilane. The inhibitory effect of silicon-based materials on DNA amplification was already mentioned by Northrup et al., who presented one of the first micro-PCR systems [10]. They found silanization as an adequate method to hold up reaction processes. This was also confirmed by Shoffner and co-workers [51], even though there have been doubts on sufficient reproducibility of efficiency [51,136]. Munro et al. presented an evaluation of different silica coatings, especially silanization, for PCR systems tested on reliability in electrophoretic microchips. The preferred coating consisted of a silanization with chlorodimethyloctylsilane dissolved in toluene and combined with a polymer layer of poly(vinylpyrrolidone) [137]. The silanization treatment was also applied to continuous flow-through amplification systems [5, 20, 23, 24] as well as oscillating systems [8]. Another method to create biocompatible environments is the use of silicon dioxide-coated silicon chips also allowing an DNA amplification in microchambers [14, 35, 36]. Static coating with polyethylene glycol (PEG) and its positive effect of preventing protein adsorption have been proven by different groups. Holmberg and co-workers have grafted PEG chains to polystyrene [138]. Österberg et al. used the same substrate material but, additionally, they compared PEG with dextran and differentiated between end-on and side-on polymer configurations [139]. They observed a more effective reduction of protein adsorption using end-on PEG and side-on dextran coatings. Furthermore, packing density seems to be more important in preventing protein adsorption than layer thickness. Surface modifications of PDMS have been performed by Hu and co-workers [140]. In a one-step procedure, they covalently linked different polymers to the channel walls of PDMS microdevices. A so far uncommon surface passivation has been presented by Shin et al. [75]. They

Nakayama et al. placed a fluorinated oil plug in front of the sample volume [81]. It helped to increase the internal pressure of the latter solution and led to stabilized flow conditions. Circumventing the restriction on the number of cycles, Liu et al. developed a rotary device, also made from PDMS [74]. A 12 nL sample was introduced into a loop and can be circulated by an integrated pump at a rate of 2–3 revolutions/min. Reaction temperatures were provided by three integrated heaters. A numeric simulation with regard to fluid transport mechanism was presented by Gui and Ren [99]. Instead of using a pressure-driven flow, they presented an alternative approach based on electrokinetic flow.

### 3.2.3. Special Realization Methods

The majority of microfluidic PCR systems is based either on wells or on continuous flow. But new approaches for DNA amplifications have also proven functional efficiency. Instead of pumping continuously only in one direction, a sample in the shape of a plug can be transported back and forth in an oscillating manner while the microchannel is mounted on three separately controlled temperature zones, cf. Fig. 5. The setting of different PCR protocols regarding plug transport, number of cycles, temperature ramps, residence times in each temperature zone, and, of course, the temperature of each zone itself can be easily performed by electronic control units and is not fixed by the design structure itself. Similar to continuous flow systems, the setup provides a rapid heat transfer from the channel walls to the plug due to the three independent thermostatic zones. Computer fluid dynamics (CFD) simulations of the heating and cooling process indicate a temperature equalization of the plug only within 80 ms [7]. This is not only caused by the short thermal diffusion path but also by the recirculating flow conditions within the plug, which speeds up the heat transfer by convection.

The oscillating plug flow system presented by Münchow and co-workers consisted of a structured polymeric chip that was placed on three heating blocks [7]. To avoid crosscontamination, it is used as a disposable. The sample plug was driven by a ferrofluidic actuator pneumatically connected to the PCR chip. It provided a precise, pulsation-free, and reliable micropump that is essential for an accurate positioning of the plug during cycling. A complete amplification including 40 cycles can be performed within 5 min. Instead, Auroux et al. used a glass capillary mounted on copper heating blocks [21]. The sample was sandwiched between two mineral oil plugs and transported back and forth using a syringe pump.

covered the PDMS wall surface of a PCR microchip with parylene. The polymer is known as an inert coating for medical devices and was used here to inhibit protein adsorption as well as sample evaporation.

In contrast, mainly three different substances have been associated with dynamic coatings of miniaturized PCR systems. The most frequently used is the protein bovine serum albumin (BSA), which sticks to almost all surfaces and is typically used as a blocking agent to reduce nonspecific binding in enzyme-linked immunosorbent assay (ELISA) systems [141]. It acts as a stabilizer for other proteins and enzymes and as protein base controls and calibrators. The molecular weight of BSA is 66 kDa, its density 1.36 g $cm^{-3}$ [142], and the isoelectric point approximately pH 4.9 [143]. The concentration of BSA solutions used for pretreatment of microchambers or channels varies from 2 mg $mL^{-1}$ [47] to 10 mg $mL^{-1}$ [64, 77]. Often these solutions also contain additional compounds like Tween [33, 64], polyethylene glycol (PEG) [94], or PCR buffer itself [33, 63]. However, either the solution is flushed through the device [63] or the microstructures are flooded for 2 [30] or up to 30 min before use [77]. An additional temper step for 30 min at 80°C during the BSA coating process has been performed by Zou et al. [47]. Not specifically examined for PCR systems but performed at different flow rates, Chun and co-workers studied the protein adsorption on PEG-modified surfaces [130]. It was determined that specific proteins were adsorbed more at higher flow rates compared to lower flow rates. This could be important for oscillating or continuous flow-through PCR systems. A reduced adsorption of Taq polymerase due to PEG pretreatment of polymeric microchannels was also presented by Münchow et al. and Hennig et al. [7, 94]. To create a more dense PEG molecular brush-like coating on PMMA surfaces, Bi and co-workers modified PEG with butyl methacrylate (BMA), which also led to less nonspecific protein adsorption [144]. A PEG vapor-phase coating of glass surfaces and its reducing effect on nonspecific bindings has been presented by Popat et al. [145]. Epoxy polymers coatings have been used for either capillary electrophoresis or microchamber PCR systems [31,146]. Ferrance et al. have coated their glass chambers by flushing them with 1M NaOH solution for 10 min and a subsequent flush with 0.4% (w/v) epoxy-polydimethylacrylamide (EPDMA) solution [31]. In systems based on glass, EPDMA is only effective as a pretreatment when adsorbed to the wall surface before use. The coating does not work when only dissolved in the PCR mixture itself [136].

Although dynamic coatings can be performed as an initial operation step, surface treatments can also be applied during the PCR process itself. Once again, BSA represents one of the most used PCR mixture additives preventing inhibition of PCR processes by dynamic coating. Compared to a pretreatment solution, the BSA concentration within the PCR mixture is reduced and varies between 0.2 ng $mL^{-1}$ [61] and 2.5 mg $mL^{-1}$ [46]. The average concentration lies between 0.1 and 0.5 mg $mL^{-1}$ [19, 56, 59, 62]. The addition of BSA has also been combined with Triton X-100, PEG, and Tween [19, 59, 62, 90]. For Erill et al., the addition of BSA became essential for obtaining a PCR product [46]. Kaack and co-workers presented a three times higher yield of PCR product once BSA was added [61] and Nagai et al. also reported a strong increase of fluorescence intensity corresponding to the amplified DNA amount after adding BSA [14]. Overall, the application of BSA as a surface coating has been presented for all kinds of miniaturized PCR systems consisting of silicon, glass, or polymers. However, polyvinylpyrrolidone (PVP) is also known as a supporter of PCR reactions and has been used for dynamic coating of glass. An applied concentration of 2.5% (w/v) in the PCR mixture leads to 78% of the product amount relative to the amount observed in standard polypropylene inserts used for conventional thermocyclers [136]. For $SiO_2$ microstructures, Krishnan et al. also make use of Teflon coating or magnesium added to the reaction mixture [53].

Not specifically examined for PCR but for continuous biochemical flow-through systems in serial mode (droplets within a carrier liquid), Roach and co-workers presented the use of surfactants influencing the protein adsorption at the liquid–liquid interface. The adsorption of proteins can be either prevented or induced [147].

Concluding, a further enhancement of preventing protein adsorption can be reached by the combination of surface pretreatment and PCR mixture additives [5, 11, 14, 20, 23, 24, 30, 33, 63, 64, 77]. A good overview with regard to polymer wall coatings is given by Horvath et al. [148].

### 3.4. Detection of Amplified DNA

Still, the most famous target detection procedure is to remove the reaction mixture from the microchip after amplification and analyze it by gel [7, 20, 33] or capillary electrophoresis (endpoint detection) [12, 55, 61]. Because of utilized sieving matrices, these techniques allow a separation by size and the ability to identify different DNA segments of a PCR product. Visualization is performed by intercalators, a type of dye molecule that binds to double-stranded DNA (dsDNA) and inserts itself into the DNA structure. After binding, the intercalators show a strong increase in fluorescence intensity. Because of the transparent nature of glass and different kinds of polymers, the use of intercalators also enables real-time detection during the amplification process itself [43, 96]. However, since the intercalators do not only bind to specific but also to nonspecific products, there will be no differentiation between them. The most popular examples for intercalators are ethidium bromide and SYBR® Green.

Besides the above-mentioned sequence-independent detection methods, there have been several approaches applying sequence-dependent detection methods. Mainly these methods utilize the ssDNA–ssDNA hybridization (ssDNA, single-stranded DNA). After DNA is heated to denaturation temperature and has formed single strands, these single strands will reform to double helices at regions of sequence complementarity after cooling. This technique has been used for real-time PCR [149] as well as for subsequent detections on microarrays. Another example for sequence-dependent real-time detection is the use of TaqMan® probes consisting of a fluorophore, an oligonucleotide, and a fluorescence quencher, which are added to the PCR mixture. They are designed to hybridize between the two primers, meaning that a third part of the target sequence is scrutinized. During PCR, when the Taq polymerase replicates a template on which a TaqMan® probe is bound, the 5′ nuclease activity of the polymerase cleaves the probe that leads to the separation of the quencher and fluorophore. The corresponding fluorescence signal, which is generated during the amplification and increased in each cycle, indicates the total amount of the target sequence. Further real-time PCR chemistries are Molecular Beacons and Scorpions®, all allowing multiple DNA species to be measured in the same sample (multiplex PCR). Therefore, fluorescence dyes with different emission spectra are attached to different probes.

A detection method in combination with capillary electrophoresis (CE), which is not based on fluorescence, has been presented by Abad-Villar et al. [150]. With an external contactless conductivity detector, they not only detected DNA but also peptides or proteins.

General overviews with regard to DNA detection are given by various authors [113, 121, 151], especially Kubista et al., who presented a broad overview of real-time polymerase chain reactions [152].

### 3.5. Integrated Micro-PCR Systems

The polymerase chain reaction is only one part in a row of steps toward a complete genetic analysis. In the first step, the nucleic acids have to be extracted from cells. In most cases, the released DNA strands have to be purified before subsequent amplification steps. After amplification, the PCR products have to be purified again before using them in CE or real-time detection systems. Depending on the assay, there are also differences in the number of necessary analytical steps. For example, in the case of a slab gel electrophoresis, the second purification step can often be omitted. In general, this kind of conventional genetic analysis lasts up to 12 h [153]. Using appropriate microsystems, this time can be reduced down to between 20 and 70 min. The great challenge of fully integrated microsystems is to combine various analytical steps which are prior and subsequential to the PCR process. The advantages of integrated systems are the reduction of analysis times, sample and reagents volumes, as well as elimination of manual transfer steps between individual analysis steps. Hence, integrated systems require microfluidic connections between different microfluidic units as well as chip-to-world connections for sample loading and removal. Additionally, the directed transport of the sample has to be carried out by devices such as micropumps and microvalves. Corresponding microsystems can be classified according to existing analytical steps, i.e., systems including a subsequent detection based on, e.g., separation or hybridization, systems including presteps like sample preparation, and finally fully integrated systems that include both. But even the embedding of microheaters and sensors, which was presented for most microchamber-based systems, represents the first step toward integrated PCR systems.

In most cases, PCR devices have been first extended by a detection unit. Here, one possibility is real-time fluorescence measurement during PCR in microchambers [18, 27, 43, 69, 154] or microarrays [14, 16, 76] via addition of a DNA fluorescence dye. An online laser-induced fluorescence detection system combined with a segmented continuous flow-through PCR system was presented by Curcio et al. [58]. The combination of PCR and a subsequent capillary electrophoresis is another alternative for an integrated detection and was already presented by Woolley et al. in 1996 [87]. Here, a PCR microchamber was attached to a CE chip made of glass. After 15 min of amplification, the sample was electrokinetically injected directly into the CE chip and analyzed within 120 s. A further example for integrated systems including PCR and CE has been performed by Lagally and co-workers [22]. Also, microfabricated heaters, temperature sensors, valves, and hydrophobic vents were included, which allowed fast heating and cooling rates and sensorless positioning of samples. The amplification and integrated detection of samples has been achieved within 15 min and was later prepared for pathogen detection and genotyping [29]. The PCR device for nanoliter samples presented by Burns et al. also included heaters and CE channels [54]. The sample loading itself was controlled by capillary forces, integrated hydrophobic stop valves, and air vents.

To reach a higher level of integration, optical components also find their way into integrated PCR systems. Schabmueller and co-workers [45] positioned a diode detector inside a PCR microchamber. An optical fibre guided the light of a HeNe laser into the chamber, and resulting fluorescence emission has been detected by a built-in detector. The integrated system presented by Huang and co-workers consists of not only a PCR chamber with a connected CE channel, but also a chip-integrated optical fiber for on-line detection [80]. Further optical detection systems, not associated with PCR but with microfluidic bioanalysis systems, have been presented by several authors [155–158]. An alternative label-free PCR product quantification, also applicable for integrated PCR systems, has been introduced by Hou and co-workers [159]. The microelectronic sensor is based on field effect measurements and can monitor the product concentration at various stages of PCR. The readout is similar to real-time fluorescent measurements.

Besides real-time fluorescence measurement inside a microchamber or subsequent capillary electrophoresis, hybridization is a further possibility to detect amplified DNA fragments and enable sequence-specific analysis. A polycarbonate PCR chip including such a hybridization channel was introduced by Liu et al. [65]. After successful PCR, the amplicons were pumped through an attached hybridization channel by a syringe pump where oligonucleotides were previously deposited. After hybridization, the readout can be performed by laser scanning operating with desired wavelengths. A hybridization system that was not based on fluorescence but on sequence-specific electrochemical detection was presented by Lee et al. [44]. After reaction, single-stranded and amplified target DNA hybridized to probes immobilized on a working electrode, that was also located inside the PCR microchamber. The binding of additional indicators led to an electrochemical transduction and to a final detection.

The highest level of integration is reached as soon as only the raw sample has been injected followed by final results after a specific time span. Hence, to minimize the number of analysis steps done off-chip, further tasks, especially sample preparation, have to be integrated. Yuen and co-workers designed a microdevice containing a filter structure inside the PCR microchamber, which retained white blood cells from whole blood samples with volumes below 3 μL [71]. After filtration, the chamber was flooded with PCR reagents to complete the cell isolation process and amplification was carried out. A dielectrophoresis (DEP) sample pretreatment enabling cell capture and simultaneous removal of PCR inhibiting components has also been integrated in PCR systems [88]. After purification, the cells were transported into the microreaction chamber and PCR was performed. A system consisting of a cell lysis reactor, different reservoirs for reagents, an active mixer, and PCR chamber was presented by Lee et al. [79]. After thermal lysis of the introduced cells, the extracted DNA samples as well as the PCR reagents were driven electroosmotically into the mixing section. By applying a high external voltage to shielded electrodes, the solution was electrokinetically mixed and driven into the PCR chamber where the final amplification was performed. One of the first PCR systems, which combined preceding as well as subsequent genetic analysis steps, was performed by Waters and co-workers in 1998 [115]. The chip design was based on the previous work of Woolley et al. but included a thermal step for cell lysis [87]. The PCR chamber was adapted to a CE chip. After amplification, the products were loaded electrokinetically into the separation path and an additional reservoir provided DNA markers for electrophoretic separation. A strong integrated system, which combined sample preparation from whole blood (including bead-based target cell capture, cell preconcentration, purification, and cell lysis), PCR and a final DNA microarray detection, was presented by Liu and co-workers [66]. Additional microfluidic mixers, valves, pumps, and heaters completed the self-contained analysis system. A different device but also containing multiple genetic analysis steps was presented by Anderson [67]. It allowed the extraction and enrichment of nucleic acids, amplification, and a final hybridization. The device was similar in size of a credit card and can handle more than 60 sequential operations. A capillary-based system offering an analysis directly from cheek cells has been presented by He and co-workers [113]. The device included all necessary DNA analysis steps followed by the separation and detection of the PCR product.

An overview of integrated systems is given in Table 3. Additional reviews can also be found in the literature [162–164].

**Table 3.** Overview of integrated PCR systems

| | |
|---|---|
| PCR + capillary electrophoresis | 29–31, 54, 78, 80, 82, 87, 115, 116 |
| PCR + fluorescence detection of PCR mixture (including real-time) | 18, 34, 43, 45, 58, 69, 96, 100, 104, 154, 160 |
| PCR + hybridization | 44, 65–67, 96 |
| PCR + other detection methods | 159 |
| sample preparation + PCR | 71, 79, 161 |
| sample preparation + PCR + detection | 66, 67, 115, 116 |

## 4. Alternative Protocols to PCR

Very often, amplification is equated with PCR. But PCR represents a class of amplification methods and does not, by far, include all amplification principles. In the following, the large field of PCR and further possibilities for amplifications are presented, without being exhaustive.

A variant of PCR, called *multiplex PCR*, enables the simultaneous amplification of many targets of interest in one reaction by using more than one pair of primers. First presented by Chamberlain et al. in 1988 [165], the method has been applied to many areas of DNA testing. Sometimes the target sequence is not well known. In these cases, degenerated primers, meaning a set of primers whose base sequence varies within the unknown sequence section of the target, are used. This technique is called degenerate oligonucleotide-primed PCR (*DOP PCR*) and can also be applied to whole genome amplification if random primers are used [166–168]. Human Alu sequences are short interspersed elements that are found at about one million sites in our genomes. If the goal of amplification is a genetic fingerprint of bands from an uncharacterized human DNA, these Alu sequences are amplified using PCR primers specific to them. The corresponding PCR is called *Alu-PCR* [169].

Another variation of amplification uses two pairs of primers, both of which belong to the same gene. The amplification starts, as usual, with only one primer pair amplifying a large amplicon. After 15–20 cycles, the second pair of primers (nested primers) is added. During the following 20–25 cycles, these primers bind within the first amplified amplicon and produce a second PCR product that will be shorter than the first one. As a result, this *nested PCR* has a higher specificity and sensitivity.

One last example of the variability is represented by the *asymmetric PCR*. This kind of PCR is used to preferentially amplify one strand of the original DNA. Compared to normal PCR, the concentration of primers for the chosen strand is significantly higher, or the second primer is not added at all. Accordingly, either the forward or backward reaction is highly enhanced, which is especially interesting for, for example, a subsequent microarray hybridization where the presence of only two complementary strands is ideal. It is noteworthy that the advantage of an exponential amplification disappears, at least in later cycles, after the limiting primer type has been consumed and the reaction passes into a linear amplification.

However, another central ingredient, the polymerase, also offers quite some variability. There are more than four different polymerases to choose from: Taq-, Tth-, Pwo-, and Pfu-polymerase. Tth- and Taq-polymerase show a high 5′-3′-activity in addition to a reverse transcriptase activity. This allows a good forward and backward amplification as well as RNA generation. Besides their 5′-3′-activity, Pwo- and Pfu-polymerase also possess proof-reading activity (5′-3′-exonuclease-activity). Thus, the choice of the polymerase type leads to differences in the performance that can be decisive for an application.

To prevent nonspecific amplification during the setup of PCR, hot-start polymerases have been developed. This special kind of polymerase is activated only after a certain temperature is exceeded. An alternative option is to use thermally activated primers developed by TriLink Bio Technologies Inc.

If the initial template is not DNA but RNA, the amplification process starts with the translation of RNA into cDNA using reverse transcriptase, e.g., MMLV-RTase, AMV-RTase, or Tth-polymerase. This type of PCR is called reverse transcriptase PCR (*RT-PCR*).

For whole genome amplification, in particular, there are a number of other methods, such as ligase-mediated PCR (*LM-PCR*) [170–173], linker-adaptor PCR (*LA-PCR*) [174–177], primer extension preamplification PCR (*PEP-PCR*) [169, 178, 179] or *Tagged-PCR* [180], all of which use different approaches according to primers. In the first two approaches, sequences are ligated to the DNA allowing specific primers to anneal. The other two approaches, PEP-PCR and Tagged-PCR, are characterized by special random primers. However, it must be noted that the above-mentioned methods are not only limited to whole genome amplifications.

An alternative amplification method to PCR, which also uses high temperature for denaturation and a lower temperature for hybridization, is the ligase chain reaction (*LCR*). This reaction achieves a similar sensitivity compared to PCR [181]. After denaturation, two oligonucleotides, called probes, are hybridized to the single-stranded DNA and then ligated by a thermally stable ligase. After ligation, the cycle is completed and starts again with the denaturation step. A more elaborated method is multiplex ligation probe amplification (*MLPA*) [182]. Utilizing the same denaturation, hybridization, and ligation steps, the difference lies in the applied probes ending with a sequence that is used for PCR amplification afterward. One of them also includes a stuffer sequence between the primer and hybridization sequence and is used for product determination. Therefore, the assay can easily be multiplexed using different hybridization sequences all equipped with stuffer sequences of varying length but with the same primer. The reason for this is that a subsequent PCR can be performed with only one primer pair and the different amplified and ligated probes can be separated using capillary electrophoresis.

So far, all presented amplification methods use thermal cycling by some means, but isothermal methods have also been developed. The first isothermal transcription-based amplification system was described by Guatelli imitating the retroviral replication [183]. Two commercial variants of this basic methodology are the nucleic acid sequence-based amplification (*NASBA*) and the transcription-mediated amplification (*TMA*) [184]. NASBA requires Ribonuclease H as well as reverse transcriptase and T7-RNA-polymerase, whereas TMA only needs two different kinds of enzymes, namely reverse transcriptase and T7-RNA-polymerase. Furthermore, these two methods are very similar since they both rely on the same principal amplifying RNA. Compared to RT-PCR methods, the amount of product shows an exponential increase leading to much higher amplifications [185].

Another isothermal amplification method is the strand displacement amplification (*SDA*) [186]. This method works at 40°C and results in an amplification of eight orders of magnitude within 2 h. The technique is based upon the ability of restriction endonuclease to nick a hemimodified recognition site and of exonuclease-deficient DNA polymerase to displace a downstream DNA strand during replication. SDA is a cyclic process in which modified target sequences are repeatedly nicked. The strand is displaced and the displaced strands are primed [186–188].

Besides the above-mentioned methods using PCR for whole genome amplification, the strand displacement, used by SDA, can also be applied for whole genome amplification. The technique is called multiple strand displacement (*MSD*) and uses phi29, a highly processive mesophillic DNA polymerase, to replicate up to 100,000 bp without dissociating from the genomic DNA [189, 190]. The first step is a random priming followed by the phi29 DNA polymerase movement along the DNA template displacing the complementary strand at 30°C. The displaced strand becomes a template for replication allowing for the generation of high yields of high-molecular-weight DNA.

Supported by its processivity, the same phi29 polymerase can also be used for the rolling cycle amplification (*RCA*) amplifying circular DNA, e.g., plasmids [191, 192]. The use of a single primer leads to a linear amplification that can be transformed into an exponential method by adding a second primer.

During helicase-dependent amplification (*HDA*), the DNA is denaturated by the enzyme helicase operating at a constant temperature of 37°C. After a primer pair anneals to each end of the target DNA, it is extended by a polymerase leading again to double-stranded DNA representing the starting point for the next cycle [193].

A further amplification method is the Q-beta-replicase amplification using the homonymous enzyme q-beta-replicase. This enzyme has the unique feature not to amplify a target DNA sequence but to replicate RNA having a specific secondary structure. This RNA hybrid consists of a sequence called MDV-1 and a target probe complementary to the DNA sequence of interest. After binding to the DNA, the RNA hybrid contorts activating the replicase that starts to copy the MDV-1 and its tag-along RNA sequence [194]. As the RNA products serve as new templates, this method allows a target amplification of nine orders of magnitude within 30 min.

All these amplification methods presented demonstrate the wide range of possibilities to choose from. Methods that have not been mentioned here are methods normally based on the variation of process parameters like temperature, e.g., touchdown PCR [195], and methods subsequent to the amplification, like melting curve analysis or microarrays.

Corresponding to the aforementioned scaling properties nonisothermal methods always have a certain potential to be accelerated by miniaturization. On the other hand, for many applications, e.g., point of care, and relating to power consumption of micro devices, isothermal amplifications might be more advantageous compared to thermal cycling.

## 5. Conclusion

Since the first miniaturized PCR devices were presented in the early 1990s, miniaturized PCR systems represent one of the most important applications of lab-on-chip technologies. Showing a huge range of variations, they have developed from feasibility studies toward microchip-based products [116, 149, 160, 196] and handheld systems [197–199]. Continuous flow devices in particular have been commercialized by different companies, e.g., IQuum Inc., Thermal Gradient Inc., and microfluidic ChipShop GmbH. Current efforts primarily aim bioanalytical point of care systems, which also comprises additional analytical steps, such as sample preparation and target detection. But, in addition to that, the automation of microfluidic control including sample loading and liquid handling (e.g., fluid transport, mixing, and valving) has to be addressed to obtain reliable and easy-to-use systems in the future. Micro-PCR systems, which allow reduction of amplification time and less reagent consumption, can provide, together with other analytical processes, a new level of standardization of high throughput gene analysis.

## References


1. Zhao R (2005) From single cell gene-based diagnostics to diagnostic genomics: Current applications and future perspectives. Clin Lab Sci 18: 254–262
2. Hodinka RL (1998) The clinical utility of viral quantitation using molecular methods. Clin Diagn Virol 10: 25–47
3. Mueller JD, Pütz B, Höfler H. (1997) Self-sustained sequence replication (3SR): An alternative to PCR. Histochem Cell Biol 108: 431–437
4. Wuesten JJAM, Carpay WM, Oosterlaken TAM, van Zuijlen MCA, van de Wiel PA (2002) Principles of quantitation of viral loads using nucleic acid sequence-based amplification in combination with homogeneous detection using molecular beacons. Nucleic Acids Res 30: e26
5. Kopp MU, deMello AJ, Manz A (1998) Chemical amplification: Continuous-flow PCR on a chip. Science 280: 1046–1048
6. Schneegaß I, Köhler JM (2001) Flow-through polymerase chain reactions in chip thermocyclers. Rev Mol Biotechnol 82: 101–121
7. Münchow G, Dadic D, Doffing F, Hardt S, Drese KS (2005) Automated chip-based device for simple and fast nucleic acid amplification. Expert Rev Mol Diagn 5: 613–620
8. Wang W, Li ZX, Luo R, Lü SH, Xu AD, Yang YJ (2005) Droplet-based micro oscillating-flow PCR chip. J Micromech Microeng 15: 1369–1377
9. Sasaki N, Izawa M, Shimojo M, Shibata K, Akiyama JI, Itoh M, Nagaoka S, Carninci P, Okazaki Y, Moriuchi T, Muramatsu M, Watanabe S, Hayashizaki Y (1997) A novel control system for polymerase chain reaction using a RIKEN GS384 thermocycler. DNA Res. 4: 387–391
10. Northrup MA, Ching MT, White RM, Watson RT (1993) DNA amplification with a microfabricated reaction chamber. Proceedings of Transducers '93: 924–926
11. Daniel JH, Iqbal S, Millington RB, Moore DF, Lowe CR, Leslie DL, Lee MA, Pearce MJ (1998) Silicon microchambers for DNA amplification. Sensor Actuator Phys 71: 81–88
12. Northrup MA, Gonzalez C, Hadley D, Hills RF, Landre P, Lehew S, Saiki R, Sninsky JJ, Watson R, Watson R Jr (1995) A MEMS-based miniature DNA analysis system. Proceedings of Transducers '95: 764–767
13. Poser S, Schulz T, Dillner U, Baier V, Köhler JM, Schimkat D, Mayer G, Siebert A (1997) Chip elements for fast thermocycling. Sensor Actuator Phys 62: 672–675
14. Nagai H, Murakami Y, Morita Y, Yokoyama K, Tamiya E (2001) Development of a microchamber array for picoliter PCR. Anal Chem 73: 1043–1047
15. Nagai H, Murakami Y, Yokoyama K, Tamiya E (2001) High-throughput PCR in silicon-based microchamber array. Biosens Bioelectron 16: 1015–1019
16. Nagai H, Murakami Y, Wakida S, Niki E, Tamiya E (2001) High throughput single cell PCR on a silicon microchamber array. Proceedings of Micro Total Analysis System 268–270

17. Felbel J, Bieber I, Pipper J, Köhler JM (2004) Investigations on the compatibility of chemically oxidized silicon ($SiO_x$)-surfaces for applications towards chip-based polymerase chain reaction. Chem Eng J 101: 333–338
18. Neuzil P, Pipper J, Hsieh TM (2006) Disposable real-time microPCR device: Lab-on-a-chip at a low cost. Mol BioSyst 2: 292–298
19. Obeid PJ, Christopoulos TK (2003) Continuous-flow DNA and RNA amplification chip combined with laser-induced fluorescence detection. Anal Chim Acta 494: 1–9
20. Sun K, Yamaguchi A, Ishida Y, Matsuo S, Misawa H (2002) A heater-integrated transparent microchannel chip for continuous-flow PCR. Sensor Actuator Chem 84: 283–289
21. Auroux PA, Day PJR, Niggli F, Manz A (2003) PCR micro-volume device for detection of nucleic acids. Proceedings of Nanotec, 55–58
22. Lagally ET, Emrich CA, Mathies RA (2001) Fully integrated PCR-capillary electrophoresis microsystem for DNA analysis. Lab Chip 1: 102–107
23. Li S, Fozdar DY, Ali MF, Li H, Shao D, Vykoukal DM, Vykoukal J, Floriano PN, Olsen M, McDevitt JT, Gascoyne PRC, Chen S (2006) A continuous-flow polymerase chain reaction microchip with regional velocity control. J Microelectromechyst S 15: 223–236
24. Park N, Kim S, Hahn JH (2003) Cylindrical compact thermal-cycling device for continuous-flow polymerase chain reaction. Anal Chem 75: 6029–6033
25. Friedmann NA, Meldrum DR (1998) Capillary tube resistive thermal cycling. Anal Chem 70: 2997–3002
26. Oda RP, Strausbauch MA, Huhmer AFR, Borson N, Jurrens SR, Craighead J, Wettstein PJ, Eckloff B, Kline B, Landers JP (1998) Infrared-mediated thermocycling for ultrafast polymerase chain reaction amplification of DNA. Anal Chem 70: 4361–4368
27. Baker J, Strachan M, Swartz K, Yurkovetsky Y, Rulison A, Brooks C, Kopf-Sill A (2003) Single molecule amplification in a continuous flow lab-chip device. Proceedings of Micro Total Analysis Systems, 1335–1338
28. Legendre LA, Bienvenue JM, Roper MG, Ferrance JP, Landers JP (2006) A simple, valveless microfluidic sample preparation device for extraction and amplification of DNA from nanoliter-volume samples. Anal Chem 78: 1444–1451
29. Lagally ET, Scherer JR, Blazej RG, Toriello NM, Diep BA, Ramchandani M, Sensabaugh GF, Riley LW, Mathies RA (2004) Integrated Portable Genetic analysis microsystem for pathogen/infectious disease detection. Anal Chem 76: 3162–3170
30. Khandurina J, McKnight TE, Jacobson SC, Waters LC, Foote RS, Ramsey JM (2000) Integrated system for rapid PCR-based DNA analysis in microfluidic devices. Anal Chem 72: 2995–3000
31. Ferrance JP, Wu Q, Giordano B, Hernandez C, Kwok Y, Snow K, Thibodeau S, Landers JP (2003) Developments towards a complete mirco-total analysis system for Duchenne muscular dystrophy diagnosis. Anal Chim Acta 500: 223–236

32. Zhang Q, Wang W, Zhang H, Wang Y (2002) Temperature analysis of continuous-flow micro-PCR based on FEA. Sensor Actuator Chem 82: 75–81
33. Schneegaß I, Bräutigam R, Köhler JM (2001) Miniaturized flow-through PCR with different template types in a silicon chip thermocycler. Lab Chip 1: 42–49
34. Zhao Z, Cui Z, Cui D, Xia S (2003) Monolithically integrated PCR biochip for DNA amplification. Sensor Actuator Phys 108: 162–167
35. Schabmueller CGJ, Lee MA, Evans AGR, Brunnschweiler A, Ensell GJ, Leslie DL (2000) Closed chamber PCR chips for DNA amplification. Eng Sci and Educ J 9: 259–264
36. Cheng J, Shoffner MA, Hvichia GE, Kricka LJ, Wilding P (1996) Chip PCR. II. Investigation of different PCR amplification systems in microfabricated silicon–glass chips. Nucleic Acids Res 24: 380–385
37. Bu M, Melvin T, Ensell G, Wilkinson JS, Evans AGR (2003) Design and theoretical evaluation of a novel microfluidic device to be used for PCR. J Micromech Microeng 13: S125-S130
38. Yoon DS, Lee YS, Lee Y, Cho HJ, Sung SW, Oh KW, Cha J, Lim G (2002) Precise temperature control and rapid thermal cycling in a micromachined DNA polymerase chain reaction chip. J Micromech Microeng 12: 813–823
39. Wilding P, Shoffner MA, Kricka LJ (1994) PCR in a silicon microstructure. Clin Chem 40: 1815–1818
40. Köhler JM, Dillner U, Mokansky A, Poser S, Schulz T (1998) Micro channel reactors for fast thermocycling. Proceedings of 2nd International Conference on Microreaction Technology, 241–247
41. Poser S, Ehricht R, Schulz S, Uebel S, Dillner U, Köhler JM (1999) Rapid PCR in flow-through Si chip thermocyclers. Proceedings of 3rd International Conference on Microreaction Technology, 411–419
42. Poser S, Schulz T, Dillner U, Baier V, Köhler JM, Mayer G, Siebert A, Schimkat D (1997) Temperature-controlled chip reactor for rapid PCR. Proceedings of 1st International Conference on Microreaction Technology, 294–301
43. Oh KW, Park C, Namkoong K, Kim J, Ock KS, Kim S, Kim YA, Cho YK, Ko C (2005) World-to-chip microfluidic interface with built-in valves for multichamber chip-based PCR assays. Lab Chip 5: 845–850
44. Lee TMH, Carles MC, Hsing IM (2003) Microfabricated PCR-electrochemical device for simultaneous DNA amplification and detection. Lab Chip 3: 100–105
45. Schabmueller CGJ, Pollard JR, Evans AGR, Wilkinson JS, Ensell G, Brunnschweiler A (2001) Integrated diode detector and optical fibres for in situ detection within micromachined polymerase chain reaction chips. J Micromech Microeng 11: 329–333
46. Erill I, Campoy S, Rus J, Fonseca L, Ivorra A, Navarro Z, Plaza JA, Aguiló J, Barbé J (2004) Development of a CMOS-compatible PCR chip: Comparison of design and system strategies. J Micromech Microeng 14: 1558–1568

47. Zou ZQ, Chen X, Jin QH, Yang MS, Zhao JL (2005) A novel miniaturized PCR multi-reactor array fabricated using flip-chip bonding techniques (2005) J Micromech Microeng 15: 1476–1481
48. Kim SH, Noh J, Jeon MK, Kim KW, Lee LP, Woo SI (2006) Micro-Raman thermometry for measuring the temperature distribution inside the microchannel of polymerase chain reaction chip. J Micromech Microeng 16: 526–530
49. Pal D, Venkataraman V (2002) A portable battery-operated chip thermocycler based on induction heating. Sensor Actuator Phys 102: 151–156
50. Gulliksen A, Solli L, Karlsen F, Rogne H, Hovig E, Nordstrøm T, Sirevåg R (2004) Real-time nucleic acid sequence-based amplification in nanoliter volumes. Anal Chem 76: 9–14
51. Shoffner MA, Cheng J, Hvichia GE, Kricka LJ, Wilding P (1996) Chip PCR. I. Surface passivation of microfabricated silicon-glass chips for PCR. Nucleic Acids Res 24: 375–379
52. Erill I, Campoy S, Erill N, Barbé J, Aguiló J (2003) Biochemical analysis and optimization of inhibition and adsorption phenomena in glass-silicon PCR-chips. Sensor Actuators Chem 96: 685–692
53. Krishnan M, Burke DT, Burns MA (2004) Polymerase chain reaction in high surface-to-volume ratio $SiO_2$ microstructures. Anal Chem 76: 6588–6593
54. Burns MA, Johnson BN, Brahmasandra SN, Handique K, Webster JR, Krishnan M, Sammarco TS, Man PM, Jones D, Heldsinger D, Mastrangelo CH, Burke DT (1998) An integrated nanoliter DNA analysis device. Science 282: 484–487
55. Giordano BC, Ferrance J, Swedberg S, Hühmer AFR, Landers JP (2001) Polymerase chain reaction in polymeric microchips: DNA amplification in less than 240 seconds. Anal Biochem 291: 124–132
56. Chen Z, Qian S, Abrams WR, Malamud D, Bau HH (2004) Thermosiphon-based PCR reactor: Experiment and modeling. Anal Chem 76: 3707–3715
57. Daly J, Davies M, Dalton T (2005) A fully controlled natural convection polymerase chain reaction device. Proceedings of ECI International Conference on Heat Transfer and Fluid flow in Microscale
58. Curcio M, Roeraade J (2003) Continuous segmented-flow polymerase chain reaction for high-throughput miniaturized DNA amplification. Anal Chem 75: 1–7
59. Nakano H, Matsuda K, Yohda M, Nagamune T, Endo I, Yamane T (1994) High speed polymerase chain reaction in constant flow. Biosci Biotech Biochem 58: 349–352
60. Chiou J, Matsudaira P, Sonin A, Ehrlich D (2001) A closed-cycle capillary polymerase chain reaction machine. Anal Chem 73: 2018–2021
61. Kaack RM, Jung A, Wenz HM, Zengerle R, Daub M (2004) PCR-slide: A modular and cascadable platform for DNA sample processing with integrated nanolitre dosage. Proceedings of 17th IEEE International Conference on Micro Electro Mechanical Systems MEMS, 338–342

62. Yang J, Liu Y, Rauch CB, Stevens RL, Liu RH, Lenigk R, Grodzinski P (2002) High sensitivity PCR assay in plastic micro reactors. Lab Chip 2: 179–187
63. Hashimoto M, Chen PC, Mitchell MW, Nikotopoulos DE, Soper SA, Murphy MC (2004) Rapid PCR in a continuous flow device. Lab Chip 4: 638–645
64. Wang J, Chen Z, Corstjens PLAM, Mauk MG, Bau HH (2006) A disposable microfluidic cassette for DNA amplification and detection. Lab Chip 6: 46–53
65. Liu Y, Rauch CB, Stevens RL, Lenigk R, Yang J, Rhine DB, Grodzinski P (2002) DNA amplification and hybridization assays in integrated plastic monolithic devices. Anal Chem 74: 3063–3070
66. Liu RH, Yang J, Lenigk R, Bonanno J, Grodzinski P (2004) Self-contained, fully integrated biochip for sample preparation, polymerase chain reaction amplification, and DNA microarray detection. Anal Chem 76: 1824–1831
67. Anderson RC, Su X, Bogdan GJ, Fenton J (2000) A miniature integrated device for automated multistep genetic assays. Nucleic Acids Res 28: e60
68. Chen J, Wabuyele M, Chen H, Patterson D, Hupert M, Shadpour H, Nikitopoulos D, Soper SA (2005) Electrokinetically synchronized polymerase chain reaction microchip fabricated in polycarbonate. Anal Chem 77: 658–666
69. Lee DS, Park SH, Yang H, Chung KH, Yoon TH, Kim SJ, Kim K, Kim YT (2004) Bulk-micromachined submicroliter-volume PCR chip with very rapid thermal response and low power consumption. Lab Chip 4: 401–407
70. Krishnan M, Ugaz VM, Burns MA (2002) PCR in a Rayleigh-Bénard convection cell. Science 298: 793
71. Yuen PK, Kricka LJ, Fortina P, Panaro NJ, Sakazume T, Wilding P (2006) Microchip module for blood sample preparation and nucleic acid amplification reactions. Genome Res 11: 405–412
72. Niu ZQ, Chen WY, Shao SY, Jia XY, Zhang WP (2006) DNA amplification on a PDMS-glass hybrid microchip. J Micromech Microeng 16: 425–433
73. Liao CS, Lee GB, Liu HS, Hsieh TM, Wang CH, Fan CL, Luo CH (2004) Micro reverse-transcription polymerase chain reaction system for clinical diagnosis. Proceedings of Micro Total Analysis Systems, 87–89
74. Liu J, Enzelberger M, Quake S (2002) A nanoliter rotary device for polymerase chain reaction. Electrophoresis 23: 1531–1536
75. Shin YS, Cho K, Lim SH, Chung S, Park SJ, Chung C, Han DC, Chang JK (2003) PDMS-based micro PCR chip with parylene coating. J Micromech Microeng 13: 768–774
76. Yu X, Zhang D, Li T, Hao L, Li X (2003) 3-D microarrays biochip for DNA amplification in polydimethylsiloxane (PDMS) elastomer. Sensor Actuator Phys 108: 103–107
77. Prakash AR, Adamia S, Sieben V, Pilarski P, Pilarski LM, Backhouse CJ (2006) Small volume PCR in PDMS biochips with integrated fluid control and vapour barrier. Sensor Actuator Chem 113: 398–409

78. Lee CY, Lee GB, Liu HH, Huang FC (2002) MEMS-based temperature control systems for DNA amplification. Int J Nonlinear Sci Numer Simul 3: 215–218
79. Lee CY, Lee GB, Lin JL, Huang FC, Liao CS (2005) Integrated microfluidic systems for cell lysis, mixing/pumping and DNA amplification. J Micromech Microeng 15: 1215–1223
80. Huang FC, Liao CS, Lee GB (2006) An integrated microfluidic chip for DNA/RNA amplification, electrophoresis separation and on-line optical detection. Electrophoresis 27: 3297–3305
81. Nakayama T, Kurosawa Y, Furui S, Kerman K, Kobayashi M, Rao SR, Yonezawa Y, Nakano K, Hino A, Yamamura S, Takamura Y, Tamiya E (2006) Circumventing air bubbles in microfluidic systems and quantitative continuous-flow PCR applications. Anal Bioanal Chem DOI 10.1007/s00216-006-0688-7
82. Hong JW, Fujii T, Seki M, Yamamoto T, Endo I (2001) Integration of gene amplification and capillary gel electrophoresis on a polydimethylsiloxane-glass hybrid microchip. Electrophoresis 22: 328–333
83. Gulliksen A, Solli LA, Drese KS, Sörensen O, Karlsen F, Rogne H, Hovig E, Sirevåg (2005) Parallel nanoliter detection of cancer markers using polymer microchips. Lab Chip 5: 416–420
84. Cheng JY, Hsieh CJ, Chuang YC, Hsieh JR (2005) Performing microchannel temperature cycling reactions using reciprocating reagent shuttling along a radial temperature gradient. Analyst 130: 931–940
85. Zou Q, Miao Y, Chen Y, Sridhar U, Chong CS, Chai T, Tie Y, The CHL, Lim TM, Heng CK (2002) Micro-assembled multi-chamber thermal cycler for low-cost reaction chip thermal multiplexing. Sensor Actuator Phys 102: 114–121
86. Wheeler EK, Benett B, Stratton P, Richards J, Christian A, Chen A, Weisgraber T, Ness K, Ortega J, Milanovich F (2003) Convectively driven polymerase chain reaction thermal cycler. Proceedings of Micro Total Analysis Systems, 1133–1135
87. Woolley AT, Hadley D, Landre P, de Mello AJ, Mathies RA, Northrup MA (1996) Functional integration of PCR amplification and capillary electrophoresis in a microfabricated DNA Analysis Device. Anal Chem 68: 4081–4086
88. El-Ali J, Perch-Nielsen IR, Poulsen CR, Jensen M, Tellemann P, Wolff A (2003) Microfabricated DNA amplification device monolithically integrated with advanced sample pre-treatment. Proceedings of Transducers '03: 214–217
89. West J, Karamata B, Lillis B, Gleeson JP, Alderman J, Collins JK, Lane W, Mathewson A, Berney H (2002) Application of magnetohydrodynamic actuation to continuous flow chemistry. Lab Chip 2: 224–230
90. Leamon JH, Lee WL, Tartaro KR, Lanza JR, Sarkis GJ, deWinter AD, Berka J, Lohman KL (2003) A massively parallel PicoTiterPlate™ based platform for discrete picoliter-scale polymerase chain reactions. Electrophoresis 24: 3769–3777

91. Hühmer AFR, Landers JP (2000) Noncontact infrared-mediated thermocycling for effective polymerase chain reaction amplification of DNA in nanoliter volumes. Anal Chem 72: 5507–5512
92. Fermér C, Nilsson P, Larhed M (2003) Microwave-assisted high-speed PCR. Eur J Pharm Sci 18: 129–132
93. Chaudhari AM, Woudenberg TM, Albin M, Goodson KE (1998) Transient liquid crystal thermometry of microfabricated PCR vessel arrays. J Microelectromech Syst 7: 345–355
94. Hennig M, Braun D (2005) Convective polymerase chain reaction around micro immersion heater. Appl Phys Lett 87: 183901
95. Chen Z, Wang J, Quian S, Bau HH (2005) Thermally-actuated, phase change flow control for microfluidic systems. Lab Chip 5: 1277–1285
96. Guttenberg Z, Müller H, Habermüller H, Geisbauer A, Pipper J, Felbel J, Kielpinski M, Scriba J, Wixforth A (2005) Planar chip device for PCR and hybridization with surface acoustic wave pump. Lab Chip 5: 308–317
97. Litborn E, Curcio M, Emmer Å, Roeraade J (2000) Synthesis and analysis of chemical components in nanoscale. Proceedings of Micro Total Analysis Systems, 447–454
98. Barrett B, Sirr N, Sayers M, Dalton T, Davies M (2005) Liquid properties for droplet or plug-flow microfluidic polymerase chain reaction devices. Proceedings of ECI International Conference on Heat Transfer and Fluid flow in Microscale
99. Gui L, Ren CL (2006) Numeric simulation of heat transfer and electrokinetic flow in an electroosmosis-based continuous flow PCR chip. Anal Chem 78: 6215–6222
100. Pollack MG, Paik PY, Shenderov AD, Pamula VK, Dietrich FS, Fair RB (2003) Investigation of electrowetting-based microfluidics for real-time PCR applications. Proceedings of Micro Total Analysis Systems, 619–622
101. Griffith EJ, Akella S (2005) Coordination multiple droplets in planar array digital microfluidic systems. Int J Rob Res 24: 933–949
102. Griffith EJ, Akella S, Goldberg MK (2006) Performance characterization of a reconfigurable planar-array digital microfluidic system. IEEE Trans Comput Aided Des Integrated Circuits Syst 25: 340–352
103. Zhang T, Chakrabarty K, Fair RB (2002) System performance evaluation with SystemC for two PCR microfluidic systems. Proceedings of Nanotech, 48–53
104. Braun D, Goddard NL, Libchaber A (2003) Exponential DNA replication by laminar convection. Phys Rev Lett 91: 158103
105. Krishnan M, Agrawal N, Burns MA, Ugaz VM (2004) Reactions and fluidics in miniaturized natural convection systems. Anal Chem 76: 6254–6265
106. Walsh EJ, King CR, Grimes RH, Davies MR, Dalton TM (2005) Passive movement and manipulation of microfluidic using buoyancy. Proceedings of ECI International Conference on Heat Transfer and Fluid flow in Microscale
107. Leamon JH, Link DR, Egholm M, Rothberg JM (2006) Overview: Methods and applications for droplet compartmentalization of biology. Nat Methods 3: 541–543

125. Kricka LJ, Wilding P (2003) Microchip PCR. Anal Bioanal Chem 377: 820–825
126. Obeid PJ, Christopoulos TK (2004) Microfabricated systems for nucleic acid analysis. Crit Rev Clin Lab Sci 41: 429–465
127. Zhang C, Xu J, Ma W, Zheng W (2006) PCR microfluidic devices for DNA amplification. Biotechnol Adv 24: 243–284
128. Lassen B, Malmsten M (1997) Competitive protein adsorption at plasma polymer surfaces. J Colloid Interface Sci 186: 9–16
129. Sukhishvili SA, Granick S (1999) Adsorption of human serum albumin: Dependence on molecular architecture of the oppositely charged surface. J Chem Phys 110: 10153–10161
130. Chun CJ, Lenghaus K, Riedel L, Bhalkikar A, Hickman JJ (2006) A behaviour of protein adsorption on poly(ethylene glycol)-modified surfaces under flow conditions at relatively low concentrations for microfluidics systems. Proceedings of Nanotech 2: 674–677
131. Kai J, Sohn YS, Ahn CH (2002) Study on protein (IgG) adsorption in terms of surface modification of cyclic olefin copolymer (COC) for protein biochip. Proceedings of Micro Total Analysis Systems, 419–421
132. Felbel J, Bieber I, Köhler JM (2002) Chemical surface management for micro PCR in silicon chip thermocyclers. Proceedings of SPIE 4937: 34–40
133. Lacher NA, de Rooij NF, Verpoorte E, Lunte SM (2003) Comparison of the performance characteristics of poly(dimethylsiloxane) and Pyrex microchip electrophoresis devices for peptide separations. J Chromatogr A 1004: 225–235
134. Vreeland WN, Barron AE (2002) Functional materials for microscale genomic and proteomic analyses. Curr Opin Biotechnol 13:87–94
135. Hjertén S (1985) High-performance electrophoresis. Elimination of electroendosmosis and solute adsorption. J Chromatogr 347: 191–198
136. Giordano BC, Copeland ER, Landers JP (2001) Towards dynamic coating of glass microchip chambers for amplifying DNA via the polymerase chain reaction. Electrophoresis 22: 334–340
137. Munro NJ, Hühmer AFR, Landers JP (2001) Robust polymeric microchannel coating for microchip-based analysis of neat PCR products. Anal Chem 73: 1784–1794
138. Holmberg K, Bergström K, Brink C, Österberg E, Tiberg F, Harris JM (1993) Effects on protein adsorption, bacterial adhesion and contact angle of grafting PEG chains to polystyrene. Contact Angle, Wettability and Adhesion, Ed. Mittal KL, 813–827
139. Österberg E, Bergström K, Holmberg K, Schuman TP, Riggs JA, Burns NL, van Alstine JM, Harris JM (1995) Protein-rejecting ability to surface-bound dextran in end-on and side-on configurations: Comparison to PEG. J Biomed Mater Res 29: 741–747
140. Hu S, Ren X, Bachman M, Sims CE, Li GP, Allbritton N (2002) Surface modification of poly(dimethylsiloxane) microfluidic devices by ultraviolet polymer grafting. Anal Chem 74: 4117–4123

141. Eteshola E, Leckband D (2001) Development and characterization of an ELISA assay in PDMS microfluidic channels. Sensor Actuator Chem 72: 129–133
142. Lodish H, Baltimore D, Berk A, Zipursky SL, Matsudaira P, Darnell J (1996) Molekulare Zellbiologie, Berlin, de Gruyter
143. Putnam FW (1975) The Plasma Proteins: Structure, Function and Genetic Control. Vol. 1, New York, Springer-Verlag
144. Bi H, Meng S, Li Y, Guo K, Chen Y, Kong J, Yang P, Zhong W, Liu B (2006) Deposition of PEG onto PMMA microchannel surface to minimize nonspecific adsorption. Lab Chip 6: 769–775
145. Popat KC, Desai TA (2004) Poly(ethylene glycol) interfaces: An approach for enhanced performance of microfluidic systems. Biosens Bioelectron 19: 1037–1042
146. Chiari M, Cretich M, Damin F, Ceriotti F, Ceriotti L, Consonni R (2000) New adsorbed coatings for capillary electrophoresis. Electrophoresis 21: 909–916
147. Roach LS, Song H, Ismagilov RF (2005) Controlling nonspecific protein adsorption in a plug-based microfluidic system by controlling interfacial chemistry using fluorous-phase surfactants. Anal Chem 77: 785–796
148. Horvath J, Dolnik V (2001) Polymer wall coatings for capillary electrophoresis. Electrophoresis 22: 644–655
149. Lee JG, Cheong KH, Huh N, Kim S, Choi JW, Ko C (2006) Microchip-based one step DNA extraction and real-time PCR in one chamber for rapid pathogen identification. Lab Chip 6: 886–895
150. Abad-Villar EM, Kubáň P, Hauser PC (2005) Determination of biochemical species on electrophoresis chips with an external contactless conductivity detecto. Electrophoresis 26: 3609–3614
151. Kan CW, Fredlake CP, Doherty EAS, Barron AE (2004) DNA sequencing and genotyping in miniaturized electrophoresis systems. Electrophoresis 25: 3564–3588
152. Kubista K, Andrade JM, Bengtsson M, Forootan A, Jonák J, Lind K, Sindelka R, Sjöback R, Sjögreen B, Strömbom L, Ståhlberg A, Zoric N (2006) The real-time polymerase chain reaction. Mol Aspects Med 27: 95–125
153. Lagally ET, Soh ET (2005) Integrated genetic analysis microsystems. Crit Rev Solid State Mater Sci 30: 207–233
154. Belgrader P, Benett W, Hadley D, Richards J, Stratton P, Mariella R Jr, Milanovich F (1999) PCR detection of bacteria in seven minutes. Science 284: 449–450
155. Namasivayam V, Lin R, Johnson B, Brahmasandra S, Razzacki Z, Burke DT, Burns MA (2004) Advances in on-chip photodetection for applications in miniaturized genetic analysis systems. J Micromech Microeng 14: 81–90
156. Roulet JC, Völkel R, Herzig HP, Verpoorte E, de Rooij NF, Dändliker R (2002) Performance of an integrated microoptical system for fluorescence detection in microfluidic systems. Anal Chem 74: 3400–3407
157. Chabinyc ML, Chiu DT, McDonald JC, Stroock AD, Christian JF, Karger AM, Whitesides GM (2001) An integrated fluorescence detection system in

poly(dimethylsiloxane) for microfluidic applications. Anal Chem 73: 4491–4498
158. Kamei T, Paegel BM, Scherer JR, Skelley AM, Street RA, Mathies RA (2003) Integrated hydrogenated amorphous Si photodiode detector for microfluidic bioanalytical devices. Anal Chem 75: 5300–5305
159. Hou CSJ, Milovic N, Godin M, Russo PR, Chakrabarti R, Manalis SR (2006) Label-free microelectronic PCR quantification. Anal Chem 78: 2526–2531
160. Cho YK, Kim J, Lee Y, Kim YA, Namkoong K, Lim H, Oh KW, Kim S, Han J, Park C, Pak YE, Ki CS, Choi JR, Myeong HK, Ko C (2006) Clinical evaluation of micro-scale chip-based PCR system for rapid detection of hepatitis B virus. Biosens Bioelectron 21: 2161–2169
161. Wilding P, Kricka LJ, Cheng J, Hvichia G, Shoffner MA, Fortina P (1998) Integrated cell isolation and polymerase chain reaction analysis using silicon microfilter chambers. Anal Biochem 257: 95–100
162. Burke DT, Burns MA, Mastrangelo C (1997) Microfabrication technologies for integrated nucleic acid analysis. Genome Res 7: 189–197
163. Paegel BM, Blazej RB, Mathies RA (2003) Microfluidic devices for DNA sequencing: Sample preparation and electrophoretic analysis. Curr Opin Biotechnol 14: 42–50
164. Erickson D, Li D (2004) Integrated microfluidic devices. Anal Chim Acta 507: 11–26
165. Chamberlain JS, Gibbs RA, Ranier JE, Nguyen PN, Caskey CT (1988) Deletion screening of the Duchenne muscular dystrophy locus via multiplex DNA amplification. Nucleic Acids Res 16: 11141–11156
166. Telenius H, Carter NP, Bebb CE, Nordenskjold M, Ponder BA, Tunnacliffe A (1992) Degenerate oligonucleotide-primed PCR: General amplification of target DNA by a single degenerate primer. Genomics 13: 718–725
167. Pich U, Houben A, Fuchs J, Meister A, Schubert I (1994) Utility of DNA amplified by degenerate oligonucleotide-primed PCR (DOP-PCR) from the total genome and defined chromosomal regions of field bean. Mol Gen Genet 243: 173–177
168. Liu B, Segal G, Vega JM, Feldmann M, Abbo S (1997) Isolation and characterization of chromosome-specific DNA sequences from a chromosome-arm genomic library of common wheat. Plant J 11: 959–965
169. Zhang L, Cui X, Schmitt K, Hubert R, Navidi W, Arnheim N (1992) Whole genome amplification from a single cell: Implications for genetic analysis. Proc Natl Acad Sci USA 89: 5847–5851
170. Mueller PR, Wold B (1989) In vivo footprinting of a muscle specific enhancer by ligation mediated PCR. Science 246: 780–786
171. Schlissel M, Constantinescu A, Morrow T, Baxter M, Peng A (1993) Double-strand signal sequence breaks in V(D)J recombination are blunt, 5'-phosphorylated, RAG dependent, and cell cycle regulated. Genes Dev 7: 2520–2532

172. Staley K, Blaschke AJ, Chun J (1997) Apoptotic DNA fragmentation is detected by a semiquantitative ligation-mediated PCR of blunt DNA ends. Cell Death Differ 4: 66–75
173. Liu D, Liu C, DeVries S, Waldman F, Cote RJ, Datar RH (2004) LM-PCR permits highly representative whole genome amplification of DNA isolated from small number of cells and paraffin-embedded tumor tissue sections. Diagn Mol Pathol 13: 105–115
174. Ludecke HJ, Senger G, Claussen U, Horsthemke B (1989) Cloning defined regions of the human genome by microdissection of banded chromosomes and enzymatic amplification. Nature 338: 348–350
175. Kao FT, Yu JW (1991) Chromosome microdissection and cloning in human genome and genetic disease analysis. Proc Natl Acad Sci USA 88: 1844–1848
176. Klein CA, Schmidt-Kittler O, Schardt JA, Pantel K, Speicher MR, Riethmüller G (1999) Comparative genomic hybridization, loss of heterozygosity, and DNA sequence analysis of single cells. Proc Natl Acad Sci USA 96: 4494–4499
177. Zhang FY, Yin WB, Shi R, Hu YK, Yan YM, Chen YH, Zhou YH, Hu J, Wang R, Hu ZM (2005) Construction and characterization of chromosome 1b specific DNA library of wheat. Can J Plant Sci 85: 309–316
178. Snabes MC, Chong SS, Subramanian SB, Kristjansson K, DiSepio D, Hughes MR (1994) Preimplantation single-cell analysis of multiple genetic loci by whole-genome amplification. Proc Natl Acad Sci USA 91: 6181–6185
179. Barrett MT, Reid BJ, Joslyn G (1995) Genotyping analysis of multiple loci in somatic cells by whole genome amplification. Nucleic Acids Res 23: 3488–3492
180. Grothues D, Cantor CR, Smith CL (1993) PCR amplification of megabase DNA with tagged random primers (T-PCR). Nucleic Acids Res 21: 1321–1322
181. Wu DY, Wallace RB (1989) The ligation amplification reaction (LAR)-amplification of specific DNA sequences using sequential rounds of template-dependent ligation. Genomics 4: 560–569
182. Schouten JP, McElgunn CJ, Waaijer R, Zwijnenburg D, Diepvens F, Pals G (2002) Relative quantification of 40 nucleic acid sequences by multiplex ligation-dependent probe amplification. Nucleic Acid Res 30: e57
183. Guatelli JC, Whitfield KM, Kwoh DY, Barringer KJ, Richman DD, Gingeras TR (1990) Isothermal, in vitro amplification of nucleic acids by a multienzyme reaction modeled after retroviral replication. Proc Natl Acad Sci USA 87: 1874–1878
184. Kwoh DY, Davis GR, Whitfield KM, Chappelle HL, DiMichele LJ, Gingeras TR (1989) Transcription-based amplification system and detection of amplified human immunodeficiency virus type 1 with a bead-based sandwich hybridization format. Proc Natl Acad Sci USA 86: 1173–1177

185. Chan AB, Fox JD (1999) NASBA and other transcription-based amplification methods for research and diagnostic microbiology. Rev Med Microbiol 10: 185–196
186. Walker GT, Fraiser MS, Schram JL, Little MC, Nadeau JG, Malinowski DP (1992) Strand displacement amplification – An isothermal, in vitro DNA amplification technique. Nucleic Acids Res 20:1691–1696
187. Persing DH, Smith TF, Tenover FC, White TJ (1993) Diagnostic Molecular Microbiology Principles and Applications. Americal Society for Microbiology, Washington D.C.
188. Walker GT, Little MC, Nadeau JG, Shank DD (1992) Isothermal in vitro amplification of DNA by restriction enzyme/DNA polymerase system. Proc Natl Acad Sci USA 89: 392–396
189. Lizardi PM (2000) Multiple displacement amplification. United States Patent 6124120
190. Dean FB, Hosono S, Fang L, Wu X, Faruqi AF, Bray-Ward P, Sun Z, Zong Q, Du Y, Du J, Driscoll M, Song W, Kingsmore SF, Egholm M, Lasken RS (2002) Comprehensive human genome amplification using multiple displacement amplification. Proc Natl Acad Sci USA 99: 5261–5266
191. Nelson JR, Cai YC, Giesler TL, Farchaus JW, Sundaram ST, Ortiz-Rivera M, Hosta LP, Hewitt PL, Mamone JA, Palaniappan C, Fuller CW (2002) TempliPhi, phi29 DNA polymerase based rolling circle amplification of templates for DNA sequencing. Biotechniques 32: S44–S47
192. Lizardi PM, Huang X, Zhu Z, Bray-Ward P, Thomas DC, Ward DC (1998) Mutation detection and single-molecule counting using isothermal rolling-circle amplification. Nat Genet 19: 225–232
193. Vincent M, Xu Y, Kong H (2004) Helicase-dependent isothermal DNA amplification. EMBO Rep 5: 795–800
194. Lizardi PM, Guerra CE, Lomeli H, Tussie-Luna I, Kramer FR (1988) Exponential amplification of recombinant-RNA hybridization probes. Bio/Technology 6: 1197–1202
195. Hecker KH, Roux KH (1996) High and low annealing temperatures increase both specificity and yield in touchdown and stepdown PCR. Biotechniques 20: 478–485
196. Conoci S, Di Pietro P, Petralia S, Amore MG, San Biagio F, Alaimo G, Iacono G, Alessi E, Ricceri D, Di Trapani G, Di Francesco F, Palmieri M (2006) Fast and efficient nucleic acid testing by ST's In-Check™ lab-on-chip platform. Proc Nanotech: 562–565
197. Higgins JA, Nasarabadi S, Karns JS, Shelton DR, Cooper M, Gbakima A, Koopman RP (2003) A handheld real time thermal cycler for bacterial pathogen detection. Biosens Bioelectron 18: 1115–1123
198. Lee TMH, Hsing IM (2006) DNA-based bioanalytical microsystems for handheld device applications. Anal Chim Acta 556: 26–37
199. Belgrader P, Smith JK, Weedn VW, Northrup MA (1998) Rapid PCR for identity testing using a battery-powered miniature thermal cycler. J Forensic Sci 43: 315–319

# Chapter 14
# Cytometry on Microfluidic Chips

Petra S. Dittrich[1], Fedor Malik[2] and Petra Schwille[2]

[1]ISAS – Institute for Analytical Sciences, Bunsen-Kirchhoff-Str. 11, 44263 Dortmund (Germany)
[2]Biophysics/BIOTEC, TU Dresden, Tatzberg 47-51, 01309 Dresden (Germany)

## 1. Introduction

Cell counting and sorting have become important standard methods in cell biology. It plays an increasing role in basic science, as well as in drug discovery and in clinical applications. Fundamental cellular processes such as apoptosis, cell growth, or protein expression can be studied, as well as the influences of specific compounds (e.g., drugs) on these processes [3, 34, 40, 46, 51, 69].

After the first developments in the 1970s [39], cytometry devices have undergone a dramatic improvement with respect to speed, accuracy, and applicability. Modern commercial devices are capable to detect, analyze and sort several tens of thousands of cells within one second, after multi-parametric analysis usually based on fluorescence and light scattering. The main features of a standard cell sorter are sketched in Fig. 1. After pre-treatment (e.g., tagging with fluorescent markers), the cells are suspended in a buffer solution and introduced into the device. The cells are aligned in a thin stream using a sheath flow, and then analyzed serially (one-by-one) by a focused laser beam with respect to their fluorescence and/or light scattering signal. To individually address them according to their properties, the stream is broken into droplets containing one cell each at a small nozzle, which is facilitated by slight vibrations of the nozzle tip. Sorting of the droplets is achieved by deflection in an electrical field, such that target cells can be collected and enriched in a particular well, while undesired cells are disposed.

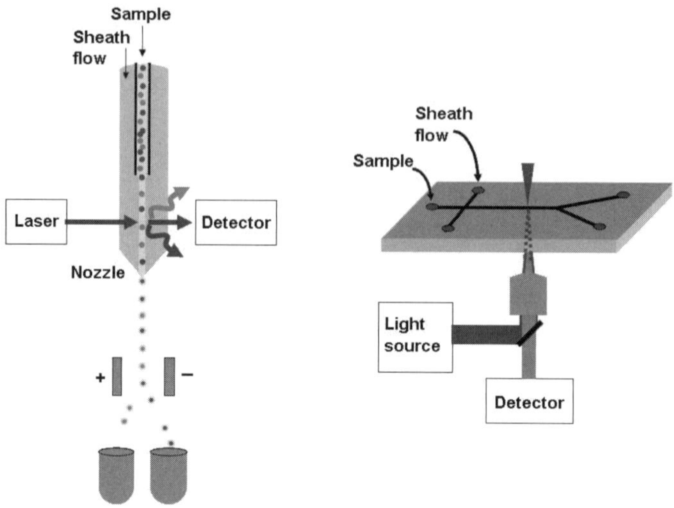


**Fig. 1.** Sketches (not to scale) of a conventional fluorescence-activated cell sorter FACS (left) and a microfabricated FACS device, e.g. positioned on a microscope (right)

Conventional fluorescence-activated cell sorters (FACS machines) are extremely powerful analytical devices because of their high throughput and the direct access to the cells. However, they are expensive equipment with high maintenance requirements, and specialized personnel are usually required to operate the device. The entire device is bulky and consumes large sample volumes to be analyzed. The high demand of cell analyses and sorting for biological and medical purposes has triggered many efforts to improve the existing technology, and to develop simpler and less expensive devices. Of particular interest is, on the other hand, the miniaturization of FACS applications, and the possibility to combine the actual sorting with other preparation steps, such as cell lysis and component purification, for proteomics studies.

Since the early 1990s, new opportunities for cell treatment and analysis were investigated using microfabrication technologies. The employment of microfluidic platforms for cellular studies promises a couple of persuasive advantages. The size of living cells fit very well with dimensions of microstructures that are typically in the range from a few up to hundreds of micrometers. Thus, microfluidic structures are valuable tools to transport cells, and, moreover, to manipulate such small objects individually, by integration of certain functional units (e.g., miniaturized electrodes) into the microchips. Planar microfluidic chips are typically inexpensive devices that are nowadays easy to fabricate. They can be used as disposable chips, rendering time-consuming cleaning procedures unnecessary. They are small and can be self-contained. Therefore, only small sample volumes are required, and the use of hazardous materials is more feasible. The small dimensions in microfluidic chips provide short transport times along the channels. Moreover, the high surface-to-volume ratio enables fast heating and cooling procedures in chemical reactions. Considering that cell assays are often complex, and multistep reactions and diverse treatment of the cells are often needed, the probably most promising aspect of microfluidic systems lays in its ability to integrate several functionalities on a single chip. Nowadays, microfluidic structures are widely utilized for various cell studies [2, 21, 26, 107], including cell lysis [66, 77], cell electroporation [71], cell culture for long term cell treatment [35, 91], tissue formation [52], formation of cell culture analogues [85], and for the creation of biohybrid devices [75, 89, 97].

Since the introduction of microstructured systems, the development of integrated systems for sample preparation, analysis, and separation is of major interest – also referred to as micrototal analysis systems (µTAS [64]). For separation of a sample mixture, most often devices are designed, in which the separation mechanism relies on physical properties of the sample, such as size and charge in capillary electrophoresis. Separation of particles and cells with different sizes is often achieved by exploiting a certain channel/chip design, e.g. by use of chip-based filtering structures employing sieves and sieves-like obstacles [7, 12, 44], or by variations in the flow profile (by widening of the channels) in a pinched microchannel [104]. Large particles move slower compared to small particles, also enabling a passive filtering perpendicular to the laminar flow direction. Such schemes can be utilized to separate particles with different sizes [98] as well as to separate motile cells from less motile cells [10].

Particle and cell separation devices that use magnetic forces (magnetic-activated cell sorting, MACS) were likewise miniaturized [5, 47, 72]. In such devices, magnetic particles with certain surface modifications attach to target cells. In a magnetic field, they are attracted by a magnet and deflected perpendicular to the flow, or caged by the magnetic field. Other separation techniques utilize ultrasound [74], dielectrophoresis [4, 11, 57], or optical forces [61] to separate cells with certain characteristics from a bulk cell suspension.

In contrary to these separation devices based on passive sorting, "intelligent" active sorting devices can be designed to detect, analyze and, sort cells based on any sizeable parameter, one after another. This operation mode on the individual cell level enables a sorting routine that is uncoupled from physical properties and amenable to various kinds of secondary control parameters, e.g., fluorescence tagging. This larger flexibility with respect to molecular or cellular properties requires highly dynamic systems for manipulation of individual cells or small liquid quantities. For precisely controlling cell sorting in such a device, valves and switches are needed, which are fast and reliable on the one hand, and easy to operate on the other hand. In the past years, several cell counting and sorting devices have been developed on microfluidic platforms, and are commercialized by several companies such as Micronics Inc., Evotec Technologies, or Agilent Technologies. The main features will be described in the following, and are sketched in Fig. 1. Like commercial cell sorting machines, microfluidic cell sorters comprise three main components: the focusing unit, the detection area, and the sorting switch with typical geometries of Y, T, or X shaped channels.

## 2. Design of Microfluidic Flow Cytometers

### *2.1. Transport and Focusing of Cell Suspensions*

Different pumping mechanisms have been realized in microfluidic analytical devices. In passive transport systems, driven by capillary forces, gravity, or evaporation, no external pumps are required. A different height of the liquid reservoirs connected to the (open) ends of the channels induces a hydrodynamic flow. Continuous, active transport is induced by external devices. For pressure-driven flow, the flow rate is controlled by syringe pumps connected via tubing systems, or by applying an air pressure onto the liquid reservoirs. Pressure-driven flow is typically fast, and enables flow velocities up to a few $m\,s^{-1}$. For electroosmotic flow (EOF), electrodes are inserted into the channel end reservoirs, or integrated into the channels themselves during microfabrication. The fluid can be moved by applying a voltage between the electrodes. In contrast to pressure-driven flow, the flow profile is flat rather than parabolic, which can be advantageous for concentration control. Also, switching the flow direction, e.g., for sorting, is faster and more reliable by EOF. Although all methods are applicable for miniaturized flow cytometry, usually gentle pumping and transport mechanisms, e.g., slow hydrodynamic flow, are favorable because they avoid the danger of cell rupture due to shear stress, or cell death due to high voltage.

After introduction into the microfluidic device, the cells need to be aligned by a focusing unit. Focusing is an important task of the device to ensure that cells are sequentially passing the detection unit, i.e. to guarantee detection on single cell level. Moreover, the quality of focusing is a critical issue for the quality of optical detection, because the detection sensitivity is often inhomogeneous across the microfluidic channel, e.g., in confocal detection schemes as described in the next section. Therefore, it is important that every cell passes the detection window on the same transit pathway.

The cell thread can be generated by different focusing techniques. Narrowing the channel itself in the area of detection would be a straightforward solution. However, clogging of the narrow neck sections by dispersed dust particles, or larger cell and buffer components, can occur easily. Furthermore, a higher initial pressure would be required to pump the sample solution through the channel.

Hydrodynamic focusing is certainly the most popular technique, since it is a simple approach and easy adaptable to any planar microfluidic chip. For hydrodynamic focusing, the liquid in the main channel containing the cell suspension is merged with the streams of two side channels (Fig. 2).

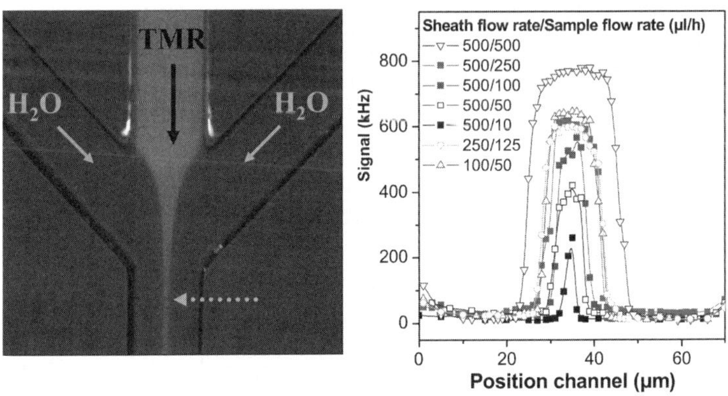


**Fig. 2.** Hydrodynamic focusing of a TMR (Tetramethylrhodamine) solution in a microfluidic chip by two side streams containing water. Left: Fluorescence image of the merging section. Right: The width of the focused stream depends on the flow rates of the sheath flow and the TMR flow, and was measured utilizing a confocal fluorescence setup directly after merging of the three streams as specified by the arrow in the left image

Hereby, the widths of the three streams decrease, and their flow speed increases accordingly. An appropriate choice of flow rates of the side channels and the main channel (the flow speed in the main channel should preferably be lower than in the side channels) results in a narrow stream of cells, confined on two sides by the side streams. The cross section of the focused stream can be made as small as a 1–2 μm by such an arrangement. By improvement and development of nonplanar microchips in the last years, it is nowadays possible to obtain confinement of the cell stream in two dimensions, if necessary [15, 84, 101]. Focusing induced by different flow rates of merging streams was also achieved using electroosmotic flow [78], or by use of an air sheath flow [45].

### 2.2. The Sorting Unit: Active Microfluidic Switches

The sorting unit facilitates the selection of target cells. The basic requirement for a sorting unit is its ability either to direct the target cell individually or to direct a small liquid volume, in which one target cell is suspended, into the desired output reservoir. A number of switching schemes were developed for cell sorting. In the following, several examples of microfluidic sorting devices shall be given, which are listed in Table 1.

One of the first cell sorting systems with an active switch was developed by Fuhr and co-workers in the 1990s [28]. The manipulation of cells was realized by forces generated by AC fields under conditions of negative dielectrophoresis. Differently arranged, miniaturized electrode arrays were inserted into the microchip. They functioned as funnel and aligner to concentrate and narrow the cell stream. A cage was used to trap the cells for detection and analysis, and a (manually activated) electrode array covering half of the channel width directs the cells into one of two output channels. Cell manipulation and sorting devices based on dielectrophoresis were further improved during the last years, and commercialized by Evotec Technologies (Fig. 3).

In 1999, Quake and coworkers presented a simple disposable device, made of silicon elastomer that comprised a T-shaped channel [29]. Bacterial cells were transported and sorted by means of electroosmotic flow towards the T-junction. Instead of sheath flow, the channels were narrowed at the T-junction to align and detect cells individually. Electrodes were activated to direct the cells either into the waste channel or into the collection channel. However, electroosmotic flow and electroosmotic switches suffer from the drawback that living cells are sensible to high voltages; hence the viability rate of the cells in such a system is low. In a subsequent study of the same

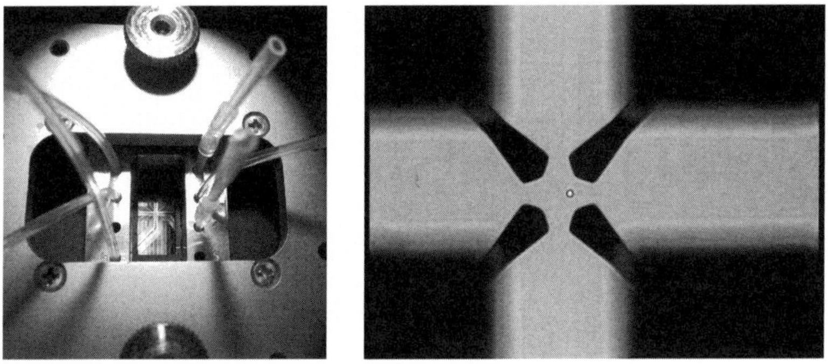

**Fig. 3.** Left: Microchip Cytocon 400 (Evotec Technologies) integrated in an interface for electronic and fluidic connections. Right: Trapping of a baker's yeast cell by positive dielectrophoresis inside the microchip enabling long observation times. In this image, a weak flow from left to right deflects the yeast cells slightly from the centre of the electrode array. (Courtesy of H. Kortmann, L. Blank and A. Schmid, University of Dortmund, Germany)

group, the device was thus improved. The cells were pumped by hydrodynamic flow, and the switch was realized by use of pneumatic valves [30]. The valves were designed by multilayer soft lithography, a technique that exploits the elasticity and surface chemistry of silicone elastomers. In a two-layer system, the bottom layer consisted of the fluidic network to transport the cells, while the top layer was operated as control channels to pneumatically activate the valves. When pressurized gas was introduced into the control line, the thin membrane of silicone elastomer that separates the two layers was deflected downward, and seals the fluidic channels in the bottom channel. Different cell sorting algorithms were applied, and enrichment factors up to 83% could be achieved.

Other kind of valves can be directly inserted into the microfluidic chip. A mechanical valve actuated via air bubbles that were generated electrochemically was fabricated, and demonstrated for cell sorting [41]. In 2006, Shirasaki et al. created valves formed by a thermoreversible gelation polymer that was mixed with the cell suspension, and introduced into a Y-shaped microchannel [83]. If the polymer was exposed to infrared laser light, it formed a water-impermeable plug. If target cells were identified (by their fluorescence), a laser beam was used to plug the waste channel, allowing the cell to enter the collection channel. Otherwise the collection channel was closed, and the waste channel was open. The response time of the sol–gel transformation to form a plug was 3 ms, and a flow switching time of 120 ms was achieved.

**Table 1.** List of microfluidic signal-activated cell sorting devices listed by publication year

| Pumping method/sorting method | Focusing | Detection | Speed of sorting | Comments | References |
|---|---|---|---|---|---|
| Hydrodynamic flow/DEP | Integrated funnel, aligner and cell trap | Microscopy | – | Commercialized by Evotec Technologies | [28] |
| Electrokinetic flow/electrokinetic switch | Narrow channel section | Fluorescence (LIF, CCD camera) | 10 beads, cells $s^{-1}$ | – | [29] |
| Hydrodynamic flow/integrated pneumatic valves | Narrow channel sections | Fluorescence (LIF, PMT) | Max. 44 cells $s^{-1}$ | – | [30] |
| Hydrodynamic flow/dhydrodynamic switching or off-chip valves | Hydrodynamic focusing | Flurescence (laser diode or LEDs, APD) | – | – | [53] |
| Hydrodynamic flow/cell trap and release array based on DEP | – | Fluorescence | – | – | [95] |
| Hydrodynamic flow/optical trap | – | Microscopy | – | – | [70] |
| Hydrodynamic flow/electrokinetic switch | Hydrodynamic focusing | Fluorescence (confocal setup; LIF, APD) | ~1 cell $s^{-1}$ | Integrated (labeling) reaction | [18] |
| Hydrodynamic flow/off-chip valve | Hydrodynamic focusing | Fluorescence (Hg lamp, PMT) | 12,000 cell $s^{-1}$ 100-fold enrichment | Integrated fucntionalities (cell culture chamber) | [101] |
| Electrokinetic flow/electrokinetic switch | Eletrokinetic focusing | Scattering | – | Integrated optical components (waveguides) | [31] |
| Gravity/electric switch | Focusing unit | Flurescence (diode laser, PMT) | 42–64 cells/30s run time | – | [106] |

**Table 1.** (cont.)

| | | | | | |
|---|---|---|---|---|---|
| Electroosmotic flow/electrokinetic switch super-imposed with hydrodynamic flow | – | Microscopy | – | – | [50] |
| Electrokinetic flow/electrokinetic switch | Electrokinetic focusing | Fluorescence (confocal setup; LIF, APD) | – | – | [55] |
| Hydrodynamic flow/integrated pneumatic valves | – | Fluorescence (CCD camera) | – | – | [87] |
| Hydrodynamic flow/optical switch | Hydrodynamic focusing | Fluorescence (LIF, PMT) | Max. 105.9 cells $s^{-1}$ | – | [96] |
| Hydrodynamic flow/DEP | Funnel-formed electrode array for focusing | Dielectrophoretic amplitude response | 100,000 cells $s^{-1}$ (200-fold enrichment) | – | [43] |
| Hydrodynamic flow/Mechanical switch | – | Microscopy | – | – | [41] |
| Hydrodynamic flow/Selection through sorting slit | – | Fluorescence (LIF, CCD camera) | – | Integrated (labeling) reaction, commercialized by Micronics | [56] |
| Hydrodynamic flow/DEP | Hydrodynamic | Digital camera | – | Two sample streams could be sorted in parallel | [59] |
| Hydrodynamic flow/Pneumatic valves | Hydrodynamic focusing | Fluorescence (Laser diode, APD) | – | – | [105] |
| Hydrodynamic flow/Valves made by thermoresponsive gel | – | Fluorescence (LIF, PMT) | Max. 25 cells $s^{-1}$ | – | [83] |

DEP, Dielectrophoresis; APD, Avalanche photodiode; LIF, Laser-induced fluorescence; PMP, Photomultipler tube.

High-speed valves with response times of a few ms that are inserted in the tubing system outside the chip were introduced by several groups [53, 101]. The off-chip valve systems have the advantage that simple planar chips can be used, and the cells are not directly in contact with the valves (e.g., no rupture of the cells can occur).

In the approaches utilizing valves, the liquid stream is temporally redirected into different channels, by blocking one channel while another one is opened. Our group developed a cell sorting device without use of a valve [18]. Pressure-driven flow was used to pump the cells through the main channel, and to hydrodynamically focus the cell suspension. Electroosmotic flow was applied in a perpendicular channel to sort the cells (Fig. 4). The perpendicular channel was located directly before the main channel branches into the two output channels. Only a short voltage pulse was required to induce electroosmotic flow and to minimally deflect the cells from their former path. This minimal deflection was sufficient to alter the selection probability of previously 50% to almost 100% accuracy, and to achieve efficient sorting into the desired outlet channel. Detection was performed using a confocal fluorescence setup, where the detection sensitivity (on molecular level) and efficiency (approximately 100%) was on almost ultimate scale. This procedure is a gentle way to sort cells, and the recovery rate of (bacterial) cells is high (up to 95%).

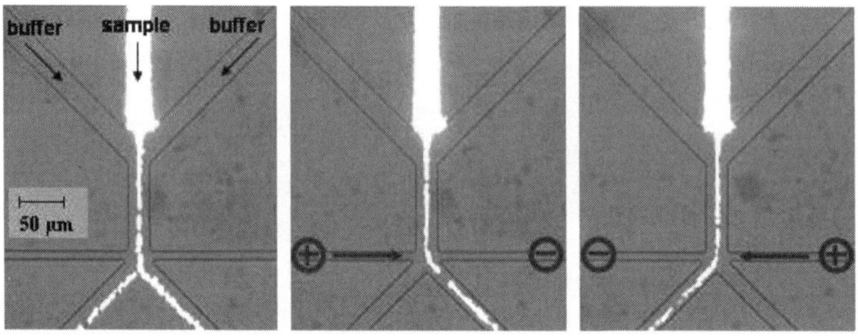


**Fig. 4.** Microfluidic sorting device. The sample suspension (here fluorescent beads) is hydrodynamically focused by the stream of two side channels, and then equally directed into two output channels. Instead of a buffer solution, the side channels could be filled with reactive agents, e.g. fluorescently labeled molecules to tag the sample. Detection of the sample would be performed in the region in front of the Y-branch of the microchannel. In the middle and right images, the principle of the sorting procedure is shown. A potential is applied in the small channel that induces a weak electroosmotic flow (towards the negative pole). This additional flow deflects the sample into the respective output channel. The electroosmotic flow can be temporally controlled. A short pulse would be sufficient for sorting of individual particles

The lateral displacement of the cells is also a common strategy used in several other sorting devices [31, 59, 96]. In 2005, a study by Wang et al. was published that describes the microfluidic sorting of cells by optical force switching [96]. In a Y-shaped sorting unit, passing cells were displaced within the laminar flow utilizing a focused infrared laser beam that acts as an optical trap. The laser spot was translated at a speed matching the flow velocity, and at a small angle relative to the axis of the flow, to maximize the interaction time between the laser and the cell. A temporally controllable magnetic switch was introduced by another group in 2006, which facilitates sorting of magnetic particles, and can probably be transferred for single cell sorting in near future [49].

Development and integration of switching units into microfluidic chips are (still) challenging. Particularly for automated and high speed sorting routines, the demand on the quality of the switching unit is enormous. Besides the flow rate that can be realized to pump the cells through the channels, there are two critical issues of the sorting speed with respect to the switching unit. First, the response time of the switching unit that is typically very short (e.g, the most common valves open/close within ms). Second, the time needed to redirect the particle or the liquid into the desired output channels, which is among others a matter of distance of the displacement of particles or liquid.

In an automated sorting routine, the sorting switch has to be actuated by the analyzing unit, i.e., dependent on the outcome of the analysis. This feedback system needs careful adjustment to the speed and concentration of the cell suspension, so that the switching and direction of a single cell towards the specific outlet channel is finished before the next cell is passing the switching area of the microchannel. Very high concentrated cell suspensions, in which cells are flowing back-to-back, naturally cause a reduction in accuracy of the sorting process.

### 2.3. Integration of Several Functionalities on Microchips

Cell sorting is typically a multistep process rather than an isolated procedure. Pretreatment of cells such as fluorescence labeling or addition of drugs must be performed before sorting; in specific applications, cell treatment or analysis is desired after the selection of target cell. In contrary to conventional FACS machines, planar microfluidic chips are advantageous to integrate these additional reaction steps, since any functional unit can in principle be implemented. The integration of cell treatment processes can significantly reduce preparation times, since no pipetting procedures are necessary, and since incubation times can be reduced [18, 56]. In chip designs with a focusing unit, a reaction step prior to sorting can simply be performed by addition of a reagent into the sheath flow (Figs. 4 and 7). For example, small molecules for tagging of the cells could be supplied by the sheath flow. The molecules are diffusing fast towards the cell train that is focused to a small stream in the channel's centre, and can tag the target cells by binding to the cell membrane within ms [18].

There are other examples of treatment and analysis of cells in a high throughput mode on microfluidic chips that could be used in combination with cytometry devices. The approaches include mechanical and chemical cell lysis and subsequent analysis of the cytosol [66, 103], as well as cell electroporation [82]. Wolff et al. integrated a microchamber inside a sorting chip to culture and store cells at the end of the collection channel [101]. The chamber was surrounded by a microchannel. The connection between cell culture chamber and microchannel was achieved by small gold dots that were placed at the chamber wall. The gold dots prevent complete sealing of the wall with the glass cover, and hence, act as filter that allow supplying fresh medium to the cells, and draining of excess liquid during sorting. In this way, losses of cells were diminished, which is in particular valuable for rare cell sorting.

## 3. Detection Concepts for Ultrasensitive Cytometry

### *3.1. Single Molecule Fluorescence Spectroscopy in Microfluidic Channels*

One main feature of flow cytometry device is its ability to detect and identify target cells with high speed and accuracy. The speed of cell counting and analysis that can be achieved depends on the time required for data acquisition and analysis, and on the flow rate. The requirements of fast and specific detection are perfectly met by fluorescence spectroscopy [1, 20]. Since most microfluidic devices are made by transparent materials, their adaptation to any optical setup is usually easily achieved.

Fluorescence spectroscopy is a noninvasive technique that does not alter the sample after detection (e.g., cells remain viable). The high specificity of fluorescence spectroscopy is achieved by attaching fluorophores of defined spectral properties to defined positions of the sample. Hence, inter- and intramolecular reactions can be studied for in vivo and in vitro applications. A wide variety of staining assays for biological cells are available that enable the identification of cells with certain functionalities by fluorescence spectroscopy. Fluorophores introduced into cells facilitate

108. Nakano M, Komatsu J, Matsuura SI, Takashima K, Katsura S, Mizuno A (2003) Sigle-molecule PCR using water-in-oil emulsion. J Biotechnol 102: 117–124
109. Haase AT, Retzel EF, Staskus KA (1990) Amplification and detection of lentiviral DNA inside cells. Proc Natl Acad Sci USA 87: 4971–4975
110. Komminoth P, Werner M (1997) Target and signal amplification: Approaches to increase the sensitivity of in situ hybridization. Histochem Cell Biol 108: 325–333
111. Brehm-Stecher BF, Johnson EA (2004) Single-cell microbiology: Tools, technologies, and applications. Microbiol Mol Biol Rev 68: 538–559
112. Bagasra O, Harris T (2006) Latest developments in in situ PCR. Methods Mol Biol 334: 221–240
113. He Y, Zhang YH, Yeung ES (2001) Capillary-based fully integrated and automated system for nanoliter polymerase chain reaction analysis directly from cheek cells. J Chromatogr A 924: 271–284
114. Lagally ET, Simpson PC, Mathies RA (2000) Monolithic integrated microfluidic DNA amplification and capillary electrophoresis analysis system. Sensor Actuator Chem 63: 138–146
115. Waters LC, Jacobson SC, Kroutchinina N, Khandurina J, Foote RS, Ramsey JM (1998) Microchip device for cell lysis, multiplex PCR amplification, and electrophoretic sizing. Anal Chem 70: 158–162
116. Raisi F, Blizard BA, Shabari AR, Ching J, Kintz GJ, Mitchell J, Lemoff A, Taylor MT, Weir F, Western L, Wong W, Joshi R, Howland P, Chauhan A, Nguyen P, Petersen KE (2004) Human genomic DNA analysis using a semi-automated sample preparation, amplification, and electrophoresis separation platform. J Sep Sci 27: 275–283
117. Liao CS, Lee GB, Wu JJ, Chang CC, Hsieh TM, Huang FC, Luo CH (2005) Micromachined polymerase chain reaction system for multiple DNA amplification of upper respiratory tract infectious disease. Biosens Bioelectron 20: 1341–1348
118. Wang H, Chen J, Zhu L, Shadpour H, Hupert ML, Soper SA (2006) Continuous flow thermal cycler microchip for DNA cycle sequencing. Anal Chem 78: 6223–6231
119. Agrawal N, Ugaz VM (2006) A buoyancy-driven compact thermocycler for rapid PCR. JALA 11: 217–221
120. Marcus JS, Anderson WF, Quake SR (2006) Parallel picoliter RT-PCR assays using microfluidics. Anal Chem 78: 956–958
121. Auroux PA, Koc Y, deMello A, Manz A, Day PJR (2004) Miniaturised nucleic acid analysis. Lab Chip 4: 534–546
122. Krishnan M, Namasivayam V, Lin R, Pal R, Burns MA (2001) Microfabricated reaction and separation systems. Curr Opin Biotechnol 12: 92–98
123. Vilkner T, Janasek D, Manz A (2004) Micro total analysis systems. Recent developments. Anal Chem 76: 3373–3386
124. Lee SJ, Lee SY (2004) Micro total analysis system (μ-TAS) in biotechnology. Appl Microbiol Biotechnol 64: 289–299

the differentiation of live and dead cells. Certain organelles can be stained individually and specifically; fluorescently labeled reagents, e.g., antibodies or ligands, can be attached to the outer cell membrane. Fluorescent proteins like GFP and mutants can be introduced into cells by standard techniques of molecular biology. Furthermore, cells and proteins exhibit intrinsic fluorescence (autofluorescence), which is caused by components such as NADH, FAD in cells, or aromatic amino acids in proteins.

Fluorescent markers are available over a wide spectral range from UV, visible, and IR light, which allows the simultaneous detection of fluorophores with different colors. Many commercial fluorophores are optimized with respect to their absorption coefficient, quantum yield, and photostability, so that they provide a strong, easily detectable signal utilizing inexpensive standard detection setups for concentrations of mM to μM.

Fluorescence spectroscopy is a very sensitive technique, since a single fluorophore can undergo the excitation–emission cycle repeatedly, thus continuously providing a detectable signal. Standard fluorophores such as Rhodamine dyes absorb and emit up to a few hundred thousand photons before a photobleaching reaction occurs that results in an irreversible loss of fluorescence. In this context, it should also be noted that photobleaching is a minor issue in flowing systems. Since the sample passes the excitation/detection area rapidly, and the sample is analyzed only once (i.e., bleaching after detection is irrelevant), high power light sources (lasers) at relatively high intensities can be used.

The high sensitivity of fluorescence spectroscopy can be greatly enhanced in specific optical setups providing very small detection volumes, which significantly reduces the background signal. The background light, mostly produced by scattering, scales linearly with the detection volume, while the signal arising from a single fluorophore remains constant. In a confocal setup, e.g., detection volumes of less than a femtoliter can be achieved. A typical experimental configuration is shown in Fig. 5. Laser light is focused by a microscope objective with high numerical aperture by coupling it in through the back aperture. The detection volume is axially defined by the tight focusing of the laser light (diffraction limited), and laterally by a field diaphragm (pinhole) to be positioned in the image plane, in front of the detectors. Combined with sensitive photodiodes or intensified CCD cameras, detection with high sensitivity down to the single-molecule level is feasible.

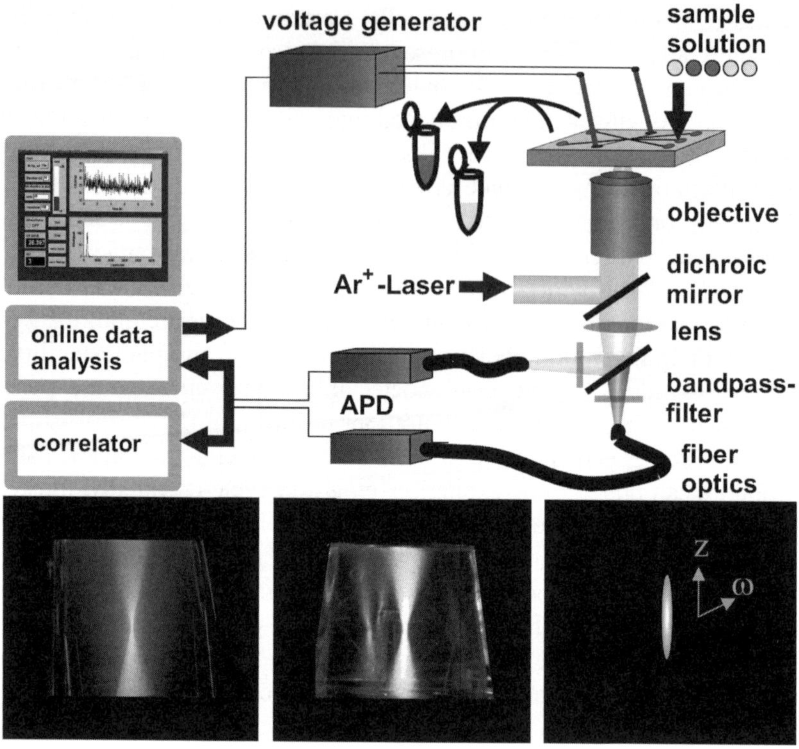


**Fig. 5.** Top: Confocal setup used for cell analysis and sorting and sorting as described in Fig. 4. Laser light is adjusted into an inverted microscope and focused by a microscope objective. Fluorescence light emitted from the sample is collected by the same objective and directed towards the pinhole. Appropriate optical filter blocks stray light and Raman scattering. Photons are counted by avalanche photodiodes (APD), and analyzed by computer. A positive signal activates the sorting module (here: a voltage generator to induce electroosmotic flow). The images on the bottom show the shape of the focused laser light (visualized with scattering particles, left). The sample fluoresces along the laser light (middle). The use of the pinhole in front of the detector limits the volume from which fluorescence is detected (right). The dimension of the detection volume in a confocal setup is typically less than 1 μm in radial direction (ω) and 1–5 μm in axial direction ($z$)

The maximum number of photons that can be detected from a passing fluorophore that is either dissolved in buffer or bound to particles or cells, depends on its dwell time inside the detection volume, as well as on the time needed for a fluorophore to undergo an excitation–emission cycle, which is mainly determined by the fluorescence lifetime $\tau_F$ (typically 1–10 ns). For a dwell time of 0.5 ms, the number of emitted photons is $5\times10^4$ – $5\times10^5$, which is diminished due to photophysical reactions and photobleaching. However, only a small percentage of ~5% emitted photons can be collected and registered by the detector, so that finally a maximum of 2,500 photons could be expected for one single molecule transit through the illuminating laser spot.

Figure 6 shows typical raw data for molecules passing the detection volume in a confocal setup. Detected photons are integrated over a certain time ("bin"). Photon bursts, arising out of a background trace, can be assigned to passing fluorophores. High signal to background ratios are desirable to clearly differentiate between signal and background values. Furthermore, the integration time should be adapted to the dwell time of the fluorophore to ensure that the counted photons will lay in one or few consecutive integration intervals. For a real-time analysis of the signals – necessary in sorting devices – a threshold is set in advance to differentiate between background and signal.

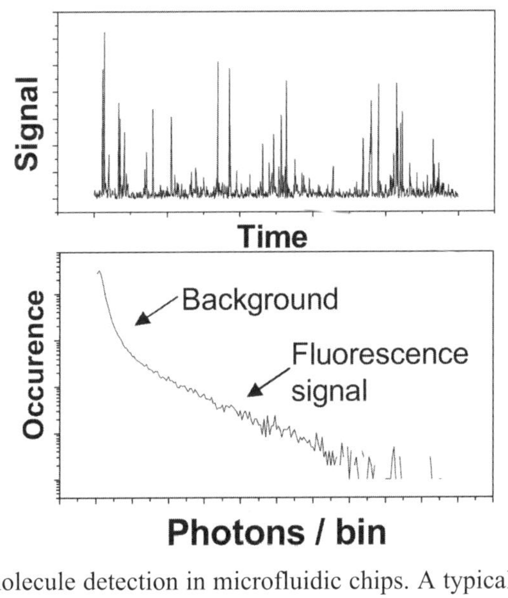


**Fig. 6.** Single molecule detection in microfluidic chips. A typical raw data trace is shown with the signal bursts that indicate the transit of a fluorescent dye through the detection volume (here Rhodamine Green). The data are displayed in a histogram in the graph below

An overview about individual registered fluorescence photons per unit time is usually displayed as a histogram. Since the background signal is low and only subject to statistical fluctuations, it can be described by a Poissonian distribution, giving the following probability $P$ to detect a certain number of photons per integration time ($n$):

$$P(n) = \frac{\mu^n}{n!} e^{-\mu} \tag{1}$$

where $\mu$ is the expectation value with the standard deviation $\sigma = \sqrt{\mu}$. Often, $3\sigma$ is chosen as a threshold [23, 54].

The fluorescence signal in the histogram, on the other hand, can be described by an exponential equation:

$$y = A \cdot \exp\left(\frac{-(x-t)}{H}\right) \tag{2}$$

with y, number of events; A, normalization factor; x, photons per bin; t, threshold; and H, brightness.

In such histograms, every time bin is recorded. For determination of the brightness, however, it should be considered that the photons from one event (transit of a molecule) could be represented in two or even more bins. Instead of a histogram that considers the values for *every* integration time, the number of detected photons per photon burst can be plotted, which directly corresponds to the brightness of a molecule or particle [94, 108]. In devices where two or more fluorescence colors can be detected simultaneously, the brightness of molecules with respect to a certain color is accessible, as well as the ratio of the colors, which often improves the identification of molecules (Fig. 7).

Histograms as described above are influenced by several parameters. Aggregation within the sample, or a small distance of two consecutive photon bursts may corrupt the data analysis. Additionally, the flow rate must be constant for the time of data acquisition to ensure that the all particles/molecules remain for equal time inside the detection volume. Moreover, in a confocal setup, the detection efficiency is usually inhomogeneous, i.e., the excitation intensity as well as the efficiency to detect emitted photons are highest in the center of the focused laser spot and drop down at a distance from the center. Therefore, the sample should preferably pass the center of the detection spot on highly reproducible tracks.

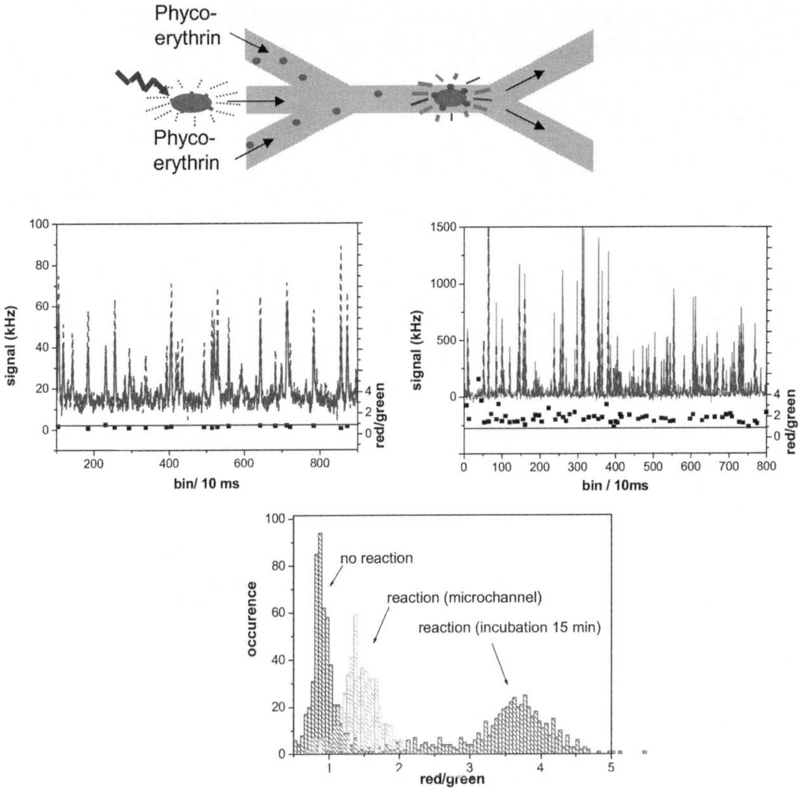


**Fig. 7.** Reaction in a microfluidic chip. Top: Bacterial cells expressing a membrane protein that binds to the fluorescent marker phycoerythrin are introduced into the central channel on a microfluidic sorting device. From the side channels, a buffer solution with phycoerythrin is supplied. Shortly afterwards (~ a few ms), the cells are exposed to the spot of blue laser light and green, and red fluorescence is detected. The cells attached to phycoerythrin exhibit a strong red signal (middle, *solid line*), whereas cells without phycoerythrin produce a green signal due to scattering light and autofluorescence (*dotted line*). For each fluorescent burst, the ratio of the red and green signal is calculated and plotted in the histogram (*bottom*). Cells with and without membrane protein, i.e. with or without attached phycoerythrin could be differentiated in this way. For comparison, the results for an external incubation in a test tube for 15 min are also shown

### 3.2. Determination of Flow Velocity by Fluorescence Correlation Spectroscopy (FCS)

In microfluidic systems of any kind, it is crucial to have extremely sensitive control units and output parameters to characterize the flow inside the microfluidic channels. For instance, in cell sorting devices, the determination of local flow velocity supports the temporal control of the switching unit.

Different methods have been proposed to characterize flow velocities in microchannels, such as μPIV [76], optical Doppler tomography [8], as well as two-laser-spot configurations. In configurations with two laser spots, the first laser spot initiates a reaction, e.g., photobleaching of a dye-containing fluid [79] or photoactivation of a fluorophore [73, 81], while a second laser functions as probe beam to detect the bleached, or photoactivated sample stream with spatial and temporal delay. Using a standard confocal setup, the transit time of a molecule through the detection volume in flowing systems is mainly defined by the flow velocity. Only a small contribution can be assigned to diffusion of the molecule, which can be neglected for high flow rates. Thus, the duration of the burst corresponds directly to the flow rate [24].

A very fast and precise determination of flow velocities of a fluorescent solution inside a microfluidic channel with high spatial resolution can be performed by fluorescence correlation spectroscopy (FCS), a single molecule fluorescence technique [63]. The principle of FCS is the quantification of fluorescence fluctuations in the signal recorded from a very small spatial volume element, usually the confocal setup (Fig. 5). In bulk solutions or cells, FCS can be utilized to characterize single molecule dynamics such as diffusion properties or rate constants of chemical reactions, but also to reveal intramolecular fluctuations as induced by conformational changes [62, 80]. By autocorrelation analysis of the fluorescence countrate, the spontaneous fluctuations of the fluorescence intensity are temporally evaluated by multiplying the time signal with itself at different delay times ($\tau$), which results in an autocorrelation curve that decays from a maximum value at $\tau \rightarrow 0$ to zero for large delay times (Fig. 8). Multiple parameters are accessible on single molecule level by FCS (Table 2). The amplitude of the correlation function is inversely proportional to the concentration of fluorescent molecules. The characteristic half-value decay time of such a curve corresponds to the average residence time of the particles that induced the fluctuations (usually single fluorescent molecules) in the focal spot. If additional dynamics such as blinking or flickering on faster time scales occur, the autocorrelation curve shows additional characteristic decay times that reflect on the underlying fast dynamics.

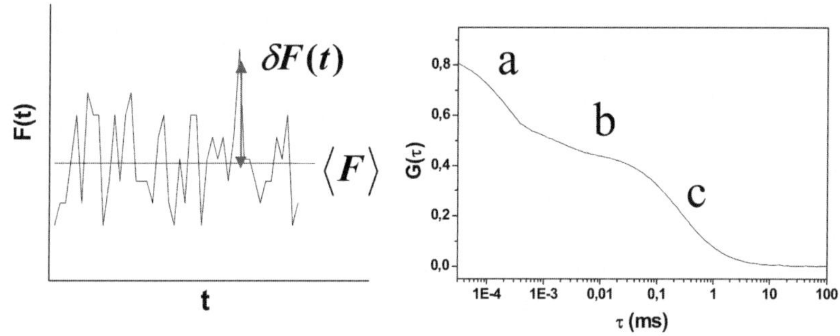


**Fig. 8.** Principle of Fluorescence Correlation Spectroscopy. Left: Fluorescent molecules entering and leaving the excitation/detection volume produce a fluctuating signal with the average value [$F$]. Right: From the fluctuations $\delta F(t)=F(t)$-[$F$] the autocorrelation curve can be derived. The decay time of the autocorrelation curve corresponds to the time scale of the process that causes the signal fluctuation. Diffusion into and out of the detection volume is the slowest process to be analyzed (**c**), while intramolecular reactions (**b**) or rotational movement (**a**) typically occur on faster time scales.

**Table 2.** Parameters that can be derived by correlation analysis. The sketches show the pathway of a molecule through the focused laser beam (*bright line*, fluorescent state; *dark line*, not fluorescent).

| | | |
|---|---|---|
| 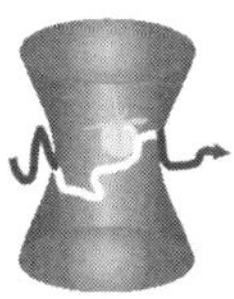 | Diffusion analysis | • Diffusion coefficient<br>• Local concentration |
| 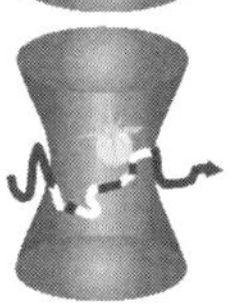 | Intramolecular reactions (changes in fluorescence intensity) | • Triplet transition<br>• Conformational changes<br>• Rotational movement |
| 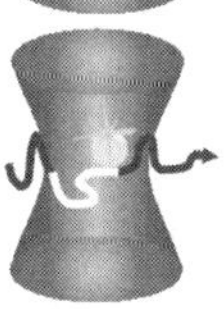 | Irreversible reactions | • Photobleaching |
| 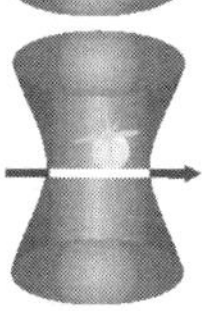 | Mode of movement | • Three-/two-dimensional diffusion<br>• Directed transport, e.g. flow |

Hereby, photophysical transitions [99], reversible intramolecular reactions [37], or photobleaching reactions [100] can be characterized. FCS enables also to distinguish random (Brownian) diffusion in one, two, or three dimensions from directed transport, e.g., to analyze the flow of fluorescent molecules through the laser focus in microfluidic channels, as well as along molecular or cytoskeletal tracks in living cells [86]. Figure 9 demonstrates how the FCS autocorrelation curve is modified for varying flow velocities, if a stream of fluorescent molecules is analyzed in a microchannel.

FCS has not only been employed to determine the flow velocity and thus, the flow profile, in microfluidic channels, but also to evaluate the directionality of flow [6, 17] by utilizing a cross-correlation scheme. For this technique, two detection volumes, i.e., laser foci, with a defined distance from each other are generated, and the respective emission signals are cross-correlated. In case of directed transport, the fluorescent dye molecules are passing the detection volumes successively, resulting in a correlation curve with a distinct maximum that corresponds to the transition time between both volumes. By using two-photon excitation, both the suppression of background signal and the spatial resolution are improved (Fig. 10).

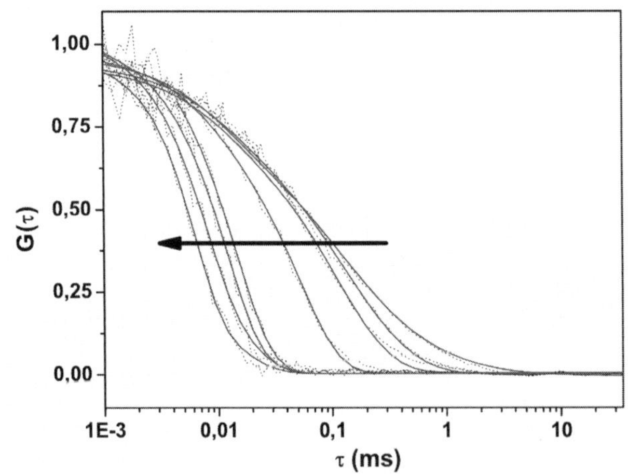


**Fig. 9.** FCS autocorrelation curves (*dotted line*) with fit (*solid line*) for increasing flow velocities (direction of arrow; 1.5–70 mm $s^{-1}$)

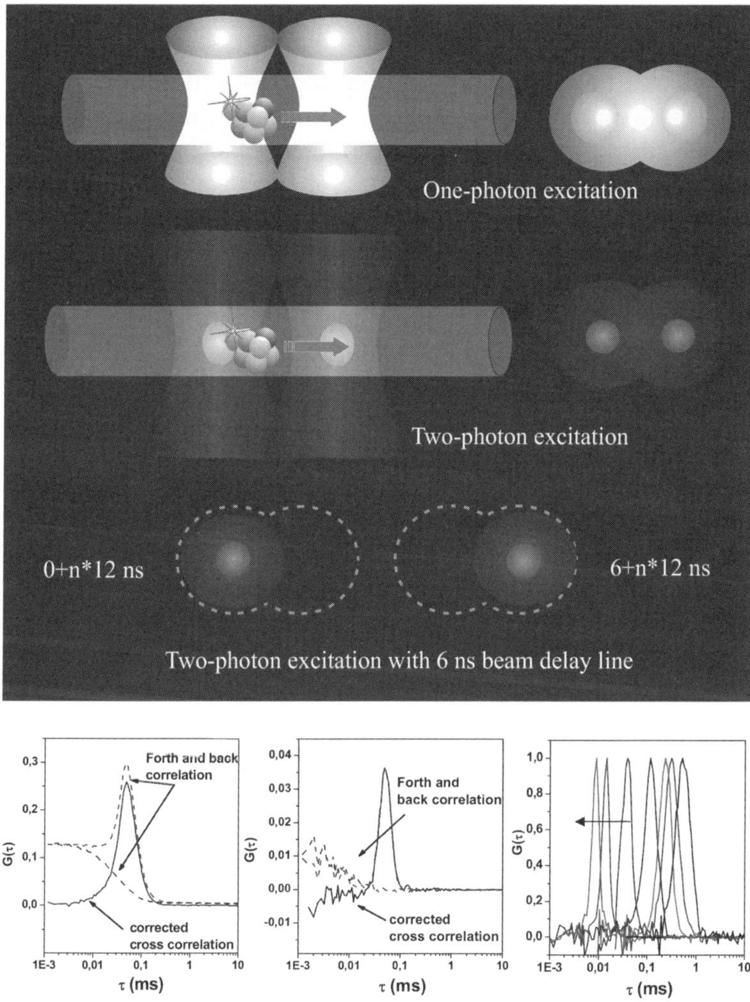


**Fig. 10.** Top: Principle of dual-beam cross-correlation measurements of flow in microstructures. Two-photon excitation with pulsed lasers enables the possibility to time-gate the excitation–detection cycle, which allows a significant reduction of cross-talk between the two detection channels. Bottom: in contrast to FCS autocorrelation curves, the spatial cross-correlation recorded in the dual-beam setup exhibits a clear maximum that corresponds to the time a particle needs to travel between the two volume elements. Two correlation curves are calculated (signal 1 correlated to signal 2 and vice versa), from which the corrected curves are derived. Shown are typical cross-correlation curves after one- and two-photon excitation (left and middle), as well as the movement of the maximum to shorter times $\tau$, when the flow velocity increases, which is indicated by the arrow (right)

### 3.3. Integration of Optical Components into Microfluidic Chips

Most microfluidic flow cytometry devices are designed in a way that the microchip can be placed on top of an optical setup, which is commonly a microscope, equipped with a light source (mercury lamp or laser) and a detector (photomultiplier tubes, photodiodes, or CCD cameras). In this configuration, the entire device is still bulky, and can not be used as stand-alone device.

A first step towards the size reduction of the entire devices was made by integration of optical fibers [58, 92]. Tung et al. fabricated microfluidic chips with grooves [92]. Optical fibers fit tightly in these grooves and were directed to the microchannel that contained the sample. Two fibers perpendicular to the microchannel were used to expose blue and red excitation light to the sample, two other fibers in an angle of 45° collected the fluorescence.

The integration of optical fibers reduces alignment procedures of the optical setup, while additionally, no microscope is required. Although such microchips, in principle, can be designed as small and easily transportable devices, standard light sources and detectors are still bulky, and a further reduction in the overall size would be desirable. Some efforts are therefore made to miniaturize optical components such as gratings and lenses [16], light sources such as light-emitting diodes [25] and lasers [32], but also detectors such as organic photodiodes [42] or avalanche photodiodes [38] and to integrate them to the microchips.

### 3.4. Other Detection Techniques

Fluorescence spectroscopy is certainly the most important and widely applicable method to detect passing cells or particles. However, fluorescence spectroscopy is frequently based on staining or labeling of cells, i.e. at least one reaction step is necessary prior to analysis, although autofluorescence of cells could be a helpful tool to distinguish certain cell types [27]. For certain applications such as sizing of cell populations, or to probe mechanical properties of cells, alternative detection methods have been developed that render cell labeling unnecessary.

Among other optical detection techniques, light scattering is maybe most popular to extract the size and shape of cells. The required optical setup is similar to the requirements for fluorescence spectroscopy, and hence often applied in combination. Furthermore, absorption measurements were performed in microfluidic chips. In standard applications, the absorption of the sample is proportional to the length of the absorption pathway, which is naturally extremely short in a micrometer-sized channel. Thus, specific high sensitivity techniques were developed for the measurement of single cell/particle absorption in capillaries and microfluidic channels, such as thermal lens microscopy [65], or cavity ring-down spectroscopy [90].

Using an optical stretcher, the elasticity of cells can be determined, which is altered in certain diseases such as cancer, or for specific cell types such as stem cells [36, 60]. In these setups, cells are stably trapped between two opposing divergent laser beams, and stretched out along the laser axis, while the degree of deformation can be analyzed by microscopy. The interaction of optical gradient fields with live cells, which depends on the refractive index, composition, size, and morphology of the cells, were studied using time-of-flight optophoresis [109]. In this method, travel times of cells along a fixed distance are determined, while the cells are irradiated by laser light.

Among the nonoptical detection techniques for cytometry, impedance spectroscopy permits rapid dielectric characterization of label-free cells. It provides information on cell size, membrane capacitance, and cytoplasm conductivity as a function of frequency of the electrical field. For impedance spectroscopy, electrodes are integrated in the microfluidic channel, and an AC field with frequencies ranging between 350 kHz and 20 MHz is applied. The measurement consists of an impedance change due to particles or cells passing an electrode array. Submicrometer sizing of beads and cells, as well as discrimination of different cell populations with throughput rates of 1,000 cells $s^{-1}$ have been reported utilizing impedance spectroscopy [9, 33]. The method was also combined with fluorescence measurements [68]. As an alternative to integrated metal electrodes, Chun et al. used polyelectrolytic salt bridge-based electrodes arranged perpendicular to the main channel that contains the cell suspension, to measure the impedance signal of passing cells and particles [14].

Wood et al. presented a high-bandwidth radio-frequency counter [102]. They determined the reflected RF power measured with integrated coplanar RF electrodes of passing 15 μm sized beads with counting rates up to 30 kHz.

In this context, it is imaginable to devise small fluidic compartments as microreactors, where interactions among molecules (e.g., intermolecular reactions, aggregation) could be studied. Microfluidic methods could thus be exploited to isolate or enrich molecules of interest, e.g., proteins or enzymes with desired structures or properties, which is subject of the next section.

### *4.2. Cell-Free Protein Expression in Microfluidic Chips*

In a sorting device that aims for the selection of proteins with desired properties, e.g., in biotechnological screening applications [67], it is mandatory to couple the protein phenotype to its genetic code. This is possible and has been successfully demonstrated by our group [19], by generating small aqueous compartments in microfluidic structures that contain a so-called "cell-free protein expression machinery," which can meanwhile be purchased from several companies.

The use of cell-free systems for the in vitro expression of proteins is a rapidly growing area, with applications in basic research, molecular diagnostics, and high-throughput screening. In vitro expression encompasses two general strategies. The first is to use isolated RNA synthesized in vivo or in vitro as a template for the translation reaction (e.g., using Rabbit Reticulocyte Lysate or Wheat Germ Extract Systems). The second is to use a coupled transcription/translation system in which DNA is used as a template (e.g., *E. coli* S30 Extract Systems). This DNA may be either a gene cloned into a plasmid vector (cDNA) or a PCR-generated template.

By now, *E. coli* extracts seem to be the most practical and efficient tools for cell-free translation and transcription–translation of functional proteins in vitro. In recent years, detailed knowledge of the protein-synthesizing machinery of the *E. coli* cell (including structure, genetics, metabolism, and regulation of its components and reactions) has produced remarkable improvement in *E. coli* extract cell-free systems. Virtually any genetic information can be translated into a polypeptide in *E. coli* extract cell-free systems. However, proper protein folding remains a substantial challenge, especially for the synthesis of multidomain and disulfide-bonded eukaryotic proteins. *E. coli* extract systems are quite tolerant of various additives, including cofactors, metabolites, unnatural amino acids, and even small amounts of detergents. This implies that one can modify synthesis and folding conditions to maximize the yield of soluble and functionally active proteins. Unfortunately, cotranslational and post-translational glycosylation and other complex modifications of eukaryotic proteins are not yet feasible in *E. coli* extract cell-free systems.

In our own approach to realize cell-free protein expression on a microchip, we generated droplets of buffer solution enriched with *E.coli* extract, and the genetic code of green fluorescent protein (GFP) as an easily detectable protein system, in soft polymer (PDMS) microchips. The separation of the different buffer compartments was realized by using mineral oil as a carrier solution, and by dispensing the buffer droplets at built-in narrow section (Fig. 11) at an intersection of different channels. Flow speed in the two channels could be used to adjust the exact size of the droplets, usually sizes around 5 μm in diameter were chosen because of their straightforward handling and detection. Figure 12 displays the principle of the setup. After generation of the droplets and filling with both, genetic information (i.e., DNA) and the cell-free expression kit, the droplets had to be stored for some time to allow GFP expression and maturation, and could then be analyzed by fluorescence spectroscopy in the channel system. In principle, this setup can now be easily combined with a selection step as described in earlier parts of this section.

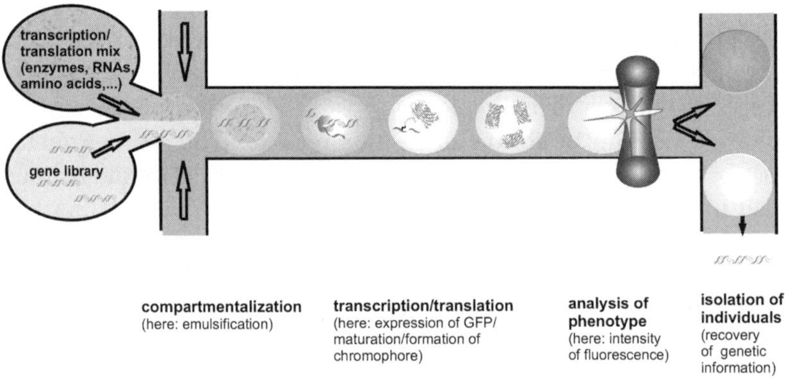


**Fig. 12.** Principle of in vitro protein expression in emulsion droplets generated on a microfluidic chip. In combination with fluorescent readout, a FACS analogon with artificial cells can be realized

Whether water-in-oil emulsions will be in the end the proper "artificial cell" system for biotechnological screening applications on a chip depends crucially on the homogeneity of size and thc speed by which the droplets can be generated, but also on the possibility to store large numbers of them for a certain time in appropriate devices, because the in vitro protein expression, especially at low gene copies, is usually a rather slow process. Ideally, the vision is to insert only single DNA molecules (e.g., out of a large DNA library) in each of the droplets, which can naturally only be realized statistically, i.e. results in a high number of "empty" droplets.

In our ongoing work, we so far succeeded in constructing low-cost PDMS microstructures for the creation of water-in-oil emulsions with reproducible droplet sizes between 1 and 50 µm. Although the possibility of having only single DNA molecules in a droplet is larger at small droplet sizes, diameters below 1 µm are usually undesirable because of the impossibility of proper quality control by optical techniques. Droplets were formed by supplying aqueous solution from a side channel into a flow of a hydrophobic liquid. As long as the flow rates (1 nl $h^{-1}$– 10 µl $h^{-1}$) of water and oil were constant, the dimensions of the formed droplets were homogeneous. The frequency of the droplet formation according to the flow rates ranged from 1 to 30 Hz and dimension of the droplets ranged from 5 to 50 µm. Figure 13 shows some of the formed water-in-oil droplets at different sizes, as well as a histogram of their size distribution.

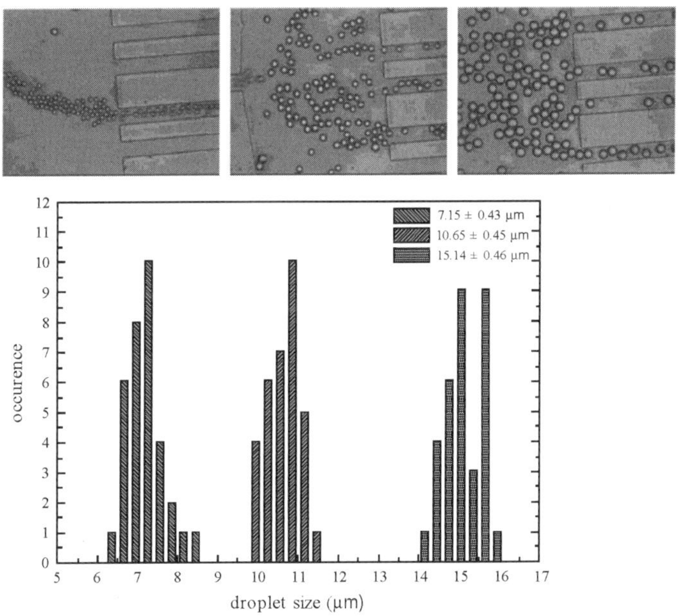


**Fig. 13.** Top row: water-in-oil droplets made at different flow speeds of mineral oil. Bottom: sistribution of droplet sizes of three sets of particles illustrated above. Thirty-three droplets of each set were analyzed with a standard deviation of less than 10%. The carrying media was prepared by dissolving *L*-α-Phosphatidylcholine (Egg PC) in mineral oil at the final concentration of 2.5 mg $ml^{-1}$

The frequency of the droplet formation could be increased to up to 100 Hz. With respect to a proper storing of the droplets to allow the in vitro expression to take place, the storage area has to have a capacity of several thousands of droplets, where at the same time a fusion or collapsing of the droplets need to be ruled out. After reaching enough high droplet density, the temperature of the chip was raised to 30°C. After two hours, the fluorescence of the expressed GFP could be observed by standard wide-field fluorescence microscopy (Fig. 14).

**Fig. 14.** Top: emulsion droplets filling the strage area of the PDMS chip at different magnifications. Bottom: expression of GFP displayed with a fluorescence arc lamp after 30 min. In our chip design, the storage area is 10 μm high and supported by columns 20 μm in diameter, in order to prevent the chip ceiling to collapse. After connecting the syringe pumps to the channel inlets, droplets of expression solution were produced over the next 3h

## *4.3. Perspectives of Generating Membrane Vesicles in Microstructures*

The described water-in-oil emulsions display in many respects a valid artificial cell system for drug screening applications, combining confinement and separation of biological compounds with the ability for protein expression. However, since many important drug targets are actually membrane-bound proteins, the possibility of creating compartments with a true lipid bilayer boundary is a highly attractive alternative. High-throughput creation of large unilamellar vesicles on a microchip would therefore be a quite attractive option, and some recent examples demonstrate the feasibility.

Small lipid vesicles with nm dimensions were formed in a X-shaped microchannel [48]. In this approach, an organic solvent containing the lipids was focused hydrodynamically in a crossed channel by two aqueous streams from two side channels. Lee and co-workers generated first aqueous droplets in a liquid lipid phase on a microchip. In a second step, the emulsion was transferred into an aqueous/alcohol solution to form vesicles [88].

The design and principle functioning of the microdevice for on-chip extrusion of lipid membranes to form three-dimensional lipid structures with various shapes is shown in Fig. 15 [22]. The central component of the device is a silicon microchip that contains small, micrometer-sized apertures that are coated with a lipid membrane. Microchannels above and below the apertures are filled with the sample solution or buffer solution. Upon applying a flow in the top channel, or sucking the aqueous solution out of the bottom channel, the lipid membrane is extruded through the apertures. In dependence of the conditions (in particular, flow rates), spherical structures (vesicles) or tubular structures are formed.

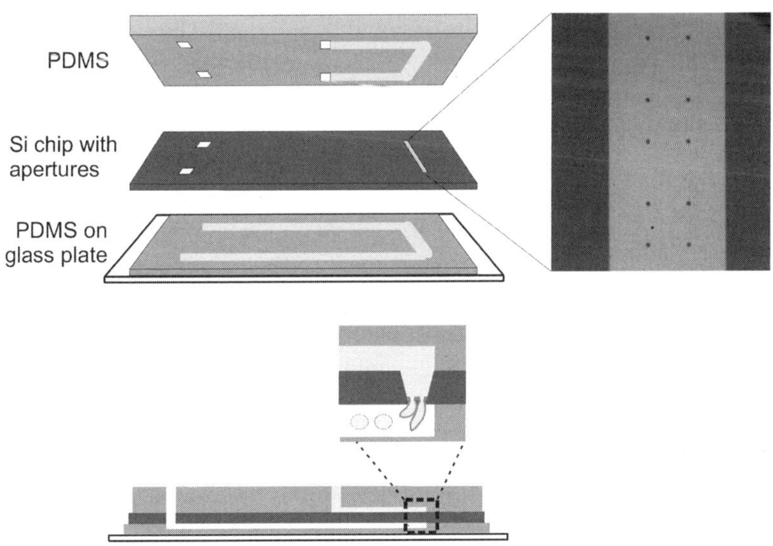


**Fig. 15.** Principle of on-chip vesicle formation. The central piece of the device is the silicon slide comprising an area with small apertures (with a diameter of ~3 μm). The apertures are coated with a lipid membrane, and the device is assembled and flooded with buffer solution. Flow either in the top or in the bottom channel causes the extrusion of the membrane through the apertures, and vesicles are released in the bottom channel of the device

Lipid vesicles are mimicking cells and cell organelles, lipid tubes could be a model for neuronal axons. The cell membranes could be simulated very precisely by choosing appropriate lipid compositions and by insertion of specific membrane proteins. After formation, the artificial lipid structures could be handled like living cells, and directly analyzed by a microcytometer. Fusion and fission in a controlled way, or electroporation to further introduce compounds into the interior of the vesicles have been shown, which opens new opportunities to simulate and study simple intracellular reactions as well as reactions at the cell membrane.

## 5. Conclusion

Many examples of recent research demonstrate that microfluidic systems are indeed a valuable alternative to conventional flow cytometers. Screening of suspended cells could be performed with the same speed as in commercial devices, while the detection sensitivity of such microfluidic systems could be dramatically increased by using confocal fluorescence spectroscopy. Although the sorting speed in the miniaturized systems is typically very low or could not run in an automated routine at all (most devices are single copies made only for research purpose), the benefits are clearly convincing, in particular, the possibility to implement many functional modules on a single chip.

Besides the usage for cell screening, miniaturized flow cytometers are yet capable to analyze and sort single (macro)molecules. An exciting novel aspect is the formation of artificial cells inside the microchip that are capable of performing basic cellular reaction such as protein expression.

## References

1. Ambrose WP, Goodwin PM, Jett JH, Van Orden A, Werner JH, Keller RA (1999) Single molecule fluorescence spectroscopy at ambient temperature. Chem Rev 99: 2929–2956
2. Andersson H, van den Berg A (2003) Microfluidic devices for cellomics: A review. Sens Act B 92: 315–325
3. Ashcroft RG, Lopez PA (2000) Commercial high speed machines open new opportunities in high throughput flow cytometry (HTFC). J Immunol Meth 243: 13–24

# 4. Perspectives for Biotechnology

## *4.1. Sorting of Single Molecules*

Microfluidic devices provide excellence performance for cell analysis and sorting. As described, fast and sensitive detection can be achieved in micrometer-sized channels. Moreover, both the small dimensions and the possibility of integrated detection with single-molecule sensitivity may even facilitate sorting procedures on the molecular level. The investigation of molecules passing the detection volume can give information about characteristic molecular properties, according to which a selection could be performed.

Several approaches aiming at molecular selection and sorting have been published. Chou et al. presented a microfluidic chip to determine the size of DNA fragments, and to sort the fragments in dependence of the length [13]. In a microfluidic chip similar to the device in Fig. 4, molecular steering of microtubules in a kinesin-coated microchannel was realized [93]. The microtubules moved with a velocity of 0.75 μm along the kinesin motor proteins. In an electrical field of 70 V per 2 mm, the microtubules' pathway was oriented towards the positive electrode. The electrophoretic movement of the microtubules was exploited to sort microtubules with red and green fluorescent tags.

To selectively handle single molecules in the fluid environment, the molecules can be attached to beads or encapsulated in vesicles or bubbles (Fig. 11).

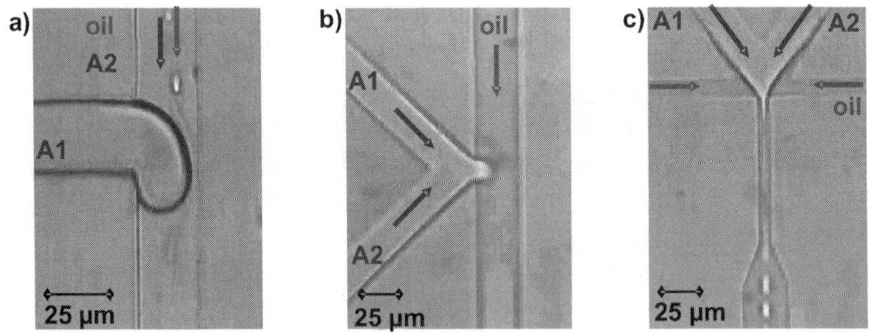


**Fig. 11.** Different channel designs for generation of water-in-oil emulsions in a microstructure. Components ("A1" and "A2") to be confined in the buffer droplets can be ideally mixed by using a preemulsion (**a**), or by merging two streams of the solutions during droplet generation (**b** and **c**)

4. Barbulovic-Nad I, Xuan X, Lee JSH, Li D (2006) DC-dielectrophoretic separation of microparticles using an oil droplet obstacle. Lab Chip 6: 274–279
5. Berger M, Castelino J, Huang R, Shah M, Austin RH (2001) Design of a microfabricated magnetic cell separator. Electrophoresis 22: 3883–3892
6. Brinkmeier M, Dörre K, Stephan J, Eigen M (1999) Two-beam cross-correlation: A method to characterize transport phenomena in micrometer-sized structures. Anal Chem 71: 609–616
7. Carlson RH, Gabel CV, Chan SS, Austin RH, Brody JP, Winkelman JW (1997) Self-Sorting of White Blood Cells in a Lattice. Phys Rev Lett 79: 2149–2152
8. Chen Z, Milner TE, Srinivas S, Wang XD, Malekafzali A, van Gemert MJC, Nelson JS (1997) Noninvasive imaging of in vivo blood flow velocity using optical Doppler tomography. Opt Lett 22: 1119–1121
9. Cheung K, Gawad S, Renaud P (2005) Impedance spectroscopy flow cytometry: On-chip label-free cell differentiation. Cytometry A 65A: 124–132
10. Cho BS, Schuster TG, Zhu XY, Chang D, Smith GD, Takayama S (2003) Passively driven integrated microfluidic system for separation of motile sperm. Anal Chem 75: 1671–1675
11. Choi S, Park JK (2005) Microfluidic system for dielectrophoretic separation based on a trapezoidal electrode array. Lab Chip 5: 1161–1167
12. Chou CF, Bakajin O, Turner SWP, Duke TAJ, Chan SS, Cox EC, Craighead HG, Austin RH (1999) Sorting by diffusion: An asymmetric obstacle course for continuous molecular separation. Proc Natl Acad Sci USA 96: 13762–13765
13. Chou HP, Spence C, Scherer A, Quake SR (1999) A microfabricated device for sizing and sorting of DNA. Proc Natl Acad Sci USA 96: 11–13
14. Chun HG, Chung TD, Kim HC (2005) Cytometry and velocimetry on a microfluidic chip using polyelectrolytic salt bridges. Anal Chem 77: 2490–2495
15. Chung S, Park SJ, Kim JK, Chung C, Han DC, Chang JK (2003) Plastic microchip flow cytometer based on 2- and 3-dimensional hydrodynamic flow focusing. Microsystem Technol 9: 525–533
16. Damean N, Sia SK, Linder V, Narovlyansky M, Whitesides GM (2005) Space- and time-resolved spectrophotometry in microsystems. Proc Natl Acad Sci USA 102: 10035–10039
17. Dittrich PS, Schwille P (2002) Spatial two-photon fluorescence cross-correlation spectroscopy for controlling molecular transport in microfluidic structures. Anal Chem 74: 4472–4479
18. Dittrich PS, Schwille P (2003) An integrated microfluidic system for reaction, high-sensitivity detection, and sorting of fluorescent cells and particles. Anal Chem 75: 5767–5774
19. Dittrich PS, Jahnz M, Schwille P (2005) A new embedded process for compartmentalized cell-free protein expression and on-line detection in microfluidic devices. Chembiochem 6: 811–814

20. Dittrich PS, Manz A (2005) Single-molecule fluorescence detection in microfluidic channels – The Holy Grail in μTAS? Anal Bioanal Chem 382: 1771–1782
21. Dittrich PS, Manz A (2006) Lab-on-a-chip: Microfluidics in drug discovery. Nature Rev Drug Disc 5: 210–218
22. Dittrich PS, Heule M, Renaud P, Manz A (2006) On-chip extrusion of lipid vesicles and tubes through microsized apertures. Lab Chip 6: 488–493
23. Dörre K, Stephan J, Eigen M (2001) Highly efficient single molecule detection in different micro- and submicrometer channels with cw excitation. Single Mol 2: 165–175
24. Edel JB, de Mello AJ (2003) Single particle confocal fluorescence spectroscopy in microchannels: Dependence of burst width and burst area distributions on particle size and flow rate. Anal Sci 19: 1065–1069
25. Edel JB, Beard NP, Hofmann O, DeMello JC, Bradley DDC, deMello AJ (2004) Thin-film polymer light emitting diodes as integrated excitation sources for microscale capillary electrophoresis. Lab Chip 4: 136–140
26. El-Ali J, Sorger PK, Jensen KF (2006) Cells on chips. Nature 442: 403–411
27. Emmelkamp J, Wolbers F, Andersson H, DaCosta RS, Wilson BC, Vermes I, van den Berg A (2004) The potential of autofluorescence for the detection of single living cells for label-free cell sorting in microfluidic systems. Electrophoresis 25: 3740–3745
28. Fiedler S, Shirley SG, Schnelle T, Fuhr G (1998) Dielectrophoretic sorting of particles and cells in a microsystem. Anal Chem 70: 1909–1915
29. Fu AY, Spence C, Scherer A, Arnold F, Quake SR (1999) A microfabricated fluorescence-activated cell sorter. Nat Biotechnol 17: 1109–1111
30. Fu AY, Chou HP, Spence C, Arnold FH, Quake SR (2002) An integrated microfabricated cell sorter. Anal Chem 74: 2451–2457
31. Fu LM, Yang RJ, Lin CH, Pan YJ, Lee GB (2004) Electrokinetically driven micro flow cytometers with integrated fiber optics for on-line cell/particle detection. Anal Chim Acta 507: 163–169
32. Galas JC, Torres J, Belotti M, Kou Q, Chen Y (2005) Microfluidic tunable dye laser with integrated mixer and ring resonator. Appl Phys Lett 86: 264101
33. Gawad S, Schild L, Renaud P (2001) Micromachined impedance spectroscopy flow cytometer for cell analysis and particle sizing. Lab Chip 1: 76–82
34. Georgiou G (2001) Analysis of large libraries of protein mutants using flow cytometry. Adv Protein Chem: Evol Protein Des 55: 293–315
35. Gu W, Zhu XY, Futai N, Cho BS, Takayama S (2004) Computerized microfluidic cell culture using elastomeric channels and Braille displays. Proc Natl Acad Sci USA 101: 15861–15866
36. Guck J, Schinkinger S, Lincoln B, Wottawah F, Ebert S, Romeyke M, Lenz D, Erickson HM, Ananthakrishnan R, Mitchell D, Kas J, Ulvick S, Bilby C (2005) Optical deformability as an inherent cell marker for testing malignant transformation and metastatic competence. Biophys J 88: 3689–3698
37. Haupts U, Maiti S, Schwille P, Webb WW (1998) Dynamics of fluorescence fluctuations in green fluorescent protein observed by fluorescence correlation spectroscopy. Proc Natl Acad Sci USA 95: 13573–13578

38. Hayden O, Agarwal R, Lieber CM (2006) Nanoscale avalanche photodiodes for highly sensitive and spatially resolved photon detection. Nature Materials 5: 352–356
39. Herzenberg LA, Sweet RG, Herzenberg LA (1976) Fluorescence-activated cell sorting. Sci Am 234: 108–117
40. Herzenberg LA, Parks D, Sahaf B, Perez O, Roederer M, Herzenberg LA (2002) The history and future of the fluorescence activated cell sorter and flow cytometry: A view from Stanford. Clin Chem 48: 1819–1827
41. Ho CT, Lin RZ, Chang HY, Liu CH (2005) Micromachined electrochemical T-switches for cell sorting applications. Lab Chip 5: 1248–1258
42. Hofmann O, Miller P, Sullivan P, Jones TS, deMello JC, Bradley DDC, deMello AJ (2005) Thin-film organic photodiodes as integrated detectors for microscale chemiluminescence assays. Sens Act B 106: 878–884
43. Hu XY, Bessette PH, Qian JR, Meinhart CD, Daugherty PS, Soh HT (2005) Marker-specific sorting of rare cells using dielectrophoresis. Proc Natl Acad Sci USA 102: 15757–15761
44. Huang LR, Cox EC, Austin RH, Sturm JC (2004) Continuous particle separation through deterministic lateral displacement. Science 304: 987–990
45. Huh D, Tung Y-C, Wei H-H, Grotberg JB, Skerlos SJ, Kurabayashi K, Takayama S (2002) Use of air–liquid two-phase flow in hydrophobic microfluidic channels for disposable flow cytometers. Biomed Microdev 4: 141–149
46. Ibrahim SF, Van den Engh G (2003) High-speed cell sorting: Fundamentals and recent advances. Curr Opinion Biotechnol 14: 5–12
47. Inglis DW, Riehn R, Austin RH, Sturm JC (2004) Continuous microfluidic immunomagnetic cell separation. Appl Phys Lett 85: 5093–5095
48. Jahn A, Vreeland WN, Gaitan M, Locascio LE (2004) Controlled vesicle self-assembly in microfluidic channels with hydrodynamic focusing. J Am Chem Soc 126: 2674–2675
49. Jiang Z, Llandro J, Mitrelias LT, Bland JAC (2006) An integrated microfluidic cell for detection, manipulation, and sorting of single micron-sized magnetic beads. J Appl Phys 99: 08S105
50. Johann R, Renaud P (2004) A simple mechanism for reliable particle sorting in a microdevice with combined electroosmotic and pressure-driven flow. Electrophoresis 25: 3720–3729
51. Katsuragi T, Tani Y (2000) Screening for microorganisms with specific characteristics by flow cytometry and single-cell sorting. J Biosci Bioeng 89: 217–222
52. Khademhosseini A, Langer R, Borenstein J, Vacanti JP (2006) Microscale technologies for tissue engineering and biology. Proc Natl Acad Sci USA 103: 2480–2487
53. Krüger J, Singh K, O'Neill A, Jackson C, Morrison A, O'Brien P (2002) Development of a microfluidic device for fluorescence activated cell sorting. J Micromech Microeng 12: 486–494

54. Kung C, Banes M, Lermer N, Whitten W, Ramsey J (1999) Single-molecule analysis of ultradilute solutions with guided streams of 1-μm water droplets. Appl Opt 38: 1481–1487
55. Kunst BH, Schots A, Visser A (2004) Design of a confocal microfluidic particle sorter using fluorescent photon burst detection. Rev Sci Instrum 75: 2892–2898
56. Lancaster C, Kokoris A, Nabavi M, Clemmens J, Maloney P, Capadanno J, Gerdes J, Battrell CF (2005) Rare cancer cell analyzer for whole blood applications: Microcytometer cell counting and sorting subcircuits. Methods 37: 120–127
57. Lapizco-Encinas BH, Simmons BA, Cummings EB, Fintschenko Y (2004) Dielectrophoretic concentration and separation of live and dead bacteria in an array of insulators. Anal Chem 76: 1571–1579
58. Lee GB, Lin CH, Chang GL (2003) Micro flow cytometers with buried SU-8/SOG optical waveguides. Sens Act A 103: 165–170
59. Lee GB, Lin CH, Chang SC (2005) Micromachine-based multichannel flow cytometers for cell/particle counting and sorting. J Micromech Microeng 15: 447–54
60. Lincoln B, Erickson HM, Schinkinger S, Wottawah F, Mitchell D, Ulvick S, Bilby C, Guck J (2004) Deformability-based flow cytometry. Cytometry A 59A: 203–209
61. MacDonald MP, Spalding GC, Dholakia K (2003) Microfluidic sorting in an optical lattice. Nature 426: 421–424
62. Magde D, Elson E, Webb WW (1972) Thermodynamic fluctuations in a reacting system – Measurement by fluorescence correlation spectroscopy. Phys Rev Lett 29: 705–708
63. Magde D, Webb WW, Elson EL (1978) Fluorescence correlation spectroscopy. III. Uniform translation and laminar flow. Biopolymers 17: 361–376
64. Manz A, Graber N, Widmer HM (1990) Miniaturized total chemical analysis systems: A novel concept for chemical sensing. Sens Act B 1: 244–248
65. Mawatari K, Kitamori T, Sawada T (1998) Individual detection of single-nanometer-sized particles in liquid by photothermal microscope. Anal Chem 70: 5037–5041
66. McClain MA, Culbertson CT, Jacobson SC, Allbritton NL, Sims CE, Ramsey JM (2003) Microfluidic devices for the high-throughput chemical analysis of cells. Anal Chem 75: 5646–5655
67. Miller OJ, Bernath K, Agresti JJ, Amitai G, Kelly BT, Mastrobattista E, Taly V, Magdassi S, Tawfik DS, Griffith AD (2006) Directed evolution by in vitro compartimentalization. Nat Methods 3: 561–570
68. Morgan H, Holmes D, Green NG (2006) High speed simultaneous single particle impedance and fluorescence analysis on a chip. Curr Appl Phys 6: 367–370
69. Nolan JP, Sklar LA (1998) The emergence of flow cytometry for sensitive, real-time measurements of molecular interactions. Nature Biotechnol 16: 633–638

70. Oakey J, Allely J, Marr DWM (2002) Laminar-flow-based separations at the microscale. Biotechnol Progr 18: 1439–1442
71. Olofsson J, Nolkrantz K, Ryttsen F, Lambie BA, Weber SG, Orwar O (2003) Single-cell electroporation. Curr Opinion Biotechnol 14: 29–34
72. Pamme N, Manz A (2004) On-chip free-flow magnetophoresis: Continuous flow separation of magnetic particles and agglomerates. Anal Chem 76: 7250–7256
73. Paul PH, Garguilo MG, Rakestraw DJ (1998) Imaging of pressure- and electrokinetic driven flows through open capillaries. Anal Chem 70: 2459–2467
74. Petersson F, Nilsson A, Holm C, Jonsson H, Laurell T (2004) Separation of lipids from blood utilizing ultrasonic standing waves in microfluidic channels. Analyst 129: 938–943
75. Roos W, Ulmer J, Gräter S, Surrey S, Spatz JP (2005) Microtubule gliding and cross-linked microtubule networks on micropillar interfaces. Nano Lett 5: 2630–2634
76. Santiago JG, Wereley ST, Meinhart CD, Beebe DJ, Adrian RJ (1998) A particle image velocimetry system for microfluidics. Exp Fluids 25: 316–319
77. Schilling EA, Kamholz AE, Yager P (2002) Cell lysis and protein extraction in a microfluidic device with detection by a fluorogenic enzyme assay. Anal Chem 74: 1798–1804
78. Schrum DP, Culbertson CT, Jacobson SC, Ramsey JM (1999) Microchip flow cytometry using electrokinetic focusing. Anal Chem 71: 4173–4177
79. Schrum KF, Lancaster JM, Johnston SE, Gilman SD (2000) Monitoring electroosmotic flow by periodic photobleaching of a dilute, neutral fluorophore. Anal Chem 72: 4317–4321
80. Schwille P, Bieschke J, Oehlenschläger F (1997) Kinetic investigations by fluorescence correlation spectroscopy: The analytical and diagnostic potential of diffusion studies. Biophys Chem 66: 211–228
81. Shelby JP, Chiu DT (2003) Mapping fast flows over micrometer-length scales using flow-tagging velocimetry and single-molecule detection. Anal Chem 75: 1387–1392
82. Shin YS, Cho K, Kim JK, Lim SH, Park CH, Lee KB, Park Y, Chung C, Han DC, Chang JK (2004) Electrotransfection of mammalian cells using microchannel-type electroporation chip. Anal Chem 76: 7045–7052
83. Shirasaki Y, Tanaka J, Makazu H, Tashiro K, Shoji S, Tsukita S, Funatsu T (2006) On-chip cell sorting system using laser-induced heating of a thermoreversible gelation polymer to control flow. Anal Chem 78: 695–701
84. Simonnet C, Groisman A (2005) Two-dimensional hydrodynamic focusing in a simple microfluidic device. Appl Phys Lett 87: 14104
85. Sin A, Chin KC, Jamil MF, Kostov Y, Rao G, Shuler ML (2004) The design and fabrication of three-chamber microscale cell culture analog devices with integrated dissolved oxygen sensors. Biotechnol Prog 20: 338–345
86. Stagi M, Dittrich PS, Frank N, Iliev AI, Schwille P, Neumann H (2005) Breakdown of axonal synaptic vesicle precursor transport by microglial nitric oxide. J Neuroscience 25: 352–362

87. Studer V, Jameson R, Pellereau E, Pepin A, Chen Y (2004) A microfluidic mammalian cell sorter based on fluorescence detection. Microelectronic Eng 73-74: 852–527
88. Tan Y-C, Hettiarachchi K, Siu M, Pan Y-R, Lee AP (2006) Controlled microfluidic encapsulation of cells, proteins, and microbeads in lipid vesicles. J Am Chem Soc 128: 5656–5658
89. Tanaka Y, Morishima K, Shimizu T, Kikuchi A, Yamato M, Okano T, Kitamori T (2006) An actuated pump on-chip powered by cultured cardiomyocytes. Lab Chip 6: 362–368
90. Tarsa PB, Wist AD, Rabinowitz P, Lehmann KK (2004) Single-cell detection by cavity ring-down spectroscopy. Appl Phys Lett 85: 4523–4525
91. Tourovskaia A, Figueroa-Masot X, Folch A (2005) Differentiation-on-a-chip: A microfluidic platform for long-term cell culture studies. Lab Chip 5: 14–19
92. Tung YC, Zhang M, Lin CT, Kurabayashi K, Skerlos SJ (2004) PDMS-based opto-fluidic micro flow cytometer with two-color, multi-angle fluorescence detection capability using PIN photodiodes. Sens Act B 98: 356–367
93. van den Heuvel MGL, de Graaff MP, Dekker C (2006) Molecular Sorting by electrical steering of microtubules in kinesin-coated channels. Science 312: 910–914
94. van Orden A, Machara NP, Goodwin PM, Keller RA (1998) Single molecule identification in flowing sample streams by fluorescence burst size and intraburst fluorescence decay rate. Anal Chem 70: 1444–1451
95. Voldman J, Gray ML, Toner M, Schmidt MA (2002) A microfabrication-based dynamic array cytometer. Anal Chem 74: 3984–3990
96. Wang MM, Tu E, Raymond DE, Yang JM, Zhang HC, Hagen N, Dees B, Mercer EM, Forster AH, Kariv I, Marchand PJ, Butler WF (2005) Microfluidic sorting of mammalian cells by optical force switching. Nature Biotechnol 23: 83–87
97. Weibel DB, Garstecki P, Ryan D, Diluzio WR, Mayer M, Seto JE, Whitesides GM (2005) Microoxen: Microorganisms to move microscale loads. Proc Natl Acad Sci USA 102: 11963–11967
98. Weigl BH, Yager P (1999) Microfluidic diffusion-based separation and detection. Science 283: 346–347
99. Widengren J, Rigler R, Mets U (1995) Triplet-state monitoring by fluorescence correlation spectroscopy – A theoretical and experimental study. J Phys Chem 99: 13368–13379
100. Widengren J, Rigler R (1996) Mechanisms of photobleaching investigated by fluorescence correlation spectrocopy. Bioimaging 4: 149–157
101. Wolff A, Perch-Nielsen IR, Larsen UD, Friis P, Goranovic G, Poulsen CR, Kutter JP, Telleman P (2003) Integrating advanced functionality in a microfabricated high-throughput fluorescent-activated cell sorter. Lab Chip 3: 22–27
102. Wood DK, Oh SH, Lee SH, Soh HT, Cleland AN (2005) High-bandwidth radio frequency Coulter counter. Appl Phys Lett 87: 184106
103. Wu HK, Wheeler A, Zare RN (2004) Chemical cytometry on a picoliter-scale integrated microfluidic chip. Proc Natl Acad Sci USA 101: 12809–12813

104. Yamada M, Nakashima M, Seki M (2004) Pinched Flow Fractionation: Continuous size separation of particles utilizing a laminar flow profile in a pinched microchannel. Anal Chem 76: 5465–5471
105. Yang SY, Hsiung SK, Hung YC, Chang CM, Liao TL, Lee GB (2006) A cell counting/sorting system incorporated with a microfabricated flow cytometer chip. Meas Sci Technol 17: 2001–2009
106. Yao B, Luo GA, Feng X, Wang W, Chen LX, Wang YM (2004) A microfluidic device based on gravity and electric force driving for flow cytometry and fluorescence activated cell sorting. Lab Chip 4: 603–607
107. Yi C, Li C-W, Ji S, Yang M (2006) Microfluidics technology for manipulation and analysis of biological cells. Anal Chim Acta 560: 1–23
108. Zander C, Drexhage KH, Han KT, Wolfrum J, Sauer M (1998) Single-molecule counting and identification in a microcapillary. Chem Phys Lett 286: 457–465
109. Zhang HC, Tu E, Hagen ND, Schnabel CA, Paliotti MJ, Hoo WS, Nguyen PM, Kohrumel JR, Butler WF, Chachisvillis M, Marchand PJ (2004) Time-of-flight optophoresis analysis of live whole cells in microfluidic channels. Biomed Microdev 6: 11–21

# Index

## *ABOUT THE COVER*

The editors gratefully acknowledge the cover photograph by Felice Frankel, Artist in Residence at the Massachusetts Institute of Technology and coauthor of *On the Surface of Things: Images of the Extraordinary in Science.*

This particular image, taken with Nomarski optics, presents a wafer-bonded piezoresistive pressure sensor. It is fabricated in the sealed-cavity process developed by Professor Martin Schmidt of the Massachusetts Institute of Technology with his graduate students, Lalitha Parameswaran and Charles Hsu. The piezoresistors are clearly visible, and the slight contrast across the central diaphragm region shows that the diaphragm is actually slightly bent by the pressure difference between the ambient and the sealed cavity beneath.

Printed in the United States of America.